Entomology Monographs

Series Editor

Hideharu Numata, Kyoto University, Kyoto, Japan

Insects are the most diverse group of organisms, and many biological advances have been derived from this group. Although entomology is considered to be a classic science, recent developments in molecular methods, application of modern theoretical concepts and collaboration with related sciences have opened new directions in entomology. Japanese scientists play a significant role in these fields, and this book series will focus on such developments. The book series Entomology Monographs publishes refereed volumes on all aspects of entomology, including ecology, ethology, physiology, taxonomy, systematics, morphology, evolutionary developmental biology, genetics, biochemistry, and molecular biology in insects and related arthropods. Authors are not restricted to Japanese entomologists, and other international experts will also be considered on the basis of their recent contribution to these fields.

Kyoko Sawabe • Chizu Sanjoba • Yukiko Higa
Editors

Medical Entomology in Asia

Springer

Editors
Kyoko Sawabe
Department of Medical Entomology
National Institute of Infectious Diseases (NIID)
Tokyo, Japan

Chizu Sanjoba
Graduate School of Agricultural and Life Sciences
The University of Tokyo
Tokyo, Japan

Yukiko Higa
Department of Medical Entomology
National Institute of Infectious Diseases (NIID)
Tokyo, Japan

ISSN 2522-526X ISSN 2522-5278 (electronic)
Entomology Monographs
ISBN 978-981-97-4405-3 ISBN 978-981-97-4406-0 (eBook)
https://doi.org/10.1007/978-981-97-4406-0

This Springer imprint is published by the registered company Springer Nature Singapore Pte Ltd.
The registered company address is: 152 Beach Road, #21-01/04 Gateway East, Singapore 189721, Singapore

Foreword 1

Vector-borne diseases (VBDs) account for more than 17% of all infectious diseases, causing more than 700,000 deaths annually. They require an arthropod vector, insect, or tick to transmit the pathogen from one vertebrate host to another. VBDs have shaped the history of mankind; plague, yellow fever, dengue fever, malaria, typhus, sleeping sickness, and many others are transmitted by insect vectors. Therefore, gaining more knowledge of insect biology and ecology is essential to understanding transmission mechanisms and disease epidemiology and to developing vector control methods.

Medical entomology was introduced in Asia in 1877, when a British doctor, Patrick Manson, described the role of mosquitoes in the biological cycle of filarial parasites. It was in Amoy (now Xiamen), in China that he observed a worm growing in a mosquito. With Pasteur's discoveries and the contributions from his eminent colleagues, microbiology has emerged as a new field of research, thus turning the page on the theory of spontaneous generation. Medical entomology has established itself as a discipline in its own right. There are several examples of this phenomenon: Alphonse Laveran described the malaria parasite *Plasmodium*, in Constantine, Algeria, in November 1880 (Nobel Prize in 1907). In 1897, the British scientist Ronald Ross demonstrated in India that *Anopheles* mosquitoes were responsible for *Plasmodium* transmission (Nobel Prize in 1902). Another emblematic example is yellow fever, which caused devastating outbreaks in its original cradle in sub-Saharan Africa and also in tropical America where it was introduced in slave ships. The idea that mosquitoes spread the disease was proposed in 1854 by the Guadeloupean Louis-Daniel Beauperthuy and then taken over by Cuban Carlos Finlay, who clearly suggested in 1881 the role of the mosquito *Aedes aegypti* in the transmission of yellow fever. Walter Reed from the American Army mission provided decisive proof. Plague is a scourge dating back to the middle Ages; in 1892, a plague epidemic hit Yunnan and spread to Canton, followed by Hong Kong and Amoy. Alexandre Yersin discovered the bacterium responsible for the disease, hereafter called *Yersinia pestis*. In 1898, Paul-Louis Simond demonstrated that the flea is a vector. Typhus is another emblematic example; it is considered to be one of the most devastating diseases in humans, with mortality rates exceeding 50%. In 1909, Charles Nicolle demonstrated that the bacteria were transmitted by the flea (Nobel Prize in 1928). Without vaccines and therapeutics, vector control is required to control the diseases. In Switzerland, in 1939, the chemist Paul Hermann Müller discovered the insecticidal properties of DDT, a highly efficient and stable insecticide (Nobel Prize in 1948). With DDT, we entered a new phase of vector control through the extensive use of insecticides. The future will tell us that this is not a panacea, as vectors have developed resistance to chemical insecticides, and more environmentally friendly methods and new technologies are needed. Medical entomology also deals with insects because of the various disorders that they can directly cause. They can be pathogenic arthropods, such as mites causing scabies, scorpions injecting their venom, urticants, or allergenic butterflies causing lepidopterism.

Asian countries have already placed a heavy burden on VBDs. Dengue fever is the most important VBD, with an incidence 30-fold higher than that in the past 50 years, and Asia accounts for almost 75% of human cases worldwide. Malaria, chikungunya, Japanese encephalitis, tick-borne diseases, and other diseases have also taken their toll worsened due to environmental changes (land use and agricultural practices), climate change, and the globalization of trade and travel, which create favorable conditions for the development and expansion of mosquitoes. This book is timely given the global context and unique as it comprises more than 520 pages with all you must know about medical entomology. It focuses on VBDs in Japan and, more broadly, Asia, which has been the scene of many discoveries in medical entomology.

Institute Pasteur Paris, Department of Virology
Paris, France

Anna-Bella Failloux

Foreword 2

Medical entomology is a field of science that primarily deals with the arthropods that transmit pathogens to humans and cause infectious diseases. Arthropods play key roles in the maintenance and circulation of human pathogens in nature. Medical entomology research includes various subjects, including the biology and ecology of arthropods and their interactions with pathogens and vertebrate hosts. Medical entomology also addresses the effects of the environment on arthropods in this era of global warming. Medical entomology studies collaborate accordingly with multiple research fields, including biology, medicine, zoology, microbiology, and climatology.

Vector-borne diseases (VBDs) have posed serious threats to humans for many years. Most of these diseases were local and confined to limited regions of the world in the past because human activities were rather regional and not worldwide. However, the global economic situation has changed rapidly. In contrast to the pre-modernization days, the movements of vectors and pathogens are rapid and worldwide along with humans and trade goods. VBDs are no longer restricted to local regions and can quickly become a worldwide threat. For example, dengue fever has spread from Asia to the Americas and Africa, and epidemics have also occurred in temperate countries. These changes have raised great concerns about VBDs worldwide.

Countermeasures against VBDs are complex and multi-targeted and focus on arthropod vectors, vertebrate hosts, and pathogens. The development and use of pesticides are effective measures. However, the emergence of pesticide resistance hampers our strategies against VBDs and shows that arthropods are complex creatures and that battles against them are not easy. Advances in other fields of science, such as molecular biology, genetics, microbiology, and immunology, have been introduced into medical entomology. Classical medical entomology has been re-established as "modern" medical entomology in the twenty-first century; this trend has made medical entomology one of the most exciting fields of medical science.

In Asia, vector-borne pathogens are the major cause of morbidity and mortality. Medical entomology studies have been extensively conducted, and many important scientific achievements have been made in Asia, including Japan. Thus, it is appropriate and timely to publish the book titled *Medical Entomology in Asia*. I recommend this textbook as a must-read for those at various stages of research careers, particularly graduate and Ph.D. students and postdoctoral fellows in entomology, as well as for those in basic and clinical infectious diseases. I am confident that the readers will be truly fascinated by scientific achievements and exciting advances in the current medical entomology.

National Institute of Infectious Diseases
Tokyo, Japan

Ichiro Kurane

Preface

Medical entomology is the study of animals directly and indirectly involved in the physical and mental health of humans, and the science of the mechanisms of pathogenesis, prevention, and treatment of diseases caused by these arthropods. A large number of arthropods, such as mosquitoes and ticks, are known to transmit pathogens to humans, and many are capable of infecting other mammals with pathogens. The physical and psychological harm caused by stinging, parasitic, allergenic, and unpleasant blood-feeding arthropod species is compounded by their role as vectors of debilitating zoonotic diseases, particularly in livestock.

With the globalization of the world, the scale and speed of movement of people and goods has increased dramatically. This has led to a corresponding increase in the spread and global distribution of arthropods and the infectious agents they carry. The rapid global spread of the COVID-19 epidemic that we are currently experiencing is evidence of this. The number of invasive species entering Japan has also increased significantly, and there is an urgent need to strengthen pest control. Of course, when considering pest control from the perspective of globalization, focusing only on domestic pests is a one-sided approach, and a holistic perspective that includes neighboring Asian countries that are geographically and economically close to Japan is also necessary. On the other hand, Japan has experience in eradicating malaria and lymphatic filariasis, drastically reducing the number of Japanese encephalitis cases, and successfully controlling flies in the Tokyo metropolitan area. This history of our past should be valuable information for the Asian region, which is still struggling with pest infestations today.

This book focuses on Japan, where invasive arthropods are on the rise, and also provides a comprehensive overview of the recent situation in neighboring Asian countries. In addition, by integrating the latest Medical Entomology with past successes and experiences in this field, this book will be an indispensable reference for many professionals in planning sustainable pest control. We hope that this publication will facilitate the access of interested doctoral students and early career researchers to the rudiments and latest findings in medical entomology and ensure the continued development of the field.

Shinjuku, Tokyo, Japan
November 26th, 2023

Kyoko Sawabe

Acknowledgments

The authors are deeply indebted to the many mentors and colleagues who have encouraged, supported, and provided valuable advice over the years. We were able to continue our research with them who have guided and encouraged us in our studies in the field of medical entomology. In addition, they deeply understood the importance of making this book and gave us valuable advice. Many thanks to Matsumoto, Y., Goto, Y. (The Univ. Tokyo), Miyagi, I., Toma, T. (Univ. Ryukyus), Minakawa, N., Sunahara, T., Kawada, H., Takagi, M., Hasebe, F., Horio, M. (NEKKEN, Nagasaki Univ.), Takai, K., Tadano, T. (SMR Univ.), Tsukamoto, M., Makiya, K. (UOEH), Okazawa, T., Tuno, N. (Kanazawa Univ.), Mori, Y. (Kobe Univ.), Kano, S. (NCGM), Agui, N., Kobayashi, M., Tsuda, Y., and Kurihara, T. (NIID) from Japan.

Over the years, we have planned several projects and received grants to conduct joint research collaborations. We believe this ongoing international exchange has helped us obtain information from abroad, especially from Asian countries, at an early stage. Many contributors have written texts for this book. We hope that this exchange will continue for many years to come. We also thank our international research collaborators, Nguyen, T.Y., Nga, P.T. (NIHE, Vietnam), Siew, F.W., Leh, M.U. (Sarawak Museum, Malaysia), Lucida, M.I., Mulyatno, K.C., Subekti, S. (UNAIR, Indonesia), Filho, J.L.L., Alves, L.C. (LIKA, Brazil), Suwonkerd, W. (Thailand), Pemba, D. (Malawi), Bertuso, A.G. (UP-Manila, Philippines), Brey, P. (IPL, Lao PDR), Roychoudhury, S. (UK), Amoa-Bosompem, M. (UTN, USA), Bhutto, A.M. (SMBBMU, Pakistan), Rahman, F. Shyaml, P. (Bangladesh), Karunaweera, N., Silva, H. (Sri Lanka), Volf, P. (Czech), Depaquit, J. (France). For financial support, we thank the Ministry of Health, Labour and Welfare, Japan (MHLW), Ministry of Education, Culture, Sports, Science and Technology, Japan (MEXT), Japan Agency for Medical Research and Development (AMED), and Japan International Cooperation Agency (JICA).

Finally, many thanks to Professor Hideharu Numata (Graduate School of Science, Kyoto University and Series Editor of Entomology Monographs) for inviting us to write this book as a contribution to this Series and Fumiko Yamaguchi, Jayesh Kalleri, Sindhuja Aroumougame, and the staff at Springer Nature of Springer for their patience and editing of this manuscript.

Special thanks to the families of KS, CS, and YH for their long support of our scientific activities.

Contents

About the Editors

Yukiko Higa began her career as a mosquito scientist in the 1990s. She obtained a bachelor's and a master's degrees in health sciences on mosquito taxonomy for *Anopheles saperoi* under the supervision of Professors Ichiro Miyagi and Takako Toma of the University of the Ryukyus, Okinawa, Japan, and a doctoral degree in Medicine on habitat selection of dengue vectors, *Aedes aegypti* and *Aedes albopictus* under the supervision of Professors Masahiro Takagi and Yoshio Tsuda of the Institute of Tropical Medicine (NEKKEN), Nagasaki University, Nagasaki, Japan. She became interested in mosquitoes after learning about malaria from her grandparents' generation. Her motto is "Say yes and challenge" and "Learn from the old, know the new." She currently works for the National Institute of Infectious Diseases in Japan. Japanese sake and cocktail lover.

Chizu Sanjoba has been focused on the epidemiology and molecular biology of parasitic diseases, especially leishmaniasis, to understand the host–parasite relationship. She received a master's degree in epidemiological research of leishmaniasis and obtained a doctoral degree in the characterization of leishmania amastigotes under the supervision of Professor Yoshitugu Matsumoto of the University of Tokyo, Japan. Since 2002, her research has also focused on vector ecology towards integrated control for vector-borne parasitic diseases. Lab to field and field to lab is her motto. She has conducted field research in Mongolia, Taiwan, Sri Lanka, Bangladesh, Turkey, Zambia, etc. She is an associate professor at the Laboratory of Molecular Immunology, Graduate School of Agricultural and Life Sciences, The University of Tokyo, Japan.

Kyoko Sawabe studied insect physiology under Professor Sumio Tojo at the Faculty of Agriculture, Saga University, Japan. She received her Ph.D. in 1991 on the topic "Long-distance migration of brown planthoppers" with the support of Professor Fusao Nakasuji of Okayama University. She entered the "mosquito world" in 1985 and conducted physiological studies on mosquitoes at two universities. In 2002, she joined the National Institute of Infectious Diseases as chief and later director of the department. Her research interests expanded to include all arthropods such as mosquitoes, ticks, flies, spiders, and bed bugs. In 2022, her group published the presence of long-distance migratory mosquitoes. Her current interests include predicting vector-borne disease (VBD) outbreaks based on the biology of arthropod vectors such as mosquitoes and ticks.

Chapter 1
Mosquito

Pradya Somboon, Kyoko Futami, and Yukiko Higa

Abstract There are approximately 3700 species of mosquitoes worldwide, but the number of species involved in transmitting pathogens is limited. Asian countries have historically suffered from mosquito-borne diseases such as malaria, filariasis, dengue fever, and Japanese encephalitis. However, because of the efforts of researchers and medical/health professionals, some countries have overcome those diseases. This chapter shares significant experiences and studies of mosquitoes in Asian countries.

Keywords Asia · Cryptic species · DNA · Fossil · Human vector contact · Mosquitoes · New genotype · New serotype · Pathogens · Simian malaria · Taxonomy · Tsunami

1.1 Introduction

Over 3700 species of mosquitoes are currently recognized worldwide (Harbach 2023a). Many species are nuisance pests or act as vectors transmitting pathogens and parasites to humans and animals. Since different mosquito species exhibit differences in distribution, biting preference, bionomics and vectorial capacity, it is not the aim of this chapter to present exhaustive details of specific mosquitoes but rather

P. Somboon (✉)
Department of Parasitology, Faculty of Medicine, Chiang Mai University, Chiang Mai, Thailand
e-mail: pradya.somboon@cmu.ac.th

K. Futami
Department of Environment and Vector Ecology, Institute of Tropical Medicine, Nagasaki University, Nagasaki, Japan
e-mail: futami@nagasaki-u.ac.jp

Y. Higa
Department of Medical Entomology, National Institute of Infectious Diseases, Tokyo, Japan
e-mail: saperoi@niid.go.jp

K. Sawabe et al. (eds.), *Medical Entomology in Asia*, Entomology Monographs,
https://doi.org/10.1007/978-981-97-4406-0_1

to provide basic knowledge of their evolution, diversification, taxonomy, classification, morphology, identification, biology and role as disease vectors in Asian countries.

1.2 Origins of Mosquitoes

Estimation of the divergence times at which insects began to evolve is primarily based on fossil records, fossil calibration points, historical biogeography, morphological characteristics and sequence data of mtDNA and nuclear genes of extant species in relation to the sister group. Available fossils and molecular information estimated that the earliest and closest known relatives of flies appeared in the Permian (298.9–251.9 Mya), but the early dipterans began to diversify in the mid- to Late Triassic (230–210 Mya), including the earliest fossils of dipterans with a long proboscis specialized for blood-sucking (Blagoderov et al. 2007). It was not until the Jurassic that definitive insect ectoparasites appeared, and not until the Cretaceous that most of definitive hematophagous insect groups, including mosquitoes (Culicidae), biting midges (Ceratopogonidae), sandflies (Psychodidae), blackflies (Simuliidae), horseflies (Tabanidae), tsetse flies (Glossinidae), kissing bugs (Reduviidae), and louse flies (Hippoboscidae) began to diversify (Grimaldi and Engel 2005; Bertone et al. 2008; Wiegmann et al. 2011; Wiegmann and Yeates 2017). These findings support Edwards (1932), who proposed that mosquitoes probably evolved in the Mesozoic Era (252–66 Mya).

A number of fossil species have been assigned to Culicidae, but most of which were badly preserved and poorly figured, and only 28 can be placed in the family with confidence. Hence, the available fossils are insufficient for conclusions about the actual ages of the Culicidae and extant taxa. Of the Culicidae fossils (only adult stage found) currently available, three were from the Cretaceous period, and the rest were from the Cenozoic Era (Harbach 2023b). Two oldest fossils of extinct mosquitoes were found in mid-Cretaceous Myanmar amber (aged 110–97 Mya) from Kachin Province, northern Myanmar, i.e., *Burmaculex antiquus* Borkent & Grimaldi (Borkent and Grimaldi 2004, 2016) and *Priscoculex burmanicus* Poinar, Zavortink & Brown (Poinar et al. 2019). *Burmaculex antiquus* is very small (wing length 1.6–1.9 mm) and bears several plesiomorphic features, including a short proboscis, which is intermediate between modern mosquitoes and other midges. *Priscoculex burmanicus* is larger (wing length 2.6 mm) and possesses basic morphological features that align it with the subfamily Anophelinae. However, some features, such as the setaceous proboscis and palpus and the long setae on the wing veins, are not typical of modern anophelines and are regarded as ancestral characters. Their proboscis, which consists of a maxillary stylet, is capable of piercing and sucking blood from the host skin. The fossil *Priscoculex burmanicus* is older than *Paleoculicis minututs* Poinar, Zavortink & Johnston from Late Cretaceous Canadian amber (100–66 Mya) (Poinar et al. 2000), which has morphological features that indicate a closer affinity with culicine than with anopheline mosquitoes. It is believed that most primitive mosquitoes were extinct in the Cretaceous–Paleogene extinction event (66 Mya), which caused a sudden mass extinction of three-quarters of the plant and animal species (including dinosaurs) on Earth. This is probably a reason why 90% of Culicidae fossils found have been dated to the Cenozoic Era (66 Mya to present). Fossils found in the Cenozoic Era consist of three extinct genera, *Aetheapnomyia* Harbach & Greenwalt, *Eoaedes* Harbach & Greenwalt, and *Neoculicites* Evenhuis, and six extant genera of Culicidae, i.e., *Aedes*, *Anopheles*, *Coquillettidia*, *Culex*, *Culiseta* and *Toxorhynchites*, which were found mainly in fossils dated to the Eocene (56–33.9 Mya) and Oligocene Epochs (33.9–23 Mya), to a lesser extent Miocene (23.03–5.333 Mya) (Gilka et al. 2021). Based on the available fossils, it is possible that modern mosquitoes may have evolved in the early Cenozoic Era, and their morphological features have exhibited little change at the generic level since then. For example, *Anopheles dominicus* Zavortink & Poinar, contained in Dominican amber from the Eocene (Zavortink and Poinar 2000), has spotted wings similar to the subgenus *Nyssorhynchus*; *Culiseta kishenehn* Harbach & Greenwalt and *Cs. lemniscata* Harbach & Greenwalt (Harbach and Greenwalt 2012) from compression fossils from the 46-million-year-old Kishenehn shale deposits in Montana, USA, appear to share features with extant species of subgenera *Climacura* and *Culicella*, respectively.

Several studies have used molecular methods, morphology and fossil calibration points to estimate the divergence time of Culicidae (e.g., Borkent and Grimaldi 2004; Bertone et al. 2008; Reidenbach et al. 2009; Moreno et al. 2010; Neafsey et al. 2015; da Silva et al. 2020; Lorenz et al. 2021). Although variable depending on taxon samplings, genes and methods used, the results generally agree that the Culicidae and Chaoboridae (non-biting midges) shared a common ancestor and split at ~200 Mya. Although the oldest Culicidae fossils are dated to the Cretaceous, molecular studies suggest that the Culicidae probably evolved in the Jurassic (200–145 Mya), along with the early mammals, first birds and first flowering plants. The Culicinae and Anophelinae probably split in the early Jurassic (~190 Mya). The diversification between *Culex* and Aedini taxa occurred at 130 Mya (da Silva et al. 2020). Neafsey et al. (2015) concluded that *Anopheles* began to diversify 100 Mya during the mid-Cretaceous. Harbach (2013), who advanced the theory of Christophers (1933), hypothesized that the ancestral lineage of *Anopheles* existed before the breakup of Pangaea and subsequently diversified into the extant subgenera after the separation of the continents. The last 50 million years showed a rise in the number of species of both Anophelinae and Culicinae (Lorenz et al. 2021), coincident with the dominance of mammals, birds and flowering plants, which were potentially key elements (e.g., blood meals, nectars and breeding habitats) driving culicid diversification. This may explain why Culicidae fossils were more abundant in the Eocene. Diversification of species within groups of mosquitoes has occurred in the last million years. For example, Morgan et al. (2009) reported that divergent times of species within the Annularis and Maculatus Groups of *Anopheles* occurred during the Pliocene Epoch (5.333–2.58 Mya). More details and discussions of fossil Culicidae, phylogeny, and historical biogeography can be found in Edwards (1932), Lukashevich and Mostovski (2003), Harbach (2013), and Harbach (2023b).

1.3 Evolution of Mosquitoes and Parasitic Infections

The early association of bloodsucking dipterans and haemoprotozoa, e.g., malarial parasites, leishmania and trypanosomes, has long been a subject of interest and speculation. Numerous published documents related to these subjects are available with variable results, so it is not possible to include all of them here. Rather, we create an evolutionary history of haemoprotozoa and their vectors after a comprehensive review of the literature related to the topics.

Fossils of an ancient sandfly, *Palaeomyia burmitis* (Diptera: Phlebotomidae), infected with a *Leishmania* parasite, *Paleoleishmania proterus* (Trypanosomatidae) (Poinar and Poinar 2004; Poinar 2004), and an ancient biting midge (Diptera: Ceratopogonidae) infected with a *Haemoproteus*-like malarial parasite, *Paleohaemoproteus burmacis* (Haemospororida: Plasmodiidae), were detected in Myanmar amber dating back to the Cretaceous (Poinar and Telford 2005), providing a minimum age for the dipteran-haemoprotozoa association. Based on this evidence, both bloodsucking dipterans and haemoprotozoa probably coevolved in the Mesozoic Era, when early birds and reptiles were abundant. Both groups of vertebrates were probably the major blood sources and acted as hosts of the parasites. Today, some species of sandflies are vectors of leishmaniasis, a zoonotic parasitic disease caused by the protozoan of the genus *Leishmania* (Trypanosomatidae), and vectors of *Plasmodium* spp. in reptiles. Some species of biting midges are known to be vectors of the *Haemoproteus* (Haemoproteidae) and *Trypanosoma* spp. parasites in birds, *Hepatocystis* (Plasmodiidae) in bats, *Plasmodium* spp. in reptiles and *Leishmania* spp. in vertebrates including humans (Garnham 1966; Coatney et al. 1971; Sunantaraporn et al. 2021).

Fossils of ancestral Culicidae with haemoprotozoa infections dating to the Cretaceous have not been found, and hence, their historical association with this group of parasites in that period is unknown. Poinar (2005) reported the presence of *Plasmodium* oocysts and sporozoites in a fossil of a female mosquito of the genus *Culex* found in Dominican amber dating to the Eocene, providing a minimum age for the mosquito-malarial parasite association. This finding agrees with molecular studies showing that *Plasmodium* lineages began to diversify in the early Cenozoic Era (Silva et al. 2015; Pacheco et al. 2018, 2020). Today, some species of *Culex* and other culicines are known to be vectors of avian and reptile malarial parasites (Garnham 1966). Culicine mosquitoes are also vectors of *Trypanosoma culicavium* and *T. thomasbancrofti* in avians (Votýpka et al. 2012; Fialová et al. 2021) and *T. rotatorium* in frogs (Desser et al. 1975). The route of transmission of the trypanosomes to the vertebrate hosts is not by injection during biting as for the malarial sporozoites but by ingestion of the infective stage developed in the alimentary tract of infected mosquitoes. From the above evidence, the ancient haemoprotozoa appeared to parasitize a wide range of bloodsucking dipterans, and there might be multiple switches among the insect vectors and vertebrate hosts.

Molecular studies (e.g., Martinsen et al. 2008; Pacheco et al. 2018, 2020) on the extant haemosporidia showed that the genus *Leucocytozoon* (Haemosporida:

Leucocytozoidae), which infects a wide diversity of birds and is vectored by blackflies (Simuliidae), is the oldest lineage but is not monophyletic. Some species form a clade in a sister relationship to other genera of haemosporidian parasites, *Haemoproteus* (including two subgenera *Haemoproteus* and *Parahaemoproteus*), *Plasmodium* and *Hepatocystis*. The genus *Haemoproteus* is polyphyletic within the clade of *Plasmodium* taxa. The subgenus *Haemoproteus* is a clade of parasites that infect doves worldwide and is vectored by louse flies (Hippoboscidae). The subgenus *Parahaemoproteus* is highly divergent, infects a wide range of birds and is transmitted by biting midges of the family Ceratopognidae. The genus *Plasmodium* is paraphyletic and vectored primarily by mosquitoes, and it is relative to *Haemoproteus* and *Hepatocystis*. *Hepatocystis* is closely related to the *Plasmodium* of mammals and vectored by biting midges. However, there are discrepancies in the divergence times and their origins, which depend on taxon samplings, the genes and the methods used.

Based on current knowledge, it is probable that the ancestral form of the parasite groups was most likely an avian parasite similar to *Leucocytozoon,* which evolved in the Cretaceous and was vectored by blackflies (Pacheco et al. 2020). *Haemoproteus* and *Parahaemoproteus* originated in the early Cenozoic with vector shifts from blackflies, first to hippoboscid flies (*Haemoproteus*), and then to biting midges, Ceratopogonidae (*Parahaemoproteus*). *Plasmodium* began to evolve later, in the early Cenozoic (64–40 Mya) (Silva et al. 2015; Pacheco et al. 2018, 2020), and members of this genus are paraphyletic, involving three major lineages, i.e., ungulate, avian and reptile, and mammalian *Plasmodium* spp. Species of the ungulate *Plasmodium* primarily infect bovines, but the vector is not known. The *Plasmodium* species of birds exploit mosquitoes from the genera *Culex*, *Aedes*, and *Culiseta*, with a few species found in *Anopheles*, *Psorophora*, and *Mansonia* (Valkiūnas 2005) (split 40–20 Mya) and the *Plasmodium* species of squamate reptiles (e.g., lizards), which are transmitted by sandflies (split 24–15 Mya) (Pacheco et al. 2018, 2020). There is evidence for repeated switching *Plasmodium* between birds and squamate hosts (Martinsen et al. 2008). The *Plasmodium* species of mammals began to diversify approximately 40 Mya and evolved contemporaneously with their hosts, with little evidence for parasite host-switching on an evolutionary scale, coincident with a switch from Culicine to Anopheline mosquitoes. Today, all *Plasmodium* of mammals, including humans, non-human primates and rodents, are transmitted by *Anopheles* mosquitoes, indicating that the shift of *Plasmodium* into mammals was associated with specialization on anopheline vectors. The age of the split of rodent malarial parasites, which are confined to only Africa, was estimated to be approximately 20–12 Mya (Pacheco et al. 2020). *Hepatocystis* also infects mammals (e.g., bats) but is vectored by biting midges. It is the closest relative to the *Plasmodium* of primates and rodents and split 20–11 Mya (Pacheco et al. 2020). One possible explanation for the placement of *Hepatocystis* is the introgression of the organellar genomes from *Plasmodium* to the ancestral *Hepatocystis* (Martinsen et al. 2008).

The evolution of *Plasmodium* is of particular interest because of the health and economic significance of the parasites. Species of *Plasmodium* that infect mammals are classified into three subgenera, i.e., *Laverania*, *Plasmodium*, and *Vinckeia.* The subgenus *Laverania* is a monophyletic group, including *P. falciparum*, which infects humans, and other *Plasmodium* spp. that infect great apes (e.g., chimpanzees and gorillas). The subgenus *Plasmodium* includes a monophyletic group of species that infect humans, monkeys and apes (e.g., *P. malariae, P. vivax, P. ovale* and *P. knowlesi*). The subgenus *Vinckeia* includes a monophyletic group of species that infect mammals other than higher primates (e.g., rodent and bovine malarias). The subgenus *Laverania* has a sister phylogenetic relationship with the subgenera *Plasmodium* and *Vinckeia* (except bovine malaria). The *Laverania* subgenus split at 24–14 Mya (Pacheco et al. 2018, 2020). This suggests that the diversification of these parasites is contemporaneous with that of the Homininae subfamily (containing humans, chimpanzees, bonobos, and gorillas), which took place in the past 15 Mya (Raaum et al. 2005). It is widely accepted that *P. falciparum*, similar to other species of the subgenus *Laverania*, originated in Africa. However, the divergence of *P. falciparum* has been controversial and frequently revised. Early studies based mainly on sequences of rDNA, mtDNA (partial or genome), and selected genes suggested that *P. falciparum* shared a recent common ancestor with malarial parasites infecting avian (e.g., Waters et al. 1991), chimpanzees (Escalante and Ayala 1995), bonobos (Krief et al. 2010) or rodents (Blanquart and Gascuel 2011). Previously, *P. falciparum* and *P. reichenowi*, which infect chimpanzees, were believed to share a common ancestor and split approximately 5.5–3.0 Mya (Silva et al. 2015). Based on genome sequencing of *Plasmodium* spp., however, it is much more likely that human *P. falciparum* shares a recent common ancestor with the gorilla parasite *P. praefalciparum*, which split 60,000–40,000 years ago followed by a population bottleneck approximately 6000–4000 years ago (Liu et al. 2010; Loy et al. 2017; Otto et al. 2018). Within the subgenus *Laverania*, *P. falciparum* is the only parasite that has successfully adapted to humans after a transfer from gorillas and subsequently spread all over the world.

The *Plasmodium* subgenus, which includes three other species of human malaria parasites (*P. malariae*, *P. ovale* and *P. vivax*) and other primates, appears to have originated in Old World monkeys in Africa (Escalante and Ayala 1995). Rutledge et al. (2017) estimated the split of *P. ovale* and *P. malariae* at 20.3 and 3.5 Mya, respectively. Radiation of Asian non-human primate malaria parasites (e.g., *P. knowlesi* and *P. cynomolgi*) occurred 10–7 Mya (Pacheco et al. 2020). The split between *P. vivax* and *P. cynomolgi* was approximately 5 Mya. All previous models assumed that *P. vivax* originated in humans in Asia, following the cross-species transmission of a monkey parasite, and that humans then brought the parasite to African apes (e.g., Escalante et al. 2005; Neafsey et al. 2012). However, the discovery of *P. vivax* in large numbers of chimpanzees and gorillas provides compelling evidence for an African, rather than an Asian, origin of human *P. vivax* (Liu et al. 2014; Loy et al. 2017). Unlike the ape *Laverania* parasites, which do not normally cause blood stage infections in humans, ape *P. vivax* is much less host-specific and has the potential to infect Duffy-positive humans, suggesting that human and ape *P. vivax* parasites represent a single species (Loy et al. 2017).

1.4 Taxonomy

Mosquito taxonomy encompasses the description, identification, nomenclature, systematics, and classification of insects in the family Culicidae. It is the fundamental knowledge for studying various aspects of mosquitoes, including genetics, fauna, diversity, molecular biology and recognition of sibling or cryptic species. In the past, most species were described primarily based on morphological identification. Cytogenetic, protein electrophoresis, and cross-mating methods were popularly applied to help the recognition of sibling species for a few decades since the 1960s. These methods, however, have several limitations. For example, expertise is required to prepare and interpret polytene chromosome spreads, and they are only available from semi-gravid females or fourth-instar larvae, making them an impractical option for large-scale applications. Protein electrophoresis requires fresh or cryopreserved specimens. The laboratory cross-mating method determines postzygotic isolation, not prezygotic isolation, and requires mosquito colonies at the same time for artificial mating. Therefore, these methods have received less attention after the availability of DNA sequencing, molecular methods, and phylogenetic analyses. Despite a long history of study since the earliest classification of mosquitoes by Linnaeus (1758) and the establishment of the family Culicidae Meigen, 1818, with numerous published documents, the taxonomy of mosquitoes is far from complete.

To date, many genetic species of sibling species complexes await formal names or discovery. Many subspecies and variant forms, identified morphologically, genetically or molecularly, await further investigations for their specific status. A large number of species lack descriptions of adults or immature stages (Wilkerson et al. 2021). Many species do not have a certain taxonomic position in the classification. The phylogenetic relationship of some groups of mosquitoes based on morphological characters does not agree with DNA data, e.g., *Culex* and *Lutzia* (Vesgueiro et al. 2011). Moreover, many unknown species are expected to occur in many areas, especially in isolated and unique ecosystems (e.g., high mountains and islands), where surveys have never been conducted or have not been thoroughly conducted. In the current situation, however, there is a critical shortage of mosquito taxonomists because many senior mosquito taxonomists have retired or passed away, while the number of young researchers who are interested in this field has dramatically decreased.

1.5 Classification

Mosquitoes are organisms classified in the phylum Arthropoda, class Insecta, order Diptera, suborder Nematocera, and family Culicidae. A total of 3721 extant species of mosquitoes are currently formally recognized (as of December 2023). The family is classified into two subfamilies: Anophelinae (523 species) and Culicinae (3198 species) (Harbach 2023a). The Anophelinae are divided into three genera: *Anopheles*, *Bironella* and *Chagasia*.

The genus *Anopheles* consists of 510 formally recognized species, which are placed in eight subgenera:

Anopheles (206 species) (cosmopolitans)
Baimaia (1) (the Thai-Myanmar border)
Cellia (233) (the Afrotropical, Australian, Oriental Regions and southern areas of the Palaearctic)
Christya (2) (sub-Sahara)
Kerteszia (12) (Neotropical Region)
Lophopodomyia (6) (Panama and northern South America)
Nyssorhynchus (45) (Argentina to the southern USA.)
Stethomyia (5) (Southern Central America (Costa Rica, Panama) and northern South America)

The genus *Bironella* includes eight species divided into three subgenera, restricted to the Australasian Region:

Bironella (2 species)
Brugella (3)
Neobironella (3).

The genus *Chagasia* is a small genus of only five species, restricted to South and Central Americas. Phylogenetic analyses revealed that the subfamily Anophelinae is a monophyletic lineage basal to all other Culicidae, and the genus *Chagasia* is an ancestral relationship to all other anophelines.

Based on a classification of the Tribe Aedini proposed by Reinert et al. (2004, 2006, 2009), the subfamily Culicinae has 110 genera and two groups of *incertae sedis*, i.e., *'Aedes' sensu auctorum* and *'Ochlerotatus' sensu auctorum*. However, Wilkerson et al. (2015) argued that Reinert et al.'s classification of Aedini may cause confusion and is not applicable for many reasons. To stabilize the Aedini classification and maximize its usefulness to the operational community, the authors encouraged using the traditional classification and reduced the new genera proposed by Reinert et al. to the subgeneric level. We follow the traditional classification suggested by Wilkerson et al. (2015). As a result, the number of genera in subfamily Culicinae is reduced to 38, which are segregated into 11 tribes as follows:

Aedeomyiini: *Aedeomyia* (7 species).
Aedini: *Aedes* (966), *Armigeres* (58), *Eretmapodites* (49), *Haemagogus* (28), *Heizmannia* (40), *Opifex* (2), *Psorophora* (49), *Udaya* (3), *Verrallina* (95), *Zeugnomyia* (4).
Culicini: *Culex* (817), *Deinocerites* (18), *Galindomyia* (1), *Lutzia* (9).
Culisetini: *Culiseta* (41).
Ficalbiini: *Ficalbia* (8), *Mimomyia* (46).
Hodgesiini: *Hodgesia* (11).
Mansoniini: *Coquillettidia* (58), *Mansonia* (25).

Orthopodomyiini: *Orthopodomyia* (36).
Sabethini: *Isostomyia* (4), *Johnbelkinia* (3), *Kimia* (5), *Limatus* (9), *Malaya* (12), *Maorigoeldia* (1), *Onirion* (7), *Runchomyia* (8), *Sabethes* (42), *Shannoniana* (4), *Topomyia* (68), *Trichoprosopon* (17), *Tripteroides* (126), *Wyeomyia* (141).
Toxorhynchitini: *Toxorhynchites* (101).
Uranotaeniini: *Uranotaenia* (276).

Details of the formal and internal classifications, characteristics, phylogenetic relationships, bionomics and disease relations, distribution, principal references, and updated lists of all species are available at the website "Mosquito Taxonomic Inventory" (https://mosquito-taxonomic-inventory.myspecies.info/), which is highly recommended for further study. Another website that is useful for taxonomic study is https://www.wribu.si.edu/index.php/, compiled and maintained by The Walter Reed Biosystematics Unit.

1.6 Morphology

1.6.1 Adult

Mosquitoes belong to Diptera, which are insects with two wings. Among dipteran insects, insects with the following morphological characteristics can be identified as Culicidae, which we call mosquitoes. Mosquitoes generally have long antennae with 13 flagellomeres or more, wings with scales, and long proboscis for blood and sugar feeding. The head contains two compound eyes, pedicels, antennae, and maxillary palpi, and one proboscis. The color and shape of the scales on the vertex behind the compound eyes are important morphological characteristics for identification. The thorax comprises a variety of body parts, such as wings, scutum, scutellum, antepronotum, postpronotum, spiracles, postspiracular area, paratergite, meskatepisternum, mesanepimeron, coxae, and legs. Different species have different colors, shapes, and patterns of setae and scales on those parts. The abdomen comprises 10 segments and genitalia. Segments IX and X are small. Therefore, dissection is required for morphological observation. Hairy antennae and large genitalia of males compared to females are good characteristics for separating sex.

1.6.2 Pupa

The pupae actively move but do not feed during the pupal period, and their appearance is a comma-like shape (Fig. 1.1). The number of branches and shape of setae on the body are sometimes used for identification but overlap among species compared to larvae. Pupa breathes through two trumpets on its head. Males and females can be separated by the size of the genital lobe in the terminal part of the abdomen.

Fig. 1.1 Pupae of *Aedes albopictus*. (Photo by Kasai, S.)

Fig. 1.2 Larvae of *Aedes albopictus*. (Photo by Kasai, S.)

1.6.3 Larva

The larvae, as well as pupae, breed in aquatic environments and have no legs, antennae on both sides of the head, an abdomen with 9 apparent segments, and a siphon (absent in Anophelinae) for respiration (Fig. 1.2). There are various types of setae and criticized scales on its body. The number of branches and shapes of certain setae and scales are species-specific and well-used for species identification. Larvae such as the genera *Toxorhynchites* and *Lutzia* have developed mouth parts and prey interspecific or intraspecific larvae. The larvae of the genera *Mansonia* and *Coquillettidia* have a modified siphon tapering to the apex to pierce the roots of water plants for respiration.

Details of the terminology used for the morphology of mosquitoes in all stages should be referred to Harbach and Knight (1980) with illustrations.

1.7 Molecular Identification

This section is not aimed at presenting an exhaustive review of molecular methods of identification used in mosquitoes but rather to illustrate the usefulness of approaches with a few examples. Excellent reviews of molecular identification are available elsewhere (e.g., Walton et al. 1999a; Garros and Dujardin 2013; WHO 2017). Molecular techniques have proved very useful in helping to identify mosquito species accurately, which is essential for not only the surveillance and control of vectors transmitting pathogens to humans and animals but also the understanding of mosquito diversity and fauna. Members of species complexes may differ in several ways, including biting behavior, vectorial capacity, distribution and insecticide resistance. Members of species complexes are usually indistinguishable or difficult to identify with certainty by traditional morphological methods due to their similarity or overlapping characters. In addition, morphological identification can also be problematic when trying to identify young instars, eggs and damaged specimens. Of the 22 currently recognized *Anopheles* species complexes, 16 have been detected in the Oriental Region, as listed below. *Anopheles* mosquitoes often exhibit variations in morphology (intraspecific variation) due to genetics, environmental conditions or geographical isolation. For example, the larvae and adults of *An. yaeyamaensis,* a species of the Minimus Complex occurring in the Ryukyu Archipelago, become much larger and darker in the winter (Somboon et al. 2001, 2010). The larvae of the Fluviatilis and Minimus Complexes are difficult to differentiate, while the wings of adults of a variant form of *An. minimus* (varuna-like form) and *An. fluviatilis* are similar (Harrison 1980). To overcome these limitations, the identification of genetic markers in the nucleus and ribosomal and mitochondrial DNAs has the potential to identify isomorphic species complexes. DNA sequences can be obtained by sequencing methods and are available in international databases, e.g., GenBank (available at https://www.ncbi.nlm.nih.gov/nucleotide/) or BOLD (https://www.boldsystems.org/index.php). The similarity of sequences may be determined by using the Basic Local Alignment Search Tool (BLAST, available at http://blast.ncbi.nlm.nih. gov/Blast.cgi). Apart from species identification, the sequences are also used for the study of phylogenetic relationships.

One of the molecular approaches popularly used to obtain taxonomic information from unknown specimens is "DNA barcoding." The method initially involves sequencing a short fragment of the mitochondrial DNA (mtDNA), *cytochrome c oxidase* subunit I (*COI* or *COX1*) gene, "DNA barcodes," and performing comparisons with a library of DNA databases of known taxonomy as a reference (Herbert et al. 2003). The *COI* gene is often used because of its valuable characteristics, including the availability of conserved regions for designing universal primers, the ease of amplification and enough variability to distinguish intra- and interspecific variations. In addition to the *COI* gene, the *COII* gene, *cytochrome c oxidase* subunit II, is also used as a barcode. In general, sequence divergence of >2% at the *COI* and *COII* genes corresponds to variation at the interspecific level in most species of *Anopheles* mosquitoes (e.g., Wang et al. 2012, 2017). However, maximum intraspecific variation in some species groups may be greater than 3% [e.g., species of the *Anopheles lindesayi* Complex (Namgay et al. 2020)]. Interspecific variation may be lower than <2% [e.g., species of subgenus *Metalutzia* of genus *Lutzia* (Phanitchakun et al. 2019) and species of *Culex mimeticus* Subgroup (Somboon et al. 2023a)]. Partial sequences of nuclear genes and ribosomal DNA (rDNA), e.g., the ITS2 region, the internal transcribed spacer 2 located between the 5.8 and 28S subunits, are also often used as barcodes. The spacer regions are particularly suitable for distinguishing closely related species, as they are subject to few functional constraints and are likely to evolve rapidly. While significant differentiation at ITS2 is a clear indication of discrete species (Walton et al. 1999a), the opposite is not necessarily true, as recently diverged species can show little or no divergence at these markers. For example, in comparing ITS2 sequences of five species in the *An. gambiae* complex, differences ranged from a low of 0.4% to a high of 1.6% (Paskewitz et al. 1993). Low divergence at the ITS2 sequences was also found in *An. dirus* Peyton & Harrison and *An. scanloni* Sallum & Peyton of the Oriental Dirus Complex (Walton et al. 1999b) and *An. sundaicus* and *An. epiroticus* of the Sundaicus Complex (Linton et al. 2005). The D3 region of the 28S subunit is also useful for the identification of *An. minimus* complex (Somboon et al. 2001). However, the D3 sequences of *An. fluviatilis* species S and *An. harrisoni* (formally *An. minimus* species C) are identical (Chen et al. 2006), but their ITS2 and 28S-D2 sequences are distinct (Singh et al. 2006). *Anopheles paraliae* in Thailand is considered conspecific to *An. lesteri* from Korea, based on their similar DNA sequences and cross-mating experiments showing no postzygotic isolation barrier, although they are morphologically distinct (Taai et al. 2013).

Species differentiation using data from both mtDNA and rDNA sequences was generally consistent in topology. In some cases, however, while rDNA sequences are distinct, their mtDNA sequences may be very similar and vice versa. *COI* sequences may fail to distinguish some taxa of mosquitoes, e.g., some members of the Hyrcanus Group, which was attributed to mtDNA introgression because recent hybridization transfers mtDNA from one species to another and can lead to mtDNA variation (Fang et al. 2017a, b). Some mosquito populations with identical or similar ITS2 sequences may exhibit significantly different mtDNA sequences. Mitochondrial DNA is maternally inherited and independently evolved from nuclear genes with a faster mutation rate (Castro et al. 1998; Wang et al. 2017). In this case, interpretation should be made with caution, and more supportive evidence is required before a definite conclusion can be made. For example, *An. subpictus* populations in Chiang Mai, Thailand, exhibited three distinct *COI* sequences, but their analyzed ITS2 sequences were very similar (Wilai et al. 2020a). Phylogenetic analyses of *COI* sequences produced a tree with a different topology than the ITS2 tree. However, cross-mating experiments among the three *COI* variants of *An. subpictus* revealed genetic compatibility, supporting the occurrence of nuclear gene flow and mtDNA introgression between them. Therefore, the three *COI* variants are considered a single biological species. Another example is the *An. maculatus* Complex in the Indonesian Archipelago (Ali et al. 2019a). The *COII* sequences of *An. maculatus* from mainland Asia were clearly distinct from those found on several islands of the Indonesian Archipelago, forming six clades, but their ITS2 sequences were recovered in only two clades. Molecular identification should be supported by other evidence, e.g., morphology and, if possible, cross-mating experiments. A new species of *An. maculatus* was successfully identified in central Java using these methods (Ali et al. 2019b).

Since DNA sequencing is rather expensive and time-consuming, using allele-specific PCR and multiplex PCR methods for species identification are much cheaper and suitable for identifying many specimens. The methods are useful for the identification of known species of the complex or closely related species occurring in the study area. Allele-specific primers are designed for specific nucleotide segments of mtDNA or rDNA sequences of each species. Optimization of PCR conditions is required to obtain good amplification DNA products or amplicons, which are visualized as bandings on agarose gel electrophoresis. Multiplex PCR methods based on *COI*, ITS2 or D3 sequences for the identification of *Anopheles* species complexes or species groups were successfully developed, e.g., the Annularis Group (Walton et al. 2007a; Swain et al. 2010), the Barbirostris Complex (Brosseau et al. 2019; Wilai et al. 2020b), the Culicifacies Complex (Manonmani et al. 2007, 2013), the Dirus Complex (Walton et al. 1999b), the Fluviatilis Complex (Manonmani et al. 2001; Singh et al. 2004a), the Hyrcanus Group (Wijit et al. 2013; Fang et al. 2017a, b), the Maculatus Group (Walton et al. 2007b), the Minimus Group (Phuc et al. 2003; Garros et al. 2004) and the Sundaicus Complex (Dusfour et al. 2007a). In some cases, specific nucleotide segments of different species may be very close, so the size of amplification products may be difficult to distinguish in the gel. In addition, non-specific bandings may be difficult to eliminate. To solve these problems, the mismatch method and addition of a long GC tail may be applied for some primers to increase the product size so that the PCR products are clearly distinguishable, e.g., identification of the Barbirostis Complex in Thailand (Wilai et al. 2020b) and the Lindesayi Complex in Bhutan (Zhou et al. 2023). The developed assay should be validated by testing several known species before it is applied to wild-caught mosquitoes. If there is no PCR product in some tested specimens, which could be due to poor DNA quality, intraspecific variation or an unknown species, DNA sequencing would be necessary.

Morphological-based identification is essential for mosquito research. It is not practical to identify all collected specimens by molecular methods. Therefore, a portion of collected specimens should be selected at random for molecular identification to help confirm species of doubtful specimens or when cryptic species are expected to be present. It should be noted that DNA sequences from misidentified specimens are not uncommon in GenBank and other databases. In addition, sequences listed as a particular species may not represent the species from its type locality. For example, *Cx. mimulus* from Bhutan has the *COI* sequence differing from the specimens from Sarawak, Malaysia (its type locality), Singapore and India, which are also different from *Cx. mimulus* from China (Somboon et al. 2023a). Therefore, sequences from databases not derived from holotypic or paratypic specimens should be used cautiously. If they are not available, sequences from specimens newly collected from the same location of the holotype can be representative. It should be noted that outdated or obsolete names, forms, variants, or *sensu lato* listed for some sequences have not been corrected to currently valid names, e.g., *An. campestris* (Paredes-Esquivel et al. 2009) and *An. campestris*-like (Thongsahuan et al. 2011), which was more recently described as *An. wejchoochotei* by Taai and Harbach (2015). These issues are likely to cause confusion and difficulties for researchers who have no knowledge of recent formal taxonomic changes and fail to apply integrated approaches for the collection, rearing, morphological study, and molecular analysis of mosquitoes.

1.8 Species Complexes and Recent Taxonomic Changes

The most recent review of *Anopheles* species complexes in Asian countries was published by WHO (2017). However, more knowledge of *Anopheles* and other genera of mosquitoes has been gained since then. As a result, it is recommended to follow the updated valid species, classification of species groups, subgroups and species complexes, which can be found on the website https://mosquito-taxonomic-inventory.myspecies.info/. The current species complexes of *Anopheles* in Asia with current taxonomic changes are listed below.

Anopheles
Baileyi Complex (Somboon et al. 2020, 2022a)
  baileyi Edwards (species A)
  baileyi (species D)
  bhutanensis Somboon, Namgay & Harbach (species C)
  monticola Somboon, Namgay & Harbach (species B)
  simlensis James
Gigas Complex (redefined by Somboon et al. 2020)
  crockeri Colless
  danaubento Mochtar & Walandouw
  formosus Ludlow
  gigas Giles
  prachongae Rattanarithikul & Harrison
  refutans Alcock
  sumatrana Swellengrebel & Rodenwaldt
Lindesayi Complex (Namgay et al. 2020; Somboon et al. 2022b)
  benguetensis King
  cameronensis Edwards
  druki Somboon, Namgay & Harbach (lindesayi species C)
  himalayensis Somboon, Namgay & Harbach (lindesayi species D)
  japonicus Yamada
  lindesayi Giles (species A)
  lindesayi (species B)
  pleccau Koidzumi

thimphuensis Somboon, Namgay & Harbach (*lindesayi* species E)
Barbirostris Complex (Satoto 2001; Taai and Harbach 2015)
barbirostris van der Wulp
campestris Reid
dissidens Taai & Harbach (species III of Paredes-Esquivel et al. 2009)
saeungae Taai & Harbach (species IV of Paredes-Esquivel et al. 2009)
vanderwulpi Townson & Harbach
wejchoochotei Taai & Harbach
(note: *Anopheles donaldi* Reid is now included in the complex; a new species, *An. sarpangensis* Somboon, Namgay & Harbach, was recently described by Somboon et al. 2024)
Culicifacies Complex (Green and Miles 1980)
culicifacies Giles (species A, B, C, D and E)
Fluviatilis Complex (Sarala et al. 1994)
fluviatilis James (species S, T and U)
Minimus Complex (Green et al. 1990; Harbach et al. 2007a; Somboon et al. 2001, 2010)
harrisoni Harbach & Manguin (species C)
minimus Theobald (species A)
yaeyamaensis Somboon & Harbach (species E) (Fig. 1.3)
Annularis Complex (Atrie et al. 1999)
annularis van der Wulp (species A and B)
pallidus Theobald
philippinensis Ludlow
schueffneri Stanton
Nivipes Complex (Green et al. 1985; Harrison et al. 1990)
nivipes (Theobald) (2 cytogenetic species in Thailand)
Maculatus Complex (Ali et al. 2019b)
maculatus Theobald
maculatus Javanese form (Ali et al. 2019a)

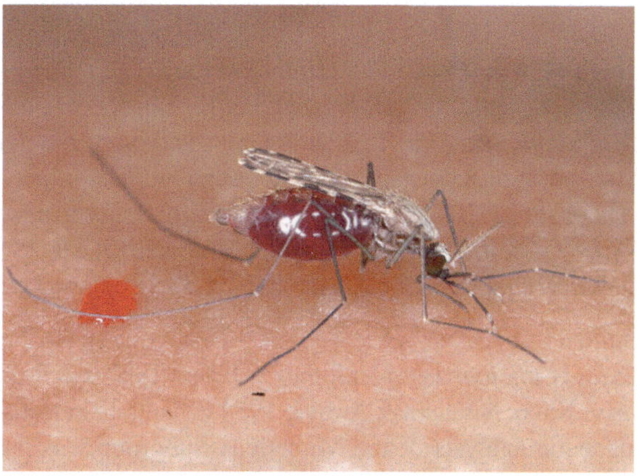

Fig. 1.3 *Anopheles yaeyamaensis*, a vector of *Plasmodium falciparum*. (Photo by Kasai, S.)

Dirus Complex (Sallum et al. 2005b)
  aff. *takasagoensis* (Takano et al. 2010)
  baimaii Sallum & Peyton
  cracens Sallum & Peyton
  dirus Peyton & Harrison
  elegans (James)
  nemophilous Peyton & Ramalingam
  scanloni Sallum & Peyton
  takasagoensis Morishita
Leucosphyrus Complex (Sallum et al. 2005a)
  balabacensis Baisas
  introlatus Colless
  latens Sallum & Peyton
  leucosphyrus Dönitz
Farauti Complex (Schmidt et al. 2003)
  farauti Laveran
  farauti 4 and 5 (Foley et al. 1993)
  farauti 8 (Bower et al. 2008)
  hinesorum Schmidt
  irenicus Schmidt
  oreios Bangs & Harbach, 2014
  torresiensis Schmidt
Tessellatus Complex (Bourke et al. 2021)
  tessellatus Theobald (species A, B, C, D, E and F)
Subpictus Complex (Suguna et al. 1994)
  subpictus Grassi (species A, B, C and D)
  subpictus Chiang Mai form (Wilai et al. 2020a)
Sundaicus Complex (Sukowati et al. 1999)
  epiroticus Linton & Harbach
  sundaicus (Rodenwaldt)
  sundaicus (species B, C, D and E) (Dusfour et al. 2007b)

Aedes

Three subspecies of *Aedes* (*Hulecoeteomyia*) *japonicus* have been formally elevated to species rank: *Ae.* (*Hul.*) *amamiensis* Tanaka, Mizusawa & Saugstad, 1979, *Ae.* (*Hul.*) *shintienensis* Tsai & Lien, 1950 and *Ae.* (*Hul.*) *yaeyamensis* Tanaka, Mizusawa & Saugstad, 1979, together with nominotypical *Ae.* (*Hul.*) *japonicus* (Theobald, 1901) (Wilkerson et al. 2022). Two new subgenera of *Aedes*, i.e., *Reinertia* Somboon, Namgay & Harbach, 2021, (Somboon et al. 2021), and *Orohylomyia* Somboon & Harbach, 2023, (Somboon et al. 2023b) were described.

Culex

Culex (*Culiciomyia*) *spiculothorax* Bram, 1967 was formally synonymized with *Cx. sasai* Kano, Nitahara & Awaya, 1954 (Phanitchakun et al. 2017). A review and phylogenetic analysis of the Mimeticus Subgroup was published by Somboon et al.

(2023a). In this publication, *Cx. gaugleri* Suman and *Cx. kodaikanalensis* Suman, Veer & Chandra, from India (Suman et al. 2021, 2022, respectively), are implicit members of the Mimeticus Complex and are recognized as junior subjective synonyms of *Cx. jacksoni* Edwards, 1934. Moreover, *Culex katezari* Karlekar, Andrew & Deshpande from India (Karlekar et al. 2020), described as a species of the Mimulus Complex, is formally regarded as a *nomen dubium.*

Lutzia
A new species of *Lutzia* (*Metalutzia*), i.e., *Lt. chiangmaiensis* Somboon & Harbach, 2019 was described from Chiang Mai, Thailand (Somboon and Harbach 2019). The larvae and adults of this species were previously misidentified as *Lt. vorax* (Rattanarithikul et al. 2005a; Phanitchakun et al. 2019).

Tripteroides
Tripteroides (*Tripteroides*) *bambusa yaeyamensis* Tanaka, Mizusawa & Saugstad, 1979 was raised to a specific status by Toma et al. (2019).

Recently, in addition, many subspecies of mosquito genera have been elevated to specific status (Harbach and Wilkerson 2023).

1.9 Important Ecological and Biological Features of Mosquito Vectors for Disease Transmission

Mosquitoes' egg, larval and pupal stages spend their life in an aquatic environment, and that of adults is terrestrial. Eggs hatch within a few days after oviposited. Larvae develop for approximately a week in tropical regions and become pupae. Adults emerge and mate. Females start searching for host animals for blood feeding, suck blood to produce eggs, digest blood and oviposit eggs to favorable water habitats. After the oviposition, the females try to find new hosts for eggs. One gonotrophic cycle from blood feeding to oviposition usually takes about three days.

The ecology and biology of vectors related to human-vector contact are essential for transmitting pathogens. Biting time, biting place (endophagy=blood fed indoors or exophagy=blood fed outdoors), resting place (endophily=resting indoors or exophily=resting outdoors), host preference, and larval breeding place are well studied to understand the degree of the human-vector contact (Silver 2008). The mosquito species with high human contact (coincidence of human activity with mosquito biting time, high biting and resting indoor density, and larval breeding vicinity of human dwellings) have high potential for the disease vectors.

Females get infected with pathogens from infected hosts when they feed on blood and transmit to susceptible hosts after the pathogens are amplified enough in their bodies. The period during which fed females become infectious is the extrinsic incubation period (EIP). EIPs of malaria parasites and dengue viruses are affected by temperature and are usually shortened (Chan and Johansson 2012; Stopard et al. 2021). Chan and Johansson (2012) analyzed 146 EIP observations of dengue viruses from 8 studies with censored Bayesian time-to-event models and their best-fitting temperature-dependent EIP model estimated that 95% of EIPs are between 5 and 33 days at 25 °C, and 2 and 15 days at 30 °C, with means of 15 and 6.5 days, respectively. Stopard et al. (2021) also estimated that increasing temperature from 21 to 34 °C decreased the EIP. Therefore, mosquito species with longevity longer enough than the EIP have the potential to be efficient vectors.

1.9.1 Malaria Vectors

The unique feature of malaria vectors is that *Anopheles* species responsible for transmission vary among regions or areas (see Sect. 1.10. Mosquito-borne Diseases). Because several species with different vectorial capacities are involved in transmission, understanding their ecology and biology is significant in controlling malaria. Because most of the malaria vector's feeding activities are high at night, using a bed net effectively avoids the infection. Mosquito species known as effective vectors in Asia, e.g., *Anopheles dirus* and *An. minimus* are generally endophagy and prefer humans as a blood source to other animals.

1.9.2 Dengue Fever/Chikungunya Fever/Zika Vectors

In contrast to malaria parasites transmitted by various local *Anopheles* species, dengue, chikungunya and Zika viruses are transmitted mainly by *Aedes aegypti* (Fig. 1.4) and *Ae. albopictus* (Fig. 1.5) distributed worldwide. They are day-biters. The distribution of the two species overlaps around the world. However, biting place, resting place, and host preference related to human vector contact are

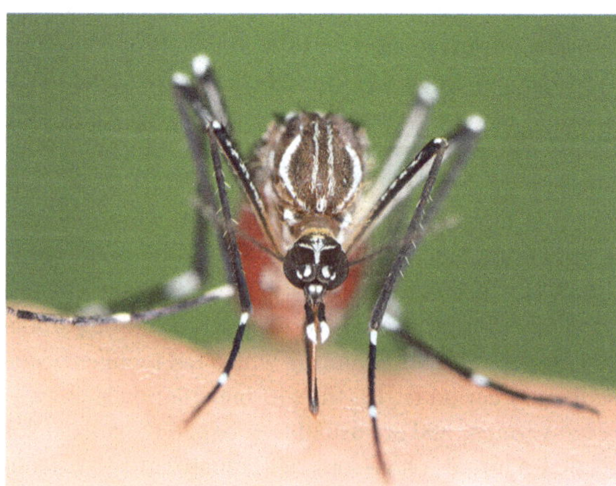

Fig. 1.4 *Aedes aegypti,* a vector of dengue, chikungunya and Zika virues. (Photo by Kasai, S.)

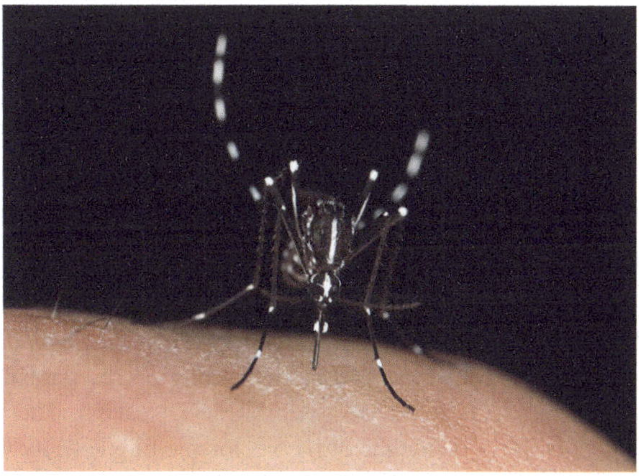

Fig. 1.5 *Aedes albopictus*, a vector of dengue and chikungunya virues. (Photo by Kasai, S.)

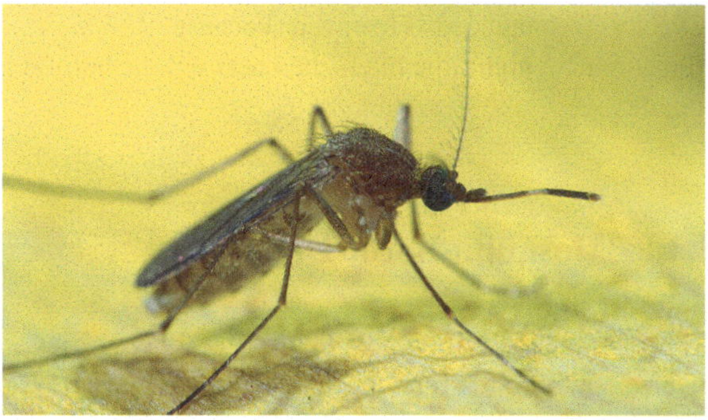

Fig. 1.6 *Culex tritaeniorhynchus*, a vector of Japanese encephalitis virus. (Photo by Kasai, S.)

somewhat different. The former is endophagus, endophilic and abundant in urban areas, and the latter is exophagus, exophilic and abundant in vegetation from urban to rural areas (Rodhain and Rosen 1997). McBride et al. (2014) successfully showed *Ae. aegypti* genetically had a strong preference for humans and responded to human odor in their experiments. Blood meal identification by several studies indicated that *Ae. albopictus* often took blood from humans and also fed on various animals, from mammals to amphibians, depending on host availability (Sawabe et al. 2010; Tamashiro et al. 2011). Those differences make *Ae. aegypti* a more efficient and vital vector than *Ae. albopictus*.

1.9.3 Japanese Encephalitis Vectors

The most important biological feature of the primary vector, *Culex tritaeniorhynchus* (Fig. 1.6), in the Japanese encephalitis virus (JEV) transmission is a robust host preference for pigs, which is the most effective amplifier of JEV. Humans became accidentally bitten by the infected *Cx. tritaeniorhynchus* female and infected with JEV. Because JEV does not amplify enough in humans for the vector to be infective, humans are dead-end hosts. In rural areas, fed females oviposit eggs to large water bodies, e.g., paddy fields, ponds, and ditches. Therefore, it is challenging to control JE using vector control.

1.9.4 Lymphatic Filariasis Vectors

Mosquitoes in the genera *Aedes*, *Anopheles*, *Culex*, *Coquilletidia* and *Mansonia* are known to be filariasis vectors, depending on the parasite species and geographical areas (see Sect. 1.10. Mosquito-borne Diseases). The most interesting ecological aspect is the coincidence of the vector-feeding activity with the daily activity of the parasites (Ichimori 2021). Microfilariae appear in blood at a particular time of day when the feeding density of mosquito vectors is high. In this case, transmission efficacy is enhanced. However, because *Anopheles*, *Culex*, *Coquilletidia*, and Mansonia are nighttime biters, bed nets will be helpful in avoiding bites by infected females.

1.10 Mosquito-Borne Diseases

Approximately 4 billion people in over 129 countries are at risk of mosquito-borne diseases, with an estimated 500,000 deaths every year (WHO 2023a). Principal mosquito-borne diseases in Asia include malaria, filariasis, and viral diseases, e.g., dengue, chikungunya, Zika and Japanese encephalitis.

1.10.1 Malaria

Malaria is one of the most important mosquito-borne diseases caused by *Plasmodium* parasites transmitted to humans by infected *Anopheles* female mosquitoes. Four human malaria species include *P. falciparum*, *P. vivax*, *P. malariae* and *P. ovale*. Most species of malarial parasites are common in tropical and subtropical regions, but *P. vivax* can be found in temperate areas where the summer isotherm is >15 °C (Garnham 1966). Several species of *Anopheles* mosquitoes in Asia are vectors of malarial parasites, but their vectorial capacity and role in malaria transmission vary depending on local epidemiological conditions. Mosquitoes of the subgenus *Cellia* are generally susceptible to all human malaria parasites. In contrast, mosquitoes of subgenus *Anopheles*, e.g., *An. sinensis* (Fig. 1.7) and *An. barbirostris s.l.* are generally refractory to *P. falciparum* and may also be less susceptible to some strains of *P. vivax* (Somboon et al. 1994; Thongsahuan et al. 2011). In Japan, *An. yaeyamaensis* (*An. minimus* species E) (Fig. 1.3) was the primary vector of *falciparum*

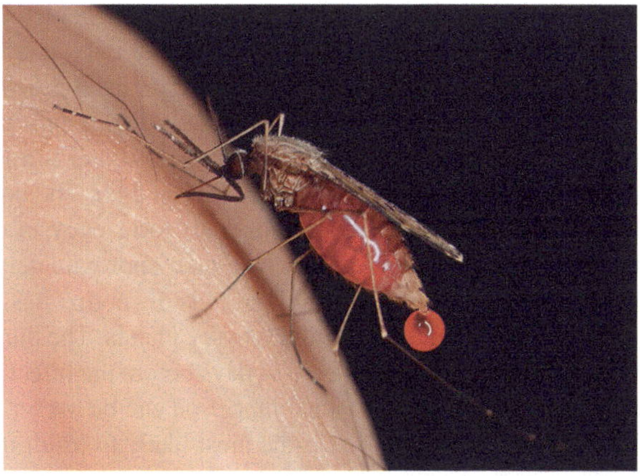

Fig. 1.7 *Anopheles sinensis*, a vector of *Plasmodium vivax*. (Photo by Kasai, S.)

malaria in the southern islands of the Ryukyu Archipelago, particularly after World War II until 1962, when transmission was completely interrupted (WHO 1966). Members of the Hyrcanus Group, e.g., *An. sinensis* (Fig. 1.7) and *An. lesteri* are the primary vectors of *vivax* malaria in China and north Korea (Sinka et al. 2011). Species of the Leucosphyrus Group, e.g., *Anopheles dirus* complex and *An. balabacensis,* the Minimus Subgroup, e.g., *An. minimus* and *An. fluviatilis s.l.,* and the Maculatus Group, e.g., *An. maculatus, An. pseudowillmori, An. sawadwongporni* and *An. willmori* are the primary or secondary vectors in rural and forest areas across Asian countries. Species of *Anopheles culicifacies* and *An. subpictus* complexes are the primary vectors in India and Sri Lanka. The *Anopheles sundaicus* complex is an important vector along the coastal areas and islands in south and southeast Asia. In addition, humans infected with *P. knowlesi*, a species of malarial parasite that naturally infects macaques in Southeast Asia, have been increasingly reported. Mosquitoes of the Leucosphyrus Group are known to be vectors of this parasite.

The life cycle of malarial parasites consists of sporogony in female *Anopheles* mosquitoes and schizogony in vertebrate hosts. In the infective stage, sporozoites are injected into blood circulation via the saliva of infected mosquitoes during blood feeding. Within one hour, the sporozoites enter hepatocytes and develop into merozoites, which take a minimum of seven days or more for maturation, depending on the species and strains of the parasites. Some sporozoites of *P. vivax* and *P. ovale* do not develop once entering the liver cells but retain a stage called hypnozoites, which begin to develop into merozoites several weeks or months post-infection. Each liver stage forms several thousand exoerythrocytic merozoites, which are ultimately released into the blood circulation and infect erythrocytes for further development. In erythrocytes, the parasites develop into ring forms (early trophozoites) and consume hemoglobin to grow. Each parasite produces erythrocytic merozoites (6–16 or more), which take approximately 48 or 72 h for maturation, depending on the parasite species. The merozoites released from broken host cells are ready to infect new erythrocytes for the next cycle of merozoite development. Some trophozoites develop into micro- (male) and macro- (female) gametocytes waiting for a mosquito to pick up upon blood feeding. Shortly (5–20 min) after entering the midgut of mosquitoes, each microgametocyte develops producing 6–8 microgametes, while macrogametocytes develop to macrogametes without multiplication. The microgametes are ultimately liberated in the midgut lumen and fertilized with macrogametes to become zygotes and then ookinetes within 24 h. The ookinetes are motile and penetrate the epithelium of the midgut to form oocysts on the midgut wall. Inside oocytes, sporozoites develop which take approximately one week for maturation. Each mature oocyte produces thousands of sporozoites, which are released into the haemocoel and finally enter the salivary glands. The duration in the mosquito phase is generally approximately 12–14 days before the infected mosquito becomes infectious, but this extrinsic incubation period can be longer in cool conditions.

Plasmodium falciparum is the most likely to result in severe infections and may lead to death. WHO estimated 241 million cases in 2020 in 85 malaria-endemic countries located in tropical and subtropical regions, and malaria deaths stood at 627,000 (WHO 2023b). Most cases occur in Africa and account for approximately 95% of malaria cases globally. The global trends of malaria incidence and deaths have decreased over the last two decades. WHO sets global targets to eliminate malaria transmission in 2030. For large-scale deployment, WHO recommends pyrethroid-only long-lasting insecticidal nets (LLINs) for preventing and controlling malaria in areas with ongoing malaria transmission but no pyrethroid resistance of vectors. In areas with ongoing malaria transmission where the principal malaria vector(s) exhibit pyrethroid resistance conferred mainly or partially by a monooxygenase-based resistance mechanism, pyrethroid-PBO nets is recommended. PBO or piperonyl butoxide is an oxidase enzyme inhibitor that can reduce the resistance level of enzyme-based resistant mosquitoes. However, it has no effect on resistance caused by knockdown-resistance (*kdr*) genes. Malaria vaccines are not available for wide use in the community. However, a malaria vaccine, RTS,S/AS01 (Mosquirix™), is under evaluation (Laurens 2020).

1.10.2 Dengue

Dengue is a flavivirus that causes a wide spectrum of diseases, ranging from no or mild symptoms (e.g., flu-like symptoms) to severe symptoms, e.g., dengue hemorrhagic fever and dengue shock syndromes. The disease is a significant public health problem in tropical and subtropical areas. The virus is transmitted by female mosquitoes mainly of the species *Aedes aegypti* (Fig. 1.4) and, to a lesser extent, *Ae. albopictus* (Fig. 1.5). There are four distinct, but closely related, serotypes of the virus that cause dengue (DENV-1, DENV-2, DENV-3 and DENV-4). Recovery from infection is believed to provide lifelong immunity against that serotype. Subsequent infections (secondary infection) by other serotypes increase the risk of developing severe dengue. Severe dengue was first recognized in the 1950s during dengue epidemics in the Philippines and Thailand. The global incidence of dengue has grown dramatically, with approximately half of the world's population now at risk. Although an estimated 100–400 million infections occur each year, over 80% are generally mild and asymptomatic (WHO 2023c).

Japan is not a dengue-endemic country, although dengue cases were continuously observed from 1942 to 1945. Approximately 50 to 200 imported and confirmed cases are annually reported (Wang and Nishiura 2021). In the summer of 2014, a large local outbreak occurred in Tokyo, involving 160 confirmed dengue cases caused by DENV-1. Most of the cases were associated with Yoyogi Park in Tokyo, where the potential vector *Ae. albopictus* was abundant. The virus was probably introduced by tourists from Southeast Asia. In Europe, over 3000 cases have been reported in recent years, principally due to imported cases. However, local dengue outbreaks and other mosquito-borne viral diseases have occurred in several European countries (WHO 2023c). The invasive species, i.e., *Ae. albopictus* and *Ae. japonicus* (Fig. 1.8), which are native Asian mosquitoes, are potential vectors responsible for autochthonous transmission (WHO 2013).

In endemic countries, dengue transmission is associated with the rainy season due to the increased availability of breeding sites in water-collecting containers. Humans are known to be the major reservoir of dengue viruses. In addition, dengue positivity was detected in bats (10.1%), non-human primates (27.3%), birds (11%), bovid (4.1%), dogs (1.6%), horses (5.1%), pigs (34.1%), rodents (3.5%), marsupials (13%) and other small animals (7.3%), as reviewed by Gwee et al. (2021). However, the role of these reservoir animals in transmitting the virus to humans is not clearly known and requires more investigation. Control of dengue transmission is difficult for several reasons: (1) no effective dengue vaccine is available for use in the community, and there is no specific drug for treating dengue infection; (2) the mosquito *Ae. aegypti* vector has developed resistance to several classes of insecticides used as larvicides or adulticides. It hence reduces the efficacy of vector control using insecticide-based methods, (3) people usually collect water in containers in their properties to provide breeding sites for the vectors, (4) community participation to

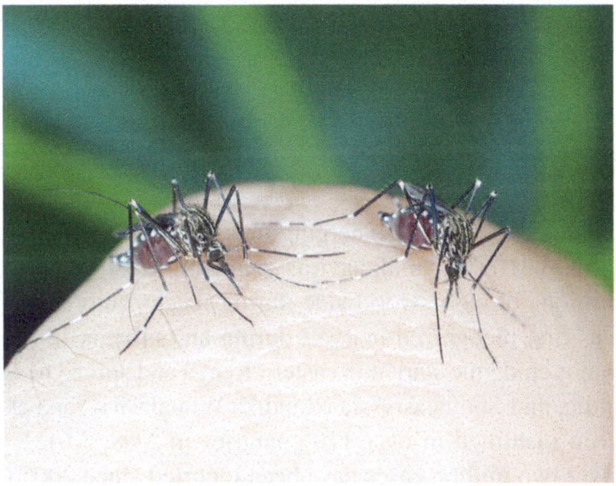

Fig. 1.8 *Aedes japonicus*. (Photo by Kasai, S.)

prevent and eliminate the breeding sites of the vector is often active for a short period, and (5) the virus can pass from the female to their eggs (vertical transmission). The emerging females can readily be infected without taking any infected blood meal. Male mosquitoes that emerge from infected eggs can also be infected and transfer the virus to virgin females via copulation. This can maintain the virus in the vector population in the absence of human and other vertebrate carriers.

1.10.3 Japanese Encephalitis

Japanese encephalitis (JE) is a zoonotic viral disease caused by the Japanese encephalitis virus (JEV), which was first documented in 1871 in Japan. Today, it is the leading cause of encephalitis in Southeast Asia and the Western Pacific region. The annual incidence of clinical disease varies both across and within endemic countries, ranging from <1 to >10 per 100,000 population or higher during outbreaks. It is estimated that nearly 68,000 clinical cases of JE occur globally each year, with approximately 13,600 to 20,400 deaths (WHO 2023d). JE primarily affects children. Most JEV infections are mild (fever and headache) or without apparent symptoms, but approximately 1 in 250 infections results in severe clinical illness. Severe disease is characterized by rapid onset of high fever, headache, neck stiffness, disorientation, coma, seizures, spastic paralysis and ultimately death. Of those who survive, 20–30% suffer permanent intellectual, behavioral or neurological sequelae such as paralysis, recurrent seizures or the inability to speak.

Safe and effective JE vaccines are available to prevent disease. Pigs are the primary reservoir host of JEV, while birds may also be important in transmission. *Culex tritaeniorhynchus* (Fig. 1.6), *Cx. vishnui, Cx. gelidus* and *Cx. fuscocephala* are known to be vectors of the virus. They breed primarily in rice fields. In Thailand, the peak of larvae of vector populations occurred during ploughing activity and rapidly declined after transplanting rice towards harvesting (Somboon et al. 1989). In Japan, however, the larvae of *Cx. tritaeniorhynchus* are abundant throughout the transplanting period (Mogi and Wada 1973).

1.10.4 Chikungunya

Chikungunya is a re-emerging mosquito-borne arbovirus caused by Chikungunya virus (CHIKV), an alphavirus genus of the family Togaviridae, which has caused millions of cases worldwide during recent outbreaks (WHO 2023e). The disease was first recognized in 1952 during an outbreak in southern Tanzania. In 2004, a major epidemic started in eastern Kenya and spread to several Indian Ocean Islands, India, and Southeast Asia (Randika Wimalasiri-Yapa et al. 2019). CHIKV has now been identified in over 110 countries in Asia, Africa, Europe, and the Americas. Over two million cases have been reported since 2005 (WHO 2023e). Three major

CHIKV genotypes have been identified: West African, East/Central/South African and Asian. Joint pain in both arms and legs is common in infected patients and may last for a few months. The virus is transmitted by the same vectors of dengue virus, *Ae. aegypti* (Fig. 1.4) and *Ae. albopictus* (Fig. 1.5), and as a result, co-infection of the two viruses is not uncommon.

1.10.5 Zika Virus Infection

Zika virus infection is a viral disease caused by Zika virus (ZIKV). It is also an emerging mosquito-borne virus that was first isolated from a sentinel rhesus monkey in the Zika Forest in Uganda in 1947. From the 1960s to the 1980s, sporadic human infections were detected across Africa and Asia. However, since 2007, outbreaks of Zika virus disease have been recorded in Africa, the Americas, Asia and the Pacific. To date, a total of 89 countries and territories have reported evidence of mosquito-transmitted Zika virus infection, but the incidence is not known because infection in humans usually causes either mild disease or no symptoms (WHO 2023f). However, ZIKV infection can be associated with the development of Guillain Barré syndrome, and infection during the early stages of pregnancy has been linked to congenital microcephaly syndrome. *Aedes aegypti* (Fig. 1.4) and *Ae. albopictus* (Fig. 1.5) are the vectors of ZIKV.

1.10.6 Lymphatic Filariasis

Lymphatic filariasis is caused by infections with filarial parasites through mosquitoes. Infection is usually acquired in childhood, causing damage to the lymphatic system. Chronic infections result in profoundly disfiguring visible manifestations of the disease, lymphoedema, elephantiasis and scrotal swelling occurring later in life and can lead to permanent disability. In 2020, approximately 800 million people in 50 countries were living in endemic areas, with an estimated 25 million men with hydrocele and over 15 million people with lymphoedema, and at least 36 million people remain with these chronic disease manifestations (WHO 2023g).

The parasites that cause lymphatic filariasis are classified as nematodes (roundworms) of the family Filariodidea. There are three types of these thread-like filarial worms: *Wuchereria bancrofti,* which is responsible for 90% of the cases, mainly in Africa, Asia and South America. *Brugia malayi* which causes most of the remainder of the cases, mainly in Asia. *Brugia timori* is confined to only the Lesser Sunda Islands of Indonesia.

Adult worms inhabit the lymphatic vessels and disrupt the normal function of the lymphatic system. The worms can live for approximately 6–8 years and, during their lifetime, produce millions of microfilariae (immature larvae) that circulate in the blood. Mosquitoes are infected with microfilaria by ingesting blood when biting

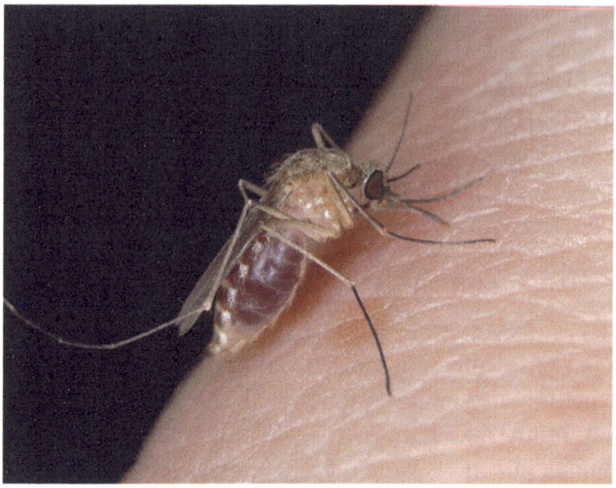

Fig. 1.9 *Culex quinquefasciatus*, a vector of filaria parasite. (Photo by Kasai, S.)

an infected host. The microfilariae shed sheaths, penetrate the midgut, and migrate to the thoracic muscles where the microfilariae increase in size, molt, and develop into infective larvae (L1 to L3) over a span of 7–21 days. Third-stage infective larvae move to the head and proboscis of the mosquito. When infected mosquitoes bite people, mature parasite larvae are deposited on the skin and migrate through the wound to the lymphatic vessels, where they develop into adult worms.

Several species of the genera *Aedes*, *Anopheles*, *Culex*, *Coquilletidia* and *Mansonia* are known to be vectors of filariasis, depending on the parasite species and geographical areas (Iyengar 1953; Wharton 1962). *Culex quinquefasciatus* (Fig. 1.9) is a major vector of *W. bancrofti* across urban and semi-urban areas in Africa, Asia and South America. Some species of *Aedes* and *Anopheles* are vectors of *W. bancrofti* in rural and forest areas. Some species of *Mansonia* and *Coquilletidia* are vectors of *B. malayi*. Some species of *Aedes* are vectors of *W. bancrofti* in endemic islands in the Pacific.

1.11 Checklists of Mosquitoes in Asia and Principal References

Morphological identification of mosquitoes is fundamentally important for studying mosquitoes, especially when working on wild-caught specimens. Misidentification may badly affect the validity of the results. Adult and larval stages are generally used for morphological identification. For beginners, it is necessary to learn the external anatomical characters of adults and larvae and practice identifying specimens that were already identified or confirmed by an expert. If available, keys to the mosquitoes in that country should be used. If this is not possible, keys to the mosquitoes found in neighboring countries may be applied. In this case, it is important to check if the keys used include all species occurring in the country or not. If not, additional keys which include the missing species are required.

Checklists of mosquitoes occurring in most Asian countries have been published, as shown below. It should be noted that new species of mosquitoes may be described after the checklists of each country were published.

Bangladesh
  Irish et al. (2016)
Bhutan
  Namgay et al. (2018)
Cambodia
  Maquart et al. (2021)
China
  Lu et al. [*sic*]. (1997)
Hong Kong
  Rueda et al. (2004)
India
  Rao (1984), Nagpal and Sharma (1995), Bhattacharyya et al. (2014)
Indonesia
  Elyazar et al. (2013)
Japan
  Tanaka et al. (1979)
  Toma and Miyagi (1986)
Korea
  Ree (2003)
Lao PDR
  Oo et al. (2004), Motoki et al. (2019)
Pakistan
  Aslamkhan (1971)
Philippines
  Delfinado (1966), Cagampang-Ramos and Darsie (1970), Rattanarithikul and Harbach (1990)
Singapore
  Lam-Phua et al. (2019)
Southwestern Asia
  Harbach (1988)
Thailand
  Rattanarithikul and Panthusiri (1994), Rattanarithikul et al. (2005a, b, 2006a, b, 2007, 2010), Taai and Harbach (2015)
Vietnam
  Bui and Darsie (2008)

1.12 Principal and Suggested Taxonomic References (Alphabetical Order)

Abercrombie (1977), Barraud (1934), Belkin (1962), Bonne-Wepster and Swellengrebel (1953), Bram (1967), Christophers (1933), Colless (1958, 1959), Delfinado (1967, 1968), Edwards (1932), Gutsevich et al. (1974), Harbach et al. (2007b), Harbach and Kitching (1998), Harbach and Knight (1980, 1981), Harrison (1980), Harrison and Scanlon (1975), Harrison et al. (1990), Huang (1972, 1977a, b, 1979), Klein (1973), Knight (1968), Knight and Stone (1977), La Casse and Yamaguti (1950), Leicester (1908), Lien (1968), Macdonald (1960), Mattingly (1957a, b, 1958a, b, 1959, 1960, 1961, 1970, 1971, 1981), Peyton (1977), Reid (1968), Reinert (1970, 1972, 1973a, b, c, 1974, 1976a, b, 1979, 1981, 1985, 1989, 1990, 1999, 2000a, b, 2003, 2006, 2009a, b), Reinert et al. (2004, 2006, 2008, 2009), Sallum et al. (2005b), Sirivanakarn (1972, 1976, 1977), Thurman (1959), Toma and Miyagi (1986), Tyson (1970), Wharton (1962), Wilkerson et al. (2015), Wilkerson and Peyton (1990), Zavortink (1971)

1.13 New Diseases

1.13.1 The New Serotype and Genotype of Endemic Viruses

1.13.1.1 Dengue Virus Serotype 5 (DENV-5)

This new serotype of the dengue virus isolated during outbreaks in Malaysia Sarawak in 2007 was first taken for DENV-4. However, it did not respond to DENV-4 diagnostic tests, and its entire genome sequence constructed a new branch of the dengue family tree (Normile 2013). DENV-5 is considered to be the sylvatic strain maintained in macaques in the forest of Sarawak. A spillover of this strain into humans resulted in the outbreak in Sarawak (Mustafa et al. 2015). Phylogenetic analysis revealed that the other four serotypes evolved from the sylvatic strain of each serotype and adapted to humans (Weaver and Vasilakis 2009). The possibility of an evolution of an urban strain from this sylvatic strain due to suspected little or no adaptive barrier in DENV-5 to limit the infection to primates, including humans, and the impact of frequent spillover into humans on vaccine development call for a more integrated control program (Mustafa et al. 2015).

Japanese Encephalitis Virus Genotype V (JEV genotype V) Detailed genetic analysis of the Muar strain isolated in 1952 from a patient in Malaysia revealed that its genotype is clearly different from the other four genotypes and branched from them in the 1500s, indicating that it is the oldest strain of JEV (Mohammed et al. 2011). This new genotype, genotype V, had not been reported for 57 years but was isolated from *Cx. tritaeniorhynchus* collected in Tibet in 2009 (Li et al. 2011). JEV genotypes IV and V circulate in the Indonesia-Malaysia region, while genotypes

I-III circulate throughout Asia (Solomon et al. 2003; Mohammed et al. 2011). In South Korea, after the first report of genotype V isolated from *Cx. bitaeniorhynchus* in the northern part of the country, including Kyeongii-do, the new genotype was repeatedly isolated from *Cx. pipiens*, *Cx. orientalis*, *Cx. bitaeniorhynchus*, *Cx. tritaeniorhynchus* (Takhampunya et al. 2011; Woo et al. 2020; Lee et al. 2022). This repeated isolation of genotype V in the past decade suggests the occurrence of a genetic shift that is frequently reported from JEV (Lee et al. 2022). Genotype V has not yet been reported in Japan and Southeast Asia (Maeki et al. 2018; Kuwata et al. 2020). However, considering its high pathogenesis in mice (de Wispelaere et al. 2015) and the low immunity of the present JEV vaccine against genotype V (Cao et al. 2016), the expansion of this new genotype should be monitored with caution.

1.13.2 Risk of Yellow Fever and West Nile Fever

There has been no report of yellow fever (YF) outbreaks in Asian countries, and the reason has not yet been determined (Weaver and Reisen 2010). However, the recent increase in global human transportation has raised the possibility of an alarming YF epidemic in Asia (Wasserman et al. 2016; Wilder-Smith and Leong 2017; Wilder-Smith et al. 2019). During the YF outbreak in Africa in 2016, more than ten travelers returning from Angola to China presented symptoms, which was the first report of YF in Asia (Wasserman et al. 2016; Wilder-Smith and Leong 2017). The patients were reported from Fujian Province, an area where *Ae. aegypti*, the major vector of YF, is distributed. Additionally, it is suggested that Asian *Ae. aegypti* shows high competency to YFV (Lataillade et al. 2020). These factors suggest the possibility of autochthonous transmission of YF in Asia and prior control measures, including virus surveillance (Wasserman et al. 2016).

West Nile Virus (WNV) has been reported sporadically in Africa, the Middle East, and Europe, but it has expanded to the Western hemisphere since the outbreak in New York in 1999 (Kilpatrick 2011). However, there are no reports from Southeast and East Asian countries. The reason why WNF outbreaks have not occurred in these areas is unknown, although the presence of cross-immunity between WNV and other viruses, such as JEV, is estimated. However, JEV outbreaks have occurred in India, including Assam, the northeastern area where JEV is distributed (Paramasivan et al. 2003; Khan et al. 2011). Considering the presence of *Culex* vectors and viral adaptation to hosts, we need to keep looking out for the distribution of WNV in Asia (Kilpatrick 2011; Dash et al. 2013). In Thailand, WNV was not reported during mosquito surveys from 2009 to 2010 (Changbunjong et al. 2012). On the other hand, in South Korea, the virus has been detected in domestic pigeons (Kim et al. 2016). In Cambodia, there is serological evidence of WNV from domestic birds (Auerswald et al. 2020). Additionally, the cause of an outbreak of viral encephalitis in Xinjian, China in 2004 is assumed to be WNV based on serological evidence (Liang et al. 2018). In Japan, there has not been any report of WNV circulation, but detailed vector research is based on particular attention to WNV vectors, considering the epidemics in Russia. *Culex pipiens* group and *Cx. pipiens* form *molestus*, the major vectors of WNV, are distributed in urban areas in Japan with high density, frequently feeding on birds and humans and developing insecticide resistance; Japan must be alert to signs of the virus invasion (Kobayashi et al. 2008).

1.13.3 New Plasmodium Parasites for Humans

One of the most important topics of malaria infection in the 2000s is the discovery of a high prevalence of simian malaria in human populations. *Plasmodium knowlesi* (Pk) is a simian malaria in Southeast Asia, and its human infection has rarely occurred. However, in 2004, a large-scale infection with Pk in a human population was reported in Sarawak, Malaysian Borneo (Singh et al. 2004b). In the Kapit division, a nested PCR assay showed that 106 (51%) of 208 malaria cases investigated were solely infected with Pk and 14 (7%) were coinfected with Pk and other *Plasmodium* species (Singh et al. 2004b). After this report, it was revealed that Pk was morphologically confused with *P. falciparum* (Pf) and *P. malariae* (Pm) and serologically confused with *P. vivax* (Pv) due to similar antigens and that Pk is distributed in a wide range of Southeast Asia, mainly in Malaysia (Jongwutiwes et al. 2004; Cox-Singh et al. 2008; Anderios et al. 2010; Jeyaprakasam et al. 2020). Recently, imported Pk cases from Southeast Asian countries have been reported in Europe, the Americas, other Asian areas and Oceania. The significant increase in the number of Pk cases in recent years in Southeast Asia may be partly due to more diagnostic tests than before. Other possible reasons may include an increase in contact chance due to deforestation, a decrease in cross-immunity due to a decrease in human malaria cases, a relaxation of the malaria control programme and the possibility of human-to-human transmission. To date, six Anopheline species (*An. hackeri, An. cracens, An. dirus, An. balabacensis, An. introlatus, An. latens*) are known as Pk vectors, and all species belong to the Leucosphyrus group. From these species, multiple infections of *Plasmodium* spp. including human malaria have been found, suggesting that they are effective bridge vectors because of the feeding preference of both monkeys and humans.

In simian populations, the infection rates of *P. cynomolgi* (Pcy) and *P. inui* are higher than that of Pk, suggesting that we must cautiously consider their epidemiology (Jeyaprakasam et al. 2020). Humans naturally infected with Pcy were reported for the first time in 2014 in Malaysia (Ta et al. 2014), followed by reports from Thailand and Cambodia (Kojom Foko et al. 2022). The number of species determined or considered its vector mosquitoes is more than that of Pk (19 species, Table 3 in Kojom Foko et al. 2022) and *An. minimus*, a famous human malaria vector in Southeast Asia, is also included. Although there are few reports of natural infection of humans with Pcy, silent epidemics could occur due to the overlap of vector species with human malaria, the nonsevere symptoms of humans infected with Pcy, the production of hypnozoite and microscopic and genetic similarity to Pv (Ta et al. 2014). Thus, the development of an effective and rapid test kit for Pk and

Pcy is required for quick diagnosis (Kojom Foko et al. 2022); an erythrocytic cycle of Pk is 24 h, which is shorter than the four human malarial cycles, and a delay in diagnosis possibly causes severe symptoms in patients.

One of the attractive topics of Pk is the periodicity of gametocyte production (Anderios et al. 2010). In the laboratory experiments, Pk showed that the nocturnal production of gametocytes at 24:00–6:00 overlapped with the feeding time (19:00–6:00) of the vector mosquito, *An. latens* (Anderios et al. 2010). Such interaction between the parasite and vector mosquitoes has been observed between lymphatic filariasis and various vector species, which expects the coevolution of simian *Plasmodium* and its bridge vector mosquitoes.

1.14 Current Topics

1.14.1 Effect of Tsunami

During the 2000s, large earthquakes occurred repeatedly in Asia, which caused tsunami disasters in the coastal areas. The intrusion of seawater onto land by the tsunami has resulted in both negative and positive effects on the habitats of vector mosquitoes in the suffering areas.

On 26th December 2004, a large earthquake hit off the west coast of northern Sumatra, Indonesia, and the tsunami disaster attacked the surrounding coastal areas such as India, Sri Lanka, Indonesia and Thailand. The intrusion of the tsunami far inland resulted in the establishment of new seawater marshes and stagnant brackish water bodies at rice fields and fallow paddy fields. These new water bodies became favorable habitats for *An. sundaicus* and caused an increase in malaria cases in the Adaman and Nicobar Islands, India (Krishnamoorthy et al. 2005); this malaria outbreak had been suppressed by the proper control program (Kumari et al. 2009). On the other hand, malaria outbreaks have not been reported in mainland India and Sri Lanka because there are no *Anopheles* species that emerge from brackish water (Balaraman et al. 2005; Briët et al. 2006). Furthermore, it was suggested that the intrusion of seawater into rice fields, which are the main habitats of the JEV vector, *Cx. tritaeniorhynchus* (Fig. 1.6), may increase salinity and lead to a decrease in the risk of vector-borne diseases, including JE (Balaraman et al. 2005). In Thailand, the increase in brackish habitats led to more collections of *Cx. sitiens*, which prefers brackish water such as *An. sundaicus*, but the data on the prevalence of mosquito-borne diseases were not reported (Prummongkol et al. 2012).

The 2009 earthquake in Padang did not lead to a tsunami. However, the rainy season started soon after the earthquake, and the residents stored water in containers instead of the broken pipeline, which led to the dengue outbreak due to the increase in the vector habitats (Fanany 2011, 2012).

In Japan, the 2011 earthquake off the Pacific coast of Tohoku on 11th March led to a tsunami disaster followed by a change in the mosquito population. During the mosquito survey in Iwate, Miyagi and Fukushima prefectures in 2011 after the tsunami disaster, more mosquitoes were collected in the affected areas than in nonaffected areas; *Cx. pipiens pallens*, *Cx. tritaeniorhynchus* and *Cx. inatomii* were the dominant species in the adult collection, and *Ae. togoi* and *Cx. inatomii* were in the larvae (Watanabe et al. 2012; Tsuda et al. 2012). On the other hand, *Ae. albopictus* was abundant in residential areas without tsunami damage (Tsuda et al. 2012). The observed habitats of *Ae. togoi* and *Cx. inatomii* were brackish water (average salinity in the range of 0.07–0.88 %, Watanabe et al. 2012), which suggests that the tsunami produced new habitats for those mosquitoes. However, the identification of feeding hosts of those mosquitoes revealed that they contacted humans with low frequency, leading to a low risk of mosquito-borne infectious diseases (Watanabe et al. 2012). Furthermore, *Cx. inatomii* does not have enough flying ability to reach residential areas and rarely contacts humans (Tsuda et al. 2012). Interestingly, *Ae. vexans nipponii*, which was the dominant species in the area in 1965, was not collected in the survey in 2011. This may occur because the tsunami flushed out its overwintered eggs in the rice fields (Tsuda et al. 2012).

Increasing sea levels due to global warming may possibly expand brackish/saline water bodies and induce more severe tsunami disasters, which could provide effects similar to those of the previous tsunami on mosquito populations (Ramasamy and Surendran 2011; Coalson et al. 2021). Recently, land subsidence has also led to coastline expansion. Such expansion of brackish water bodies has a different impact on each mosquito species (Coalson et al. 2021) and may facilitate the adaptation of *Ae. aegypti* and *Ae. albopictus* in brackish water (Ramasamy et al. 2015). Because the effects of earthquakes and tsunamis are not only for increasing the abundance of species living in brackish water bodies but also for increasing freshwater habitats reported in debris in a destroyed residential area, establishing a research methodology and prediction model after such natural disasters is also required for the future (Ramasamy and Surendran 2011; Ramasamy et al. 2015; Coalson et al. 2021).

1.14.2 Cryptic Species of Aedes albopictus

Because cryptic species may prevent the effect of vector control of mosquito-borne infectious diseases, the ecology of cryptic species must be clarified (Zheng 2020). Culicidae has many cryptic species, especially *Anopheles* spp., but new cryptic species of *Ae. albopictus* has recently been reported in Vietnam and China (Minard et al. 2017; Guo et al. 2018).

Aedes albopictus was collected from the forest reserve in southern Vietnam and included a population that was clearly different from *Ae. albopictus* by *COI* sequences (K2P difference >10%) (Minard et al. 2017). Because intraspecific variation in *COI* sequences is generally approximately 2% in mosquitoes, the population is identified as a new cryptic species. Minard and colleagues also found that the same sequences were registered in GenBank as *Ae. albopictus* (Minard et al. 2017). After this report, a survey was conducted in a wide area of China, and two populations isolated from *Ae. albopictus* were found; one belonging to the same genetic clade as the Vietnamese population was collected from Baoting in Hainan Province and Wuzhou in Guangxi Province on the opposite coast of Hainan Island, and another belonging to a distant clade was collected from Baisha in Hainan Province (Guo et al. 2018). Survey of cryptic species of *Ae. albopictus* by analyzing *COI* sequences in a large area of the Indo-Pacific region has been carried out, but there are still no reports except for the above two countries (Misbah et al. 2022; Yang et al. 2022).

Wolbachia infection is an important subject related to the cryptic species. To use cytoplasmic incompatibility (CI) for mosquito vector control, we need to understand the *Wolbachia* infection status of *Ae. albopictus* and its cryptic species. Although most *Ae. albopictus* is infected with wAlbA and wAlbB strains of *Wolbachia*, the cryptic species found in Vietnam and China were infected with low levels or not infected with them (Minard et al. 2017; Guo et al. 2018). One study suggested that the suppression of the expansion of *Wolbachia* in association with reproductive isolation is linked to speciation (Zheng 2020). In the future, *Ae. albopictus* and its cryptic species could be model species for evolutionary theory.

1.14.3 Evidence for Two Genetically Different Types in the Japanese Encephalitis Mosquito Vector, **Culex tritaeniorhynchus**

Japanese encephalitis, caused by the Japanese encephalitis virus (JEV), causes severe encephalitis in humans, and even now, approximately 68,000 people a year get infected with JEV in Asia, mainly in non-vaccinated areas. There are five genotypes (GI to GV) of JEV. Although the number of Japanese encephalitis cases has decreased dramatically in Japan since 1992, JEV is still active and isolated from mosquitoes and pigs every summer. Pigs, boars, and waterfowl are amplifying animals, and humans are infected by infected mosquitoes (mainly *Culex*) that suck blood from these animals (humans are the dead-end hosts). During the study for the mechanism of annual occurrence and the genotype of JEV from the entomological aspect in Japan, the existence of a genetically unique population by mtDNA *COI* barcoding region was found in the mosquito vector, *Culex tritaeniorhynchus* in Japan (Arai et al. 2022) (Fig. 1.6).

Phylogenetic analysis showed that *Cx. tritaeniorhynchus* from Japan, Vietnam, the Philippines, Indonesia, China, South Korea, Taiwan, and Thailand were classified into two genetically independent populations, the population found in temperate Japan (the Japanese type, *Ct*-J) and the other found from the subtropical Japan and Asian region except for the temperate Japan (the Continental type, *Ct*-C). The study also showed that approximately 10% of *Cx. tritaeniorhynchus* trapped during the summer in western Kyushu were *Ct*-C, and they could fly for up to 38 hours continuously, suggesting the potential for long-distance migration from overseas to

Japan. Morphological comparisons were made for 56 characteristics with larval, pupal, and adult samples from Japan, Vietnam, and Thailand (YH, unpublished data). The morphological characteristics examined overlapped among specimens from Japan, Vietnam, and Thailand. The study confirmed the existence of two genetically independent groups in *Cx. tritaeniorhynchus*. Because the morphological characteristics examined overlapped among specimens from Japan, Vietnam, and Thailand, a cross-mating experiment is required to confirm the taxonomic status of *Ct*-J.

1.14.4 Existence of Cryptic Species in **Anopheles lindesayi**

The recent utilization of molecular techniques in mosquito taxonomy has led to several new findings. One of these is the discovery of a cryptic species in *Anopheles lindesayi*. The species was first described from India in 1900 (Giles 1900) and subsequently included five subspecies, *lindesayi*, *japonicus*, *pleccau*, *cameronensis* and *benguetensis* (now all are elevated to species). Larvae of these subspecies are collected from the ground or rock pools beside streams in mountainous areas (Koidzumi 1921; King 1931; LaCasse and Yamaguti 1950; Reid 1968; Harrison et al. 1990). *Anopheles l. pleccau* and *japonicus* are capable of forming sporozoites of *Plasmodium falciparum* and *P. vivax*. However, their actual roles in the malaria parasites transmission in nature are unclear.

In Japan, Imanishi et al. (2018) found that morphologically and genetically three distinct groups existed in *An. l. japonicus*. The first one is the Eastern Japan group which is regarded as *An. l. japonicus*, the second one is the Western Japan group without any identical groups reported so far, and the third one is the Ryukyu group genetically identical to *An. l. pleccau* from Taiwan. The three groups were distinguished by combinations of larval seta 1-S, pupal setae 5 through segments IV-VII, and pupal setae 6 on segments IV-VII. K2P divergence of the *COI* barcoding region among the three groups was genetically isolated >3% (3.04–9.73%). The results suggested that the Western Japan group was an unrecognized species.

Namgay et al. (2020) reported five genetically (mtDNA *COI* barcoding region) distinct clades as species A, B, C, D, and E of *An. lindesayi* from Bhutan. Larvae of Clades B, C, D and E are similar in having seta 4-C branched rather than single. The adults of Clades C, D and E were distinguishable from those of Clade A and other subspecies. Specimens of Clade C were unique in having a long pale spot on wing vein R and the subcosta, scattered pale scales on several veins and a dark spot at the tip of vein R_2. The adults of Clades D and E were similar in having a dark spot at the tip of vein R_2 and no scattered pale scales on all other veins. The authors concluded that species A was *An. lindesayi* sensu stricto and the others are unnamed species. Somboon et al. (2022b) formally described and named species C as *Anopheles druki* Somboon, Namgay & Harbach, species D as *Anopheles himalayensis* Somboon, Namgay & Harbach and species E as *Anopheles thimphuensis* Somboon, Namgay & Harbach.

1.14.5 Discovery of Insect Flaviviruses from Mosquitoes

Until the 2000s, cell fusing agent (CFA) (Stollar and Thomas 1975) and Kamiti River virus (KRV) (Crabtree et al. 2003; Sang et al. 2003) were found in mosquitoes. These two insect flaviviruses are phylogenetically placed in the most outgroup lineage of flaviviruses and are considered prototypes of flaviviruses (Cook and Holmes 2006). In addition, another DNA sequence related to CFA, known as the cell silent agent (CSA), has been incorporated into the *Ae. aegypti* genome (Crochu et al. 2004). These findings suggest that these insect flaviviruses are closely related to their mosquito hosts and that other insect flaviviruses associated with mosquitoes may exist in the natural environment. Subsequently, new insect flaviviruses named Culex flavivirus (CxFV) (Hoshino et al. 2007) and Aedes flavivirus (AEFV) (Hoshino et al. 2009) were isolated from *Culex* and *Aedes* mosquitoes, respectively, during nationwide WNV surveillance in mosquitoes conducted in Japan since 2003. This has led to worldwide research on insect-specific viruses (ISVs).

Recent advances in next-generation sequencing (NGS) technology have made it possible to investigate a wide range of viral diversity (Atoni et al. 2018), and virome analysis of not only mosquitoes but also many other vector arthropods has been significantly and rapidly accelerated, as has arbovirus surveillance. Currently, numerous arthropod-specific viruses (ASVs) have also been detected and isolated from mosquitoes, horseflies, ticks, and other arthropods (Calisher and Higgs 2018; Faizah et al. 2020; Kobayashi et al. 2021). Most ASVs are unique to their arthropod hosts and are not thought to have a direct impact on human health (Hall et al. 2017). However, some of these ASVs may act as natural competitors to various arboviruses within the host arthropod, making them promising as potential biocontrol agents. Taking mosquitoes as an example, the relationship between WNV and CxFV and DENV and AEFV has been highlighted. These ASVs are known to be vertically transmitted, and it has been suggested that they eliminate co-infection with arboviruses and enhance antiviral immune responses (Öhlund et al. 2019). The potential ability of ASVs to modulate virus transmission to vertebrates by reducing the ability of vector arthropods to transmit arboviruses and altering the host immune response to subsequent arbovirus infections has also been reported (Bolling et al. 2015; Vasilakis and Tesh 2015; Hall-Mendelin et al. 2016). As these studies have demonstrated specificity in the relationship between ASVs and specific arboviruses (Xiao et al. 2018), virome research against medically important arthropods will continue to evolve.

The ASVs presented here not only contribute to research on the evolution of pathogenic viruses, but we also hope that they can be used as a new method to control vector arthropods, especially vector mosquitoes.

References

Abercrombie J (1977) Medical entomology studies - IX. The subgenus *Christophersiomyia* of the genus *Aedes* (Diptera: Culicidae). Contrib Am Entomol Inst (Ann Arbor) 14(2):1–72

Ali RSM, Wahid I, Saeung A, Wannasan A, Harbach RE, Somboon P (2019a) Genetic and morphological evidence for a new species of the Maculatus Group of *Anopheles* subgenus *Cellia* (Diptera: Culicidae) in Java, Indonesia. Parasit Vectors 12(1):107

Ali RSM, Wahid I, Saingamsook J, Saeung A, Wannasan A, Walton C et al (2019b) Molecular identification of mosquitoes of the *Anopheles maculatus* group of subgenus *Cellia* (Diptera: Culicidae) in the Indonesian Archipelago. Acta Trop 199:105124

Anderios F, NoorRain A, Vythilingam I (2010) In vivo study of human *Plasmodium knowlesi* in *Macaca fascicularis*. Exp Parasitol 124(2):181–189

Arai S, Kuwata R, Higa Y, Maekawa Y, Tsuda Y, Roychoudhury S, Bertuso AG, Tran VP, Nguyen TY, Etoh T, Otsuka A, Matsumura M, Nabeshima T, Taya KT, Okabe N, Kobayashi M, Sawabe K (2022) Two hidden taxa in the Japanese encephalitis vector mosquito, *Culex tritaeniorhynchus*, and the potential for long-distance migration from overseas to Japan. PLoS Negl Trop Dis 16(6):e0010543

Aslamkhan M (1971) The mosquitoes of Pakistan: I. A checklist. Mosq Syst Newsl 3(4):147–159

Atoni E, Wang Y, Karungu S, Waruhiu C, Zohaib A, Obanda V, Agwanda B, Mutua M, Xia H, Yuan Z (2018) Metagenomic virome analysis of *Culex* mosquitoes from Kenya and China. Viruses 10(1):30

Atrie B, Subbarao SK, Pillai MKK, Rao SRV, Sharma VP (1999) Population cytogenetic evidence for sibling species in *Anopheles annularis* (Diptera: Culicidae). Ann Entomol Soc Am 92(2):243–249

Auerswald H, Ruget AS, Ladreyt H, In S, Mao S, Sorn S, Tum S, Duong V, Dussart P, Cappelle J, Chevalier V (2020) Serological evidence for Japanese encephalitis and West Nile virus infections in domestic birds in Cambodia. Front Vet Sci 7:15

Balaraman K, Sabesan S, Jambulingam P, Gunasekaran K, Boopathi Doss PS (2005) Risk of outbreak of vector-borne diseases in the tsunami hit areas of southern India. Lancet Infect Dis 5(3):128–129

Barraud PJ (1934) The fauna of British India, including Ceylon and Burma. Diptera. Volume V. Family Culicidae. Tribes Megarhinini and Culicini. Taylor and Francis, London, p 463

Belkin JN (1962) Mosquitoes of the South Pacific (Diptera, Culicidae), vol I and II. University of California Press, Berkeley and Los Angeles, p 608

Bertone MA, Courtney GW, Wiegmann BM (2008) Phylogenetics and temporal diversification of the earliest true flies (Insecta: Diptera) based on multiple nuclear genes. Syst Entomol 33(4):668–687

Bhattacharyya DR, Rajavel AR, Mohapatra PK, Jambulingam P, Mahanta J, Prakash A (2014) Faunal richness and the checklist of Indian mosquitoes (Diptera: Culicidae). Check List 10(6):1342–1358

Blagoderov V, Grimaldi DA, Fraser NC (2007) How time flies for flies: diverse Diptera from the Triassic of Virginia and early radiation of the order. Am Mus Novit 3572:1–39

Blanquart S, Gascuel O (2011) Mitochondrial genes support a common origin of rodent malaria parasites and *Plasmodium falciparum*'s relatives infecting great apes. BMC Evol Biol 11(1):70

Bolling BG, Weaver SC, Tesh RB, Vasilakis N (2015) Insect-specific virus discovery: significance for the arbovirus community. Viruses **7**(9):4911–4928

Bonne-Wepster J, Swellengrebel NH (1953) The anopheline mosquitoes of the Indo-Australian Region. J.-H. de Bussy, Amsterdam, p 504

Borkent A, Grimaldi DA (2004) The earliest fossil mosquito (Diptera: Culicidae), in mid-Cretaceous Burmese amber. Ann Entomol Soc Am 97(5):882–888

Borkent A, Grimaldi DA (2016) The Cretaceous fossil *Burmaculex antiquus* confirmed as the earliest known lineage of mosquitoes (Diptera: Culicidae). Zootaxa 4079(4):457–466

Bourke BP, Wilkerson RC, Linton YM (2021) Molecular species delimitation reveals high diversity in the mosquito *Anopheles tessellatus* Theobald, 1901 (Diptera, Culicidae) across its range. Acta Trop 215:105799

Bower JE, Dowton M, Cooper RD, Beebe NW (2008) Intraspecific concerted evolution of the rDNA ITS1 in *Anopheles farauti* sensu stricto (Diptera: Culicidae) reveals recent patterns of population structure. J Mol Evol 67(4):397–411

Bram RA (1967) Contributions to the mosquito fauna of Southeast Asia. II. The genus *Culex* in Thailand (Diptera: Culicidae). Contrib Am Entomol Inst 2(1):1–296

Briët OJT, Galappaththy GNL, Amerasinghe PH, Konradsen F (2006) Malaria in Sri Lanka: one year post-tsunami. Malar J 5:42

Brosseau L, Udom C, Sukkanon C, Chareonviriyaphap T, Bangs MJ, Saeung A et al (2019) A multiplex PCR assay for the identification of five species of the *Anopheles barbirostris* complex in Thailand. Parasit Vectors 12(1):223

Bui P, Darsie RF Jr (2008) Tentative checklist of the mosquitoes of Vietnam employing new classification for tribe Aedini (Diptera, Culicidae). J Am Mosq Control Assoc 24(8):187–193

Cagampang-Ramos A, Darsie RF Jr (1970) Illustrated keys to the *Anopheles* mosquitoes of the Philippines Islands. In: USAF Fifth Epidemiological Flight. PACAF. Technical Report 70–1, p 49

Calisher CH, Higgs S (2018) The discovery of arthropod-specific viruses in hematophagous arthropods: an open door to understanding the mechanisms of arbovirus and arthropod evolution? Annu Rev Entomol 63:87–103

Cao L, Fu S, Gao X, Li M, Cui S, Li X, Cao Y, Lei W, Lu Z, He Y, Wang H, Yan J, Gao GF, Liang G (2016) Low protective efficacy of the current Japanese encephalitis vaccine against the emerging genotype 5 Japanese encephalitis virus. PLoS Negl Trop Dis 10(5):e0004686

Castro JA, Picornell A, Ramon M (1998) Mitochondrial DNA: a tool for population genetics studies. Int Microbiol 1(4):327–332

Chan M, Johansson MA (2012) The incubation periods of dengue viruses. PLoS One 7(11):e50972

Changbunjong T, Weluwanarak T, Toawan N, Suksai P, Sedwisai P, Chamsai T, Jirapattharasate C, Sungpradit S, Samung Y, Ratanakorn P (2012) Mosquito distribution and West Nile virus infection in zoos and in important sites of migratory and resident birds, Thailand. Asian Pacific J Trop Dis 2(4):268–272

Chen B, Butlin RK, Pedro PM, Wang XZ, Harbach RE (2006) Molecular variation, systematics and distribution of the *Anopheles fluviatilis* complex in Southern Asia. Med Vet Entomol 20(1):33–43

Christophers SR (1933) The fauna of British India, including Ceylon and Burma, Diptera, Vol. IV, Family Culicidae, Tribe Anophelini. Taylor and Francis, London, p 371

Coalson JE, Anderson EJ, Santos EM, Garcia VM, Romine JK, Dominguez B, Richard DM, Little AC, Hayden MH, Ernst KC (2021) The complex epidemiologic relationship between flooding events and human outbreaks of mosquito-borne diseases: a scoping review. Environ Health Perspect 129(9):96002

Coatney GR, Collins WE, Warren M, Contacos PG (1971) The primate malarias. U.S. Government Printing Office, Washington D.C., pp 1–366

Colless DH (1958) Notes on the culicine mosquitoes of Singapore: IV. The *Aedes niveus* subgroup (Diptera: Culicidae): Introduction and description of five new species and one new subspecies. Ann Trop Med Parasitol 52:468–483

Colless DH (1959) Notes on the culicine mosquitoes of Singapore: V. The *Aedes niveus* subgroup (Diptera: Culicidae): previously described species and keys to adults and larvae. Ann Trop Med Parasitol 53:166–179

Cook S, Holmes EC (2006) A multigene analysis of the phylogenetic relationships among the flaviviruses (family: Flaviviridae) and the evolution of vector transmission. Arch Virol 151(2):309–325

Cox-Singh J, Davis TME, Lee KS, Shamsul SSG, Matusop A, Ratnam S, Rahman HA, Conway DJ, Singh B (2008) *Plasmodium knowlesi* malaria in humans is widely distributed and potentially life threatening. Clin Infect Dis 46(2):165–171

Crabtree MB, Sang RC, Stollar V, Dunster LM, Miller BR (2003) Genetic and phenotypic characterization of the newly described insect flavivirus. Kamiti River virus. Arch Virol 148(6):1095–1118

Crochu S, Cook S, Attoui H, Charrel RN, De Chesse R, Belhouchet M, Lemasson JJ, de Micco P, de Lamballerie X (2004) Sequences of flavivirus-related RNA viruses persist in DNA form integrated in the genome of *Aedes* spp. Mosquitoes. J Gen Virol 85(7):1971–1980

da Silva AF, Machado LC, de Paula MB, da Silva Pessoa Vieira CJ, de Morais Bronzoni RV et al (2020) Culicidae evolutionary history focusing on the Culicinae subfamily based on mitochondrial phylogenomics. Sci Rep 10(1):18823

Dash AP, Bhatia R, Sunyoto T, Mourya DT (2013) Emerging and re-emerging arboviral diseases in Southeast Asia. J Vector Borne Dis 50(2):77–84

de Wispelaere M, Frenkiel M-P, Desprès P (2015) A Japanese encephalitis virus Genotype 5 molecular clone is highly neuropathogenic in a mouse model: impact of the structural protein region on virulence. J Virol 89(11):5862–5875

Delfinado MD (1966) The culicine mosquitoes of the Philippines, tribe Culicini (Diptera, Culicidae). Mem Am Entomol Inst 7:1–252

Delfinado MD (1967) Contributions to the mosquito fauna of Southeast Asia. I. The genus *Aedes*, subgenus *Neomacleaya* Theobald in Thailand. Contrib Am Entomol Inst (Ann Arbor) 1(8):1–56

Delfinado MD (1968) Contributions to the mosquito fauna of Southeast Asia. III. The genus *Aedes*, subgenus *Neomacleaya* Theobald in Southeast Asia. Contrib Am Entomol Inst (Ann Arbor) 2(4):1–74

Desser SS, McIver SB, Jez D (1975) Observations on the role of simuliids and culicids in the transmission of avian and anuran trypanosomes. Int J Parasitol 5(5):507–509

Dusfour I, Blondeau J, Harbach RE, Vythilingham I, Baimai V, Trung HD et al (2007a) Polymerase chain reaction identification of three members of the *Anopheles sundaicus* (Diptera: Culicidae) complex, malaria vectors in Southeast Asia. J Med Entomol 44(5):723–731

Dusfour I, Michaux JR, Harbach RE, Manguin S (2007b) Speciation and phylogeography of the Southeast Asian *Anopheles sundaicus* complex. Infect Genet Evol 7(4):484–493

Edwards FW (1932) Diptera, Fam. Culicidae. Fascicle 194. In: Wytsman P (ed) Genera Insectorum. Louis Desmet-Verteneuil, Bruxelles, Belgium, p 258. with illustrations

Elyazar IR, Sinka ME, Gething PW, Tarmidzi SN, Surya A, Kusriastuti R et al (2013) The distribution and bionomics of *Anopheles* malaria vector mosquitoes in Indonesia. Adv Parasitol 83:173–266

Escalante AA, Ayala FJ (1995) Evolutionary origin of *Plasmodium* and other Apicomplexa based on rRNA genes. Proc Natl Acad Sci USA 92(13):5793–5797

Escalante AA, Cornejo OE, Freeland DE, Poe AC, Durrego E et al (2005) A monkey's tale: the origin of *Plasmodium vivax* as a human malaria parasite. Proc Natl Acad Sci USA 102(6):1980–1985

Faizah AN, Kobayashi D, Isawa H, Amoa-Bosompem M, Murota K, Higa Y, Futami K, Shimada S, Kim KS, Itokawa K, Watanabe M, Tsuda Y, Minakawa N, Miura K, Hirayama K, Sawabe K (2020) Deciphering the virome of *Culex vishnui* subgroup mosquitoes, the major vectors of Japanese encephalitis, in Japan. Viruses 12(3):264

Fanany D (2011) Environmental health issues in the wake of a major earthquake in West Sumatra, Indonesia. Environ Health Biomed 15:147–157

Fanany D (2012) Dengue hemorrhagic fever and natural disaster: the case of Padang, West Sumatra. Int J Collab Res Intern Med Public Health 4(5):673–678

Fang Y, Shi WQ, Zhang Y (2017a) Molecular phylogeny of *Anopheles hyrcanus* group (Diptera: Culicidae) based on mtDNA *COI*. Infect Dis Poverty 6(1):61

Fang Y, Shi WQ, Zhang Y (2017b) Molecular phylogeny of *Anopheles hyrcanus* group members based on ITS2 rDNA. Parasit Vectors 10(1):417

Fialová M, Santolíková A, Brotánková A, Brzonová J, Svobodová M (2021) Complete life cycle of *Trypanosoma thomasbancrofti*, an avian trypanosome transmitted by culicine mosquitoes. Microorganisms 9(10):2101

Foley DH, Paru R, Dagoro H, Bryan JH (1993) Allozyme analysis reveals six species within the *Anopheles punctulatus* complex of mosquitoes in Papua New Guinea. Med Vet Entomol 7(1):37–48

Garnham PCC (1966) Malaria parasites and other haemosporidia. Blackwell Scientific, Oxford, pp 1–1114

Garros C, Dujardin JP (2013) Genetic and phenetic approaches to *Anopheles* systematics. In: Manguin S (ed) *Anopheles* mosquitoes - new insights into malaria vectors. InTech; Open Access Publisher, Rijeka, Croatia, pp 81–105

Garros C, Koekemoer LL, Coetzee M, Coosemans M, Manguin S (2004) A single multiplex assay to identify major malaria vectors within the African *Anopheles funestus* and oriental *An. minimus* groups. Am J Trop Med Hyg 70(6):583–590

Giles GMJ (1900) A handbook of the gnats or mosquitoes giving the anatomy and life history of the Culicidae. John Bale, Sons & Danielsson, Ltd., London, p 374

Gilka W, Harbach RE, Perkovsky EE (2021) Mosquitoes (Diptera: Culicidae) in Eocene amber from the Rovno region, Ukraine. Zootaxa 5016(2):257–270

Green CA, Miles SJ (1980) Chromosomal evidence for sibling species of the malaria vector *Anopheles* (*Cellia*) *culicifacies* Giles. J Trop Med Hyg 83(2):75–78

Green CA, Harrison BA, Klein T, Baimai V (1985) Cladistic analysis of polytene chromosome rearrangements in anopheline mosquitoes of subgenus *Cellia*, series Neocellia. Can J Genet Cytol 27:123–133

Green CA, Gass RF, Munstermann LE, Baimai V (1990) Population-genetic evidence for two species in *Anopheles minimus* in Thailand. Med Vet Entomol 4(1):25–34

Grimaldi D, Engel MS (2005) Evolution of the insects. Cambridge University Press, Cambridge, p 755

Guo Y, Song Z, Luo L, Wang Q, Zhou G, Yang D, Zhong D, Zheng X (2018) Molecular evidence for new sympatric cryptic species of *Aedes albopictus* (Diptera: Culicidae) in China: a new threat from *Aedes albopictus* subgroup? Parasit Vectors 11(1):228

Gutsevich AV, Monchadskii AS, Shtakel'berg AA (1974) Fauna of the U.S.S.R. Diptera. Volume 3, No. 4. Mosquitoes Family *Culicidae* [*sic*]. Keter Publishing House Jerusalem Ltd, Jerusalem, p 408

Gwee XWS, Chua PEY, Pang J (2021) Global dengue importation: a systematic review. BMC Infect Dis 21:1078

Hall RA, Bielefeldt-Ohmann H, McLean BJ, O'Brien CA, Colmant AM, Piyasena TB, Harrison JJ, Newton ND, Barnard RT, Prow NA et al (2017) Commensal viruses of mosquitoes: host restriction, transmission, and interaction with arboviral pathogens. Evol Bioinform Online 12(Suppl 2):35–44

Hall-Mendelin S, McLean BJ, Bielefeldt-Ohmann H, Hobson-Peters J, Hall RA, van den Hurk AF (2016) The insect-specific Palm Creek virus modulates West Nile virus infection in and transmission by Australian mosquitoes. Parasit Vectors 9(1):414

Harbach RE (1988) The mosquitoes of the subgenus *Culex* in southwestern Asia and Egypt (Diptera: Culicidae). Contrib Am Entomol Inst 24(1): vi, 1–240

Harbach RE (2013) The phylogeny and classipication of *Anopheles*. In: Manguin S (ed) *Anopheles* mosquitoes - new insights into malaria vectors. Infect. Dis. InTech Open Access Publisher, Rijeka, Croatia, pp 3–55

Harbach RE (2023a) Culicidae. Mosquito Taxonomic Inventory. [Accessed 27 Dec 2023] https://mosquito-taxonomic-inventory.myspecies.info/simpletaxonomy/term/6045

Harbach RE (2023b) Fossil Culicidae. Mosquito taxonomic inventory. [Accessed 27 Dec 2023] https://mosquito-taxonomic-inventory.myspecies.info/simpletaxonomy/term/6420

Harbach RE, Greenwalt D (2012) Two Eocene species of *Culiseta* (Diptera: Culicidae) from the Kishenehn Formation in Montana. Zootaxa 3530:25–34

Harbach RE, Kitching IJ (1998) Phylogeny and classification of the Culicidae (Diptera). Syst Entomol 23(4):327–370

Harbach RE, Knight KL (1980) Taxonomist' glossary of mosquito anatomy. Plexus Pub. Inc., Marlton, p 415

Harbach RE, Knight KL (1981) Corrections and additions to taxonomists' glossary of mosquito anatomy. Mosq Syst 13:201–217

Harbach RE, Wilkerson RC (2023) The insupportable validity of mosquito subspecies (Diptera: Culicidae) and their exclusion from culicid classification. Zootaxa 5303(1):1–184

Harbach RE, Garros C, Manh ND, Manguin S (2007a) Formal taxonomy of species C of the *Anopheles minimus* sibling species complex (Diptera: Culicidae). Zootaxa 1654:41–54

Harbach RE, Huong VD, Kitching IJ (2007b) Systematics of *Kimia*, a new genus of Sabethini (Diptera: Culicidae) in the oriental region. Proc Entomol Soc Wash 109(1):102–120

Harrison BA (1980) Medical entomology studies —XIII. The Myzomyia Series of *Anopheles* (*Cellia*) in Thailand, with emphasis on intra-interspecific variations (Diptera: Culicidae). Contrib Am Entomol Inst (Ann Arbor) 17(4):1–195

Harrison BA, Scanlon JE (1975) Medical entomology studies-II. The subgenus *Anopheles* in Thailand (Diptera: Culicidae). Contrib Am Entomol Inst (Ann Arbor) 12(1):1–307

Harrison BA, Rattanarithikul R, Peyton EL, Mongkolpanya K (1990) Taxonomic changes, revised occurrence records and notes on the Culicidae of Thailand and neighboring countries. Mosq Syst 22(3):196–227

Herbert PDN, Cywinska A, Ball SL, deWaard JR (2003) Biological identifications through DNA barcodes. Proc Biol Sci 270(1512):313–321

Hoshino K, Isawa H, Tsuda Y, Sawabe K, Kobayashi M (2009) Isolation and characterization of a new insect flavivirus from *Aedes albopictus* and *Aedes flavopictus* mosquitoes in Japan. Virology 391(1):119–129

Hoshino L, Isawa H, Tsuda Y, Yano K, Sasaki T, Yuda M, Takasaki T, Kobayashi M, Sawabe K (2007) Genetic characterization of a new insect flavivirus isolated from *Culex pipiens* mosquito in Japan. Virology 359(2007):405–414

Huang YM (1972) Contributions to the mosquito fauna of Southeast Asia. XIV. The subgenus *Stegomyia* of *Aedes* in Southeast Asia. I - The *scutellaris* group of species. Contrib Am Entomol Inst (Ann Arbor) 9(1):1–110

Huang YM (1977a) Medical entomology studies - VII. The subgenus *Stegomyia* of *Aedes* in Southeast Asia. II - The *edwardsi* group of species. III - The *w-albus* group of species (Diptera: Culicidae). Contrib Am Entomol Inst (Ann Arbor) 14(1):1–111

Huang YM (1979) Medical entomology studies - XI. The subgenus *Stegomyia* of *Aedes* in the Oriental Region with keys to the species (Diptera: Culicidae). Contrib Am Entomol Inst (Ann Arbor) 15(6):1–79

Huang Y-M (1977b) Medical entomology studies - VIII. Notes on the taxonomic status of *Aedes vittatus* (Diptera: Culicidae). Contrib Am Entomol Inst (Ann Arbor) 14(1):113–132

Ichimori K (2021) Human filariasis. In: Ichimori K (ed) Kagaku-Nyumon. Midori-Shobo, Tokyo, pp 133–144. (in Japanese)

Imanishi N, Higa Y, Teng HJ, Sunahara T, Minakawa N (2018) Identification of three distinct groups of *Anopheles lindesayi* in Japan by morphological and genetic analyses. Jpn J Infect Dis 71(6):427–435

Irish SR, Al-Amin HM, Alam MS, Harbach RE (2016) A review of the mosquito species (Diptera: Culicidae) of Bangladesh. Parasit Vectors 9(1):559

Iyengar MO (1953) Filariasis in Thailand. Bull World Health Organ 9(6):731–766

Jeyaprakasam NK, Liew JWK, Low VL, Wan-Sulaiman WY, Vythilingam I (2020) *Plasmodium knowlesi* infecting humans in southeast Asia: what's next? PLoS Negl Trop Dis 14(12):1–16

Jongwutiwes S, Putaporntip C, Iwasaki T, Sata T, Kanbara H (2004) Naturally acquired *Plasmodium knowlesi* malaria in human, Thailand. Emerg Infect Dis 10(12):2211–2213

Karlekar S, Andrew R, Deshpande M (2020) *Culex katezari*, a new species of *mimeticus* [sic] mosquito (Diptera: Culicidae) from the forest of Gadchiroli region of central India. Int J Mosq Res 7(6):54–58

Khan SA, Dutta P, Khan AM, Chowdhury P, Borah J, Doloi P, Mahanta J (2011) West Nile virus infection, Assam, India. Emerg Infect Dis 17(5):947–948

Kilpatrick AM (2011) Globalization, land use, and the invasion of West Nile virus. Science 334(6054):323–327

Kim CY, Oh H, Song J, Hur M, Suh JH, Jheong WH, Kim JT, Oh HS, Park JH (2016) First detection of West Nile virus in domestic pigeon in Korea. J Vet Sci 17(4):587–589

King WV (1931) The Philippine varieties of *Anopheles gigas* and *Anopheles lindesayi*. Philipp J Sci 46(4):751–757

Klein JM (1973) Contributions to the mosquito fauna of Southeast Asia. XVII. The Cambodian *Aedes* (*Neomacleaya*) species with some new descriptions (Diptera: Culicidae). Contrib Am Entomol Inst (Ann Arbor) 10(1):1–23

Knight KL (1968) Contributions to the mosquito fauna of Southeast Asia. - IV. Species of the subgroup *Chrysolineatus* of group D, genus *Aedes*, subgenus *Finlaya* Theobald. Am Entomol Inst 2(5):1–45

Knight KL, Stone A (1977) A catalog of the mosquitoes of the world (Diptera: Culicidae), vol VI, 2nd edn. The Thomas Say Foundation and entomological society of America, College Park, p 611

Kobayashi D, Watanabe M, Faizah AN, Amoa-Bosompem M, Higa Y, Tsuda Y, Sawabe K, Isawa H (2021) Discovery of a novel flavivirus (flaviviridae) from the horse fly, *Tabanus rufidens* (Diptera: Tabanidae): the possible coevolutionary relationships between the classical insect-specific flaviviruses and host dipteran insects. J Med Entomol 58(2):880–890

Kobayashi M, Kasai S, Sawabe K, Tsuda Y (2008) Distribution and ecology of potential vector mosquitoes of West Nile fever in Japan. Global Env Res 12:27–33

Koidzumi M (1921) The research on preventive medicine of mosquitoes in Taiwan. In: Research Institute of the Governor-General of Taiwan (ed) Report of the Research Institute of the Governor-Generol of Taiwan, vol. 8. Research Institute of the Governor-General of Taiwan, Taipei, pp. 34–8. (in Japanese)

Kojom Foko LP, Kumar A, Hawadak J, Singh V (2022) *Plasmodium cynomolgi* in humans: current knowledge and future directions of an emerging zoonotic malaria parasite. Infection 51(3):623–640

Krief S, Escalante AA, Pacheco MA, Mugisha L, Andre C, Halbwax M et al (2010) On the diversity of malaria parasites in African apes and the origin of *Plasmodium falciparum* from bonobos. PLoS Pathog 6(2):e1000765

Krishnamoorthy K, Jambulingam P, Natarajan R, Shriram AN, Das PK, Sehgal SC (2005) Altered environment and risk of malaria outbreak in South Andaman, Andaman & Nicobar Islands, India affected by tsunami disaster. Malar J 4:32

Kumari R, Joshi PL, Lal S, Shah W (2009) Management of malaria threat following tsunami in Andaman & Nicobar Islands, India and impact of altered environment created by tsunami on malaria situation of the islands. Acta Trop 112(2):204–211

Kuwata R, Torii S, Shimoda H, Supriyono TP, Prasertsincharoen N, Takemae H, Bautista RCJT, Ebora VDBM, Abella JAC, Dargantes AP, Hadi UK, Setiyono A, Baltazar ET, Simborio LT, Agungpriyono S, Jittapalapong S, Rerkamnuaychoke W, Hondo E, Maeda K (2020) Distribution of Japanese encephalitis virus, Japan and Southeast Asia, 2016-2018. Emerg Infect Dis 26(1):125–128

LaCasse WJ, Yamaguti S (1950) Mosquito fauna of Japan and Korea. Office of the Surgeon, Hq. 8th U.S. Army, Kyoto, p 213

Lam-Phua SG, Yeo H, Lee RML, Chong CS, Png AB et al (2019) Mosquitoes (Diptera: Culicidae) of Singapore: Updated checklist and new records. J Med Entomol 56(1):103–119

Lataillade LG, Vazeille M, Obadia T, Madec Y, Mousson L, Kamgang B, Chen CH, Failloux AB, Yen PS (2020) Risk of yellow fever virus transmission in the Asia-Pacific region. Nat Commun 11(1):5801

Laurens MB (2020) RTS,S/AS01 vaccine (Mosquirix™): an overview. Hum Vaccin Immunother 16(3):480–489

Lee A, Song JM, Seo S (2022) Emerging Japanese encephalitis virus genotype V in Republic of Korea. J Microbiol Biotechnol 32(8):955–959

Leicester GF (1908) The Culicidae of Malaya. Stud Inst Med Res Feder Malaya 3(3):18–261

Li MH, Fu SH, Chen WX, Wang HY, Guo YH, Liu QY, Li YX, Luo HM, Da W, Ji DZ, Ye XM, Liang GD (2011) Genotype v Japanese encephalitis virus is emerging. PLoS Negl Trop Dis 5(7):e1231

Liang G, Li X, Gao X, Fu S, Wang H, Li M, Lu Z, Zhu W, Lu X, Wang L, Cao Y, He Y, Lei W (2018) Arboviruses and their related infections in China: a comprehensive field and laboratory investigation over the last 3 decades. Rev Med Virol 28(1):e1959

Lien JC (1968) New species of mosquitoes from Taiwan (Diptera: Culicidae). Part V. Three new subspecies of *Aedes* and seven new species of *Culex*. Trop Med 10(4):217–262

Linnaeus C (1758) Systema Naturae per regna tria naturae, secundum classes, ordines, genera, species, cum characteribus, differentiis, synonymis, locis. Editio decima, reformata. In: Systema nature, 10th edn, vol 1. Laurentii Salvii, Holmiae, p 824

Linton YM, Dusfour I, Howard TM, Ruiz F, Minh ND, Trung HD et al (2005) *Anopheles* (*Cellia*) *epiroticus* (Diptera: Culicidae), a new malaria vector species in the Sundaicus Complex. Bull Entomol Res 95(4):329–339

Liu W, Li Y, Learn GH, Rudicell RS, Robertson JD, Keele BF et al (2010) Origin of the human malaria parasite *Plasmodium falciparum* in gorillas. Nature 467(7314):420–425

Liu W, Li Y, Shaw KS, Learn GH, Plenderleith LJ, Malenke JA et al (2014) African origin of the malaria parasite *Plasmodium vivax*. Nat Commun 5:3346

Lorenz C, Alves JMP, Foster PG, Suesdek L, Sallum MAM (2021) Phylogeny and temporal diversification of mosquitoes (Diptera: Culicidae) with an emphasis on the Neotropical fauna. Syst Entomol 46(4):798–811

Loy DE, Liu W, Li Y, Learn GH, Plenderleith LJ, Sundararaman SA, Sharp PM, Hahn BH (2017) Out of Africa: origins and evolution of the human malaria parasites *Plasmodium falciparum* and *Plasmodium vivax*. Int J Parasitol 47(2-3):87–97

Lu B et al [*sic*] (1997) Fauna Sinica, Insecta Vol. 8, Diptera: Culicidae Vols I & II. Science Press, Beijing, p 884 (in Chinese)

Lukashevich ED, Mostovski MB (2003) Hematophagous insects in the fossil record. Paleontol J 37(2):153–161

Macdonald WW (1960) Malaysian parasites. XXXVIII. On the systematics and ecology of *Armigeres* subgenus *Leicesteria* (Diptera, Culicidae). Stud. Inst. Med. Res. Feder Malaya 29:110–153

Maeki T, Tajima S, Kyaw AK, Matsumoto F, Miura K, Yamashita A, Yoshikawa A, Negishi K, Noguchi Y, Tadokoro K, Abe K, Taruya J, Koh J, Ito H, Ikegaya A, Abe F, Wada M, Nishigata T, Ikeda M, Kato F, Taniguchi S, Nakayama E, Takasaki T, Morita K, Lim CK, Saijo M (2018) Comparison of neutralizing antibody titers against Japanese encephalitis virus genotype v strain with those against genotype I and III strains in the sera of Japanese encephalitis patients in Japan in 2016. Jpn J Infect Dis 71(5):360–364

Manonmani A, Townson H, Adeniran T, Jambulingam P, Sahu S, Vijayakumar T (2001) rDNA-ITS2 polymerase chain reaction assay for the sibling species of *Anopheles fluviatilis*. Acta Trop 78(1):3–9

Manonmani AM, Sadanandane C, Sahu SS, Mathivanan A, Jambulingam P (2007) rDNA-ITS2-PCR assay for grouping the cryptic species of *Anopheles culicifacies* complex (Diptera: Culicidae). Acta Trop 104(1):72–77

Manonmani AM, Mathivanan AK, Sadanandane C, Jambulingam P (2013) Evaluation of the mtDNA-COII region based species specific assay for identifying members of the *Anopheles culicifacies* species complex. J Arthropod Borne Dis 7(2):154–163

Maquart PO, Fontenille D, Rahola N, Yean S, Boyer S (2021) Checklist of the mosquito fauna (Diptera, Culicidae) of Cambodia. Parasite 28:60

Martinsen ES, Perkins SL, Schall JJ (2008) A three-genome phylogeny of malaria parasites (*Plasmodium* and closely related genera): evolution of life-history traits and host switches. Mol Phylogenet Evol 47(1):261–273

Mattingly PF (1957a) The culicine mosquitoes of the Indomalayan Area: Part I. Genus *Ficalbia* Theobald. Bull. British Museum (Nat Hist), London, p 61

Mattingly PF (1957b) The culicine mosquitoes of the Indomalayan area. Part II. Genus *Heizmannia* Ludlow. Bull. British Museum (Nat Hist), London, p 57

Mattingly PF (1958a) The culicine mosquitoes of the Indomalayan area. Part III. Genus *Aëdes* Meigen, subgenera *Paraëdes* Edwards, *Rhinoskusea* Edwards and *Cancraëdes* Edwards. Bull. British Museum (Nat Hist), London, p 61

Mattingly PF (1958b) A revision of *Paraëdes* Edwards and *Cancraëdes* Edwards (Diptera: Culicidae). Proc R Entomol Soc Lond 27:76–83

Mattingly PF (1959) The culicine mosquitoes of the Indomalayan area. Part IV. Genus *Aëdes* Meigen, subgenera *Skusea* Theobald, *Diceromyia* Theobald, *Geoskusea* Edwards and *Christophersiomyia* Barraud. Bull. British Museum (Nat Hist), London, p 61

Mattingly PF (1960) *Aëdes* subgenus *Alanstonea* subgen. nov. (Diptera, Culicidae). Proc R Entomol Soc Lond B 29:170–171

Mattingly PF (1961) The culicine mosquitoes of the Indomalayan area. Part V. Genus *Aëdes* Meigen, subgenera *Mucidus* Theobald, *Ochlerotatus* Lynch Arribalzaga and *Neomelaniconion* Newstead. Bull. British Museum (Nat Hist), London, p 62

Mattingly PF (1970) Contributions to the mosquito fauna of Southeast Asia. VI. The genus *Heizmannia* Ludlow in Southeast Asia. Contrib Am Entomol Inst (Ann Arbor) 5:1–104

Mattingly PF (1971) Contributions to the mosquito fauna of Southeast Asia. XII. Illustrated keys to the genera of mosquitoes (Diptera, Culicidae). Contrib Am Entomol Inst (Ann Arbor) 7:1–84

Mattingly PF (1981) Medical entomology studies. XIV. The subgenera *Rachionotomyia*, *Tricholeptomyia* and *Tripteroides* (Mabinii Group) of genus *Tripteroides* in the oriental region (Diptera: Culicidae). Contrib Am Entomol Inst (Ann Arbor) 17:1–147

McBride CS, Baier F, Omondi AB, Spitzer SA, Lutomiah J, Sang R, Ignell R, Vosshall LB (2014) Evolution of mosquito preference for humans linked to an odorant receptor. Nature 515(7526):222–227

Minard G, Tran Van V, Tran FH, Melaun C, Klimpel S, Koch LK, Ly Huynh Kim K, Huynh Thi Thuy T, Tran Ngoc H, Potier P, Mavingui P, Valiente Moro C (2017) Identification of sympatric cryptic species of *Aedes albopictus* subgroup in Vietnam: new perspectives in phylosymbiosis of insect vector. Parasit Vectors 10(1):1–14

Misbah S, Low VL, Mohd Rahim NF, Jaba R, Basari N, Ya'cob Z, Abu Bakar S (2022) Mitochondrial diversity of the Asian tiger mosquito *Aedes albopictus* (Diptera: Culicidae) in Peninsular Malaysia. J Med Entomol 59(3):865–873

Mogi M, Wada Y (1973) Spatial distribution of larvae of the mosquito *Culex tritaeniorhynchus summorosus* in a rice field area. Trop Med 15(2):69–83

Mohammed MAF, Galbraith SE, Radford AD, Dove W, Takasaki T, Kurane I, Solomon T (2011) Molecular phylogenetic and evolutionary analyses of Muar strain of Japanese encephalitis virus reveal it is the missing fifth genotype. Infect Genet Evol 11(5):855–862

Moreno M, Marinotti O, Krzywinski J, Tadei WP, James AA et al (2010) Complete mtDNA genomes of *Anopheles darlingi* and an approach to anopheline divergence time. Malar J 9:127

Morgan K, O'Loughlin SM, Mun-Yik F, Linton YM, Somboon P, Min S et al (2009) Molecular phylogenetics and biogeography of the Neocellia Series of *Anopheles* mosquitoes in the Oriental Region. Mol Phylogenet Evol 52:588–601

Motoki MT, Vongphayloth K, Reuda LM, Miot EF, Hiscox A, Hertz JC, Brey PT (2019) New records and updated checklist of mosquitoes (Diptera: Culicidae) from Lao People's Democratic Republic, with special emphasis on adult and larval surveillance in Khammuane Province. J Vector Ecol 44(1):76–88

Mustafa MS, Rasotgi V, Jain S, Gupta V (2015) Discovery of fifth serotype of dengue virus (DENV-5): a new public health dilemma in dengue control. Med J Armed Forces India 71(1):67–70

Nagpal BN, Sharma VP (1995) Indian Anophelines. Oxford & IBH Publishing Co. Pvt. Ltd., New Delhi, p 416

Namgay R, Drukpa T, Wangdi T, Pemo D, Harbach RE, Somboon P (2018) A checklist of the *Anopheles* mosquito species (Diptera: Culicidae) in Bhutan. Acta Trop 188:206–212

Namgay R, Pemo D, Wangdi T, Phanitchakun T, Harbach RE, Somboon P (2020) Molecular and morphological evidence for sibling species within *Anopheles* (*Anopheles*) *lindesayi* Giles (Diptera: Culicidae) in Bhutan. Acta Trop 207:105455

Neafsey DE, Galinsky K, Jiang RH, Young L, Sykes SM, Saif S et al (2012) The malaria parasite *Plasmodium vivax* exhibits greater genetic diversity than *Plasmodium falciparum*. Nat Genet 44(9):1046–1050

Neafsey DE, Waterhouse RM, Abai MR, Aganezov SS, Alekseyev MA, Allen JE et al (2015) Highly evolvable malaria vectors: the genomes of 16 *Anopheles* mosquitoes. Science 347(6217):1258522

Normile D (2013) Surprising new dengue virus throws a spanner in disease control efforts. Science 342(6157):415

Öhlund P, Lundén H, Blomström AL (2019) Insect-specific virus evolution and potential effects on vector competence. Virus Genes 55(2):127–137

Oo TT, Storch V, Becker N (2004) Review of the anopheline mosquitoes of Myanmar. J Vector Ecol 29(1):21–40

Otto TD, Gilabert A, Crellen T, Böhme U, Arnathau C, Sanders M et al (2018) Genomes of all known members of a *Plasmodium* subgenus reveal paths to virulent human malaria. Nat Microbiol 3(6):687–697

Pacheco MA, Matta NE, Valkiunas G, Parker PG, Mello B, Stanley CE et al (2018) Mode and rate of evolution of Haemosporidian mitochondrial genomes: timing the radiation of avian parasites. Mol Biol Evol 35(2):383–403

Pacheco MA, Ceríaco LMP, Matta NE, Vargas-Ramírez M, Bauer AM, Escalante AA (2020) A phylogenetic study of Haemocystidium parasites and other Haemosporida using complete mitochondrial genome sequences. Infect Genet Evol 85:104576

Paramasivan SR, Mishra AC, Mourya DT (2003) West Nile virus: the Indian scenario. Indian J Med Res 118:101–108

Paredes-Esquivel C, Donnelly MJ, Harbach RE, Townson H (2009) A molecular phylogeny of mosquitoes in the *Anopheles barbirostris* subgroup reveals cryptic species: implications for identification of disease vectors. Mol Phylogenet Evol 50(1):141–151

Paskewitz SM, Wesson DM, Collins FH (1993) The internal transcribed spacers of ribosomal DNA in five members of the *Anopheles gambiae* species complex. Insect Mol Biol 2(4):247–257

Peyton EL (1977) Medical entomology studies—X. A revision of the subgenus *Pseudoficalbia* of the genus *Uranotaenia* in Southeast Asia (Diptera: Culicidae). Contrib Am Entomol Inst 14(3):iv + 1–273

Phanitchakun T, Wilai P, Saingamsook J, Namgay R, Drukpa T, Tsuda Y et al (2017) *Culex* (*Culiciomyia*) *sasai* (Diptera: Culicidae), senior synonym of *Cx. spiculothorax* and a new country record for Bhutan. Acta Trop 171:194–198

Phanitchakun T, Namgay R, Miyagi I, Tsuda Y, Walton C, Harbach RE, Somboon P (2019) Morphological and molecular evidence for a new species of *Lutzia* (Diptera: Culicidae: Culicini) from Thailand. Acta Trop 191:77–86

Phuc HK, Ball AJ, Son L, Hanh NV, Tu ND, Lien NG et al (2003) Multiplex PCR assay for malaria vector *Anopheles minimus* and four related species in the Myzomyia series from southeast Asia. Med Vet Entomol 17(4):423–428

Poinar G, Zavortink TJ, Brown A (2019) *Priscoculex burmanicus* n. gen. et sp. (Diptera: Culicidae: Anophelinae) from mid-Cretaceous Myanmar amber. Hist Biol 32(9):1157–1162

Poinar G Jr (2005) *Culex malariager*, n. sp. (Diptera: Culicidae) from Dominican amber: the first fossil mosquito vector of *Plasmodium*. Proc Entomol Soc Wash 107(3):548–553

Poinar G Jr, Telford SR (2005) *Paleohaemoproteus burmacis* gen. n., sp. n. (Haemospororida: Plasmodiidae) from an Early Cretaceous biting midge (Diptera: Ceratopogonidae). Parasitology 131(Pt 1):79–84

Poinar GO Jr. (2004). *Palaeomyia burmitis* gen n., sp. n. (Phlebotominidae: Diptera) a new genus of Cretaceous sandflies with evidence of blood sucking habits. Proc Entomol Soc Wash 106(3):598–605

Poinar GO Jr, Poinar R (2004) *Paleoleishmania proterus* n. gen., n. sp., (Trypanosomatidae: Kinetoplastida) from Cretaceous Burmese amber. Parasitology 155(3):305–310

Poinar GO Jr, Zavortink TJ, Pike T, Johnson PA (2000) *Paleoculicis minututs* (Diptera: Culicidae) n. gen., n. sp., from Cretaceous Canadian amber, with a summary of described fossil mosquitoes. Acta Geol Hisp 35(1–2):119–128

Prummongkol S, Panasoponkul C, Apiwathnasorn C, Lek-Uthai U (2012) Biology of *Culex sitiens*, a predominant mosquito in Phang Nga, Thailand after a tsunami. J Insect Sci 12:11

Raaum RL, Sterner KN, Noviello CM, Stewart CB, Disotell TR (2005) Catarrhine primate divergence dates estimated from complete mitochondrial genomes: concordance with fossil and nuclear DNA evidence. J Hum Evol 48(3):237–257

Ramasamy R, Surendran SN (2011) Possible impact of rising sea levels on vector-borne infectious diseases. BMC Infect Dis 11:18

Ramasamy R, Surendran SN, Jude PJ (2015) Adaptation of mosquito vectors to salinity and its impact on mosquito-borne disease transmission in the South and Southeast Asian tropics. In: Morand S, Dujardin JP, Lefait-Robin R, Apiwathnasorn C (eds) Socio-ecological dimensions of infectious diseases in southeast Asia. Springer, Berlin, pp 107–122

Randika Wimalasiri-Yapa BMC, Stassen L, Huang X, Hafner LM, Hu W, Devin GJ et al (2019) Chikungunya virus in Asia– Pacific: a systematic review. Emerg Microbes Infect 8(1):70–79

Rao TR (1984) The Anophelines of India. Indian Council Medical Research, New Delhi, p 518

Rattanarithikul R, Harbach RE (1990) *Anopheles maculatus* (Diptera: Culicidae) from the type locality of Hong Kong and two new species of the Maculatus Complex from the Philippines. Mosq Syst 22(3):160–183

Rattanarithikul R, Panthusiri P (1994) Illustrated keys to the medically important mosquitos of Thailand. Southeast Asian J Trop Med Public Health 25(Suppl 1):1–66

Rattanarithikul R, Harbach RE, Harrison BA, Panthusiri P, Jones JW, Coleman RE (2005a) Illustrated keys to the mosquitoes of Thailand, II Genera *Culex* and *Lutzia*. Southeast Asian J Trop Med Public Health 36(Suppl 2):1–97

Rattanarithikul R, Harrison BA, Panthusiri P, Coleman RE (2005b) Illustrated keys to the mosquitoes of Thailand I. Background; geographic distribution; lists of genera, subgenera, and species; and a key to the genera. Southeast Asian J Trop Med Public Health 36(Suppl 1):1–80

Rattanarithikul R, Harrison BA, Harbach RE, Panthusiri P, Coleman RE (2006a) Illustrated keys to the mosquitoes of Thailand. IV. *Anopheles*. Southeast Asian J Trop Med Public Health 37(Suppl 2):1–128

Rattanarithikul R, Harrison BA, Panthusiri P, Peyton EL, Coleman RE (2006b) Illustrated keys to the mosquitoes of Thailand III. Genera *Aedeomyia*, *Ficalbia*, *Mimomyia*, *Hodgesia*, *Coquillettidia*, *Mansonia*, and *Uranotaenia*. Southeast Asian J Trop Med Public Health 37(Suppl 1):1–85

Rattanarithikul R, Harbach RE, Harrison BA, Panthusiri P, Coleman RE (2007) Illustrated keys to the mosquitoes of Thailand V. Genera *Orthopodomyia*, *Kimia*, *Malaya*, *Topomyia*, *Tripteroides*, and *Toxorhynchites*. Southeast Asian J Trop Med Public Health 38(Suppl 2):1–65

Rattanarithikul R, Harbach RE, Harrison BA, Panthusiri P, Coleman RE, Richardson JH (2010) Illustrated keys to the mosquitoes of Thailand. VI. Tribe Aedini. Southeast Asian J Trop Med Public Health 41(Suppl 1):1–225

Ree H-Il (2003) Taxonomic review and revised keys of the Korean mosquitoes (Diptera: Culicidae). Korean J Entomol 33(1):39–52

Reid JA (1968) Anopheline mosquitoes of Malaya and Borneo. Stud Inst Med Res Malaya No. 31. Government of Malaya. Kuala Lumpur, xiii 520 pp

Reidenbach KR, Cook S, Bertone MA, Harbach RE, Wiegmann BM, Besansky NJ (2009) Phylogenetic analysis and temporal diversification of mosquitoes (Diptera: Culicidae) based on nuclear genes and morphology. BMC Evol Biol 9:298

Reinert JF (1970) Contributions to the mosquito fauna of Southeast Asia. V. Genus *Aedes*, subgenus *Diceromyia* Theobald in Southeast Asia. Contrib Am Entomol Inst (Ann Arbor) 5(4):1–43

Reinert JF (1972) Contributions to the mosquito fauna of Southeast Asia. XV. Genus *Aedes* Meigen, subgenus *Ayurakitia* Thurman. Contrib Am Entomol Inst (Ann Arbor) 9(2):1–42

Reinert JF (1973a) Contributions to the mosquito fauna of Southeast Asia. XVI. Genus *Aedes* Meigen, subgenus *Aedimorphus* Theobald in Southeast Asia. Contrib Am Entomol Inst (Ann Arbor) 9(5):1–218

Reinert JF (1973b) Contributions to the mosquito fauna of Southeast Asia. XVIII. A reconsideration of *Diceromyia* Theobald with the inclusion of *Aedes nummatus* Edwards and *Aedes pseudonummatus* new species (Diptera: Culicidae). Contrib Am Entomol Inst (Ann Arbor) 10(1):22–40

Reinert JF (1973c) Contributions to the mosquito fauna of Southeast Asia. XIX. *Bothaella*, a new subgenus of *Aedes* Meigen. Contrib Am Entomol Inst (Ann Arbor) 10(3):1–51

Reinert JF (1974) Medical entomology studies - I. A new interpretation of the subgenus *Verrallina* of the genus *Aedes* (Diptera: Culicidae). Contrib Am Entomol Inst (Ann Arbor) 11(1):iv + 1–249

Reinert JF (1976a) Medical entomology studies - IV. The subgenera *Indusius* and *Edwardsaedes* of the genus *Aedes* (Diptera: Culicidae). Contrib Am Entomol Inst (Ann Arbor) 13(1):ii + 1–45

Reinert JF (1976b) Medical entomology studies - V. The subgenus *Rhinoskusea* of the genus *Aedes* (Diptera: Culicidae). Contrib Am Entomol Inst (Ann Arbor) 13(2):1–60

Reinert JF (1979) A description of *Isoaedes*, a new subgenus of *Aedes* Meigen, and its type species, *Ae.* (*Isa.*) *cavaticus* new species (Diptera: Culicidae). Mosq Syst 11:144–162

Reinert JF (1981) Medical entomology studies - XV. A revision of the subgenus *Paraedes* of the genus *Aedes* (Diptera: Culicidae). Contrib Am Entomol Inst (Ann Arbor) 18(4):1–91

Reinert JF (1985) A description of *Scutomyia*, a subgenus resurrected for the Albolineatus Group of the genus *Aedes* (Diptera: Culicidae). Mosq Syst 17:121–131

Reinert JF (1989) *Udaya subsimilis*, a new generic combination, a redescription of the holotype, and notes on the genus *Udaya* (Diptera: Culicidae). Mosq Syst 21:206–215

Reinert JF (1990) Medical entomology studies - XVII. Biosystematics of *Kenknightia*, a new subgenus of the mosquito genus *Aedes* Meigen from the Oriental Region (Diptera: Culicidae). Contrib Am Entomol Inst (Gainesville) 26(3):iv + 1–119

Reinert JF (1999) Restoration of *Verrallina* to generic rank in tribe Aedini (Diptera: Culicidae) and descriptions of the genus and three included subgenera. Contrib Am Entomol Inst (Gainesville) 31(3):1–83

Reinert JF (2000a) Description of *Fredwardsius*, a new subgenus of *Aedes* (Diptera: Culicidae). Eur Mosq Bull 6:1–7

Reinert JF (2000b) New classification of the composite genus *Aedes* (Diptera: Culicidae: Aedini), elevation of subgenus *Ochlerotatus* to generic rank, reclassification of the other subgenera, and notes on certain subgenera and species. J Am Mosq Control Assoc 16(3):175–188

Reinert JF (2003) Description of *Bruceharrisonius*, a new subgenus of *Ochlerotatus*, and a redescription of its type species *Oc.* (*Brh.*) *greenii*. J Am Mosq Control Assoc 19(4):309–322

Reinert JF (2006) *Tewarius* Reinert, a new genus of Aedini (Diptera: Culicidae). Proc Entomol Soc Wash 108:639–645

Reinert JF (2009a) Descriptions of *Danielsia* Theobald, a genus recently resurrected from synonymy (Diptera: Culicidae: Aedini) and its type species *Dn. albotaeniata* Leicester. Zootaxa 1976:25–38

Reinert JF (2009b) List of abbreviations for currently valid generic-level taxa in family Culicidae (Diptera). Eur Mosq Bull 27:68–76

Reinert JF, Harbach RE, Kitching IJ (2004) Phylogeny and classification of Aedini (Diptera: Culicidae) based on morphological characters of all life stages. Zool J Linnean Soc 142(3):289–368

Reinert JF, Harbach RE, Kitching IJ (2006) Phylogeny and classification of *Finlaya* and allied taxa (Diptera: Culicidae: Aedini) based on morphological data from all life stages. Zool J Linnean Soc 148(1):1–101

Reinert JF, Harbach RE, Kitching IJ (2008) Phylogeny and classification of *Ochlerotatus* and allied taxa (Diptera: Culicidae: Aedini) based on morphological data from all life stages. Zool J Linnean Soc 153:29–114

Reinert JF, Harbach RE, Kitching IJ (2009) Phylogeny and classification of tribe Aedini. Zool J Linnean Soc 157(4):700–794

Rodhain F, Rosen L (1997) Mosquito vectors and dengue virus-vector relationships. In: Gubler DJ, Kuno G (eds) Dengue and dengue hemorrhagic fever. CAB International, Wallingford, pp 45–60

Rueda LM, Pecor JE, Yuen MC, Lee MW (2004) New record, habitats and updated checklist of the mosquitoes of Hong Kong. J Am Mosq Control Assoc 20(2):204–207

Rutledge GG, Bohme U, Sanders M, Reid AJ, Cotton JA, Maiga-Ascofare O et al (2017) *Plasmodium malariae* and *P. ovale* genomes provide insights into malaria parasite evolution. Nature 542(7639):101–104

Sallum MAM, Peyton EL, Harrison BA, Wilkerson RC (2005a) Revision of the Leucosphyrus group of *Anopheles* (*Cellia*) (Diptera, Culicidae). Rev Bras Entomol 49(Suppl 1):1–155

Sallum MAM, Peyton EL, Wilkerson RC (2005b) Six new species of the *Anopheles leucosphyrus* group, reinterpretation of *An. olegans* and vector implication. Med Vet Entomol 19(2):158–199

Sang RC, Gichogo A, Gachoya J, Dunster MD, Ofula V, Hunt AR, Crabtree MB, Miller BR, Dunster LM (2003) Isolation of a new flavivirus related to cell fusing agent virus (CFAV) from field-collected flood-water *Aedes* mosquitoes sampled from a dambo in central Kenya. Arch Virol 148(6):1085–1093

Sarala KS, Nutan N, Vasantha K, Dua VK, Malhotra MS, Yadav RS et al (1994) Cytogenetic evidence for three sibling species in *Anopheles fluviatilis* (Diptera: Culicidae). Ann Entomol Soc Am 87(1):116–121

Satoto TBT (2001) Cryptic species within *Anopheles barbirostris* van derWulp, 1884, inferred from nuclear and mitochondrial gene Sequence variation. Ph.D. Thesis, The University of Liverpool (UK) ProQuest Dissertations Publishing, Liverpool

Sawabe K, Isawa H, Hoshino K, Sasaki T, Roychoudhury S, Higa Y, Kasai S, Tsuda Y, Nishiumi I, Hisai N, Hamao S, Kobayashi M (2010) Host-feeding habits of *Culex pipiens* and *Aedes albopictus* (Diptera: Culicidae) collected at the urban and suburban residential areas of Japan. J Med Entomol 47(3):442–450

Schmidt ER, Foley DH, Bugoro H, Bryan JH (2003) A morphological study of the *Anopheles punctulatus* group (Diptera: Culicidae) in the Solomon Islands, with a description of *Anopheles* (*Cellia*) *irenicus* Schmidt, sp.n. Bull Entomol Res 93(6):515–526

Silva JC, Egan A, Arze C, Spouge JL, Harris DG (2015) A new method for estimating species age supports the coexistence of malaria parasites and their mammalian hosts. Mol Biol Evol 32(5):1354–1364

Silver JB (2008) Mosquito ecology, field sampling methods, 3rd edn. Springer, Dordrecht, Netherland, p 1494

Singh B, Sung LK, Matusop A, Radhakrishnan A, Shamsul SSG, Cox-singh J, Thomas A, Conway DJ (2004a) A large focus of naturally acquired *Plasmodium knowlesi* infections in human beings. Lancet 363(9414):1017–1024

Singh OP, Chandra D, Nanda N, Raghavendra K, Sunil S, Sharma SK et al (2004b) Differentiation of members of the *Anopheles fluviatilis* species complex by an allele-specific polymerase chain reaction based on 28S ribosomal DNA sequences. Am J Trop Med Hyg 70(1):27–32

Singh OP, Chandra D, Nanda N, Sharma SK, Htun PT, Adak T et al (2006) On the conspecificity of *Anopheles fluviatilis* species S with *Anopheles minimus* species C. J Biosci 31(5):671–677

Sinka ME, Bangs MJ, Manguin S, Chareonviriyaphap T, Patil AP, Temperley WH et al (2011) The dominant *Anopheles* vectors of human malaria in the Asia-Pacific region: occurrence data, distribution maps and bionomic précis. Parasit Vectors 4:89

Sirivanakarn S (1972) Contribution to the mosquito fauna of Southeast Asia. XIII. The genus *Culex* subgenus *Eumelanomyia* Theobald in Southeast Asia and adjacent areas. Contrib Am Entomol Inst (Ann Arbor) 8(6):1–86

Sirivanakarn S (1976) Medical entomology studies – III. A revision of the subgenus *Culex* in the Oriental Region (Diptera: Culicidae). Contrib Am Entomol Inst 12(2):iii, 1–272

Sirivanakarn S (1977) Medical entomology studies: VI. A revision of the subgenus *Lophoceraomyia* in the Oriental Region (Diptera: Culicidae). Contrib Am Entomol Inst (Ann Arbor) 13(4):1–245

Solomon T, Ni H, Beasley DWC, Ekkelenkamp M, Cardosa MJ, Barrett ADT (2003) Origin and evolution of Japanese Encephalitis virus in Southeast Asia. J Virol 77(5):3091–3098

Somboon P, Harbach RE (2019) *Lutzia* (*Metalutzia*) *chiangmaiensis* n.sp. (Diptera: Culicidae), formal name for the Chiang Mai (CM) form of the genus *Lutzia* in Thailand. J Med Entomol 56(5):1270–1274

Somboon P, Choochote W, Khamboonruang C, Keha P, Suwanphanit P, Sukontasan K et al (1989) Studies on the Japanese encephalitis vectors in Amphoe Muang, Chiang Mai, Northern Thailand. Southeast Asian J Trop Med Public Health 20(1):9–17

Somboon P, Suwonkerd W, Lines JD (1994) Susceptibility of Thai zoophilic anophelines and suspected malaria vectors to local strains of human malaria parasites. Southeast Asian J Trop Med Public Health 25(4):766–770

Somboon P, Walton C, Sharpe R, Higa Y, Tuno N, Tsuda Y, Takagi M (2001) Evidence for a new sibling species of *Anopheles minimus* from the Ryukyu Archipelago, Japan. J Am Mosq Control Assoc 17(2):98–113

Somboon P, Rory A, Tsuda Y, Takagi M, Harbach RE (2010) Systematics of *Anopheles* (*Cellia*) *yaeyamaensis* sp. n., alias species E of the *An. minimus* complex in southeastern Asia (Diptera: Culicidae). Zootaxa 2651:43–51

Somboon P, Phanitchakun T, Namgay R, Wangdi T, Pemo D, Harbach RE (2020) Molecular and morphological evidence of sibling species in *Anopheles baileyi* Edwards (Diptera: Culicidae) in Bhutan and Thailand. Acta Trop 209:105549

Somboon P, Phanitchakun T, Saingamsook J, Namgay R, Harbach RE (2021) *Reinertia*, a new subgenus of the genus *Aedes* Meigen and its type species *Aedes* (*Reinertia*) *suffusus* (Diptera: Culicidae), newly recorded from Bhutan. J Med Entomol 58(3):1138–1148

Somboon P, Phanitchakun T, Namgay R, Harbach RE (2022a) The *Anopheles baileyi* species complex (Diptera: Culicidae: Anophelinae) in Bhutan. Acta Trop 226:106241

Somboon P, Phanitchakun T, Namgay R, Harbach RE (2022b) The *Anopheles lindesayi* species complex (Diptera: Culicidae) in Bhutan. J Med Entomol 59(4):1236–1251

Somboon P, Phanitchakun T, Namgay R, Harbach RE (2023a) Review of the Mimeticus Subgroup of *Culex* (*Culex*) (Diptera: Culicidae), with an assessment of the specific status of three nominal species described from India. J Med Entomol 60(2):255–271

Somboon P, Zhou L, Phanitchakun T, Saingamsook J, Harbach RE (2023b) *Aedes yunnanensis* (Diptera: Culicidae): transferred from the subgenus *Hulecoeteomyia* to *Orohylomyia* subg. nov., with descriptions of the species and the new subgenus. J Med Entomol 60(5):955–967

Somboon P, Saeung A, Saingamsook J, Namgay R, Harbach RE (2024) A new species of the *Anopheles barbirostris* complex (Diptera: Culicidae) from Bhutan. J Med Entomol 61(2):377–388

Stollar V, Thomas VL (1975) An agent in the *Aedes aegypti* cell line (Peleg) which causes fusion of *Aedes albopictus* cells. Virology 64(2):367–377

Stopard IJ, Churcher TS, Lambert B (2021) Estimating the extrinsic incubation period of malaria using a mechanistic model of sporogony. PLoS Comput Biol 17(2):e1008658

Suguna SG, Rathinam KG, Rajavel AR, Dhanda V (1994) Morphological and chromosomal descriptions of new species in the *Anopheles subpictus* complex. Med Vet Entomol 8(1):88–94

Sukowati S, Baimai V, Harun S, Dusuki Y, Andri H, Efriwati M (1999) Isoenzyme evidence for three sibling species in the *Anopheles sundaicus* complex from Indonesia. Med Vet Entomol 13(4):408–414

Suman DS, De S, Sharma G, Chandra K, Banerjee D (2021) *Culex* (*Culex*) *gaugleri*, a new species (Diptera: Culicidae) from India. Rec Zool Surv India 121(4):429–439

Suman DS, Sharma G, De S, Kumar V, Veer V, Chandra K, Banerjee D (2022) Description of *Culex* (*Culex*) *kodaikanalensis* sp. nov. (Diptera: Culicidae) from India with phyletic analysis using mitochondrial cytochrome oxidase I and 16S rRNA genes. Inter J Trop Insect Sci 42:2969–2978

Sunantaraporn S, Thepparat A, Phumee A, Sor-Suwan S, Boonserm R, Bellis G et al (2021) *Culicoides latreille* (Diptera: Ceratopogonidae) as potential vectors for *Leishmania martiniquensis* and *Trypanosoma* sp. in northern Thailand. PLoS Negl Trop Dis 15(12):e0010014

Swain S, Mohanty A, Tripathy HK, Mahapatra N, Kar SK, Hazra RK (2010) Molecular identification and phylogeny of Myzomyia and Neocellia series of *Anopheles* subgenus *Cellia* (Diptera: Culicidae). Infect Genet Evol 10(7):931–939

Ta TH, Hisam S, Lanza M, Jiram AI, Ismail N, Rubio JM (2014) First case of a naturally acquired human infection with *Plasmodium cynomolgi*. Malar J 13:68

Taai K, Harbach RE (2015) Systematics of the *Anopheles barbirostris* species complex (Diptera: Culicidae: Anophelinae) in Thailand. Zool J Linnean Soc 174(2):244–264

Taai K, Baimai V, Saeung A, Thongsahuan S, Min GS, Otsuka Y et al (2013) Genetic compatibility between *Anopheles lesteri* from Korea and *Anopheles paraliae* from Thailand. Mem Inst Oswaldo Cruz 108(3):312–320

Takano KT, Nguyen NTH, Nguyen BTH, Sunahara T, Yasunami M, Nguyen MD et al (2010) Partial mitochondrial DNA sequences suggest the existence of a cryptic species within the Leucosphyrus group of the genus *Anopheles* (Diptera: Culicidae), forest malaria vectors, in northern Vietnam. Parasit Vectors 3:41

Takhampunya R, Kim HC, Tippayachai B, Kengluecha A, Klein TA, Lee WJ, Grieco J, Evans BP (2011) Emergence of Japanese encephalitis virus genotype V in the Republic of Korea. Virol J 8:449

Tamashiro M, Toma T, Mannen K, Higa Y, Miyagi I (2011) Bloodmeal identification and feeding habits of mosquitoes (Diptera: Culicidae) collected at five islands in the Ryukyu Archipelago, Japan. Med Entomol Zool 62(1):53–70

Tanaka K, Mizusawa K, Saugstad ES (1979) A revision of the adult and larval mosquitoes of Japan (including the Ryukyu Archipelago and the Ogasawara Islands) and Korea (Diptera: Culicidae). Contrib Am Entomol Inst 16:1–987

Thongsahuan S, Baimai V, Junkum A, Saeung A, Min GS, Joshi D et al (2011) Susceptibility of *Anopheles campestris*-like and *Anopheles barbirostris* species complexes to *Plasmodium falciparum* and *Plasmodium vivax* in Thailand. Mem Inst Oswaldo Cruz 106(1):105–112

Thurman EB (1959) A contribution to a revision of the Culicidae of northern Thailand. Univ Md Agric Exp Stn Bull A 100:1–182

Toma T, Miyagi I (1986) The mosquito fauna of the Ryukyu archipelago with identification keys, pupal descriptions and notes on biology, medical importance and distribution. Mosq Syst 18:1–109

Toma T, Taira K, Oosiro A, Azama Y, Fukuchi Y, Miyagi I (2019) New taxonomic status of *Tripteroides bambusa* (Yamada), 1917 from Japan, based on experimental crossing and *COI* sequence divergence. Med Entomol Zool 70(1):15–23

Tsuda Y, Haseyama M, Ishida K, Niizuma J, Kim KS, Yanagi D, Watanabe N, Kobayashi M (2012) After-effects of Tsunami on distribution and abundance of mosquitoes in rice-field areas in Miyagi Prefecture, Japan in 2011. Med Entomol Zool 63(1):21–30

Tyson WH (1970) Contributions to the mosquito fauna of Southeast Asia. VII. Genus *Aedeomyia* Theobald in Southeast Asia. Contrib Am Entomol Lnst (Ann Arbor) 6(2):1–27
Valkiūnas G (2005) Avian malaria parasites and other haemosporidia. CRC Press, Boca Raton, p 946
Vasilakis N, Tesh RB (2015) Insect-specific viruses and their potential impact on arbovirus transmission. Curr Opin Virol 15:69–74
Vesgueiro FT, Demari-Silva B, Malafronte RDS, Sallum MAM, Marrelli MT (2011) Intragenomic variation in the second internal transcribed spacer of the ribosomal DNA of species of the genera *Culex* and *Lutzia* (Diptera: Culicidae). Mem Inst Oswaldo Cruz 106(1):1–8
Votýpka J, Szabová J, Rádrová J, Zídková L, Svobodová M (2012) *Trypanosoma culicavium* sp. nov., an avian trypanosome transmitted by *Culex* mosquitoes. Int J Syst Evol Microbiol 62(Pt3):745–754
Walton C, Sharpe RG, Pritchard SJ, Thelwell NJ, Butlin RK (1999a) Molecular identification of mosquito species. Biol J Linn Soc Lond 68(1-2):241–256
Walton C, Handley JM, Kuvangkadilok C, Collins FH, Harbach RE, Baimai V et al (1999b) Identification of five species of the *Anopheles dirus* complex from Thailand, using allele-specific polymerase chain reaction. Med Vet Entomol 13(1):24–32
Walton C, Somboon P, Harbach RE, Zhang S, Weerasinghe I, O'Loughlin SM et al (2007a) Molecular identification of mosquito species in the *Anopheles annularis* group in southern Asia. Med Vet Entomol 21(1):30–35
Walton C, Somboon P, O'loughlin SM, Zhang S, Harbach RE, Linton YM et al (2007b) Genetic diversity and molecular identification of mosquito species in the *Anopheles maculatus* group using the ITS2 region of rDNA. Infect Genet Evol 7(1):93–102
Wang G, Li C, Guo X, Xing D, Dong Y, Wang Z et al (2012) Identifying the main mosquito species in China based on DNA barcoding. PLoS One 7(10):e47051
Wang G, Li C, Zheng W, Song F, Guo X, Wu Z et al (2017) An evaluation of the suitability of *COI* and *COII* gene variation for reconstructing the phylogeny of, and identifying cryptic species in, anopheline mosquitoes (Diptera Culicidae). Mitochondrial DNA A DNA Mapp Seq Anal 28(5):769–777
Wang X, Nishiura H (2021) The epidemic risk of dengue fever in Japan: Climate change and seasonality. Can J Infect Dis Med Microbiol 2021:1–13
Wasserman S, Tambyah PA, Lim PL (2016) Yellow fever cases in Asia: primed for an epidemic. Int J Infect Dis 48:98–103
Watanabe M, Watanabe H, Tabaru Y, Hirao M, Roychoudhury S, Sawabe K, Ishikawa Y, Kawabata T, Kanno K (2012) Occurrence of vector mosquitoes at Tsunami disaster areas of the Great East Japan Earthquake. Med Entomol Zool 63(1):31–43
Waters AP, Higgins DG, McCutchan TF (1991) Evolution *Plasmodium falciparum* appears to have arisen as a result of lateral transfer between avian and human hosts. Proc Nati Acad Sci USA 88(8):3140–3144
Weaver SC, Reisen WK (2010) Present and future arboviral threats. Antivir Res 85(2):328–345
Weaver SC, Vasilakis N (2009) Molecular evolution of dengue viruses: Contributions of phylogenetics to understanding the history and epidemiology of the preeminent arboviral disease. Infect Genet Evol 9(4):523–540
Wharton RH (1962) The biology of *Mansonia* mosquitoes in relation to the transmission of filariasis in Malaya. Bull Inst Med Res Kuala Lumpur 11:1–114
Wiegmann BM, Yeates DK (2017) Phylogeny of Diptera. SURICATA 4:253–265
Wiegmann BM, Trautwein MD, Winkler IS, Barr NB, Kim JW, Lambkin C et al (2011) Episodic radiations in the fly tree of life. Proc Natl Acad Sci USA 108(14):5690–5695
Wijit A, Saeung A, Baimai V, Otsuka Y, Thongsahuan S, Taai K et al (2013) DNA barcoding for the identification of eight species members of the Thai Hyrcanus Group and investigation of their stenogamous behaviour. CR Biol 336(9):449–456

Wilai P, Ali RSM, Saingamsook J, Saeung A, Junkum A, Walton C et al (2020a) Integrated systematics of *Anopheles subpictus* (Diptera: Culicidae) in the Oriental Region, with emphasis on forms in Thailand and Sulawesi, Indonesia. Acta Trop 208:105503
Wilai P, Namgay R, Ali RSM, Saingamsook J, Saeung A, Junkum A et al (2020b) A multiplex PCR based on mitochondrial *COI* sequences for identification of members of the *Anopheles barbirostris* complex (Diptera: Culicidae) in Thailand and other countries in the region. Insects 11(7):409
Wilder-Smith A, Leong WY (2017) Importation of yellow fever into China: assessing travel patterns. J Travel Med 24(4):1–4
Wilder-Smith A, Lee V, Gubler DJ (2019) Yellow fever: is Asia prepared for an epidemic? Lancet Infect Dis 19(3):241–242
Wilkerson RC, Peyton EL (1990) Standardized nomenclature for the costal wing spots of the Genus *Anopheles* and other spotted-wing mosquitoes (Diptera: Culicidae). J Med Entomol 27(2):207–224
Wilkerson RC, Linton Y-M, Fonseca DM, Schultz TR, Price DC, Strickman DA (2015) Making mosquito taxonomy useful: A stable classification of Tribe Aedini that balances utility with current knowledge of evolutionary relationships. PLoS One 10(7):e0133602
Wilkerson RC, Linton Y-M, Strickman D (2021) Mosquitoes of the world, vol 1 and 2. Johns Hopkins University Press, Baltimore, p 1332
Wilkerson RC, Somboon P, Harbach RE (2022) Reconsideration of the status of subspecies in the Japonicus Group of the subgenus *Hulecoeteomyia* Theobald of *Aedes* Meigen (Diptera: Culicidae). Zootaxa 5162(2):198–200
Woo JH, Jeong YE, Jo JE, Shim SM, Ryou J, Kim KC, Lee WJ, Lee JY (2020) Genetic characterization of Japanese encephalitis virus genotype 5 isolated from patient, South Korea, 2015. Emerg Infect Dis 26(5):1002–1006
World Health Organization (WHO) (1966) Report of the WHO evaluation team on malaria eradication in the Ryukyu Islands. WHO WPR/419/65, Geneva, Switzerland, pp 1–53
World Health Organization (WHO) (2013) Regional framework for surveillance and control of invasive mosquito vectors and re-emerging vector-borne diseases 2014–2020. In: Van den Berg H, Velayudhan R, Ejov M (eds) World Health Organization. p 18
World Health Organization (WHO) (2017) Anopheline species complexes in South and South–East Asia. SEARO Technical Publication No. 57, p 102
World Health Organization (WHO) (2023a) Vector-borne diseases. World Health Organization. [Accessed 1 January 2023] https://www.who.int/news-room/fact-sheets/detail/vector-borne-diseases
World Health Organization (WHO) (2023b) Malaria. World Health Organization. [Accessed 2 January 2023] https://www.who.int/news-room/fact-sheets/detail/malaria
World Health Organization (WHO) (2023c) Dengue and severe dengue. World Health Organization. [Accessed 3 January 2023] https://www.who.int/news-room/fact-sheets/detail/dengue-and-severe-dengue
World Health Organization (WHO) (2023d) Japanese encephalitis. World Health Organization. [Accessed 4 Jan 2023] https://www.who.int/news-room/fact-sheets/detail/japanese-encephalitis#:~:text=Japanese%20encephalitis%20virus%20(JEV)%20is,000%20clinical%20cases%20every%20year
World Health Organization (WHO) (2023e) Chikungunya. World Health Organization. [Accessed 12 Jan 2023] https://www.who.int/news-room/fact-sheets/detail/chikungunya
World Health Organization (WHO) (2023f) Zika virus. World Health Organization. [Accessed 12 Jan 2023] https://www.who.int/news-room/fact-sheets/detail/zika-virus?gclid=CjwKCAiA2fmdBhBpEiwA4CcHzSNovjWrtGxVutWzqr6ID7yMGREw1ZFFOyWL3X194TQE_RwmW3w0QxoCui4QAvD_BwE
World Health Organization (WHO) (2023g) Lymphatic filariasis. World Health Organization. [Accessed 12 Jan 2023] https://www.who.int/news-room/fact-sheets/detail/lymphatic-

filariasis#:~:text=Lymphatic%20filariasis%2C%20commonly%20known%20as,damage%20to%20the%20lymphatic%20system

Xiao P, Li C, Zhang Y, Han J, Guo X, Xie L, Tian M, Li Y, Wang M, Liu H et al (2018) Metagenomic sequencing from mosquitoes in China reveals a variety of insect and human viruses. Front Cell Infect Microbiol 8:364

Yang Q, Chung J, Robinson KL, Schmidt TL, Ross PA, Liang J, Hoffmann AA (2022) Sex-specific distribution and classification of *Wolbachia* infections and mitochondrial DNA haplogroups in *Aedes albopictus* from the Indo-Pacific. PLoS Negl Trop Dis 16(4):e0010139

Zavortink T (1971) Contributions to the mosquito fauna of Southeast Asia. IX. The genus *Orthopodomyia* Theobald in Southeast Asia. Contrib Am Entomol Inst (Ann Arbor) 7(3):1–37

Zavortink TJ, Poinar GO (2000) *Anopheles* (*Nyssorhynchus*) *dominicanus* sp. n. (Diptera: Culicidae) from Dominican amber. Ann Entomol Soc Am 93(6):1230–1235

Zheng XL (2020) Unveiling mosquito cryptic species and their reproductive isolation. Insect Mol Biol 29(6):499–510

Zhou L, Namgay R, Saeung A, Saingamsook J, Phanitchakun T, Harbach RE, Somboon P (2023) A multiplex PCR assay for the identification of 5 sibling species of the Lindesayi complex of the genus *Anopheles* (Diptera: Culicidae) in Bhutan. J Med Entomol 60(5):924–930

Chapter 2
Tick

Koray Ergunay, Daisuke Kobayashi, Takeo Yamauchi, and Kyoko Sawabe

Abstract Ticks are hematophagous arthropods of the order Ixodida within the class Arachnida (the subphylum Chelicerata) that parasitize a variety of animals, including humans. During the blood-sucking process, ticks transmit various pathogens such as viruses, bacteria, protozoa, and nematodes to vertebrates, making ticks, along with mosquitoes, important pathogen vectors. This chapter outlines the basic biology of ticks (taxonomy, distribution, and ecology) and describes the characteristics of tick distribution in Asia. It also reviews various tick-borne pathogens reported in the Asian region and describes tick-borne viral infections of medical importance in the region (Crimean-Congo hemorrhagic fever, Kyasanur Forest disease, severe fever with thrombocytopenia syndrome, and tick-borne encephalitis) are described. In addition, emerging tick-borne viruses recently discovered in various parts of Asia are also introduced in this chapter, including their clinical manifestations, prevalence status, and vector tick species. Although ticks and mosquitoes generally play the same role as vectors of pathogens, their blood-sucking ecology is quite different. This difference also affects the ecology and epidemic dynamics of the pathogens transmitted. This chapter also compares the blood-sucking ecology of ticks and mosquitoes and discusses how their respective blood-sucking characteristics affect the transmission and spread of viruses.

K. Ergunay
Walter Reed Biosystematics Unit, Suitland, MD, USA

Hacettepe University, Ankara, Turkey
e-mail: ekoray@hacettepe.edu.tr

D. Kobayashi (✉) · K. Sawabe
Department of Medical Entomology, National Institute of Infectious Diseases, Tokyo, Japan
e-mail: dkoba@niid.go.jp; sawabe@niid.go.jp

T. Yamauchi
Schools of Agriculture and Animal Science, Obihiro University of Agriculture and Veterinary Medicine, Obihiro, Hokkaido, Japan
e-mail: tyamauchi@obihiro.ac.jp

K. Sawabe et al. (eds.), *Medical Entomology in Asia*, Entomology Monographs,
https://doi.org/10.1007/978-981-97-4406-0_2

Keywords Asia · Blood-sucking · Emerging infectious diseases · Ecology · Tick · Virus

2.1 Introduction

Currently, approximately 960 species of ticks (Arachnida: Acari: Ixodida) have been recorded worldwide, 760 of which belong to the family Ixodidae and 200 to the family Argasidae. Ticks are a well-studied taxonomic group of arthropods that parasitize and transmit pathogens to humans and domestic animals.

Ticks are known to transmit several pathogens (viruses, bacteria, protozoa, and nematodes) to vertebrates during the blood-feeding process. Among these, several tick-borne viruses that cause serious infectious diseases in humans, such as Crimean-Congo hemorrhagic fever, Kyasanur Forest diseases, and tick-borne encephalitis, and certain bacteria (e.g., *Borrelia* and *Rickettsia*) are the most medically important tick-borne pathogens in the Asian region. Tick-borne protozoa such as *Babesia* and *Theileria*, which are pathogenic to domestic and companion animals, are also found in Asia. Transmission of many tick-borne pathogens occurs through the blood-sucking action of the tick. Therefore, the blood-sucking ecology of the tick (e.g., the number of blood-sucking at each developmental stage, and the time required from full engorgement to re-blood-sucking, etc.) has a significant impact on the transmission and the ecology of pathogens.

Recent studies have shown that arthropods harbor a great diversity of viruses. These studies have led to the discovery of several arboviruses, including novel viruses, from ticks. In addition to the above-mentioned human-pathogenic tick-borne viruses reported in Asia, several other tick-borne viruses are infectious to humans but whose pathogenicity is still unknown.

2.2 Taxonomy

The order Ixodida (sometimes referred to as suborder Metastigmata, hereafter referred to as ticks) belongs to the subclass Acari, within the class Arachnida, the subphylum Chelicerata, and the phylum Arthropoda. Currently, approximately 960 species of ticks have been recorded worldwide, 760 of which belong to the family Ixodidae and 200 to the family Argasidae. Approximately 50 species from these two families are distributed in Japan. The third family, Nuttalliellidae, consists of only one African species, *Nuttalliella namaqua* Bedford.

The tick's body is extremely large and conspicuous. Some species of the family Ixodidae can reach a body length of 30 mm after sucking blood. Ticks are a well-studied taxonomic group of arthropods that parasitize humans and domestic animals and transmit pathogens. Therefore, tick fauna has been studied and reported in various regions of the world.

2.3 Distribution

Table 2.1 summarizes the reports on relatively new tick fauna in Asian countries. The number of tick species cannot be easily compared because the level of faunal elucidation varies greatly from country to country. For example, the level of faunal elucidation is considered very high in Japan and Singapore and relatively low in Cambodia and Laos. Since the distribution of ticks is closely related to that of tick-borne diseases, faunal surveys are required in areas where the level of faunal elucidation is low.

Asia contains a high diversity of ticks belonging to the family Ixodidae. The genera *Amblyomma*, *Dermacentor*, *Hyalomma*, *Haemaphysalis,* and *Ixodes* are particularly abundant. *Ixodes* ticks are the most abundant in cold northern regions, whereas *Haemaphysalis* ticks are the most abundant in tropical to temperate regions. The species diversity of the genus *Hyalomma* is high in the arid regions of Central Asia. The range of ticks varied widely among species. Parasitic avian species are widely distributed. Some species, such as *Am. testudinarium* C. L. Koch, are not avian parasites but are widely distributed, especially in Southeast Asia. However, there are species whose distribution is highly localized, for example, *Am.*

Table 2.1 Number of tick species reported in Asian countries

Country	Number of species		References
	Ixodidae	Argasidae	
Afghanistan	28	11	Hoogstraal (1971), Guglielmone et al. (2023)
Bangladesh	10	1	Ghosh et al. (2007), Guglielmone et al. (2023)
Cambodia	17		Guglielmone et al. (2023)
China	111	14	Zhang et al. (2019)
India	82	18	Ghosh et al. (2007), Guglielmone et al. (2023)
Indonesia	42+	7	Kazim et al. (2021), Guglielmone et al. (2023)
Iran	37	10	Hosseini-Chegeni et al. (2019), Guglielmone et al. (2023)
Japan	42	4	Takada and Fujita (2019)
Laos	30		Guglielmone et al. (2023)
Malaysia	44	3	Kazim et al. (2022)
Mongolia	19		Guglielmone et al. (2023)
Myanmar	34		Guglielmone et al. (2023)
Nepal	55		Pun et al. (2018)
Pakistan	25+	4	Guglielmone et al. (2023)
Philippines	30	3	Kazim et al. (2021), Guglielmone et al. (2023)
Singapore	15	1	Kwak (2018), Guglielmone et al. (2023)
Sri Lanka	30	2	Seneviratna (1962), Guglielmone et al. (2023)
Taiwan	37	3	Robbins (2005), Guglielmone et al. (2023)
Thailand	58	4	Tanskul et al. (1983), Guglielmone et al. (2023)
Turkey	40	8	Bursali et al. (2012), Guglielmone et al. (2023)
Uzbekistan	28	7	Rasulov (2007), Guglielmone et al. (2023)
Vietnam	57		Guglielmone et al. (2023)

komodoensis (Oudemans) (Komodo Island, and Flores Island), *Am. kraneveldi* Anastos (Flores and Sula Islands), *Am. orlovi* Kolonin (northern Vietnam), *Am. babirussae* Schulze (Sulawesi or adjacent small islands), *Am. cyprium aeratipes* Schulze (Mindanao), and *Am. robinsoni* Warburton (Komodo Island). Many of these species are found on remote islands in Southeast Asia and their hosts are often restricted to specific species. Because parasites with local distribution and high host specificity are at high risk of extinction, actions for these ticks must focus on species conservation.

2.4 Ecology

2.4.1 Life Cycle

The tick larvae hatch from eggs, develop into nymphs and emerge as adults. The time of the year when ticks are present is generally fixed for each tick species, and the seasonal occurrence shows similar trends each year. During winter, ticks in areas with heavy snowfall are inactive, whereas those in warmer areas are active. In general, ticks suck blood at least once at each developmental stage (larva, nymph, and adult) for a total of more than three times. Ticks that spontaneously drop off the host after saturation with blood (i.e., parasitize three hosts in their lifetime) are called three-host ticks. Most of the ticks are three-host ticks.

In three-host ticks, the larvae and nymphs suck enough blood from the host animal, voluntarily leave the body, fall to the ground surface, and hide under rocks or in the shade of grass to avoid light. After a period of rest, they molt to the next stage in which the larva does not move much until the body hardens and flattens, after which it resumes host-seeking. In ticks, the non-feeding period is much longer than the feeding period. After sucking blood from the host, adult females (some species reach 30 mm in length after sucking blood) drop to the ground and lay eggs on fallen leaves, rock crevices, and burrows. However, even if they do not reach saturation with blood, they can lay eggs after a certain amount of blood has been removed. Adult females die within a few days of egg laying. In addition, various pathogens have been reported to be transmitted via transovarial transmission (Lane and Burgdorfer 1987; Silva Costa et al. 2011).

The larvae are less than 1 mm long and may form colonies on plants depending on the species. The nymphs of many species are less than 2 mm in length. There are several nymphal stages in the family Argasidae, such as protonymph and deutonymph, whereas, in the family Ixodidae, the nymph has only one stage. Ticks are generally bisexual; however, some species, such as *Hae. mageshimaensis*, are known to be parthenogenetic and reproduce only as females. The closely related species *Hae. longicornis* Neumann has both hermaphrodite (diploid chromosomes) and parthenogenetic (triploid chromosomes) reproductive systems. Male ticks of the genus *Ixodes* rarely suck blood, and the hypostomes of many tick species, especially the teeth, are degenerated. The lifespan of male ticks of the genus *Ixodes* is significantly shorter than that of males of other tick genera.

The lifecycle duration of ticks is highly variable. For example, the life cycle of *Rhipicephalus microplus* (Canestrini) is short (approximately eight weeks) while that of *I. signatus* Birula, a three-host species, is approximately three full years (four years in total), based on field observations (Asanuma and Fukuda 1957). When conditions are suitable, the life cycle of ticks of the family Argasidae ranges from 6 weeks to 2 years, otherwise it can extend to more than 10 years (Filippova 1966).

2.4.2 Blood-Sucking

All tick species are hematophagous and can transmit a variety of pathogens while sucking blood. Therefore, all tick species are considered potential vectors of tick-borne diseases. Ticks parasitize and suck blood from humans, pets, and wild animals (mammals, birds, and reptiles). The blood-sucking period of ticks varies according to their developmental stage, being 3–6 days for the larval stage of most species, 3–10 days for the nymphal stage, and 6–12 days for the adult female. However, there are several exceptions. For example, the larvae of *Hae. inermis* Birula completes blood-sucking within 2–3 h (Nuttall and Warburton 1915). In the closely related *Hae. kitaokai* Hoogstraal, larvae complete blood sucking in as little as 1 h and 50 min, and nymphs in as little as 3 h and 30 min (Kitaoka and Morii 1967).

When ticks of the family Ixodidae become engorged with blood, they do not re-parasitize a host during the same developmental stage. In contrast, ticks in the family Argasidae suck blood repeatedly and briefly during the same developmental stages. In larvae, nymphs, and adult females, the scutum is present only on the anterior part of the dorsal surface. The posterior part of the dorsal surface, which is not covered, elongates significantly while the tick sucks blood. The amount of blood sucked by a single female (7–10 mg body weight) of the genus *Dermacentor* can reach 4000 mg. The large species *I. acutitarsus* Karsch and *Am. testudinarium* can consume up to 3 g of blood (Kitaoka 1994).

2.4.3 Search for Host Animals

The sensilla of the Haller's organ in the first leg of a tick responds to heat, carbon dioxide, ammonia, hydrogen sulfide, and other host-specific odors. Ticks can be broadly classified into non-nidicolous and nidicolous species based on where they seek hosts, molt, and lay eggs. There are two types of non-nidicolous species. The first type (e.g., *Hae. longicornis* and *Hae. flava* Neumann) climbs up a plant or other object and raises their first leg, which has a Haller's organ, to ambush an incoming animal. The other type (e.g., adults of *Am. testudinarium* and *I. acutitarsus*) walk around looking for animals to parasitize. Nidicolous species parasitize host animals in their burrows or nests.

2.4.4 Tick Mobility

In general, it is believed that most ticks usually minimize their movement to save energy. For example, in the field, the larvae of the genus *Haemaphysalis* ticks walk a total of 1 m at the most (Mori and Tsunoda 1996). Thus, the horizontal migration distance of unparasitized ticks is very limited, and the presence of a tick in a habitat indicates that a blood-saturated individual from a previous developmental stage had fallen nearby. Although the tick only moves a short distance, the host animal may transport the parasitic tick over a large distance. This is particularly true when the host is avian, which results in ticks inhabiting distant areas.

2.4.5 Habitats

Some ticks are found on the undersides of leaves, including those that fall to the undergrowth of montane forests, waiting for a host animal to visit. Tick species that parasitize large animals, such as the Japanese deer (e.g., adult *Hae. longicornis* and *Hae. kitaokai*), wait for them on the plants they feed at the height of 20–110 cm. Thus, ticks are found at high densities in grasslands, montane forests, and other environments with constant temperature and humidity, which are typical habitats for mammals (especially large ones), birds, and reptiles. Factors that determine the distribution of ticks include climatic conditions (temperature, humidity, sunlight, etc.), vegetation, and the density and behavior of host animals.

Although most ticks are found in montane forests, some are found in urban environments, such as urban parks, riverbeds, and sometimes even homes.

2.4.6 Relationship with Host Animals

Some tick species are host-specific; for example, *I. turdus* Nakatsudi parasitizes birds, *Am. geoemydae* Cantor parasitizes turtles and *Am. nitidum* Hirst & Hirst parasitizes sea snakes. Moreover, some ticks have host animal preferences, for example, *R. microplus* preferentially parasitizes cattle. Additionally, host animal preferences often differ among the three developmental stages (larvae, nymphs, and adults) of the same tick species. Most resident species are highly host-specific. Other species are also opportunistic. Thus, the relationship between ticks and host animals is diverse. Ticks that parasitize endangered animals may gradually disappear with the decline in their host animal populations. For example, *Am. crenatum* Neumann, which parasitizes the Javan rhinoceros (*Rhinoceros sondaicus* Desmarest), and *Hae. pentalagi* Pospelova-Shtrom, which parasitizes the Amami rabbit (*Pentalagus furnessi* Stone), are at high risk of extinction.

In Japan, tick fauna has been reported to vary significantly depending on the presence or absence of the Japanese Sika deer *Cervus nippon pseudaxis* Gervais (Fujimoto and Yamaguti 1990). In the northeastern part of Shimane Prefecture, the collection frequency of *Hae. longicornis* and *Hae. kitaokai* decreased with distance from the Misen Mountains, which are the only habitat of the Japanese Sika deer in Shimane Prefecture (Yamauchi et al. 2009). Accordingly, the distribution of large mammals, which are important blood-sucking sources, significantly influences the formation of tick fauna. In other words, changes in the mammalian fauna can cause changes in the tick fauna. The increase in the population and distribution of Japanese deer and wild boars in Japan in recent years have led to similar changes in their parasitic ticks. Incidentally, medium to large mammals can harbor large numbers of ticks, and it has been reported that more than 3300 ticks have been collected from a single captivated Japanese deer (Mori et al. 1990).

When ticks suck blood from a host animal, they remain attached to the skin for some time. Therefore, when host animals are transported over long distances, they are transported with their hosts. The migration of host animals, whether caused by humans or otherwise, can lead to the expansion of parasitic tick species into new areas of distribution. For example, the distribution of *R. microplus* has expanded from South and Southeast Asia to almost all tropical regions and that of *Am. variegatum* Fabricius has expanded from Africa to Madagascar and the West Indies (Barré and Uilenberg 2010). The Australian species *Hae. longicornis* is believed to have been introduced from Japan in the nineteenth century as a cattle parasite and subsequently expanded to New Zealand and small islands in the Pacific. *Rhipicephalus sanguineus* Latreille has been introduced to Japan from the United States, but its native habitat is unknown. Thus, the distribution and density of parasitic ticks are strongly influenced by the dynamics of host animals.

2.4.7 Human Bites

When a human encounters a waiting tick, it attaches to the exposed skin or clothing and locates a suitable parasitic site, and begins to suck blood. When the tick reaches a suitable site, it establishes a parasite site and begins to suck blood. In other words, ticks do not suck blood from the first site they attach to. Some tick species show a preference for biting sites: *I. ovatus* Neumann tends to bite the face (especially the eyelids), *I. turdus* the head, and *Am. testudinarium* the moist areas of the lower body such as the interdigital spaces, pubic region, and anus.

The swollen body of the blood-sucking tick is gray or blackish-brown and not blood-colored. The scutum was significantly smaller than the entire swollen body, and the legs were hidden by the ventral surface, making them difficult to see. The saliva of the tick, which is injected into the host during the blood-sucking process, contains substances that reduce itching and pain, allowing the tick to suck blood for long periods without being aware of the host. Adults of the genera *Ixodes* and *Amblyomma* have long submaxillary segments that are extremely powerful at biting the host. Therefore, the tick cannot be easily removed from the bite site. When a tick is forcibly removed, its mouthparts are often torn off and left on the skin.

Damage caused by ticks during blood sucking includes (1) damage to the host's skin, (2) transmission of various infectious disease agents, (3) paralysis (injection of toxic substances), and (4) allergy to animal flesh.

2.5 Tick-Borne Diseases in Asia

Asian regions include a variety of climatic zones in which a wide range of organisms are distributed. The Asian continent includes four of the six zoogeographic regions (Australian, Ethiopian, Oriental, and Palearctic) proposed by Wallace regarding the geographical distribution of the world's fauna (Wallace 1876). Furthermore, in a recent research, six zoogeographic regions (Australian, Oceanian, Oriental, Palearctic, Saharo-Arabian, and Sino-Japanese regions) were included in Asia based on the vertebrate taxa, distribution, and phylogenetic relationships (Halt et al. 2013). Ticks are natural vectors of pathogens due to their complex interactions with the vertebrate host, and the diversity of tick-borne pathogens in a region is also associated with that of the host organisms. Therefore, various tick-borne pathogens are expected to be present in Asia, encompassing a diverse range of zoogeographic zones.

Ticks transmit various pathogens (viruses, bacteria, protozoa, and nematodes) to vertebrates through blood feeding (Table 2.2). Among these, various tick-borne viruses that cause serious infectious diseases in humans, such as Crimean-Congo hemorrhagic fever, Kyasanur Forest disease, tick-borne encephalitis, and certain bacteria (e.g., *Borrelia* and *Rickettsia*) are the major medically important tick-borne pathogens in Asia (Table 2.2) (Rochlin and Toledo 2020). Tick-borne protozoa, such as *Babesia* and *Theileria*, which are pathogenic to domestic and companion animals, are also distributed in Asia (Lempereur et al. 2017) and cases of human babesiosis, although rare, have been reported mainly in East Asia (Table 2.2). Trypanosoma parasites have also been found in ticks, but little is known about their natural history or pathogenicity in vertebrates, including humans (Kobayashi et al. 2020). There are also reports on the distribution of tick-borne nematodes (*Cercopithifilaria* spp.) that infect companions (dogs and cats) and wild animals in Asia (Uni et al. 2013; Bezerra-Santos et al. 2022). Among these diverse tick-borne pathogens, this section discusses tick-borne viruses in detail, which are a major public health concern owing to their increasing prevalence and the recent discovery of novel viruses in Asia.

Table 2.2 Representative tick-borne diseases of public health importance in Asia

Disease	Causative agent	Primary vector tick species	Area(s) of endemicity[a]
Viral[b]			
Crimean-Congo hemorrhagic fever	Crimean-Congo hemorrhagic fever virus	*Hyalomma marginatum* complex	CA, EA, SA, WA
Kyasanur Forest disease	Kyasanur Forest disease virus	*Haemaphysalis spinigera*	SA, WA
Severe fever with thrombocytopenia syndrome	Severe fever with thrombocytopenia syndrome virus	*Haemaphysalis longicornis*	EA
Tick-borne encephalitis	Tick-borne encephalitis virus	*Ixodes persulactus*	EA
Bacterial			
Lyme disease	*Borrelia bavariensis, B. garinii, B. afzalii*	*Ixodes persulactus*	EA
Tick-borne relapsing fever	*Borrelia persica, B. caucasica, B. latyschewii*	*Ornithodoros tholozani, O. verrucosus, O. tartakowskyi*	CA, WA
Borrelia miyamotoi disease	*Borrelia miyamotoi*	*Ixodes persulactus*	EA
Mediterranean spotted fever	*Rickettsia conorii*	*Rhipicephalus sanguineus*	EA, SA, SEA, WA
Japanese spotted fever	*Rickettsia japonica*	*Haemaphysalis flava, Hae. hystricis*	EA, SEA
Protozoal			
Human babesiosis	Babesia microti etc.	*Ixodes* spp.	EA, SEA

[a]*CA* Central Asia, *EA* Eastern Asia, *SA* Southern Asia, *SEA* South-eastern Asia, *WA* Western Asia. The division of the Asian region follows the standard country or area codes for statistical use (M49) prepared by the Statistics Division of the United Nations
[b]See Table 2.3 for details

2.6 Tick-Borne Viruses Associated with Diseases in Humans in Asia

Ticks are involved in the transmission of a wide variety of viruses in at least nine families (Nuttall 2014; Matthijnssens et al. 2022a, b; Kuhn et al. 2023). Among these, 23 pathogenic viruses have been reported in Asia (Table 2.3). Although the prevalence of viral diseases varies among different types, it is inevitably low for those caused by species that have been recently discovered as a result of recent advances in sequencing technology (e.g., Alongshan virus, Beiji nairovirus, Jingmen tick virus, Songling virus, Tacheng tick virus-1 and 2, Yezo virus; see Sect. 2.7 for details). Many tick-borne viruses cause flu-like symptoms, and it is difficult to differentiate between them only from clinical symptoms. In addition, for most viruses, there are no commercially available vaccines or therapeutics agents; therefore, the prevention of tick bites is of paramount importance to avoid viral infection. This

Table 2.3 Tick-borne viruses associated with diseases in humans in Asia

Virus name	Taxonomy (family, genus)[a]	Common clinical signs and symptoms	No. of reported cases in the world[b]	Known or possible vector species in Asia (isolation[c], detection[d], oral-vector competence[e])	Areas of endemicity in Asia	References
Alkhumra hemorrhagic fever virus	*Flaviviridae*, *Orthoflavivirus*	Acute febrile flu-like illness, hepatitis, hemorrhagic manifestations, encephalitis	620	*Amblyomma lepidum*[d], *Ornithodoros savignyi*[c], *Hyalomma dromedarii*[c]	Saudi Arabia	Madani and Abuelzein (2021)
Alongshan virus	Unassigned	Headache, fever	86	*Dermacentor nuttalli*[d], *Dermacentor reticulatus*[d], *Haemaphysalis concinna*[d], *Ixodes persulcatus*[c,d], *Ixodes ricinus*[d]	China	Wang et al. (2019), Kholodilov et al. (2020), Kholodilov et al. (2021).
Beiji nairovirus	*Nairoviridae*, *Norwavirus*	Fever, headache, fatigue, myalgia or arthralgia, depression, coma	129	*Dermacentor nuttalli*[d], *Haemaphysalis concinna*[d], *Haemaphysalis longicornis*[d], *Ixodes persulcatus*[d]	China	Meng et al. (2019), Wang et al. (2021), Liu et al. (2022).
Bhanja virus	*Phenuiviridae*, *Bandavirus*	Fever, photophobia, vomiting, meningoencephalitis, pareses	4	*Amblyomma variegatum*[c], *Haemaphysalis intermedia*[c], *Haemaphysalis punctata*[c], *Hyalomma detritum*[c], *Hyalomma marginatum*[c], *Rhipicephalus bursa*[c], *Rhipicephalus decoloratus*[c]	India, Iran, Sri Lanka, Pakistan, Turkmenia, Uzbekistan, Tadjikistan, Kirghistan, Kazakhstan, Azerbaijan, Armenia	Williams et al. (1972), Hubálek (1987), Matsuno et al. (2013).

Virus name	Taxonomy (family, genus)[a]	Common clinical signs and symptoms	No. of reported cases in the world[b]	Known or possible vector species in Asia (isolation[c], detection[d], oral-vector competence[e])	Areas of endemicity in Asia	References
Crimean-Congo hemorrhagic fever virus	*Nairoviridae*, *Orthonairovirus*	Fever, photophobia, myalgia, petechial rash, hemorrhagic syndrome with necrotic hepatitis	>10,000[f]	*Dermacentor marginatus*[c,d,e], *Hyalomma impeltatum*[c,e], *Hyalomma marginatum*[c,d,e], *Hyalomma rufipes*[c,d,e], *Rhipicephalus evertsi*[c,e], *Rhipicephalus rossicus*[c,d,e], many other ixodid spp.[c,d]	Countries in Central, Southern, and Western Asia, China	Turell (2007), Blair et al. (2019), Tsapko et al. (2022).
Dhori virus	*Orthomyxiviridae*, *Thogotovirus*	Fever, headache, retro-orbital pain, myalgia, general weakness, mild symptoms of cranial nerve involvement with radicular syndrome	5	*Amblyomma gemma*[c], *Hyalomma dromedarii*[c], *Hyalomma impeltatum*[c], *Hyalomma marginatum*[c]	Saudi Arabia	Anderson and Casals (1973), Williams et al. (1973), Filipe and Casals (1979),Al-Khalifa et al. (2007), Lutomiah et al. (2014).
Ganjam virus	*Nairoviridae*, *Orthonairovirus*	fever, muscular pains, lethargy	7	*Haemaphysalis intermedia*[c], *Haemaphysalis wellingtoni*[c], *Rhipicephalus haemaphysaloids*[c]	India	Sudeep et al. (2009).
Issyk-Kul virus	*Nairoviridae*, *Orthonairovirus*	Fever (Issyk-Kul fever), headache, myalgia, nausea	>60	*Argas pusillus*[c], *Argas vespertilionis*[c]	Kazakhstan, Kyrgyz, Malaysia, Tajikistan, Turkmenistan	L'vov et al. (1984), Nuttall (2014).

(continued)

Table 2.3 (continued)

Virus name	Taxonomy (family, genus)[a]	Common clinical signs and symptoms	No. of reported cases in the world[b]	Known or possible vector species in Asia (isolation[c], detection[d], oral-vector competence[e])	Areas of endemicity in Asia	References
Jingmen tick virus	unassigned	Fever, headache, nausea, vomiting, asthenia, myalgia	12	*Amblyomma javanense*[c,d], other *Amblyomma* spp.[d], *Dermacentor* spp.[d], *Haempahysalis* spp.[d], *Ixodes* spp.[d], *Rhipicephalus microplus*[c,d], other *Rhipicephalus* spp.[d]	China, Japan, Laos, Turkey	Qin et al. (2014), Jia et al. (2019), Colmant et al. (2022).
Kemerovo virus	*Sedoreoviridae*, *Orbivirus*	Fever (Kemerovo fever), rash, neurological disorders	The exact number of cases is not known. Cases of Kemerovo fever were registered in Western Siberia in the 1960s and 1970s.	*Dermacentor reticulatus*[d], *Dermacentor nuttalli*[d], *Ixodes pavlovskyi*[d], *Ixodes persulcatus*[c,d], *Ixodes ricinus*[d]	Kazakhstan	Tkachev et al. (2014), Tkachev et al. (2017).
Kyasanur Forest disease virus	*Flaviviridae*, *Orthoflavivirus*	Fever, headache, myalgia, gastrointestinal and bleeding	≈10,000 reported cases of KFD in humans, averaging 400–500 cases annually	*Amblyomma* sp.[c], *Dermacentor auratus*[c,e], *Haemaphysalis kyasanurensis*[c,e], *Haemaphysalis minuta*[c,e], *Haemaphysalis papuana kinneari*[c,e], *Haemaphysalis spinigera*[c,e], *Haemaphysalis turturis*[c,e], *Haemaphysalis wellingtoni*[c,e], other *Heamaphysalis* spp.[c], *Hyalomma marginatum isaaci*[e], *Ixodes ceylonicus*[c], *Ixodes petauristae*[e], *Rhipicephalus haemaphysaloides*[e]	China[g], India	Hoogstraal (1966), Sreenivasan et al. (1979), Mehla et al. (2009), Holbrook (2012), Chakraborty et al. (2021).

Virus name	Taxonomy (family, genus)[a]	Common clinical signs and symptoms	No. of reported cases in the world[b]	Known or possible vector species in Asia (isolation[c], detection[d], oral-vector competence[e])	Areas of endemicity in Asia	References
Langat virus	*Flaviviridae*, *Orthoflavivirus*	Fever, leukopenia, encephalitis	21 (experimental infection)Not known natural infected cases	*Dermacentor marginatus*[e], *Haemaphysalis papuana*[c], *Haemaphysalis spinigera*[e], *Ixodes granulatus*[c], *Ixodes ricinus*[e]	Malaysia, Thailand	Hoogstraal (1966), Smith (1956), Webb (1966), Bancroft et al. (1976).
Omsk hemorrhagic fever virus	*Flaviviridae*, *Orthoflavivirus*	Fever, headache, nausea, muscle pain, cough, hemorrhages	>2700[h]	*Dermacentor marginatus*[c,d], *Dermacentor reticulatus*[c,d,e], *Ixodes persulcatus*[d]	Kazakhstan	Hoogstraal (1966), Karan et al. (2014), Wagner et al. (2022).
Quaranfil virus	*Orthomyxiviridae*, *Quaranjavirus*	Fever	2	*Argas hermanni*[c], *Argas persicus*[e], *Argas vulgaris*[c]	Afghanistan, India, Iran	Williams, (1970), Taylor et al. (1966), Kaiser (1966), Klein et al. (1980), Mourya et al. (2019).
Oz virus	*Orthomyxiviridae*, *Thogotovirus*	Fever, fatigue, arthritis pain, anorexia, vomiting, myocarditis	1	*Amblyomma testudinarium*[c]	Japan	National Institute of Infectious Diseases (2023a)

(continued)

Table 2.3 (continued)

Virus name	Taxonomy (family, genus)[a]	Common clinical signs and symptoms	No. of reported cases in the world[b]	Known or possible vector species in Asia (isolation[c], detection[d], oral-vector competence[e])	Areas of endemicity in Asia	References
severe fever with thrombocytopenia syndrome virus	*Phenuiviridae*, *Bandavirus*	High fever, gastrointestinal symptoms	>15,000[h]	*Amblyomma testudinarium*[d], *Dermacentor nuttalli*[d], *Haemaphysalis flava*[d,e], *Haemaphysalis longicornis*[c,d,e], *Hyalomma asiaticum*[d], *Ixodes nipponensis*[d], *Rhipicephalus microplus*[d]	China, Japan, Myanmar, South Korea, Thailand, Taiwan, Vietnam	Luo et al. (2015), Yoo et al. (2020), Casel et al. (2021), Huang et al. (2021), Fang et al. (2023), Rattanakomol et al. (2022), National Institute of Infectious Diseases (2023b).
Songling virus	*Nairoviridae*, *Orthonairovirus*	Headache, fever, depression, fatigue, dizziness	42	*Haemaphysalis concinna*[d], *Haemaphysalis longicornis*[d], *Ixodes crenulatus*[d], *Ixodes persulcatus*[d]	China	Ma et al. (2021).
Tacheng tick virus 1	*Nairoviridae*, *Orthonairovirus*	Fever, rash	2	*Dermacentor marginatus*[d], *Dermacentor nuttalli*[d], *Dermacentor reticulatus*[d], *Dermacentor silvarum*[d], *Hyalomma asiaticum*[d], *Ixodes canisuga*[d], *Rhipicephalus turanicus*[d]	China	Liu et al. (2020), Zhang et al. (2021), Ji et al. (2022).
Tacheng tick virus 2	*Phenuiviridae*, *Uukuvirus*	Fever, chills, severe fatigue, headache, anorexia, nausea, vomiting	1	*Dermacentor marginatus*[d], *Dermacentor nuttalli*[d], *Dermacentor reticulatus*[d], *Dermacentor silvarum*[d], *Hyalomma scupense*[d], *Hyalomma anatolicum*[d]	China, Kazakhstan, Turkey	Dong et al. (2021), Jia et al. (2021), Dinçer et al. (2022).

Virus name	Taxonomy (family, genus)[a]	Common clinical signs and symptoms	No. of reported cases in the world[b]	Known or possible vector species in Asia (isolation[c], detection[d], oral-vector competence[e])	Areas of endemicity in Asia	References
Thogoto virus	*Orthomyxiviridae*, *Thogotovirus*	Fever, anorexia, dysphagia, hepatosplenomegaly, axillary and inguinal lymphadenopathy, bilateral optic neuritis, meningitis	2	*Amblyomma gemma*[c], *Amblyomma lepidum*[c], *Amblyomma variegatum*[c,e], *Haemaphysalis longicornis*[c], *Hyalomma anatolicum*[c], *Rhipicephalus bursa*[c], *Rhipicephalus decoloratus*[c], *Rhipicephalus sanguineus*[c]	Iran, Japan	Williams et al. (1972), Calisher et al. (1987), Jones et al. (1989), Yoshii et al. (2015).
Tick-borne encephalitis virus	*Flaviviridae*, *Orthoflavivirus*	Flu-like symptoms (fever, headache, joint and muscle pain, etc.), paralysis, even coma, encephalomyelitis, encephalitis	10,000–12,000/ year	*Dermacentor nuttalli*[c], *Dermacentor marginatus*[c,e], *Dermacentor reticulatus*[c,d,e], *Dermacentor silvarum*[c,d], *Haemaphysalis concinna*[c,e], *Haemaphysalis flava*[d], *Haemaphysalis japanica*[c,d,e], *Haemaphysalis longicornis*[d,e], *Hyalomma marginatum*[c], *Ixodes arboricola*[e], *Ixodes lividus*[c], *Ixodes nipponensis*[d], *Ixodes ovatus*[c], *Ixodes persulcatus*[c,d,e], *Ixodes ricinus*[c,d,e], *Rhipicephalus* sp.[c]	China, Japan, Kazakhstan, Kyrgyz, Mongolia, South Korea	Hoogstraal (1966), Takeda et al. (1998), Demina et al. (2010), Biernat et al. (2014), He et al. (2018), Süss (2011), Yun et al. (2016), Morozova et al. (2002), Yurchenko et al. (2017), Yoshii et al. (2017), Kholodilov et al. (2019), Makenov et al. (2019), WHO (2023).
Wanowrie virus	*Nairoviridae*, *Orthonairovirus*	Fever, abdominal pain, vomiting, rectal hemorrhage, vomiting blood	1	*Hyalomma asiaticum*[c], *Hyalomma isaaci*[c]	India, Iran, Sri Lanka	Pavri et al. (1976), Sureau et al. (1980).

(continued)

Table 2.3 (continued)

Virus name	Taxonomy (family, genus)[a]	Common clinical signs and symptoms	No. of reported cases in the world[b]	Known or possible vector species in Asia (isolation[c], detection[d], oral-vector competence[e])	Areas of endemicity in Asia	References
Yezo virus	*Nairoviridae*, *Orthonairovirus*	Fever, thrombocytopenia, leukopenia	8	*Haemaphysalis megaspinosa*[d], *Ixodes ovatus*[d], *Ixodes persulcatus*[d]	China, Japan	Kodama et al. (2021), Lv et al. (2023)

[a] Virus taxonomy is based on the reports of the International Committee on Taxonomy of Viruses (Matthijnssens et al. 2022a, b; Rico-Hesse et al. 2023) and (Kuhn et al. 2023)
[b] As of September 2023
[c] Tick species with virus isolation
[d] Tick species from virus genome detection
[e] Tick species which have been experimentally demonstrated to transmit the virus
[f] Only Asian region
[g] Possible laboratory contamination is indicated (Mehla et al. 2009)
[h] Including suspected cases

section provides an overview of the four most prominent and pathogenic viruses, namely, the Crimean-Congo hemorrhagic fever virus, Kyasanur Forest disease virus, severe fever with thrombocytopenia syndrome virus, and tick-borne encephalitis virus.

2.6.1 Crimean-Congo Hemorrhagic Fever Virus (CCHFV)

CCHFV was first reported in 1944–1945 based on the symptoms prevalent among Soviet soldiers in Crimea and was initially called Crimean hemorrhagic fever. It was subsequently identified as the same virus as the Congo virus, which was isolated in 1956 in a patient with hemorrhagic fever in the Congo; thus, it was named the Crimean-Congo hemorrhagic fever virus. The virus is widely distributed throughout Africa, Europe, and Asia, with more than 10,000 cases reported in Asia alone so far. Furthermore, the fatality rate due to this viral infection is 10–40%, and CCHFV is therefore one of the most medically important tick-borne viruses. Although CCHFV has been isolated and detected in various tick species (Table 2.3), the main vector of the virus is *Hy. marginatum* complex (especially *Hy. marginatum sensu stricto* and *Hy. rufipes*). Other species of the genus *Hyalomma* and some species of *Rhipicephalus* and *Dermacentor* also transmit this virus (Table 2.3). Transstadial and transovarial transmissions (described in detail in Sect. 2.8) of the virus in the infected ticks were also observed. Ticks are therefore important not only as CCHFV vectors but also as reservoir hosts. Mammals, such as cattle, goats, sheep, and hares, are also amplifiers of the virus, and even those that do not develop viremia can transmit the virus to non-infected ticks via co-feeding (non-viremic) transmission (described in detail in Sect. 2.8). Humans are infected with CCHFV by the bite of a tick harboring the virus, as well as through the blood and body fluids of patients and infected animals.

2.6.2 Kyasanur Forest Disease Virus (KFDV)

KFDV is a tick-borne virus with regional distribution in southwest India. Although various tick species are capable of transmitting the virus (Table 2.3), *Hae. spinigera* is considered the main vector of the virus owing to its high infection rate, the relative abundance of the tick, and its blood-sucking preference for both humans and non-human mammals. Transstadial and transovarial transmission of KFDV in infected ticks has also been observed. KFDV infects a wide range of hosts including humans, rodents, bats, ground-dwelling birds, porcupines, and monkeys. Transmission of KFDV to humans is established by tick bites or contact with infected animals; however, human-to-human transmission has not been reported. Cattle, goats, and sheep are also susceptible to this virus but are considered dead-end hosts, like humans. In humans, an incubation period of 3–8 days is followed by symptoms including fever, headache, severe myalgia, gastrointestinal disturbances, and bleeding tendency. Relapse occurs in many patients after one or two weeks of remission, with symptoms varying from fever to encephalitis. The fatality rate ranges from 2–10%. To date, approximately 10,000 cases have been reported in India.

2.6.3 Severe Fever with Thrombocytopenia Syndrome Virus (SFTSV)

Severe fever with thrombocytopenia syndrome (SFTS) is a relatively novel tick-borne viral infection caused by the SFTSV, first reported in China in 2009. However, given the number of cases and pathogenicity, it is a medically important tick-borne viral infection in Asia, which is equally or even more important than the Crimean-Congo hemorrhagic fever. Since the discovery of the virus in China, incidents of the disease have been reported in Japan, South Korea, Taiwan, Vietnam, Myanmar, and Thailand. After an incubation period of 6 days to two weeks, fever and gastrointestinal symptoms (loss of appetite, nausea, vomiting, diarrhea, and abdominal pain) are observed in many cases of SFTSV infection. Other symptoms include headache, myalgia, neurological symptoms such as impaired consciousness and aphasia, lymphadenopathy, and hemorrhagic symptoms such as subcutaneous hemorrhage, and the fatality rate was reported to be 6.3–30%. Although the genome of this virus has been detected in a wide range of ticks (Table 2.3), its transmission has so far been demonstrated only by *Hae. longicornis* and *Hae. flava*. Antibodies against SFTSV have been detected in various domestic, companion, and wild animals. The virus is thought to be maintained in nature through the transmission cycle between ticks and mammals, as well as through transovarial transmission (see Sect. 2.8 for details) from SFTSV-infected female ticks to their progeny. Transmission of SFTSV in humans occurs mainly via the bite of infected ticks; however, person-to-person transmission via contact with body fluids, such as the blood of infected individuals, as well as via direct transmission from infected cats to humans, has also been reported.

2.6.4 Tick-Borne Encephalitis Virus (TBEV)

Tick-borne encephalitis viruses are phylogenetically classified into the Far Eastern (also known as Russian Spring-Summer encephalitis virus), Siberian (also known as West Siberian virus), and European (also known as Central European tick-borne encephalitis virus) subtypes. All these subtypes are distributed in Asia, with the Far Eastern subtype reported in Japan and China, the Siberian subtype in Mongolia and Kyrgyzstan, and the European subtype in South Korea. Although the virus has been isolated, detected, and even transmitted by various tick species (Table 2.3), the main vector ticks in nature are *I. ricinus*, for the European subtype, and *I. persulcatus*, for the Siberian and Far Eastern subtypes. In addition, transstadial, transovarial, and co-feeding (non-viremic) transmissions of this virus (see Sect. 2.8 for details) have been observed in several tick species. Ticks are therefore important not only as vectors of TEBV but also as viruses. TBEV transmission cycles are maintained between various wild animals, such as small wild rodents and ticks. They can be transmitted to a wide range of animal species, including humans and other companion and domestic animals, through blood-sucking by infected ticks. The incidence rate of human biting by TBEV-infected ticks is reported to be 5–30%; after an incubation period of 7–14 days, symptoms such as headache, fever, and joint and muscle pain are seen in the early stages of the disease. In severe cases, central nervous system symptoms such as coma and paralysis due to encephalomyelitis are observed, with fatality rates reported in a wide range. TBEV infection in humans has been reported not only through tick bites but also through the consumption of raw milk from infected livestock animals.

2.6.5 Tick-Borne Viruses of Unknown Pathogenicity in Humans

In addition to the aforementioned human-pathogenic tick-borne viruses reported in Asia, several other tick-borne viruses are infectious to humans, but their pathogenicity remains unknown. For example, Guertu virus (GTV) (*Bandavirus* genus, *Phenuiviridae* family) is a novel virus phylogenetically related to SFTSV isolated from *D. nuttalli* in Xinjiang, China, in 2014 (Shen et al. 2018). Similar to SFTSV, the GTV strongly inhibits the induction and activity of human interferon, a factor that inhibits viral infections in vertebrates (Min et al. 2020). Moreover, as neutralizing antibodies to the virus have been detected in residents of the area where GTV has been isolated, there may be future reports of infected patients from this area and other locations. In Japan, the Kabuto mountain virus (KAMV) (*Uukuvirus* genus, *Phenuiviridae* family) was recently identified as infectious to humans (Tran et al. 2022a). KAMV is harbored by the *Hae. flava* ticks, which are ubiquitous in Japan, and it is considered that many people have been exposed to this virus. Therefore, future studies should investigate viral infections caused by KAMV.

2.7 Advances in Nucleotide Sequencing Technology and the Discovery of Novel Pathogens

Recent breakthroughs in nucleotide sequencing technology, represented by the advent of next-generation sequencing (NGS), have revealed the diversity of microbes, such as bacteria and viruses, that exist in various organisms and environments. The term "virome" refers to the entire viral sequence present in a particular species or environment, and recent studies have revealed that arthropods harbor a great diversity of viruses. These studies have led to the discovery of various arboviruses, including novel viruses, from blood-sucking arthropods such as mosquitoes and ticks. This section provides an overview of tick-borne viruses recently found in Asia by virome analysis associated with human diseases and their epidemic risk as emerging viral infections. The clinical symptoms observed in infections with these viruses are nonspecific and cannot be distinguished from those of tick-borne infections caused by established agents.

2.7.1 Jingmenviruses—Jingmen Tick Virus and Alongshan Virus

Jingmen tick virus (JMTV) is the prototype of a large group of viruses frequently referred to as jingmenviruses (Qin et al. 2014). Jingmenvirus genomes comprise four to five segments that encode structural and nonstructural viral proteins (Colmant et al. 2022). They are currently classified as a distinct group within the *Flaviviridae* family due to similarities between their nonstructural proteins. Jingmenvirus genomes phylogenetically cluster into two major clades associated with tick-vertebrate or insect hosts. JMTV was also the first jingmenvirus to be proposed as a causative agent of tick-borne infections in humans. The first cases were reported to present with mild or severe febrile diseases with skin manifestations (Jia et al. 2019). JMTV has also been documented in CCHFV-infected individuals on the Balkan Peninsula (Emmerich et al. 2018). Originally detected in *Rhipicephaus microplus* ticks, JMTV nucleic acids have been identified in many species of *Rhipicephalus*, *Amblyomma*, *Dermacentor*, *Haemaphysalis*, *Hyalomma*, and *Ixodes* ticks, as well as in bats, cattle, rodents, and tortoises. JMTV sequences were in samples distributed globally, including those from Asia (China, Japan, Laos, and Turkey), Africa (Guinea, Kenya, and Uganda), Europe (Italy, Kosovo, Romania, and Russia), and the Americas (Brazil, Colombia, French Antilles, Trinidad, and Tobago), with further serological evidence of exposure in specific locations in Asia and Europe (Colmant et al. 2022). Currently, data on JMTV ecology, distribution, and transmission are scarce and inconclusive, although several findings support vertical transmission and salivary gland infections, suggesting probable transmission via tick bites (Kobayashi et al. 2021). Portions of JMTV genome were also documented to be integrated in the tick genome, with implications for vector screening (Morozkin et al. 2022).

Alongshan virus (ALSV) is another jingmenvirus that has potential human health impacts. The viral genome has been detected in patients with febrile diseases and a history of tick bites, followed by seroconversion (Wang et al. 2019). ALSV was subsequently identified in several tick species collected in several Asian and European countries, including China, Finland, Russia, Serbia, Germany, and France, with evidence of viral exposure in sheep, cattle, and deer (Colmant et al. 2022). The main vectors of ALSV appear to be *Ixodes* ticks, which are widely distributed in Asia and Europe and serve as the main vector for tick-borne encephalitis virus transmission (Kobayashi et al. 2021).

2.7.2 Novel Nairoviruses

In addition to the CCHFV and Nairobi sheep disease virus, several orthonairoviruses (family *Nairoviridae*, genus *Orthonairovirus*) have also been associated with other tick-borne infections, and several others have been recently discovered. Among these, the Tacheng tick virus 1 (TcTV1) and Songling virus (SGLV) are phylogenetically grouped within the Tamdy orthonairoviruses. Initial cases of TcTV1 infection have been documented in China, presenting as a febrile disease with other symptoms (Liu et al. 2020; Zhang et al. 2021), and recently, in Eastern Europe (Ergünay et al. 2023). This virus, which was initially described during tick metagenomic investigations, is harbored by several *Dermacentor* species and *Hy. asiaticum*, and follow-up studies revealed exposure in local human populations, cattle, and sheep (Li et al. 2015; Liu et al. 2020). SGLV and associated human infections and exposures were recently described. The virus forms a unique clade with Tamdy orthonairoviruses (Ma et al. 2021). The Tamdy virus, the prototype strain of the species, has also been previously implicated in human infections and is widespread in Central Asia and Asia Minor, with evidence of human and animal exposure (L'vov et al. 1984; Ergünay et al. 2020; Moming et al. 2021). A novel orthonairovirus, tentatively named the Yezo virus (YEZV), has been described in patients with tick-associated febrile diseases in Japan (Kodama et al. 2021). Further investigations revealed viral nucleic acids in retrospective human samples, *Ixodes* and *Hemaphysalis* ticks, and antibodies in wild animals. Among orthonairoviruses, YEZV is closely related to the Sulina virus, which was discovered in *I. ricinus* ticks from Romania (Tomazatos et al. 2021).

Recently, the CCHFV genotype VI (Europe 2) has been reclassified as a separate virus, called the Aigai virus (Papa et al. 2022). Initially described in Greece, Aigai virus is mainly detected in *Rhipicephalus* ticks and documented infections are reported from Turkey and Iran.

Norwavirus is a recently-ratified genus in the family *Nairoviridae*, which currently includes Grotenhout virus (GRHV) in the *Norwavirus grotenhoutense* species (Kuhn et al. 2023). Norwaviruses lack the M (medium) genomic segments that encode structural glycoproteins in nairoviruses (Kuhn et al. 2016). Several norwaviruses such as Pustyn virus and Norway nairovirus 1, not classified in the current taxonomy, have been observed in ticks from Asia (China and Russia), Europe, and the United States, suggesting a broad geographic distribution (Kuhn et al. 2016; Cai et al. 2022; Sameroff et al. 2022). Beiji nairovirus, another member of the *Norwavirus* genus, was discovered in tick-borne human infections of unknown etiology in China (Wang et al. 2021). Viral isolates can induce pathologies in cell lines and experimental infections. Antibodies were detected in convalescent serum, sheep, and cattle samples from the region. Viral nucleic acids have also been observed in various *Ixodes*, *Dermacentor*, and *Hemaphysalis* tick species (Wang et al. 2021; Cai et al. 2022).

2.7.3 Novel Uukuviruses

Tacheng tick virus 2 (TcTV-2) was initially described in *D. marginatus* in a metagenomic investigation of various arthropods in China (Li et al. 2015). Classified as a member of the newly established *Uukuvirus tachengense* species (family *Phenuiviridae*, genus *Uukuvirus*), TcTV-2 was identified in several *Dermacentor* and *Hyalomma* ticks from China, Kazakhstan, Turkey, as well as in Europe (Dong et al. 2021; Bratuleanu et al. 2022; Dinçer et al. 2022; Ergünay et al. 2023). In China, it was also reported as a probable etiologic agent in individuals with tick-associated febrile disease, with evidence of person-to-person spread via contact with droplets or body fluids. (Dong et al. 2021).

2.7.4 Novel Thogotoviruses

Three viruses (Batken, Dhori, and Thogoto) of the genus *Thogotovirus* (family Orthomyxoviridae) that are human pathogenic or infectious have been reported in Asia (Table 2.3).

Recently, Oz virus (OZV), isolated from *Am. testudinarium* ticks in Japan, was discovered as a novel virus belonging to this genus (Ejiri et al. 2018). In June 2023, the first case of death due to OZV infection in humans was reported in Japan (NIID 2023a). This patient had fever, fatigue, arthritis pain, anorexia, and vomiting as the main symptoms and died from viral myocarditis (NIID 2023a). This is the only case report of OZV infection, and more information is needed to further characterize its clinical manifestations. Regarding OZV exposure in wild animals, neutralizing antibodies against this virus have been found in Japanese macaques (*Macaca fuscata*), Japanese wild boar (*Sus scrofa leucomystax*), and Japanese sika deer (*Cervus nippon*) (Tran et al. 2022b). However, the natural vertebrate hosts of this virus and the tick species that harbor and transmit the virus have not yet been identified, and future studies are required to elucidate the virus transmission cycle. In addition, the novel Thailand tick thogoto virus has been detected in *Boophilus* sp. in Thailand as a close relative of OZV (Temmam et al. 2019), but the infectivity and pathogenicity of this virus in humans are unknown to date.

2.8 Differential Roles of Ticks and Mosquitoes in the Transmission and Maintenance of Arthropod-Borne Viruses

The transmission of many arthropod-borne pathogens occurs via blood-sucking by vector arthropods. Therefore, the blood-sucking ecology of the vector (e.g., the number of developmental stages for blood-sucking, blood-sucking frequency, and the time required from full engorgement to re-blood-sucking) has a significant impact on the transmission and ecology of pathogens. Ticks and mosquitoes are important pathogen vectors. However, the blood-sucking ecologies of the two vectors are very different. In this section, the life cycles of ticks and mosquitoes are compared, and the similarities and differences in their respective blood-sucking ecology and modes of pathogen transmission are outlined, with a primary focus on arthropod-borne viruses (arboviruses). Given these differences, this section also discusses the different roles of ticks and mosquitoes as vectors and reservoirs for viruses and the relationship between mosquito blood-sucking ecology and the rapid expansion of arbovirus epidemics.

2.8.1 Comparison of Life Cycle and Blood-Sucking Styles Between Ticks and Mosquitoes

Ticks and mosquitoes differ greatly in their blood-sucking ecology, such as the number of developmental stages of blood-sucking, blood-sucking frequency, and the time required from full engorgement to re-blood-sucking, etc. (Table 2.4), resulting in significant differences in their abilities to acquire and transmit arboviruses.

Although ticks and mosquitoes are disease vector arthropods, they differ greatly in their abilities to suck blood during their life cycle (Fig. 2.1). Ticks feed exclusively on vertebrate blood and consume blood meals during all developmental stages except the egg stage. Mosquitoes, on the other hand, feed on organic matter in the water during the larval stage. Thereafter, only adult females take blood-based meals (in some species, adult females are non-blood-sucking). Many arboviruses are acquired or transmitted when vectors suck blood from vertebrates. Therefore, it is expected that the increased blood-sucking abilities of a vector will increase the likelihood of acquiring and transmitting the virus. Although adult female mosquitoes consume blood meals several times during their lifetime, ticks appear to more efficiently acquire and transmit the virus than mosquitoes because they suck blood at all developmental stages, except eggs, of both males and females.

Ticks and mosquitoes greatly differ in their blood-sucking styles. Mosquitoes require less time to reach engorgement, consume a blood meal, and re-suck blood. They penetrate the host skin with their proboscis and directly suck blood from blood vessels. Mosquitoes usually require only a few seconds to minutes to take a blood

Table 2.4 Comparison of important characteristics between tick and mosquito for virus transmission and maintenance

Vector's characteristics	Tick	Mosquito	Effects on virus transmission and maintenance
Number of developmental stages for blood-sucking	≥ 3 (larva, nymph, and adult)	1 (adult female)	The increased opportunity to suck blood increases the chance that the vector will acquire and transmit arboviruses.
Biting frequency until full engorgement	1	≥1	The higher the frequency of bites, the higher the efficiency of arbovirus transmission to the vertebrate host.
Longevity (developmental stages capable of transmitting the virus)	Long (years)	Short (days-several months)	The longevity of developmental stage vectors that can maintain and transmit arboviruses increases the efficiency of viral transmission and maintenance.
Extrinsic incubation period (EIP) of arbovirus	Long (several months)	Short (several days)	The shorter the EIP, the more efficiently the virus is transmitted.

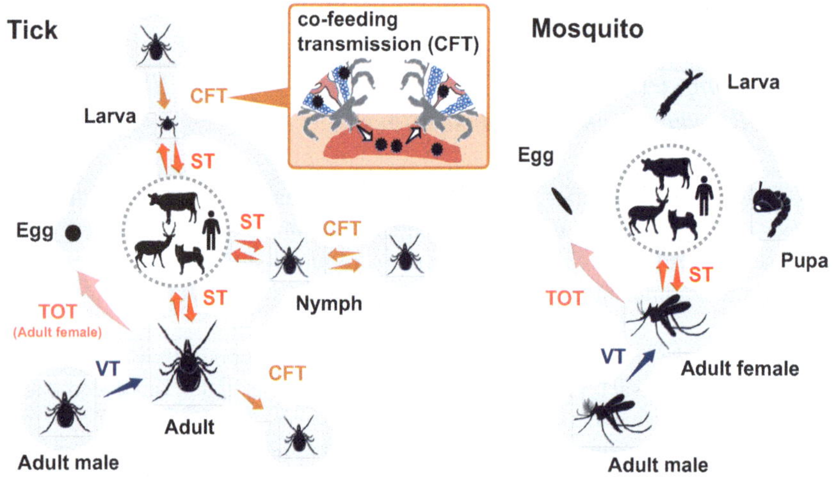

Fig. 2.1 Schematic illustration of the life cycle of ticks and mosquitoes and the route by which arthropod-borne viruses are transmitted. Ticks and mosquitoes acquire arboviruses via blood-sucking from viremic vertebrates [systemic transmission (ST)] and also transmit the viruses in the process of blood-sucking (red arrows). Both ticks and mosquitoes transmit arboviruses vertically to the next generation via transovarial transmission (TOT) (pink arrows). Horizontal transmission of the virus from virus-infected males to females via mating [venereal transmission (VN)] is indicated by the dark blue arrows. In addition, co-feeding transmission (CFT), a tick-specific mode of horizontal transmission, is shown by the orange arrows; CFT is a mode of transmission in which the virus is transmitted horizontally from virus-infected ticks to non-infected ticks between ticks that are sucking blood in close proximity on the host's body surface (Illustration in the orange square in the upper center of the figure)

meal, and they bite and suck blood repeatedly until reaching the amount required for oogenesis. In addition, mosquitoes usually require only a few days from egg-laying to blood re-sucking. In contrast, ticks are extravascular blood feeders, and it takes several days to several dozen days from biting the host to engorgement. Essentially, a tick sucks blood once at each developmental stage. Furthermore, a tick requires 1–12 months from molting after engorgement to re-sucking, depending on the developmental stage, temperature, and the presence or absence of diapause (Randolph 1998).

The period from engorgement to re-blood-sucking has a significant impact on the prevalence of arbovirus diseases. The period between virus acquisition via blood feeding and its availability for transmission is referred to as the extrinsic incubation period (EIP). The shorter the EIP, the more efficient the transmission of the virus. Ticks re-suck blood after a long period from engorgement, which inevitably results in a longer EIP. However, mosquitoes have a shorter EIP because the period from engorgement to re-sucking of blood is shorter, and viruses acquired by mosquitoes through blood-sucking can be transmitted within a period of up to ten days. Hence, mosquitoes are more efficient at transmitting arboviruses within shorter periods.

2.8.2 Arbovirus Transmission Dynamics in the Body of Ticks and Mosquitoes

Although ticks and mosquitoes differ greatly in their blood-sucking ecology, as described above, viruses acquired by these vectors via blood-sucking broadly follow the same process until they are transmitted to the next vertebrate host by the vectors (Nuttall et al. 1991). The mode of viral transmission from vertebrate to vector is called systemic transmission (ST). Viruses are acquired by ticks and mosquitoes from the blood of vertebrates with viremia. The viruses (i) infect the midgut cells, propagate in them, and (ii) escape them to disseminate into the hemocele. Thereafter, the virus propagates to the cells of various organs and tissues until (iii) infecting the salivary gland cells. Viruses that infect salivary gland cells are transmitted to the vertebrate host via saliva injected by the vector during blood-sucking. A wide variety of arboviruses transmitted by ticks and mosquitoes are currently known, and each virus has a specific vector species. This is because multiple barriers must be overcome by the virus until it is released with saliva, and as described above, the process from virus acquisition to transmission by ticks and mosquitoes is complex. Three barriers in the body of vectors are important for virus transmission, corresponding to numbers (i–iii) described above: (i) the midgut infection barrier, (ii) the midgut escape barrier, and (iii) the salivary gland infection barrier (Nuttall et al. 1991). Only viruses that can overcome these barriers are transmitted to the vertebrate host by their respective vectors, which, in this case, are defined as suitable vectors for the virus.

2.8.3 Similarities and Differences in Modes of Arbovirus Transmission Within Vector Populations

Arboviruses are transmitted not only between vectors and vertebrates but also via multiple routes within the vector population, which is thought to play a certain role in arbovirus maintenance in nature. There are two main modes of arbovirus transmission within a vector population: vertical and horizontal (Table 2.5). CFT, a form of horizontal transmission, is observed in many tick-borne viruses. Tick-borne viruses are suggested to be stably maintained in tick populations by a combination of three transmission routes: CFT, vertical transmission [transovarial transmission (TOT)] as discussed below, and ST from the vertebrate host via blood-sucking.

Two modes of horizontal transmission within a vector population are known: venereal transmission (VT) via mating from virus-infected males to virus-uninfected females and CFT (also called non-viremic transmission) (Fig. 2.1). VT has been experimentally observed in several viruses in both ticks and mosquitoes. However, its impact as a viral maintenance mechanism in nature remains unknown. CFT, on the other hand, is a well-known mode of horizontal transmission of tick-borne viruses (Randolph et al. 1996) (Fig. 2.1), in which the virus is transmitted horizontally from virus-infected ticks to uninfected ticks feeding blood in close proximity on the host's body surface. This allows horizontal transmission of the virus between ticks, even when the vertebrates are non-viremic or with low viremia, including those refractory to viral infection or infected with the virus in the past and already have antibodies against the virus (Nuttall 2014). In mosquitoes, CFT has been observed with West Nile virus (Higgs et al. 2005), but the role of this CFT in virus transmission and maintenance in natural mosquito populations is unknown.

The mode of vertical transmission within a vector population is known as TOT [transmission of the virus to the next generation by infection of the ovaries (oocytes) of adult females]. For TOT of a virus in a vector, the virus needs to pass over the midgut infection and escape barriers to reach the ovary and enter the developing oocyte before the egg is protected by the outer chorion after it is acquired via

Table 2.5 Comparison of the mode of virus transmission within vector population

Transmission mode in vector population		Tick	Mosquito	Effects or remarks on virus transmission and maintenance
Horizontal	Co-feeding transmission (CFT)	Confirmed	Confirmed (rare)	The number of virus-infected vectors increases independent of the viremia of the vertebrate host.
	Venereal transmission (VT)	Confirmed	Confirmed	Its role in the maintenance of arboviruses in nature is unknown.
Vertical	Transovarial transmission (TOT)	Confirmed (TOT rate <5%)	Confirmed (TOT rate <1%)	This vertical transmission route alone cannot maintain the virus within the vector population.

blood-sucking (Randolph 1998). For many mosquito-borne viruses, excluding a few exceptional viruses (e.g., orthobunyaviruses), the transmission rate by TOT is usually less than 1% (Kuno and Chang 2005). Therefore, the role of TOT in the maintenance of viruses in nature is likely to be limited. However, ticks are believed to have a high tolerance to TOT for various types of infectious pathogens, including viruses, due to the long time interval between blood engorgement and oogenesis (Randolph 1998). The TOT of various viruses has been reported in ticks (Nuttall et al. 1994), and generally, the transmission rate by TOT is approximately 5% or less (Nuttall 2014). At this TOT rate, the virus cannot be stably maintained in the tick population. However, mathematical modeling has shown that the combination of TOT, CFT, and ST allows the virus to be stably maintained in tick populations (Nonaka et al. 2010).

2.8.4 Relationship Between Vector Longevity and Arbovirus Transmission and Maintenance

The ability of arbovirus reservoir hosts to harbor and transmit the virus for an extended period significantly impacts the natural ecology and prevalence of arboviruses. When arboviruses infect vertebrates, the host immune system frequently and quickly eliminates the virus from the host body, resulting in a transient infection. However, arbovirus infections in arthropods such as ticks and mosquitoes are often sustained throughout their lifetime.

When comparing the longevity of mosquitoes and ticks as viral vectors, the lifespan of adult female mosquitoes is by far the shortest (a few days to a few months). In contrast, *I. ricinus,* the well-known vector tick species in Europe, completes its life cycle within three to six years (Labuda and Nuttall 2008). *Argas brumpti*, a soft tick species, has been reported to survive for 27 years (Shepherd 2022). Thus, ticks have a longer lifespan than mosquitoes. Furthermore, arboviruses acquired in the immature stages (larva or nymph) of ticks can persist even after molting that follows engorgement and remain available for further transmission. This mode of transmission between developmental stages is termed "transstadial transmission." Therefore, ticks can harbor arboviruses for extremely long periods, in contrast with mosquitoes and vertebrates. For example, several tick species of the genus *Ornithodoros* can maintain and transmit the Karshi virus (a tick-borne flavivirus) for approximately eight years after acquisition (Turell 2015). Hence, ticks are important not only as vectors but also as reservoir hosts for arboviruses.

2.8.5 The Epidemic Scale of Arbovirus Disease and Its Relationship to the Blood-Sucking Ecology of Ticks and Mosquitoes

Differences in the blood-sucking ecology of ticks and mosquitoes may affect the epidemic scale of arbovirus disease. Mosquito-borne arboviral infections in humans are by far more prevalent than tick-borne infections. This may be because mosquitoes bite humans more frequently and have a shorter EIP, which allows for a more rapid spread of arbovirus disease and possibly epidemics.

At first glance, it would appear that ticks are more efficient at transmitting arboviruses than mosquitoes because only adult female mosquitoes consume blood meals during their life cycle. In contrast, ticks feed on all developmental stages except eggs. The population of virus-infected ticks increases by horizontal transmission (e.g., CFT), independent of a suitable vertebrate host for the virus. However, endemic mosquito-borne arboviral diseases are far higher than tick-borne diseases. For example, dengue fever is caused by the mosquito-borne dengue virus, which is endemic mainly in tropical regions of the world, with more than several million reported cases each year and an estimated total of approximately 100 million (Bhatt et al. 2013; WHO 2022). In contrast, tick-borne encephalitis, the most common tick-borne viral disease, accounts for approximately 5000–12,000 cases annually (Madison-Antenucci et al. 2020). Furthermore, Lyme disease, a bacterial infection with the highest number of reported cases among tick-borne infections, has fewer than 120,000 estimated cases each year in the world (Stone et al. 2017). Differences in vertebrate hosts and other factors for each pathogen make general comparisons impossible. However, the different blood-sucking ecology between ticks and mosquitoes contributes to the differences in the disease prevalence.

In epidemiology, the infectivity of a pathogen can be evaluated using the basic reproduction number (R_0), which is the average number of new infections by one infected person in a susceptible population. As mosquitoes and ticks have different biological characteristics, as represented by their blood-sucking ecology, different formulas for calculating R_0 have been proposed for mosquito- and tick-borne pathogens (Randolph 1998). By comparing the vector-associated factors in these formulas that significantly affect the variation in R_0, the characteristics of both vectors and their impact on the magnitude of arboviral disease epidemics could be understood.

The formula for calculating R_0 for mosquito-borne pathogens shows that the variable related to mosquito bite frequency has a significant impact on epidemic dynamics, with R_0 increasing significantly as bite frequency increases (Randolph 1998). Mosquitoes, unlike ticks, repeatedly bite and suck blood until they can feed on enough blood for oogenesis; therefore, it is easy to assume that the bite frequency can increase. Recent studies have also reported instances of behavioral modifications by mosquito-borne viruses that increase the biting frequency of the vector, such as virus-infected mosquitoes biting vertebrate hosts more frequently and virus-infected vertebrates that are more likely to attract vector mosquitoes (Wei Xiang et al. 2022; Zhang et al. 2022). In addition to this increase in bite frequency, mosquitoes have a short EIP, which is expected to allow the spread of mosquito-borne viruses over a short period. However, the formula for calculating R_0 of tick-borne pathogens, which reflects the biological characteristics of ticks, suggests that the survival probability of ticks has a significant effect on R_0 (Randolph 1998). Ticks undergo diapause during the summer and winter seasons during their long lifespan. A prolonged period of arrested development due to diapause increases the probability of tick survival. However, this results in a significant slowdown in the pace of infection and transmission of pathogens (Randolph 1998). Hence, tick-borne viral diseases are less likely to cause rapid and widespread epidemics than mosquito-borne viral infections.

2.8.6 Conclusion

Although ticks and mosquitoes are the same blood-sucking arthropods, their roles in virus transmission and maintenance differ greatly because of their different life cycles and blood-sucking ecologies. Mosquitoes are highly efficient vectors of viruses between vertebrate hosts. On the other hand, ticks, in addition to being vectors, are also highly efficient reservoir hosts of viruses due to their long lifespan and mechanisms for horizontal transmission of viruses within tick populations, such as CFT. The differences in the blood-sucking ecology of the two vectors are not limited to these differences in arbovirus transmission and maintenance mechanisms but are also likely to be one factor influencing the magnitude of arbovirus disease epidemics.

References

Al-Khalifa MS, Diab FM, Khalil GM (2007) Man-threatening viruses isolated from ticks in Saudi Arabia. Saudi Med J 28(12):1864–1867

Anderson CR, Casals J (1973) Dhori virus, a new agent isolated from *Hyalomma dromedarii* in India. Indian J Med Res 61(10):1416–1420

Asanuma K, Fukuda S (1957) On the life-history of a tick, *Ixodes signatus*, infesting a black tailed gull (*Larus crassirostris*) in Japan. Jpn J Sanit Zool 8(3):147–159

Bancroft WH, Scott RM, Snitbhan R, Weaver RE Jr, Gould DJ (1976) Isolation of Langat virus from *Haemaphysalis papuana* Thorell in Thailand. Am J Trop Med Hyg 25(3):500–504

Barré N, Uilenberg G (2010) Spread of parasites transported with their hosts: case study of two species of cattle tick. Rev Sci Tech 29(1):149–160; 135–147

Bezerra-Santos MA, de Macedo LO, Nguyen VL, Manoj RR, Laidoudi Y, Latrofa MS, Beugnet F, Otranto D (2022) *Cercopithifilaria* spp. in ticks of companion animals from Asia: new putative hosts and vectors. Ticks Tick Borne Dis 13(4):101957

Bhatt S, Gething PW, Brady OJ, Messina JP, Farlow AW, Moyes CL, Drake JM, Brownstein JS, Hoen AG, Sankoh O, Myers MF, George DB, Jaenisch T, Wint GR, Simmons CP, Scott TW, Farrar JJ, Hay SI (2013) The global distribution and burden of dengue. Nature 496(7446):504–507

Biernat B, Karbowiak G, Werszko J, Stańczak J (2014) Prevalence of tick-borne encephalitis virus (TBEV) RNA in *Dermacentor reticulatus* ticks from natural and urban environment, Poland. Exp Appl Acarol 64(4):543–551

Blair PW, Kuhn JH, Pecor DB, Apanaskevich DA, Kortepeter MG, Cardile AP, Polanco Ramos A, Keshtkar-Jahromi M (2019) An emerging biothreat: Crimean-Congo hemorrhagic fever virus in Southern and Western Asia. Am J Trop Med Hyg 100(1):16–23

Bratuleanu BE, Temmam S, Chrétien D, Regnault B, Pérot P, Bouchier C, Bigot T, Savuța G, Eloit M (2022) The virome of *Rhipicephalus*, *Dermacentor* and *Haemaphysalis* ticks from Eastern Romania includes novel viruses with potential relevance for public health. Transbound Emerg Dis 69(3):1387–1403

Bursali A, Keskin A, Tekin S (2012) A review of the ticks (Acari: Ixodida) of Turkey: species diversity, hosts and geographical distribution. Exp Appl Acarol 57(1):91–104

Cai X, Cai X, Xu Y, Shao Y, Fu L, Men X, Zhu Y (2022) Virome analysis of ticks and tick-borne viruses in Heilongjiang and Jilin Provinces, China. Virus Res 323:199006

Calisher CH, Karabatsos N, Filipe AR (1987) Antigenic uniformity of topotype strains of Thogoto virus from Africa, Europe, and Asia. Am J Trop Med Hyg 37(3):670–673

Casel MA, Park SJ, Choi YK (2021) Severe fever with thrombocytopenia syndrome virus: emerging novel phlebovirus and their control strategy. Exp Mol Med 53(5):713–722

Chakraborty S, Sander WE, Allan BF, Andrade FCD (2021) Retrospective study of Kyasanur forest disease and deaths among Nonhuman Primates, India, 1957–2020. Emerg Infect Dis 27(7):1969–1973

Colmant AMG, Charrel RN, Coutard B (2022) Jingmenviruses: Ubiquitous, understudied, segmented flavi-like viruses. Front Microbiol 13:997058

Demina TV, Dzhioev YP, Verkhozina MM, Kozlova IV, Tkachev SE, Plyusnin A, Doroshchenko EK, Lisak OV, Zlobin VI (2010) Genotyping and characterization of the geographical distribution of tick-borne encephalitis virus variants with a set of molecular probes. J Med Virol 82(6):965–976

Dinçer E, Timurkan M, Oğuz B, Şahindokuyucu İ, Şahan A, Ekinci M, Polat C, Ergünay K (2022) Several tick-borne pathogenic viruses in circulation in Anatolia, Turkey. Vector Borne Zoonotic Dis 22(2):148–158

Dong Z, Yang M, Wang Z, Zhao S, Xie S, Yang Y, Liu G, Zhao S, Xie J, Liu Q, Wang Y (2021) Human Tacheng tick virus 2 infection, China, 2019. Emerg Infect Dis 27(2):594–598

Ejiri H, Lim CK, Isawa H, Fujita R, Murota K, Sato T, Kobayashi D, Kan M, Hattori M, Kimura T, Yamaguchi Y, Takayama-Ito M, Horiya M, Posadas-Herrera G, Minami S, Kuwata R, Shimoda H, Maeda K, Katayama Y, Mizutani T, Saijo M, Kaku K, Shinomiya H, Sawabe K (2018) Characterization of a novel thogotovirus isolated from *Amblyomma testudinarium* ticks in Ehime, Japan: a significant phylogenetic relationship to Bourbon virus. Virus Res 249:57–65

Emmerich P, Jakupi X, von Possel R, Berisha L, Halili B, Günther S, Cadar D, Ahmeti S, Schmidt-Chanasit J (2018) Viral metagenomics, genetic and evolutionary characteristics of Crimean-Congo hemorrhagic fever orthonairovirus in humans, Kosovo. Infect Genet Evol 65:6–11

Ergünay K, Dinçer E, Kar S, Emanet N, Yalçınkaya D, Polat Dinçer PF, Brinkmann A, Hacıoğlu S, Nitsche A, Özkul A, Linton YM (2020) Multiple orthonairoviruses including Crimean-Congo hemorrhagic fever virus, Tamdy virus and the novel Meram virus in Anatolia. Ticks Tick Borne Dis 11(5):101448

Ergünay K, Bourke BP, Reinbold-Wasson DD, Nikolich MP, Nelson SP, Caicedo-Quiroga L, Vaydayko N, Kirkitadze G, Chunashvili T, Long LS, Blackburn JK, Cleary NG, Tucker CL, Linton YM (2023) The expanding range of emerging tick-borne viruses in Eastern Europe and the Black Sea Region. Sci Rep 13(1):19824

Fang LZ, Xiao X, Lei SC, Liu JW, Yu XJ (2023) *Haemaphysalis flava* ticks as a competent vector of severe fever with thrombocytopenia syndrome virus. Ticks Tick Borne Dis 14(2):102100

Filipe AR, Casals J (1979) Isolation of Dhori virus from *Hyalomma marginatum* ticks in Portugal. Intervirology 11(2):124–127

Filippova NA (1966) Argasid ticks (Argasidae). Fauna of the USSR: Paukoobraznye. Nauka, Moskva/Leningrad 4(3):255 (in Russian)

Fujimoto K, Yamaguti N (1990) Comparison of ixodid ticks in Nara Park and surrounding regions. Jpn J Environ Entomol Zool 2(3):133–137

Ghosh S, Bansal GC, Gupta SC, Ray D, Khan MQ, Irshad H, Shahiduzzaman MD, Seitzer U, Ahmed JS (2007) Status of tick distribution in Bangladesh, India and Pakistan. Parasitol Res 101(Suppl 2):207–216

Guglielmone AA, Nava S, Robbins RG (2023) Geographic distribution of the hard ticks (Acari: Ixodida: Ixodidae) of the world by countries and territories. Zootaxa 5251(1):1–274

Halt BG, Lessard JP, Borregaard MK, Fritz SA, Araujo MB, Dimitrov D, Fabre PH, Graham CM, Graves GR, Jensson KA, Nogues-Bravo D, Wang ZH, Whittaker RJ, Fjeldsa J, Rahbek C (2013) An update of Wallace's zoological regions of the world. Science 339(6115):74–78

He X, Zhao J, Fu S, Yao L, Gao X, Liu Y, He Y, Liang G, Wang H (2018) Complete genomic characterization of three tick-borne encephalitis viruses detected along the China-North Korea border, 2011. Vector Borne Zoonotic Dis 18(10):554–559

Higgs S, Schneider BS, Vanlandingham DL, Klingler KA, Gould EA (2005) Nonviremic transmission of West Nile virus. Proc Natl Acad Sci USA 102(25):8871–8874

Holbrook MR (2012) Kyasanur forest disease. Antivir Res 96(3):353–362

Hoogstraal H (1966) Ticks in relation to human diseases caused by viruses. Annu Rev Entomol 11:261–308

Hoogstraal H (1971) Biological patterns in the Afghanistan tick fauna. In: Daniel M, Rosický B (eds) Proceedings of the 3rd international congress of acarology. Springer, Dordrecht, pp 511–514

Hosseini-Chegeni A, Tavakoli M, Telmadarraiy Z (2019) The updated list of ticks (Acari: Ixodidae & Argasidae) occurring in Iran with a key to the identification of species. Syst Appl Acarol 24(11):2133–2166

Huang X, Li J, Li A, Wang S, Li D (2021) Epidemiological characteristics of severe fever with thrombocytopenia syndrome from 2010 to 2019 in Mainland China. Int J Environ Res Public Health 18(6):3092

Hubálek Z (1987) Geographic distribution of Bhanja virus. Folia Parasitol (Praha) 34(1):77–86

Ji N, Wang N, Li F, Zhao S, Liu G, Tan W, Lu X, Wang Y (2022) Tacheng tick virus 1 in ticks, livestock, and wildlife in Border Regions of China-Kazakhstan. Authorea, May 17, 2022

Jia N, Liu HB, Ni XB, Bell-Sakyi L, Zheng YC, Song JL, Li J, Jiang BG, Wang Q, Sun Y, Wei R, Yuan TT, Xia LY, Chu YL, Wei W, Li LF, Ye JL, Lv QY, Cui XM, Guan Y, Tong YG, Jiang JF, Lam TT, Cao WC (2019) Emergence of human infection with Jingmen tick virus in China: a retrospective study. EBioMedicine 43:317–324

Jia Y, Wang S, Yang M, Ulzhan N, Omarova K, Liu Z, Kazkhan O, Wang Y (2021) First detection of Tacheng tick virus 2 in hard ticks from southeastern Kazakhstan. Kafkas Universitesi Veteriner Fakultesi Dergisi 28(1):139–142

Jones LD, Davies CR, Steel GM, Nuttall PA (1989) Vector capacity of *Rhipicephalus appendiculatus* and *Amblyomma variegatum* for Thogoto and Dhori viruses. Med Vet Entomol 3(2):195–202

Kaiser MN (1966) Viruses in ticks. II. Experimental transmission of Quaranfil virus by *Argas (persicargas) arboreus* and *A. (p.) persicus*. Am J Trop Med Hyg 15(6):976–985

Karan LS, Ciccozzi M, Yakimenko VV, Lo Presti A, Cella E, Zehender G, Rezza G, Platonov AE (2014) The deduced evolution history of Omsk hemorrhagic fever virus. J Med Virol 86(7):1181–1187

Kazim AR, Houssaini J, Ehlers J, Tappe D, Heo CC (2021) Soft ticks (Acari: Argasidae) in the island nations of Southeast Asia: a review on their distribution, associated hosts and potential pathogens. Acta Trop 223:106085

Kazim AR, Houssaini J, Tappe D, Heo CC (2022) A checklist of the ticks of Malaysia (Acari: Argasidae, Ixodidae), with lists of known associated hosts, geographical distribution, type localities, human infestations and pathogens. Zootaxa 5190(4):485–530

Kholodilov I, Belova O, Burenkova L, Korotkov Y, Romanova L, Morozova L, Kudriavtsev V, Gmyl L, Belyaletdinova I, Chumakov A, Chumakova N, Dargyn O, Galatsevich N, Gmyl A, Mikhailov M, Oorzhak N, Polienko A, Saryglar A, Volok V, Yakovlev A, Karganova G (2019) Ixodid ticks and tick-borne encephalitis virus prevalence in the south Asian part of Russia (Republic of Tuva). Ticks Tick Borne Dis 10(5):959–969

Kholodilov IS, Litov AG, Klimentov AS, Belova OA, Polienko AE, Nikitin NA, Shchetinin AM, Ivannikova AY, Bell-Sakyi L, Yakovlev AS, Bugmyrin SV, Bespyatova LA, Gmyl LV, Luchinina SV, Gmyl AP, Gushchin VA, Karganova GG (2020) Isolation and characterisation of Alongshan virus in Russia. Viruses 12(4):362

Kholodilov IS, Belova OA, Morozkin ES, Litov AG, Ivannikova AY, Makenov MT, Shchetinin AM, Aibulatov SV, Bazarova GK, Bell-Sakyi L, Bespyatova LA, Bugmyrin SV, Chernetsov N, Chernokhaeva LL, Gmyl LV, Khaisarova AN, Khalin AV, Klimentov AS, Kovalchuk IV, Luchinina SV, Medvedev SG, Nafeev AA, Oorzhak ND, Panjukova EV, Polienko AE, Purmak KA, Romanenko EN, Rozhdestvenskiy EN, Saryglar AA, Shamsutdinov AF, Solomashchenko NI, Trifonov VA, Volchev EG, Vovkotech PG, Yakovlev AS, Zhurenkova OB, Gushchin VA, Karan LS, Karganova GG (2021) Geographical and tick-dependent distribution of flavi-like Alongshan and Yanggou tick viruses in Russia. Viruses 13(3):458

Kitaoka S (1994) Ticks. In: Umeya K (ed) Outdoor harmful insects. Zenkoku Noson Kyoiku Kyokai Publishing Co. Ltd., Tokyo, pp 207–211

Kitaoka S, Morii T (1967) The biology of *Haemaphysalis* (*Alloceraea*) *ambigua* Neumann, 1901, with description of immature stages (Ixodoidea, Ixodidae). Natl Inst Anim Health Q 7:145–152

Klein JM, Sureau P, Casals J, Piazak N, Korouri C, Calvo MA (1980) Isolation of Quaranfil virus in Iran from *Argas vulgaris* ticks. Cahiers d'ORSTOM. Serie Entomologie Médicale et Parasitologie (France) 17:201–206

Kobayashi D, Faizah AN, Amoa-Bosompem M, Watanabe M, Maekawa Y, Hayashi T, Higa Y, Sawabe K, Isawa H (2020) Analysis of *Trypanosoma* sequences from *Haemaphysalis flava* (Acari: Ixodidae) and *Tabanus rufidens* (Diptera: Tabanidae) collected in Ishikawa, Japan. Med Entomol Zool 71(4):279–288

Kobayashi D, Kuwata R, Kimura T, Shimoda H, Fujita R, Faizah AN, Kai I, Matsumura R, Kuroda Y, Watanabe S, Kuniyoshi S, Yamauchi T, Watanabe M, Higa Y, Hayashi T, Shinomiya H, Maeda K, Kasai S, Sawabe K, Isawa H (2021) Detection of Jingmenviruses in Japan with evidence of vertical transmission in ticks. Viruses 13(12):2547

Kodama F, Yamaguchi H, Park E, Tatemoto K, Sashika M, Nakao R, Terauchi Y, Mizuma K, Orba Y, Kariwa H, Hagiwara K, Okazaki K, Goto A, Komagome R, Miyoshi M, Ito T, Yamano K, Yoshii K, Funaki C, Ishizuka M, Shigeno A, Itakura Y, Bell-Sakyi L, Edagawa S, Nagasaka A, Sakoda Y, Sawa H, Maeda K, Saijo M, Matsuno K (2021) A novel nairovirus associated with acute febrile illness in Hokkaido, Japan. Nat Commun 12(1):5539

Kuhn JH, Wiley MR, Rodriguez SE, Bào Y, Prieto K, Travassos da Rosa AP, Guzman H, Savji N, Ladner JT, Tesh RB, Wada J, Jahrling PB, Bente DA, Palacios G (2016) Genomic characterization of the genus *Nairovirus* (Family *Bunyaviridae*). Viruses 8(6):164

Kuhn JH, Abe J, Adkins S, Alkhovsky SV, Avšič-Županc T, Ayllón MA, Bahl J, Balkema-Buschmann A, Ballinger MJ, Kumar Baranwal V, Beer M, Bejerman N, Bergeron É, Biedenkopf N, Blair CD, Blasdell KR, Blouin AG, Bradfute SB, Briese T, Brown PA, Buchholz UJ, Buchmeier MJ, Bukreyev A, Burt F, Büttner C, Calisher CH, Cao M, Casas I, Chandran K, Charrel RN, Kumar Chaturvedi K, Chooi KM, Crane A, Dal Bó E, Carlos de la Torre J, de Souza WM, de Swart RL, Debat H, Dheilly NM, Di Paola N, Di Serio F, Dietzgen RG, Digiaro M, Drexler JF, Duprex WP, Dürrwald R, Easton AJ, Elbeaino T, Ergünay K, Feng G, Firth AE, Fooks AR, Formenty PBH, Freitas-Astúa J, Gago-Zachert S, Laura García M, García-Sastre A, Garrison AR, Gaskin TR, Gong W, Gonzalez JJ, de Bellocq J, Griffiths A, Groschup MH, Günther I, Günther S, Hammond J, Hasegawa Y, Hayashi K, Hepojoki J, Higgins CM, Hongō S, Horie M, Hughes HR, Hume AJ, Hyndman TH, Ikeda K, Jiāng D, Jonson GB, Junglen S, Klempa B, Klingström J, Kondō H, Koonin EV, Krupovic M, Kubota K, Kurath G, Laenen L, Lambert AJ, Lǐ J, Li JM, Liu R, Lukashevich IS, MacDiarmid RM, Maes P, Marklewitz M,

Marshall SH, Marzano SL, McCauley JW, Mirazimi A, Mühlberger E, Nabeshima T, Naidu R, Natsuaki T, Navarro B, Navarro JA, Neriya Y, Netesov SV, Neumann G, Nowotny N, Nunes MRT, Ochoa-Corona FM, Okada T, Palacios G, Pallás V, Papa A, Paraskevopoulou S, Parrish CR, Pauvolid-Corrêa A, Pawęska JT, Pérez DR, Pfaff F, Plemper RK, Postler TS, Rabbidge LO, Radoshitzky SR, Ramos-González PL, Rehanek M, Resende RO, Reyes CA, Rodrigues TCS, Romanowski V, Rubbenstroth D, Rubino L, Runstadler JA, Sabanadzovic S, Sadiq S, Salvato MS, Sasaya T, Schwemmle M, Sharpe SR, Shi M, Shimomoto Y, Kavi Sidharthan V, Sironi M, Smither S, Song JW, Spann KM, Spengler JR, Stenglein MD, Takada A, Takeyama S, Tatara A, Tesh RB, Thornburg NJ, Tian X, Tischler ND, Tomitaka Y, Tomonaga K, Tordo N, Tu C, Turina M, Tzanetakis IE, Maria Vaira A, van den Hoogen B, Vanmechelen B, Vasilakis N, Verbeek M, von Bargen S, Wada J, Wahl V, Walker PJ, Waltzek TB, Whitfield AE, Wolf YI, Xia H, Xylogianni E, Yanagisawa H, Yano K, Ye G, Yuan Z, Zerbini FM, Zhang G, Zhang S, Zhang YZ, Zhao L, Økland AL (2023) Annual (2023) taxonomic update of RNA-directed RNA polymerase-encoding negative-sense RNA viruses (realm *Riboviria*: kingdom *Orthornavirae*: phylum *Negarnaviricota*). J Gen Virol 104(8):001864

Kuno G, Chang GJ (2005) Biological transmission of arboviruses: reexamination of and new insights into components, mechanisms, and unique traits as well as their evolutionary trends. Clin Microbiol Rev 18(4):608–637

Kwak ML (2018) Ticks in the Lion city: a preliminary review of the tick fauna of Singapore. Exp Appl Acarol 76(2):263–267

L'vov DK, Sidorova GA, Gromashevskiĭ VL, Skvortsova TM, Aristova VA (1984) Isolation of Tamdy virus (*Bunyaviridae*) pathogenic for man from natural sources in central Asia, Kazakhstan and Transcaucasia. Vopr Virusol 29(4):487–490

Labuda M, Nuttall PA (2008) Viruses transmitted by ticks. In: Bowman AS, Nuttall PA (eds) Ticks: biology, disease and control. Cambridge University Press, Cambridge, pp 253–280

Lane RS, Burgdorfer W (1987) Transovarial and transstadial passage of *Borrelia burgdorferi* in the western black-legged tick, *Ixodes pacificus* (Acari: Ixodidae). Am J Trop Med Hyg 37(1):188–192

Lempereur L, Beck R, Fonseca I, Marques C, Duarte A, Santos M, Zúquete S, Gomes J, Walder G, Domingos A, Antunes S, Baneth G, Silaghi C, Holman P, Zintl A (2017) Guidelines for the detection of *Babesia* and *Theileria* parasites. Vector Borne Zoonotic Dis 17(1):51–65

Li CX, Shi M, Tian JH, Lin XD, Kang YJ, Chen LJ, Qin XC, Xu J, Holmes EC, Zhang YZ (2015) Unprecedented genomic diversity of RNA viruses in arthropods reveals the ancestry of negative-sense RNA viruses. elife 4:e05378

Liu X, Zhang X, Wang Z, Dong Z, Xie S, Jiang M, Song R, Ma J, Chen S, Chen K, Zhang H, Si X, Li C, Jin N, Wang Y, Liu Q (2020) A tentative Tamdy orthonairovirus related to febrile illness in Northwestern China. Clin Infect Dis 70(10):2155–2160

Liu Z, Li L, Xu W, Yuan Y, Liang X, Zhang L, Wei Z, Sui L, Zhao Y, Cui Y, Yin Q, Li D, Li Q, Hou Z, Wei F, Liu Q, Wang Z (2022) Extensive diversity of RNA viruses in ticks revealed by metagenomics in northeastern China. PLoS Negl Trop Dis 16(12):e0011017

Luo LM, Zhao L, Wen HL, Zhang ZT, Liu JW, Fang LZ, Xue ZF, Ma DQ, Zhang XS, Ding SJ, Lei XY, Yu XJ (2015) *Haemaphysalis longicornis* ticks as reservoir and vector of severe fever with thrombocytopenia syndrome virus in China. Emerg Infect Dis 21(10):1770–1776

Lutomiah J, Musila L, Makio A, Ochieng C, Koka H, Chepkorir E, Mutisya J, Mulwa F, Khamadi S, Miller BR, Bast J, Schnabel D, Wurapa EK, Sang R (2014) Ticks and tick-borne viruses from livestock hosts in arid and semiarid regions of the eastern and northeastern parts of Kenya. J Med Entomol 51(1):269–277

Lv X, Liu Z, Li L, Xu W, Yuan Y, Liang X, Zhang L, Wei Z, Sui L, Zhao Y, Hou Z, Wei F, Han S, Liu Q, Wang Z (2023) Yezo virus infection in tick-bitten patient and ticks, northeastern China. Emerg Infect Dis 29(4):797–800

Ma J, Lv XL, Zhang X, Han SZ, Wang ZD, Li L, Sun HT, Ma LX, Cheng ZL, Shao JW, Chen C, Zhao YH, Sui L, Liu LN, Qian J, Wang W, Liu Q (2021) Identification of a new orthonairovirus associated with human febrile illness in China. Nat Med 27(3):434–439

Madani TA, Abuelzein EME (2021) Alkhumra hemorrhagic fever virus infection. Arch Virol 166(9):2357–2367

Madison-Antenucci S, Kramer LD, Gebhardt LL, Kauffman E (2020) Emerging tick-borne diseases. Clin Microbiol Rev 33(2):e00083–e00018

Makenov M, Karan L, Shashina N, Akhmetshina M, Zhurenkova O, Kholodilov I, Karganova G, Smirnova N, Grigoreva Y, Yankovskaya Y, Fyodorova M (2019) First detection of tick-borne encephalitis virus in *Ixodes ricinus* ticks and their rodent hosts in Moscow, Russia. Ticks Tick Borne Dis 10(6):101265

Matsuno K, Weisend C, Travassos da Rosa AP, Anzick SL, Dahlstrom E, Porcella SF, Dorward DW, Yu XJ, Tesh RB, Ebihara H (2013) Characterization of the Bhanja serogroup viruses (*Bunyaviridae*): a novel species of the genus *Phlebovirus* and its relationship with other emerging tick-borne phleboviruses. J Virol 87(7):3719–3728

Matthijnssens J, Attoui H, Bányai K, Brussaard CPD, Danthi P, Del Vas M, Dermody TS, Duncan R, Fāng Q, Johne R, Mertens PPC, Mohd Jaafar F, Patton JT, Sasaya T, Suzuki N, Wei T (2022a) ICTV virus taxonomy profile: *Sedoreoviridae*. J Gen Virol 103(10):001782

Matthijnssens J, Attoui H, Bányai K, Brussaard CPD, Danthi P, Del Vas M, Dermody TS, Duncan R, Fāng Q, Johne R, Mertens PPC, Mohd Jaafar F, Patton JT, Sasaya T, Suzuki N, Wei T (2022b) ICTV virus taxonomy profile: *Spinareoviridae* 2022. J Gen Virol 103(11):001781

Mehla R, Kumar SR, Yadav P, Barde PV, Yergolkar PN, Erickson BR, Carroll SA, Mishra AC, Nichol ST, Mourya DT (2009) Recent ancestry of Kyasanur forest disease virus. Emerg Infect Dis 15(9):1431–1437

Meng F, Ding M, Tan Z, Zhao Z, Xu L, Wu J, He B, Tu C (2019) Virome analysis of tick-borne viruses in Heilongjiang Province, China. Ticks Tick Borne Dis 10(2):412–420

Min YQ, Shi C, Yao T, Feng K, Mo Q, Deng F, Wang H, Ning YJ (2020) The nonstructural protein of Guertu virus disrupts host defenses by blocking antiviral interferon induction and action. ACS Infect Dis 6(5):857–870

Moming A, Shen S, Fang Y, Zhang J, Zhang Y, Tang S, Li T, Hu Z, Wang H, Zhang Y, Sun S, Wang LF, Deng F (2021) Evidence of human exposure to Tamdy virus, northwest China. Emerg Infect Dis 27(12):3166–3170

Mori K, Tsunoda T (1996) Migration of larvae in the tick, *Haemaphysalis longicornis* Neumann (Acarina, Ixodidae). Jpn J Environ Entomol Zool 7(4):211–213

Mori K, Fujimagari M, Hayashi A (1990) The tick fauna on deer *Cervus nippon* in the south of Chiba Prefecture. Bull Chiba Pref Inst Public Health 14:44–47

Morozkin ES, Makenov MT, Zhurenkova OB, Kholodilov IS, Belova OA, Radyuk EV, Fyodorova MV, Grigoreva YE, Litov AG, Valdokhina AV, Bulanenko VP, Samoilov AE, Korneenko EV, Voizekhovskaya YA, Neverov AD, Karganova GG, Karan LS (2022) Integrated jingmenvirus polymerase gene in *Ixodes ricinus* genome. Viruses 14(9):1908

Morozova OV, Dobrotvorsky AK, Livanova NN, Tkachev SE, Bakhvalova VN, Beklemishev AB, Cabello FC (2002) PCR detection of *Borrelia burgdorferi* sensu lato, tick-borne encephalitis virus, and the human granulocytic ehrlichiosis agent in *Ixodes persulcatus* ticks from western Siberia, Russia. J Clin Microbiol 40(10):3802–3804

Mourya DT, Yadav PD, Nyayanit DA, Majumdar TD, Jain S, Sarkale P, Shete A (2019) Characterization of a strain of quaranfil virus isolated from soft ticks in India. Is quaranfil virus an unrecognized cause of disease in human and animals? Heliyon 5(3):e01368

National Institute of Infectious Diseases (NIID) (2023a) Oz virus infection in human. [Accessed 8 Dec 2023] https://www.niid.go.jp/niid/ja/diseases/a/ozv/2629-cepr/12134-ozv-eng.html

National Institute of Infectious Diseases (NIID) (2023b) Severe fever with thrombocytopenia syndrome. [Accessed 8 Dec 2023] https://www.niid.go.jp/niid/ja/sfts/3143-sfts.html

Nonaka E, Ebel GD, Wearing HJ (2010) Persistence of pathogens with short infectious periods in seasonal tick populations: the relative importance of three transmission routes. PLoS One 5(7):e11745

Nuttall GHF, Warburton C (1915) Ticks, a monograph of Ixodoidea. Part III. The genus *Haemaphysalis*. Cambridge University Press, Cambridge, pp 349–550

Nuttall PA (2014) Tick-borne viruses. In: Sonenshine DE, Roe RM (eds) Biology of ticks, vol 2. Oxford University Press, Oxford, pp 180–210

Nuttall PA, Jones LD, Davies CR (1991) The role of arthropod vectors in arbovirus evolution. In: Harris KF (ed) Advances in disease vector research. Springer, Heidelberg, pp 15–45

Nuttall PA, Jones LD, Labuda M, Kaufman WR (1994) Adaptations of arboviruses to ticks. J Med Entomol 31(1):1–9

Papa A, Marklewitz M, Paraskevopoulou S, Garrison AR, Alkhovsky SV, Avšič-Županc T, Bente DA, Bergeron É, Burt F, Di Paola N, Ergünay K, Hewson R, Mirazimi A, Sall AA, Spengler JR, Postler TS, Palacios G, Kuhn JH (2022) History and classification of Aigai virus (formerly Crimean-Congo haemorrhagic fever virus genotype VI). J Gen Virol 103(4):001734

Pavri KM, Anandarajah M, Hermon YE, Nayar M, Wikramsinghe MR, Dandawate CN (1976) Isolation of Wanowrie virus from brain of a fatal human case from Sri Lanka. Indian J Med Res 64(4):557–561

Pun SK, Guglielmone AA, Tarragona EL, Nava S, Maharjan M (2018) Ticks (Acari: Ixodidae) of Nepal: first record of *Amblyomma varanense* (Supino), with an update of species list. Ticks Tick Borne Dis 9(3):526–534

Qin XC, Shi M, Tian JH, Lin XD, Gao DY, He JR, Wang JB, Li CX, Kang YJ, Yu B, Zhou DJ, Xu J, Plyusnin A, Holmes EC, Zhang YZ (2014) A tick-borne segmented RNA virus contains genome segments derived from unsegmented viral ancestors. Proc Natl Acad Sci USA 111(18):6744–6749

Randolph SE (1998) Ticks are not insects: consequences of contrasting vector biology for transmission potential. Parasitol Today 14(5):186–192

Randolph SE, Gern L, Nuttall PA (1996) Co-feeding ticks: epidemiological significance for tick-borne pathogen transmission. Parasitol Today 12(12):472–479

Rasulov I (2007) Ticks status in Central Asia with a special emphasis on Uzbekistan. Parasitol Res 101(Suppl 2):183–186

Rattanakomol P, Khongwichit S, Linsuwanon P, Lee KH, Vongpunsawad S, Poovorawan Y (2022) Severe fever with thrombocytopenia syndrome virus infection, Thailand, 2019-2020. Emerg Infect Dis 28(12):2572–2574

Rico-Hesse R, Avsic-Zupanc T, Blitvich B, Bukh J, Cao-Lormeau VM, Imrie A, Kapoor A, Kramer LD, Lindenbach BD, Simmonds P, Smith DB, da Costa Vasconcelos PF (2023) The ICTV report on virus classification and taxon nomenclature *Flaviviridae* Chapter. [Accessed 8 Dec 2023] https://ictv.global/report/chapter/flaviviridae/flaviviridae

Robbins RG (2005) The ticks (Acari: Ixodida: Argasidae, Ixodidae) of Taiwan: a synonymic checklist. Proc Entomol Soc Wash 107(2):245–253

Rochlin I, Toledo A (2020) Emerging tick-borne pathogens of public health importance: a mini-review. J Med Microbiol 69(6):781–791

Sameroff S, Tokarz R, Vucelja M, Jain K, Oleynik A, Boljfetić M, Bjedov L, Yates RA, Margaletić J, Oura CAL, Lipkin WI, Cvetko Krajinović L, Markotić A (2022) Virome of *Ixodes ricinus*, *Dermacentor reticulatus*, and *Haemaphysalis concinna* ticks from Croatia. Viruses 14(5):929

Seneviratna P (1962) The ticks (Ixodoidea) of Ceylon. Part I–Argasidae. Ceylon Vet J 10(1):30–41

Shen S, Duan X, Wang B, Zhu L, Zhang Y, Zhang J, Wang J, Luo T, Kou C, Liu D, Lv C, Zhang L, Chang C, Su Z, Tang S, Qiao J, Moming A, Wang C, Abudurexiti A, Wang H, Hu Z, Zhang Y, Sun S, Deng F (2018) A novel tick-borne phlebovirus, closely related to severe fever with thrombocytopenia syndrome virus and Heartland virus, is a potential pathogen. Emerg Microbes Infect 7(1):95

Shepherd JG (2022) Record longevity and reproduction of an African tick, *Argas brumpti* (Ixodida: Argasidae). J Med Entomol 59(2):777–778

Silva Costa LF, Nunes PH, Soares JF, Labruna MB, Camargo-Mathias MI (2011) Distribution of *Rickettsia rickettsii* in ovary cells of *Rhipicephalus sanguineus* (Latreille, 1806) (Acari: Ixodidae). Parasit Vectors 4:222

Smith CE (1956) A virus resembling Russian spring-summer encephalitis virus from an ixodid tick in Malaya. Nature 178(4533):581–582

Sreenivasan MA, Bhat HR, Naik SV (1979) Experimental transmission of Kyasanur forest disease virus by *Dermacentor auratus* Supino. Indian J Med Res 69:701–707

Stone BL, Tourand Y, Brissette CA (2017) Brave new worlds: the expanding universe of lyme disease. Vector Borne Zoonotic Dis 17(9):619–629

Sudeep AB, Jadi RS, Mishra AC (2009) Ganjam virus. Indian J Med Res 130(5):514–519

Sureau P, Klein J, Casals J, Digoutte J, Salaun J, Piazak N, Calvo M (1980) Isolation of Thogoto, Wad Medani, Wanowrie and Crimean-Congo haemorrhagic fever viruses from ticks of domestic animals in Iran. Ann Virol (Inst Psteur) 131E(2):185–200

Süss J (2011) Tick-borne encephalitis 2010: epidemiology, risk areas, and virus strains in Europe and Asia-an overview. Ticks Tick Borne Dis 2(1):2–15

Takada N, Fujita H (2019) Whole classification of ticks in Japan. In: Takada N (ed) Medical acarology in Japan. Hokuryukan, Tokyo, pp 115–117

Takeda T, Ito T, Chiba M, Takahashi K, Niioka T, Takashima I (1998) Isolation of tick-borne encephalitis virus from *Ixodes ovatus* (Acari: Ixodidae) in Japan. J Med Entomol 35(3):227–231

Tanskul P, Stark HE, Inlao I (1983) A checklist of ticks of Thailand (Acari: Metastigmata: Ixodoidea). J Med Entomol 20(3):330–341

Taylor RM, Hurlbut HS, Work TH, Kingston JR, Hoogstraal H (1966) Arboviruses isolated from *Argas* ticks in Egypt: Quaranfil, Chenuda, and Nyamanini. Am J Trop Med Hyg 15(1):76–86

Temmam S, Chrétien D, Bigot T, Dufour E, Petres S, Desquesnes M, Devillers E, Dumarest M, Yousfi L, Jittapalapong S, Karnchanabanthoeng A, Chaisiri K, Gagnieur L, Cosson JF, Vayssier-Taussat M, Morand S, Moutailler S, Eloit M (2019) Monitoring silent spillovers before emergence: a pilot study at the tick/human interface in Thailand. Front Microbiol 10:2315

Tkachev S, Panov V, Dobler G, Tikunova N (2014) First detection of Kemerovo virus in *Ixodes pavlovskyi* and *Ixodes persulcatus* ticks collected in Novosibirsk region, Russia. Ticks Tick Borne Dis 5(5):494–496

Tkachev SE, Tikunov AY, Babkin IV, Livanova NN, Livanov SG, Panov VV, Yakimenko VV, Tantsev AK, Taranenko DE, Tikunova NV (2017) Occurrence and genetic variability of Kemerovo virus in *Ixodes* ticks from different regions of western Siberia, Russia and Kazakhstan. Infect Genet Evol 47:56–63

Tomazatos A, von Possel R, Pekarek N, Holm T, Rieger T, Baum H, Bialonski A, Maranda I, Erdelyi-Molnár I, Spînu M, Lühken R, Jansen S, Emmerich P, Schmidt-Chanasit J, Cadar D (2021) Discovery and genetic characterization of a novel orthonairovirus in *Ixodes ricinus* ticks from Danube Delta. Infect Genet Evol 88:104704

Tran NTB, Shimoda H, Mizuno J, Ishijima K, Yonemitsu K, Minami S, Supriyono, Kuroda Y, Tatemoto K, Mendoza MV, Takano A, Muto M, Isawa H, Sawabe K, Hayasaka D, Maeda K (2022a) Epidemiological study of Kabuto Mountain virus, a novel uukuvirus, in Japan. J Vet Med Sci 84(1):82–89

Tran NTB, Shimoda H, Ishijima K, Yonemitsu K, Minami S, Kuroda Y, Tatemoto K, Mendoza MV, Kuwata R, Takano A, Muto M, Sawabe K, Isawa H, Hayasaka D, Maeda K (2022b) Zoonotic infection with Oz virus, a novel Thogotovirus. Emerg Infect Dis 28(2):436–439

Tsapko NV, Volynkina AS, Evchenko AY, Lisitskaya YV, Shaposhnikova LI (2022) Detection of Crimean-Congo hemorrhagic fever virus in ticks collected from south Russia. Ticks Tick Borne Dis 13(2):101890

Turell MJ (2007) Role of ticks in the transmission of Crimean-Congo hemorrhagic fever virus. In: Ergonul O, Whitehouse CA (eds) Crimean-Congo hemorrhagic fever. Springer, Heidelberg, pp 143–154

Turell MJ (2015) Experimental transmission of Karshi (mammalian tick-borne flavivirus group) virus by *Ornithodoros* ticks >2,900 days after initial virus exposure supports the role of soft ticks as a long-term maintenance mechanism for certain flaviviruses. PLoS Negl Trop Dis 9(8):e0004012

Uni S, Bain O, Fujita H, Matsubayashi M, Fukuda M, Takaoka H (2013) Infective larvae of *Cercopithifilaria* spp. (Nematoda: Onchocercidae) from hard ticks (Ixodidae) recovered from the Japanese serow (Bovidae). Parasite 20:1

Wagner E, Shin A, Tukhanova N, Turebekov N, Nurmakhanov T, Sutyagin V, Berdibekov A, Maikanov N, Lezdinsh I, Shapiyeva Z, Shevtsov A, Freimüller K, Peintner L, Ehrhardt C, Essbauer S (2022) First indications of Omsk haemorrhagic fever virus beyond Russia. Viruses 14(4):754

Wallace AR (1876) The geographical distribution of animals; with a study of the relations of living and extinct faunas as elucidating the past changes of the Earth's surface, vol 1. Harper & Brothers, New York, p 503

Wang YC, Wei Z, Lv X, Han S, Wang Z, Fan C, Zhang X, Shao J, Zhao YH, Sui L, Chen C, Liao M, Wang B, Jin N, Li C, Ma J, Hou ZJ, Yang Z, Han Z, Zhang Y, Niu J, Wang W, Wang Y, Liu Q (2021) A new nairo-like virus associated with human febrile illness in China. Emerg Microbes Infect 10(1):1200–1208

Wang ZD, Wang B, Wei F, Han SZ, Zhang L, Yang ZT, Yan Y, Lv XL, Li L, Wang SC, Song MX, Zhang HJ, Huang SJ, Chen J, Huang FQ, Li S, Liu HH, Hong J, Jin YL, Wang W, Zhou JY, Liu Q (2019) A new segmented virus associated with human febrile illness in China. N Engl J Med 380(22):2116–2125

Webb HE, Wetherley-Mein G, Smith CE, McMahon D (1966) Leukaemia and neoplastic processes treated with Langat and Kyasanur Forest disease viruses: a clinical and laboratory study of 28 patients. Br Med J 1(5482):258–266

Wei Xiang BW, Saron WAA, Stewart JC, Hain A, Walvekar V, Missé D, Thomas F, Kini RM, Roche B, Claridge-Chang A, St John AL, Pompon J (2022) Dengue virus infection modifies mosquito blood-feeding behavior to increase transmission to the host. Proc Natl Acad Sci USA 119(3):e2117589119

Williams R, Abdel Wahab K, Hoogstraal H (1970) Viruses in ticks. V. Viruses isolated from Afghanistan ticks during 1968. Folia Parasitol (Prague) 17(4):359–363

Williams RE, Hoogstraal H, Casals J, Kaiser MN, Moussa MI (1973) Isolation of Wanowrie, Thogoto, and Dhori viruses from *Hyalomma* ticks infesting camels in Egypt. J Med Entomol 10(2):143–146

Williams RW, Causey OR, Kemp GE (1972) Ixodid ticks from domestic livestock in Ibadan, Nigeria as carriers of viral agents. J Med Entomol 9(5):443–445

World Health Organization (WHO) (2022) Dengue and severe dengue. [Accessed 8 Dec 2023] https://www.who.int/news-room/fact-sheets/detail/dengue-and-severe-dengue

World Health Organization (WHO) (2023) Tick-borne encephalitis. [Accessed 8 Dec 2023] https://www.who.int/health-topics/tick-borne-encephalitis#tab=tab_1

Yamauchi T, Tabara K, Kanamori H, Kawabata H, Arai S, Katayama T, Fujita H, Yano Y, Takada N, Itagaki A (2009) Tick fauna associated with sika deer density in the Shimane Peninsula, Honshu, Japan. Med Entomol Zool 60(4):297–304

Yoo JR, Heo ST, Song SW, Bae SG, Lee S, Choi S, Lee C, Jeong S, Kim M, Sa W, Lee Y, Choi H, Kee SH, Lee KH (2020) Severe fever with thrombocytopenia syndrome virus in ticks and SFTS incidence in humans, South Korea. Emerg Infect Dis 26(9):2292–2294

Yoshii K, Okamoto N, Nakao R, Klaus Hofstetter R, Yabu T, Masumoto H, Someya A, Kariwa H, Maeda A (2015) Isolation of the Thogoto virus from a *Haemaphysalis longicornis* in Kyoto City, Japan. J Gen Virol 96(8):2099–2103

Yoshii K, Song JY, Park SB, Yang J, Schmitt HJ (2017) Tick-borne encephalitis in Japan, Republic of Korea and China. Emerg Microbes Infect 6(9):e82

Yun SM, Lee YJ, Choi W, Kim HC, Chong ST, Chang KS, Coburn JM, Klein TA, Lee WJ (2016) Molecular detection of severe fever with thrombocytopenia syndrome and tick-borne encephalitis viruses in ixodid ticks collected from vegetation, Republic of Korea, 2014. Ticks Tick Borne Dis 7(5):970–978

Yurchenko OO, Dubina DO, Vynograd NO, Gonzalez JP (2017) Partial characterization of tick-borne encephalitis virus isolates from ticks of Southern Ukraine. Vector Borne Zoonotic Dis 17(8):550–557

Zhang H, Zhu Y, Liu Z, Peng Y, Peng W, Tong L, Wang J, Liu Q, Wang P, Cheng G (2022) A volatile from the skin microbiota of flavivirus-infected hosts promotes mosquito attractiveness. Cell S0092–8674(22):00641–00649

Zhang Y, Jiang L, Yang Y, Xie S, Yuan W, Wang Y (2021) A tick bite patient with fever and meningitis co-infected with *Rickettsia raoultii* and Tacheng tick virus 1: a case report. BMC Infect Dis 21(1):1187

Zhang YK, Zhang XY, Liu JZ (2019) Ticks (Acari: Ixodoidea) in China: geographical distribution, host diversity, and specificity. Arch Insect Biochem Physiol 102:e21544

Chapter 3
Fly

Siew Hwa Tan, Hiromu Kurahashi, and Kyoko Sawabe

Abstract Flies, as essential insects, exerts a dual impact on our daily lives, contributing both positive and negative aspects. They ubiquitously associate with human activities, with specific concern directed towards certain synanthropic species (e.g., *Chrysomya megacephala*, and *Musca domestica*) due to their potential to introduce medical issues and cause a nuisance. As mechanical vectors for viruses, pathogens, bacteria, and even protozoa, flies pose health risks and require scrutiny. In Asia, the families Calliphoridae, Muscidae, and Sarcophagidae, within the Diptera order, are significant in medicine, forensics, and veterinary sciences. Documented diseases associated with flies include typhoid fever, food poisoning, diarrhea, cholera, eye diseases, polio, and highly pathogenic avian influenza. Within the medical context, myiasis primarily induced by *Chrysomya bezziana* and *Lucilia sericata* poses significant challenges, leading to morbidity and mortality in both animals and humans and causing extensive economic losses in the livestock industry. Conversely, maggot therapy showcases beneficial applications for human well-being. In Japan, flies pose a considerable challenge as mechanical vectors for the transmission of *Escherichia coli* O157 and the H5N1 virus. Therefore, systematic surveillance and periodic implementation of control measures are essential to monitor the situation and preemptively mitigate the risk of potential outbreaks.

Keywords Calliphoridae · *Escherichia coli* O157 · H5N1 · Muscidae · Myiasis · Sarcophagidae

S. H. Tan (✉)
Genetics and Molecular Biology, Institute of Biological Sciences, Faculty of Science, University of Malaya, Kuala Lumpur, Malaysia

H. Kurahashi · K. Sawabe
Department of Medical Entomology, National Institute of Infectious Diseases, Tokyo, Japan
e-mail: MLB15110@nifty.com; sawabe@niid.go.jp

K. Sawabe et al. (eds.), *Medical Entomology in Asia*, Entomology Monographs,
https://doi.org/10.1007/978-981-97-4406-0_3

3.1 Introduction

In the order Diptera, flies play an important role in public health issues, particularly in medicine, forensics, and veterinary science. Among the various families of flies within this order, the families Calliphoridae (blow flies), Muscidae (house flies), and Sarcophagidae (flesh flies) are recognized as particularly important in the field of medical entomology. These families are called filth flies due to their feeding habits, including animal excrement, food waste, and garbage. Consequently, they serve as the primary vectors responsible for the transmission of pathogenic microorganisms such as bacteria, viruses, fungi, protozoa, and helminths, leading to diseases such as typhoid fever, diarrhea, cholera, and polio.

The phenomenon of fly larvae feeding on the living tissues of hosts, whether human or animal, is called myiasis. Conversely, the infestation of dead tissues of a living host by larvae, used in medical practice to treat wounds that are difficult to heal due to bacterial infestation, is known as maggot therapy. In forensic entomology, identifying fly immatures from a deceased body is valuable in criminal investigations. The collected larvae estimate the postmortem interval (PMI) by calculating their age.

This chapter focuses on three fly families of medical importance: Calliphoridae, Muscidae, and Sarcophagidae (Fig. 3.1). The chapter examines aspects of their taxonomy, role as a vector, distribution, ecological characteristics, and fly-transported diseases, particularly in the Asian context. It also explores the diverse roles that flies play in various fields, including medicine, forensics, ecology, agriculture, and animal husbandry.

3.2 Taxonomy

Flies are classified in the taxonomic order Diptera. Within this order is a rich diversity of families, genera, and species, each distinguished by unique morphological characteristics and ecological roles. The taxonomy of flies that fall within the broad order Diptera can be further elaborated, as shown in Table 3.1. Generally, a common structure consisting of seven major levels or tiers is maintained. These levels, arranged in ascending order of specificity, include (1) Kingdom, (2) Phylum, (3) Class, (4) Order, (5) Family, (6) Genus, and (7) Species. It is important to recognize that below each of these primary levels, there are a variety of additional hierarchical ranks that achieve greater precision and detail in the characterization of organisms. The hierarchical framework is a systematic support for categorizing true flies within the order Diptera, providing valuable insight into their phylogenetic relationships and distinctive characteristics at various taxonomic levels. The application of hierarchical taxonomy is of great importance in systematic studies, assisting researchers to structure and understand the interrelationships among different fly species. Each of these taxonomic levels represents a group of flies with specific shared

Fig. 3.1 Fly species with medical importance. (**a**) *Chrysomya megacephala*, (**b**) *Chrysomya rufifacies*, (**c**) *Musca domestica vicina*, and (**d**) *Parasarcophaga* sp.

characteristics and evolutionary affinities, contributing to a comprehensive understanding of fly biodiversity.

3.3 Morphology

Flies, scientifically known as Diptera, possess a pair of wings called the front wings, which give the order its name. The hind wings, on the other hand, have undergone reduction and have evolved into halteres, structures that serve primarily sensory functions and are of limited taxonomic value. The wings themselves are hyaline and transparent, allowing for clear visibility of the wing venation, which plays a crucial role in identifying fly families. Two accessory lobes, the alar and thoracic squamae, are located beneath the wings and provide valuable information for species identification.

The body of a fly is divided into three segments: the head, thorax, and abdomen. The head contains a pair of eyes and three antennal segments. The arista, a

Table 3.1 Hierarchy of fly taxonomy

Taxonomic rank	Name	Information
Domain	Eukaryota (Eukaryotes)	This is the broadest taxonomic category and encompasses all organisms with cells containing a true nucleus and membrane-bound organelles. Flies belong to this domain, like all other animals.
Kingdom	Animalia (Animals)	Animals are multicellular, heterotrophic organisms characterized by mobility and typically complex life cycles. Flies are part of the animal kingdom.
Phylum	Arthropoda (Arthropods)	Arthropods are a diverse group of invertebrates known for their jointed appendages and segmented bodies. Flies are a subgroup within this phylum.
Subphylum	Hexapoda (Hexapods)	Hexapods are a subgroup of arthropods characterized by having six legs. Insects are a prominent group within this subphylum.
Class	Insecta (Insects)	Insects are a class of arthropods characterized by three pairs of jointed legs, a segmented body, and typically one or two pairs of wings. Flies belong to the class Insecta.
Subclass	Pterygota (Winged and once-winged)	A diverse group of insects characterized by the presence of wings. The name "Pterygota" is derived from the Greek word "pteryx," which means wing, highlighting the defining feature.
Order	Diptera (True flies)	Diptera is one of the major orders within the class Insecta. True flies are distinguished by having only two wings (one pair), while other insects typically have four wings. This is the highest taxonomic level for flies.
Suborder	Brachycera (Short-horned flies)	The suborder Brachycera represents a group of flies characterized by their short antennae compared to other fly suborders. This includes many diverse fly families.
Infraorder	Muscomorpha (Muscomorpha flies)	Muscomorpha is an infraorder within the suborder Brachycera. It includes a wide range of fly families that share certain morphological characteristics.
	Cyclorrhapha (Circular-seamed flies)	Synonym or subset of Muscomorpha. Cyclorrhapha is a group of flies characterized by a particular type of pupal development called pupariation, where the pupa and adult develop within a puparium. The adult emerges from the puparium by pushing off an operculum.
Section	Schizophora	A taxonomic section within Cyclorrhapha, characterized by a distinctive feature known as the "ptilinum", which refers to a specialized inflatable membranous sac that aids in escaping from the puparium during eclosion.
Subsection	Calyptratae (Calyptrate flies)	The term "Calyptratae" originates from Greek, where "kalyptos" means "covered" and "ptera" means "wings". Flies within this taxonomic group are distinguished by a distinctive wing structure featuring calypters, small membranous flaps that shield their hind wings when at rest. This adaptation safeguards the delicate hind wings, essential for their flight.

(continued)

Table 3.1 (continued)

Taxonomic rank	Name	Information
Superfamily	Oestroidea/ Muscoidea	Oestroidea is characterized by its focus on parasitism and scavenging, with a higher likelihood of parasitic lifestyles and forensic significance. Muscoidea, on the other hand, includes a broader range of ecological roles, with some species being associated with human habitats and others fulfilling roles in decomposition, pollination, and agriculture.
Family	Calliphoridae/ Muscidae/ Sarcophagidae	These are specific families of flies within the superfamilies of Oestroidea and Muscoidea.
Subfamily	Chrysomyinae/ Muscinae/ Sarcophaginae	These subfamilies represent different branches or subdivisions within the broader family Calliphoridae, Muscidae, and Sarcophagidae.
Tribe	Chrysomyini/ Muscini/ Sarcophagini	Tribes are employed to cluster genera within a family or subfamily, with the grouping determined by shared traits, evolutionary connections, and a common ancestral lineage.
Genus	*Chrysomya/ Musca/ Parasarcophaga*	Each genus consists of one or more species that are closely related and sharing similarities in morphology, genetics, and ecological niche.
Species	*megacephala/ domestica/ dux*	Species is the fundamental unit of biological classification. Each species is unique and possesses specific characteristics that differentiate it from other species.
Scientific name	*Chrysomya megacephala/ Musca domestica/ Parasarcophaga dux*	The scientific name, or binomial nomenclature, is a dual-part naming system in italics. It comprises the genus name, with an initial capital letter, followed by the species name in lowercase.

bristle-like structure, is typically plumose. The mouth apparatus may be spongy or blood-sucking. The thorax is covered with hairs, which are useful for morphological identification. Attached to the thorax are six legs, each consisting of three main segments: femur, tibia, and tarsus. The bristles on the legs are important for morphological identification. The abdomen contains the sex organs, especially the male genitalia, which help to determine species identity.

Flies undergo a complete metamorphosis from eggs to larvae to pupae before finally emerging as adults. Most fly species are oviparous, with females laying mature eggs that hatch externally. However, certain species (e.g., *Calliphora* species) and families (e.g., Sarcophagidae) are viviparous, with females giving birth to young instar larvae instead of laying eggs. In these cases, the eggs are retained and develop within the mother's body to maturity and are then larviposited as first instar larvae. In families such as Calliphoridae, Muscidae, and Sarcophagidae, the larvae undergo three molts, corresponding to the first, second, and third instar larval stages. The duration of pupariation varies among species. In tropical climates, the life cycle of *Chrysomya* species may be as short as five or six days. Some fly species, such as

Boettcherisca peregrina Robineau-Desvoidy, diapause during the winter to survive unfavorable conditions (Moribayashi et al. 1999, 2001, 2008).

3.3.1 Family Calliphoridae

The family Calliphoridae includes flies commonly known as carrion flies, including species such as blue bottles (*Calliphora* species), green bottles (*Lucilia* species), and cluster flies (*Pollenia* species). These flies typically have a metallic appearance, with colors ranging from green and blue to purple, although certain species may lack the metallic sheen and appear brown or yellow. Their body size varies from 2.5 mm to 19.0 mm. Females of most species in this family are oviparous, although there are a few exceptions to viviparity, notably observed in *Calliphora* species.

In terms of taxonomic studies, the Calliphoridae have received more attention than the Muscidae and Sarcophagidae. In Asia, pioneering efforts in this field began as early as 1940 by Senior-White and colleagues (Senior-White et al. 1940), followed by contributions from Zumpt (1956), Kurahashi (1964, 1966, 1971, 1987a, b, 2010a) and colleagues (Kurahashi and Park 1972; Kurahashi and Tumrasvin 1979; Kurahashi et al. 1997; Kurahashi and Bunchu 2011), Fan (1965, 1992), Tumrasvin and Kano (1979), Tumrasvin et al. (1976, 1977, 1979), Inder Singh et al. (1979), and Nandi (2002a). The taxonomic studies conducted by Kurahashi were particularly significant and covered various Asian countries, including Japan (Kurahashi 1964, 1966), Korea (Kurahashi and Park 1972), Taiwan (Kurahashi 1987a; Kurahashi and Kano 1995; Yang et al. 2014), Bangladesh (Kurahashi 1989), Nepal (Kurahashi and Thapa 1994), India (Bharti and Kurahashi 2010), Pakistan (Kurahashi and Afzal 2002), Sri Lanka (Kurahashi 2001), Malaysia (Singh et al. 1979; Kurahashi et al. 1997; Kurahashi and Leh 2007, 2009), Thailand (Tumrasvin et al. 1976, 1977, 1979; Kurahashi and Tumrasvin 1979; Kurahashi and Bunchu 2011), Indochina-Vietnam, Laos, Cambodia (Kurahashi and Chowanadisai 2001), Indonesia (Kurahashi and Magpayo 1987; Kurahashi 1988; Wells and Kurahashi 1996), and the Philippines (Kurahashi and Salazar 1977; Kurahashi and Magpayo 2000). These extensive contributions to the taxonomy of Calliphoridae have greatly aided in the successful identification of species and have made it the most extensively studied family in medical entomology.

In Asia, the current classification divides Calliphoridae into nine subfamilies: (1) Ameniinae, (2) Bengaliinae, (3) Calliphorinae, (4) Chrysomyinae, (5) Luciliinae, (6) Melanodexiinae, (7) Phumosiinae, (8) Polleniinae, and (9) Rhiniinae (Kurahashi and Leh 2009). Although Calliphoridae is of great medical importance, not all species within this family are equally important. In this regard, only species from the subfamilies Chrysomyinae (*Chrysomya*) and Luciliinae (*Lucilia*) are strongly associated with disease transmission (e.g., bacteria, viruses, fungi), as well as being involved in forensic investigations and cases of myiasis (both obligate and facultative) in humans and animals. In Japan, the family Calliphoridae comprises 28 genera and 70 species, including three species of *Chrysomya* and 10 species of *Lucilia* (Kurahashi 2014a).

Almost all species within the genus *Chrysomya* are of significant medical, forensic, and veterinary importance, especially in their larval stage. The differences among these species in terms of medical, forensic, and veterinary relevance lie in their frequency and rate of occurrence. *Chrysomya* species are characterized by their metallic underglaze colors of green, blue, or purple. Their body size ranges from 5 to 13 mm, with *C. nigripes* Aubertin being the smallest and *C. thanomthini* Kurahashi and Tumrasvin the largest. They can be distinguished from other genera by the presence of distinct setulae on the postero-dorsal side at the stem vein of the wing, a subalar knob adorned with erect hairs, and largely lobulated thoracic squamae. There are 17 known *Chrysomya* species in the Oriental region (Verves 2005). Among the *Chrysomya* species, *C. bezziana* Villeneuve, commonly known as the Old-World screwworm fly, is an obligate wound parasite affecting humans, domestic animals, and wildlife. Untreated myiasis caused by *C. bezziana* larvae can lead to serious illness in humans and even death in the infested animal. However, *C. bezziana* is not present in Japan. In Japan, the *Chrysomya* species responsible for human myiasis are *C. megacephala* (Fabricius) and *C. pinguis* (Walker) (Kobayashi 1915; Kano et al. 1962).

In the Oriental region, *C. bezziana* is the most notable species causing myiasis in both animals and humans (Sreevatsa et al. 1990; Lan and Zhong 2001; Chemonges-Nielsen 2003; Chan et al. 2005; Lee et al. 2005; Faramarzi et al. 2009; Yaghoobi and Bagherani 2009; Yeung et al. 2010; Berenji et al. 2015; Han et al. 2017; Wardhana et al. 2018; Fang et al. 2019). Cases of human infestation have been reported from various body sites, including the mouth, limbs, ear, eye, nose, face, scalp, perineal and inguinal regions, trunk, intestine, tracheostomy, pharyngostomy, and urogenital region (Wadhwa et al. 2006; Rahoma and Latif 2010; Mircheraghi et al. 2016; Zhou et al. 2019). *Chrysomya megacephala,* also known as the oriental latrine fly or oriental blue fly, is very similar to *C. bezziana* and is typically associated with breeding on decomposing corpses and plays an important role in forensic entomology. However, it can also become a facultative parasite in wounds, similar to *C. bezziana*. Distinguishing between these two species is challenging due to their high degree of similarity. Initially, *C. bezziana* was originally described by Bezzi as *C. megacephala* but was later corrected by Villeneuve (1914). The differentiation of male flies of *C. bezziana* from *C. megacephala* is based on the striking differences in their hypopygia and the arrangement of their ocular facets. In *C. megacephala*, male flies have enlarged upper eye facets with a sharp transition to smaller facets in the lower third of the eye. However, distinguishing female flies of the two species can be challenging. A key distinguishing feature is the coloration of the thoracic squamae, which is white in *C. bezziana* and infuscated (darkened) in *C. megacephala* (Kurahashi et al. 1997). In addition, it is worth noting that the frons of *C. bezziana* have a generally parallel-sided configuration, in contrast to *C. megacephala*, where they narrow conspicuously towards the antennal groove. This indistinctive difference was elucidated by Zumpt (1965) and is visually depicted in a figure provided by Animal Health

Australia (Figs. 3–10, p. 28, AHA 2017). It often contributes to misidentification due to its complicated nature.

In the field of forensic entomology, *C. megacephala* (Fig. 3.1a) and *C. rufifacies* (Macquart) (Fig. 3.1b) are the two predominant species commonly found on cadavers in tropical countries, especially Malaysia and Thailand (Lee 1989, 1996a, b; Lee et al. 2004; Sukontason et al. 2007). These species have been reported in numerous forensic cases as they are early colonizers of cadavers and can be found indoors and outdoors (Kumara et al. 2012). They are easily distinguished morphologically in adults, and their larvae exhibit distinct characteristics. Adult *C. megacephala* have dark thoracic spiracles, while *C. rufifacies* have white mesothoracic spiracles. The larvae of *C. rufifacies* are notable for their hairy appearance and are known as hairy maggots, with spiny tubercles on the body, although the "hairs" are not true hairs. Both *C. megacephala* and *C. rufifacies* are considered facultative parasites capable of causing myiasis, but such occurrences are rare (Sukontason et al. 2005).

The genus *Lucilia*, commonly known as green bottle flies, has important associations with veterinary and forensic entomology. *Lucilia* species can be readily distinguished by specific characters. One distinctive feature is the posterior portion of the suprasquamal ridge, which exhibits a posterior parasquamal tuft of black setulose erect hairs on a small, well-defined black sclerite. Additionaly, the supraspiracular convexity of *Lucilia* species may either be bare or pubescent. Most *Lucilia* species have a green coloration, with the exception of one species, the Australian sheep blow fly, *L. cuprina (Wiedemann)*, which is characterized by a coppery body color. *Lucilia cuprina* is primarily responsible for causing "flystrike" in sheep in Australia, a condition that, if left untreated, can result in sheep mortality and significant production losses for livestock producers (Capinera 2008a). Another common *Lucilia* species associated with myiasis in animals and humans is *Lucilia sericata* (Meigen) (Khan and Khan 1987; Chigusa et al. 1998, 2002; Dik et al. 2012; Jang et al. 2013; Nasiri et al. 2015; Choe et al. 2016; Demirel-Kaya et al. 2016). While myiasis predominantly affects mammals and birds (Baumgartner 1988; Gomez-Puerta et al. 2021; Pezzi et al. 2022), it can occasionally occur in amphibians (Dasgupta 1962). In cases of human myiasis, species such as *L. ampullacea* Villeneuve, *L. caesar* (Linnaeus), *L. cuprina* (Wiedemann), and *L. illustris* (Meigen) have also been reported (Chigusa et al. 1996; Hashimoto et al. 1987; Ishii et al. 1988; Yoneda et al. 1998; Nazni et al. 2011; Moradi-Asl et al. 2021).

Maggot therapy, also known as larval therapy, maggot debridement therapy (MDT), biodebridement, or biosurgery, is a medical treatment method that involves the application of live and pre-sterilized larvae to non-healing wounds for debridement, disinfection, and enhanced healing. With technological advancements, maggot therapy is gaining popularity and acceptance in wound healing. The roots of maggot therapy can be traced back to the 1500s when a military surgeon used it to heal wounds in soldiers (Pechter and Sherman 1983). However, its use declined with the invention and widespread use of antibiotic treatments. In the 1990s, there was renewed interest in maggot therapy due to the rise of antibiotic resistance. Currently, *L. sericata* larvae are widely used in various countries, including China, Japan, Korea, and Thailand (Sherman et al. 1995; Thomas et al. 1996; Kim and

Kang 2008; Wang et al. 2010; Kim et al. 2011; Igari et al. 2013; Wilasrusmee et al. 2014; Bazaliński et al. 2019; Nezakati et al. 2020; Bavani et al. 2022). In Japan, Mitsui et al. (2005) reported the first successful application of this species in maggot therapy to heal a severe ischemic foot ulcer in a diabetic patient. In Malaysia and Singapore, maggot therapy using *L. cuprina* has also shown promising results in wound healing (Paul et al. 2009; Nair et al. 2021; Gunasegaran et al. 2022).

In Kyoto, Japan, the blow flies *Calliphora nigribarbis* Vollenhoven and *Aldrichina grahami* (Aldrich) were detected as carriers of the H5N1 highly pathogenic avian influenza virus (Sawabe et al. 2006). Subsequent studies have shown that blow flies have the potential to mechanically spread the H5N1 influenza virus (Tsuda et al. 2009; Sawabe et al. 2009, 2011).

3.3.2 Family Muscidae

Muscidae is a large family of flies that includes approximately 200 genera and 5000 species worldwide (Moon 2019). In Japan, 38 genera and 253 species of Muscidae have been recorded to date (Shinonaga 2014). Muscidae can be distinguished from other families of flies, especially Fanniidae, based on the wing venation of A_1+CuA_2 (or 6th vein/A1), which extends before the wing margin but does not meet A_2. Vein M_{1+2} has a distinct forward curve, the apical portion of vein *Sc* shows a distinct curve, the hind coxa typically lacks setae on the posterior surface, and interfrontal bristles are usually absent. The external morphology of house flies within the Muscidae is highly variable. Body coloration varies between genera and species; even the thorax and abdomen can differ between species. For example, *Neomyia* species are green, *Hydrotaea spinigera* (Stein) (=*Ophyra spinigera*) is black, *Musca domestica* Linnaeus is gray, and *Antherigona* species are brown. In general, however, house flies are gray with four black stripes on the thorax, the wing vein M_{1+2} is folded upright, and the abdomen is typically yellowish. An important distinction between non-biting and biting Muscidae is their mouthparts. Non-biting muscids have a normal sponging-type proboscis, similar to other saprophagous flies, whereas biting muscids develop a piercing-type style in their mouthparts that allows the proboscis to penetrate the host's skin to obtain blood. Bites from biting muscid flies can cause allergic reactions. The body size of house flies is medium, ranging from 6 to 9 mm. House flies are considered synanthropic because of their close association with humans, and the most common species is *M. domestica*. Almost all types of organic matter and waste serve as food sources for house flies, including food scraps, decaying vegetables or meat, sewage, excrement, and carrion.

Hennig (1964) divided *M. domestica* into five subspecies: *M. domestica domestica* Linnaeus, *M. domestica vicina* Macquart, *M. domestica nebulo* Fabricius, *M. domestica cuthbertsoni* Patton, and *M. domestica curviforceps* Sacca & Rivosechhi. Among these, the subspecies *domestica*, *vicina*, and *nebulo* are closely related, with the width of the frons narrowing from the former to the latter (Hori 1966). Discrimination among the five subspecies can be made by considering a combination of frons width and abdominal coloration. *Musca domestica vicina* (Fig. 3.1c) is the most important species in medical entomology because of its wide distribution, synanthropic behavior, and ability to adapt to domestic environments.

House flies, *M. domestica*, are considered important mechanical vectors of numerous pathogens, more than 100 of which cause diseases in humans and animals. This species is of particular public health concern because it feeds on decaying matter, making it an effective vector for the transmission of protozoan parasites (e.g., *Sarcocystis* spp., *Toxoplasma gondii*, *Giardia* spp.), bacteria (e.g., *Entamoeba coli*, *Salmonella* spp., *Shigella* spp.), and viruses (e.g., *enterovirus*). As a result, they contribute to the spread of diseases such as typhoid, cholera, and polio (Wallace 1971; Khan and Huq 1978; Markus 1980; Sukontason et al. 2000; Szostakowska et al. 2004). In addition, house flies can also serve as mechanical vectors for avian influenza viruses such as H1N1 and H5N1 (Nazni et al. 2013; Wanaratana et al. 2013; Tyasasmaya et al. 2016). In a series of studies conducted by Otake and colleagues, it was proposed that *M. domestica* may serve as a potential vector for porcine reproductive and respiratory syndrome virus (PRRSV), facilitating mechanical transmission of the virus between swains (Otake et al. 2003, 2004; Pitkin et al. 2009). Another study by Tan et al. (1997) found that *M. domestica* has the ability to mechanically transmit rotavirus from virus-contaminated surfaces.

According to Shinonaga and Kano (1971), *M. conducens* Walker and *M. sorbens* Wiedemann serve as mechanical vectors for stephanofilariasis and habronemiasis, respectively. *Musca conducens* can be distinguished from the other ten *Musca* species in Japan by the presence of an anterodorsal seta on the middle tibia (Shinonaga 2003). *Musca sorbens* is associated with ophthalmic diseases (Pont 1980). This species can be easily distinguished from other *Musca* species in Japan by the presence of hypopleural hairs below the metathoracic spiracle and two broad longitudinal stripes on the scutum of both sexes.

Hydrotaea leucostoma (Wiedemann), on the other hand, is considered medically important as a mechanical carrier in spreading intestinal diseases such as cholera, diarrhea, and summer dysentery (Nandi and Sinha 2004). As a hematophagous fly species, *Stomoxys calcitrans* (Linnaeus), the most prevalent blood-sucking species in Japan, can cause nuisance biting to humans and domestic animals, including dogs, cattle, and horses, by feeding on the blood of the host. It is easily distinguished from the other two *Stomoxys* species in Japan by the male having dichoptic eyes and dark spots on the abdominal tergite (Shinonaga and Kano 1971).

Musca domestica (Lata et al. 1996; Bhatt and Jayakrishnan 2000; Godhi et al. 2008; Dogra and Mahajan 2010; Kandi et al. 2013; Parwani et al. 2014) and *Musca nebulo* (Sharma et al. 2008) have been reported in cases of human myiasis cases. They have also been reported of myiasis in dogs (Dik et al. 2018) and cattle (Singh and Singh 2016).

In forensic entomology, although the frequency of occurrence of Muscidae is not as high as that of Calliphoridae and Sarcophagidae, their role in forensic investigations is equally important. Species commonly associated with forensic cases include *Synthesiomyia nudiseta* Van Der Wulp, *Hydrotaea spinigera*, *Hydrotaea* spp., and *Muscina angustifrons* (Loew) (Lee 1996a, b; Sukontason et al. 2007; Kumara et al.

2012; Kavitha et al. 2013; Grzywacz et al. 2017; Urabe et al. 2022; Yan et al. 2023). *Synthesiomyia nudiseta* is also involved in myiasis (Bernhardt et al. 2019) and is a potential mechanical vector for *E. coli* and *S. dysenteriae* (Greenberg 1973).

3.3.3 Family Sarcophagidae

Sarcophagidae, a family within the order Diptera commonly known as flesh flies, play a vital and pivotal role in forensic entomology, underscoring their considerable scientific importance. They play a crucial role in estimating the postmortem interval by assessing the species composition and life stages found on cadavers. Despite their importance as forensic indicators, the identification of sarcophagid flies can be extremely challenging due to their morphological similarities. Currently, reliable identification is limited to adult male flies because their genitalia exhibit distinct species-specific characteristics (Rohdendorf 1937; Lopes 1961; Lopes and Kano 1979; Sugiyama and Kano 1984; Kurahashi 2002). However, the identification of female flies remains unresolved due to the lack of diagnostic characters, including genitalia, which requires a thorough understanding of fly anatomy for taxonomic studies.

Since the 1950s, Japanese researchers have made notable and substantial contributions to the taxonomic studies of Sarcophagidae. They have described new species and established taxonomic keys, thereby expanding the knowledge of this family (Hori 1953, 1954a, b, 1955; Kano and Shinonaga 1964; Kano et al. 1967, Kurahashi 1970, 1972, 1973, 1975, 1994, 2004, 2005, 2010b, 2014b; Tumrasvin and Kano 1979; Kurahashi and Suenaga 1994; Kano and Kurahashi 1999; Kurahashi and Harusawa 2009, 2012; Kurahashi and Hoshino 2009; Kakinuma and Kurahashi 2010a, b, 2012a, b; Kurahashi and Kakinuma 2015).

Countries in the Asian region, including Taiwan, China, Malaysia, Thailand, and Vietnam, have been actively involved in the development of taxonomic keys for Sarcophagidae, thus contributing significantly to the progress of research in this scientific field (Fan 1965, 1992; Sugiyama et al. 1990; Kano et al. 1999; Kano and Kurahashi 2000; Kurahashi and Chaiwong 2013; Kurahashi et al. 2017, 2021). In addition, scientific investigations involving systematic analysis of Sarcophagidae, compilation of species checklists, documentation of novel locality records, and descriptions of newly discovered species have been diligently conducted (Sugiyama and Kurahashi 1988; Pape and Kurahashi 1995, 2000, 2004; Kurahashi and Pape 1996; Kurahashi and Sukontason 2004; Kurahashi and Chaiwong 2007; Kurahashi and Leh 2008; Kurahashi and Tan 2009; Tan et al. 2010; Kurahashi and Tan 2012a, b).

Flesh flies, members of the family Sarcophagidae, exhibit remarkable taxonomic diversity with three distinct subfamilies: (1) Sarcophaginae, which includes species commonly found in association with decaying carcasses and organic waste materials; (2) Miltogrammatinae, which parasitizes various insects, including bees, wasps, and butterflies; and (3) Paramacronychiinae, which remains relatively understudied in terms of its ecological context and specific contributions. In Japan, the richness of flesh fly species is represented by 43 genera and 122 species (Kurahashi and Kakinuma 2015).

Flesh flies are often confused with house flies (Muscidae) and parasitic flies (Tachinidae). Their body sizes range from 6 to 20 mm, typically larger than house flies but smaller than tachinid flies. Their characteristic appearance includes a gray body with three prominent black stripes on the thorax, which is an important distinguishing feature within this family. The abdomen has a square pattern reminiscent of a checkerboard, characterized by alternating between black and gray squares.

The life cycle of flesh flies follows a process of complete metamorphosis, typically lasting at least 14 days in tropical regions. Sarcophagid flies employ a viviparous reproductive strategy, in which the female fly does not lay eggs but rather deposits first instar larvae, known as larviposition. The eggs develop inside the female's body until she finds a suitable substrate, typically decaying organic matter rich in protein, which serves as a nutrient medium for the larvae to thrive. This intriguing behavior allows sarcophagid larvae to function as parasites or carnivorous predators, preying on smaller fly larvae. Both male and female adult flesh flies feed primarily on organic matter such as corpses, carcasses, feces, and flower nectar. The availability of protein obtained from decaying flesh serves as a primary nutrient source for female flies, facilitating the production and maturation of eggs.

In the medical field, the larvae of flesh flies have the potential to induce facultative myiasis. Several species of Sarcophagidae have been documented in human myiasis cases in Japan, including *Boettcherisca peregrina* (Yoneda et al. 1998; Chigusa et al. 2005; Miura et al. 2005), *Boettcherisca septentrionalis* (Robineau-Desvoidy) (Chigusa et al. 2006), *Helicophagella melanura* (Meigen) (Miyamoto et al. 1996; Chigusa et al. 1997), *Liopygia crassipalpis* (Macquart) (Ohashi et al. 2012), *Parasarcophaga albiceps* (Meigen) (Ito and Koshimizu 1955), *Parasarcophaga harpax* (Pandellé) (Kaneko et al. 1968), and *Parasarcophaga similis* (Meade) (Hatsushika et al. 1988, 2002; Yoneda et al. 1998). Sueyoshi (2015) conducted a comprehensive review and analysis of human myiasis cases in Japan, providing additional in-depth information and valuable insights into these cases.

In Malaysia, due to the complexity of Sarcophagidae, most human myiasis cases have only been identified at the genus level (Baharuddin et al. 1995; Ahmad et al. 2009). However, Thomas et al. (1980) successfully identified one case as *Seniorwhitea princeps* (Wiedemann) (=*Sarcophaga krameri* Böttcher). *Liopygia ruficornis* (Fabricius) has been associated with myiasis cases in comatose women in Thailand and leprosy patients in India (Sucharit et al. 1981; Sreevatsa et al. 1990). In addition, both *L. ruficornis* and *S. princeps* (=*Seniorwhitea reciproca* (Walker)) have been reported in cases of animal myiasis cases in India (Sinha 2012; Singh and Singh 2016).

Within the disciplines of medicine, forensic entomology, and veterinary science in various Asian countries, fly species belonging to Calliphoridae, Muscidae, and Sarcophagidae are of great interest. In particular, they play a key role, especially in myiasis cases and forensic investigations. In this context, Table 3.2 provides a comprehensive compilation of species relevant to these cases. These fly species have attracted attention due to their multiple roles within these scientific disciplines. Figure 3.1 showcases a selection of medically significant fly species within the families Calliphoridae, Muscidae, and Sarcophagidae in the Asian region.

3.4 Distribution and Habitat

Flies are found on every continent on Earth, from the polar regions to the tropics and from high altitudes to lowlands. Their ability to adapt to a wide range of habitats contributes to their worldwide distribution. Different species of flies occupy different habitats. Some flies are adapted to aquatic environments, while others thrive in forests, grasslands, urban areas, and agricultural landscapes. Human activities, including urbanization, agriculture, and transportation, can affect the distribution of flies. Some species are considered pests because of their association with disease transmission or crop damage. The distributions of flies are influenced by factors such as temperature, humidity, and altitude. Different species have adapted to specific climatic conditions. Some fly species have limited distributions and are found only in certain regions, known as endemism. Others may have broader ranges associated with their synanthropic behavior. It is important to note that there are many different species of flies, each with its own specific distribution patterns.

3.4.1 Calliphoridae

The family Calliphoridae plays an important role in medicine, particularly in wound care and forensic entomology. Their importance is due to their unique life cycle, behavior, and ecological interactions, which provide valuable insights and applications in various medical contexts. They are carriers of pathogens as well as agents of myiasis in humans and animals (Greenberg 1973). The presence of blow flies can provide insight into the health of ecosystems and environmental conditions. Changes in populations may indicate shifts in decomposition processes, ecological disturbances, or pollution levels, thus contributing to environmental monitoring efforts.

Chrysomya megacephala, the most important calliphorid species in medical entomology, is widely distributed throughout Asia, including countries such as India, China, Thailand, Malaysia, Indonesia, and many others (Verves 2005; Badenhorst and Villet 2018). Its distribution spans a wide range of climatic zones, from tropical to temperate regions (Ngoen-Klan et al. 2011), and even at altitudes above 2000 m (Kurahashi et al. 1997; Moophayak et al. 2014). As a synanthropic fly species, it exhibits a versatile habitat preference associated with both urban and rural environments. It often thrives in places where human activities and waste disposal are prevalent. *Chrysomya megacephala* exhibits a distinct preference for oviposition and larval development within decaying organic substrates, including animal excreta, garbage, and various decaying materials. Its ubiquitous presence is

Table 3.2 List of Calliphoridae, Muscidae, and Sarcophagidae species of medical, forensic, and veterinary significance documented in Asian countries

Species	Reported cases	Countries
Calliphoridae:		
Aldrichina grahami	Forensic	Japan
	Myiasis (human)	Japan
Calliphora erythrocephala	Forensic	India
Calliphora lata	Forensic	Korea
Calliphora nigribarbis	Forensic	Japan, Korea
	Myiasis (human)	Japan
Calliphora vicina	Forensic	Japan, Korea
	Myiasis (animal)	Turkey
Calliphora spp.	Forensic	Malaysia
	Myiasis (human)	Malaysia, Thailand
Chrysomya albiceps	Myiasis (animal)	India, Turkey
	Myiasis (human)	India
Chrysomya bezziana	Forensic	Malaysia, Thailand
	Myiasis (animal)	China, Hong Kong, Malaysia
	Myiasis (human)	China, Hong Kong, India, Indonesia, Iran, Laos, Malaysia, Pakistan, Sri Lanka, Taiwan, Thailand
Chrysomya chani	Forensic	Thailand
Chrysomya megacephala	Forensic	China, India, Korea, Malaysia, Taiwan, Thailand
	Myiasis (human)	India, Japan, Thailand
Chrysomya nigripes	Forensic	China, Malaysia, Taiwan, Thailand
Chrysomya pinguis	Forensic	Japan, Korea, Malaysia, Taiwan, Thailand
	Myiasis (human)	Japan
Chrysomya rufifacies	Forensic	China, India, Malaysia, Taiwan, Thailand
	Myiasis (human)	Thailand
Chrysomya villeneuvi	Forensic	Malaysia, Thailand
Hemipyrellia ligurriens	Forensic	Malaysia, Taiwan, Thailand
Hemipyrellia tagaliana	Forensic	Malaysia
Lucilia ampullacea	Myiasis (human)	Japan

(continued)

Table 3.2 (continued)

Species	Reported cases	Countries
Lucilia caesar	Forensic	Japan, Korea
	Myiasis (human)	Japan
Lucilia cuprina	Forensic	Thailand
	Myiasis (animal)	Zumpt (1956)[a]
	Myiasis (human)	Iran, Malaysia, Turkey
Lucilia illustris	Forensic	Japan, Korea
	Myiasis (human)	Japan
Lucilia porphyrina	Forensic	Taiwan, Thailand
Lucilia sericata	Forensic	China, Japan, Korea
	Myiasis (animal)	Korea, Turkey
	Myiasis (human)	Japan, Korea, Turkey
Protocalliphora maruyamensis	Myiasis (animal)	Japan
Muscidae:		
Hydrotaea ignava	Forensic	Japan
Musca domestica	Forensic	India
	Myiasis (human)	India, Indonesia, Japan, Thailand
Muscina stabulans	Forensic	Japan, Taiwan
	Myiasis (human)	Japan
Hydrotaea obscrifrons	Forensic	Korea
Hydrotaea spinigera (=*Ophyra spinigera*)	Forensic	Malaysia, Taiwan, Thailand
	Myiasis (human)	Japan
Hydrotaea spp. (=*Ophyra* spp.)	Forensic	Malaysia
Synthesiomyia nudiseta	Forensic	Malaysia, Taiwan, Thailand
Sarcophagidae:		
Ravinia pernix	Forensic	India
Boettcherisca formosensis	Forensic	Taiwan
Boettcherisca peregrina	Forensic	China, Japan, Korea, Malaysia, Taiwan
	Myiasis (human)	Japan
Boettcherisca septentrionalis	Forensic	Japan
	Myiasis (human)	Japan

(continued)

Table 3.2 (continued)

Species	Reported cases	Countries
Helicophagella melanura	Myiasis (human)	Japan
Liopygia ruficornis	Forensic	Malaysia, Taiwan, Thailand
	Myiasis (human)	Thailand
Liopygia crassipalpis	Forensic	Japan, Korea
	Myiasis (human)	Japan
Parasarcophaga albiceps	Myiasis (human)	Japan
Parasarcophaga dux	Forensic	Japan, Korea, Malaysia, Taiwan
	Myiasis (animal)	India
	Myiasis (human)	Thailand
Parasarcophaga harpax	Forensic	Japan
	Myiasis (human)	Japan, Taiwan
Parasarcophaga similis	Forensic	Japan
	Myiasis (human)	Japan
Sarcophaga carnaria	Forensic	India
Seniorwhitea princeps	Forensic	Malaysia
Sarcophaga spp.	Forensic	India, Malaysia

[a]No country was indicated in Zumpt (1965)

often observed around latrines, open garbage dumps, and areas of poor sanitation. Due to its feeding behavior, which involves the consumption of decaying animal carcasses and organic waste materials, this fly species is of considerable importance in several fields. These areas include forensics, its role as a myiasis agent, scavenging activities, participation in decomposition processes, and functioning as a mechanical vector of pathogens (Monzon et al. 1991; Sulaiman et al. 1988; Lee et al. 2004; Sukontason et al. 2007; Ferraz et al. 2010). According to Wallace's findings in 1971, *C. megacephala* was associated with the mechanical transmission of *Toxoplasma gondii*. The adaptability of *C. megacephala* to urban environments is highly evident. It is often associated with human settlements, particularly in areas with inadequate waste management systems. In rural areas, this species can also be found in agricultural landscapes where it has access to suitable breeding and feeding resources, especially animal dung in areas dedicated to livestock farming. However, various environmental variables can affect its distribution and population density, including altitude, rainfall patterns, temperature regimes, relative humidity levels, seasonal variations, landscape features, and habitat characteristics.

Chrysomya bezziana, commonly referred to as the Old-World screwworm fly, exhibits a close ecological association with environments characterized by live hosts, infested open wounds, and bodily orifices of animals, in both livestock and occasionally humans (Zumpt 1956). The larvae of *C. bezziana* are obligate parasites that require living tissue for sustenance and development (Zumpt 1965). The prevalence of warm-blooded animals with wounds or injuries plays an important role in influencing the distribution of *C. bezziana* within its habitat. Livestock, pets, wildlife, and even humans can serve as viable hosts for these flies, and their unchecked presence can result in significant tissue damage (Spradbery et al. 2019). Their geographic distribution spans vast areas, including regions in Africa, the Indian subcontinent, and the tropical regions of the Old World, Southeast and East Asia, as detailed in the study by Zhou et al. (2019). *Chrysomya bezziana* exhibits a clear preference for warm and tropical climates, rendering regions with elevated temperatures more conducive to its survival and reproductive success. The problem precipitated by myiasis caused by *C. bezziana* not only causes disease and mortality in both animal and human hosts but also has significant economic implications in the livestock sector.

Lucilia cuprina, the green bottle fly, is thought to have originally originated from either the Afrotropical or Oriental regions, as proposed by Aubertin (1933). However, it has subsequently been introduced into various other geographical areas, including parts of Asia, and has thus acquired a cosmopolitan distribution. In Malaysia and Thailand, there is observed evidence of *L. cuprina* has been observed to display a notable adaptation to human activities, leading to its classification as a synanthropic species (Omar et al. 2003; Chaiwong et al. 2012; Khoso et al. 2015). Notably, *Lucilia cuprina* exhibits morphological variation in different parts of its range, leading to the recognition of two distinct subspecies: *Lucilia cuprina cuprina* (Wiedmann) and *Lucilia cuprina dorsalis* Robineau-Desvoidy (Waterhouse and Paramonov 1950). In the Asian region, the subspecies *L. cuprina cuprina* is widespread. While another green bottle fly, *L. sericata*, may have originated in England and subsequently expanded its range to various countries within the continent due to its remarkable adaptability to different environments (Zumpt 1965). Both *Lucilia* species are of great economic, agricultural, and forensic importance. They are classified as saprophagous and often serve as primary facultative ectoparasites, causing myiasis in both wild and domestic animals (Zumpt 1965; Hall and Wall 1995). Consequently, their presence is a matter of considerable concern in regions while livestock production is important. In particular, *L. cuprina* tends to predominate in subtropical and warm temperate regions, such as Southeast Asia, while *L. sericata* shows a preference for cool temperate regions, including Japan, Korea, and China (Stevens and Wall 1996, 1997).

3.4.2 Muscidae

House flies are cosmopolitan insects that are found throughout the world. They are of medical importance because they potentially transmit pathogens that cause diseases like diarrhea, food poisoning, and eye infections to humans and animals. It is an important vector of several diseases and can be found in both urban and rural areas.

Musca domestica originated in the steppes of Central Asia and this ubiquitous fly species now has a cosmopolitan distribution that spans all inhabited continents (Capinera 2008b). Its ecological versatility is evidenced by its occupation of different climatic zones, including both tropical and temperate regions. It also thrives in a wide range of habitats, spanning the spectrum from rural to urban landscapes. While traditionally associated with animal excreta, its dietary preferences have adapted to include garbage, making it prolific in virtually any location inhabited by human populations. These preferences, such as those related to food choice and temperature range, explain the successful global spread of *M. domestica*. In Asia, *M. domestica* is found in both urban and rural areas, and its distribution is influenced by factors such as climate, sanitation practices, and the presence of suitable breeding sites. Among five subspecies of *M. domestica*, the subspecies of *M. domestica vicina* has a wider distribution in Asia. It is known for its close association with livestock, especially poultry and cattle. Within the Indian territory, *M. domestica* subspecies, specifically *vicina* and *nebulo*, coexist, with *nebulo* showing a markedly predominant presence (Hennig 1964).

Musca sorbens, commonly known as the African latrine fly or bazaar fly, has received attention as a potential vector of trachoma (Emerson et al. 2000). This species is primarily found in sub-Saharan Africa, but also extends its presence to tropical and subtropical regions (Shinonaga 2003). It exhibits a notable association with areas characterized by inadequate sanitation facilities and is often found in communities without access to clean water sources. A distinctive characteristic of *M. sorbens* is its aggressive behavior toward humans, as it seeks to feed on mucous membranes of the eyes, nose, or mouth, open wounds, and sweat (Robinson et al. 2021). These feeding habits have been implicated in the transmission of the debilitating eye disease trachoma in certain regions of the world.

Stomoxys calcitrans, commonly known as the stable fly or biting house fly, originated in Africa but has achieved a cosmopolitan distribution that includes regions such as Japan, Korea, Taiwan, China, Malaysia, Thailand, and Indonesia. Unlike typical house flies, stable flies exhibit primarily hematophagous behavior, with both males and females feeding on the blood of animals, including humans (Zumpt 1973; Wall and Shearer 1997). This blood-feeding propensity can lead to disease transmission, discomfort, and economic losses in livestock due to their painful bites (Baldacchino et al. 2013). Stable flies prefer habitats rich in potential blood hosts and are often found in areas frequented by livestock, such as cattle, horses, and other large mammals, where they find suitable conditions for feeding and breeding. They tend to seek shelter at night and generally avoid biting after dark, but their strong flying ability allows them to venture indoors within a radius of up to 1 km from cattle (Robinson 2005). Although stable flies occasionally and unintentionally bite humans, rare exceptions have been observed in beach environments (Hogsette and Ruff 1985). It is noteworthy that the distribution and abundance of stable flies can exhibit seasonal variations, often peaking during warmer months.

3.4.3 Sarcophagidae

Most Sarcophagidae are of considerable importance in both myiasis occurrences and forensic applications within the medical context (Fig. 3.1d). Sarcophagid flies are predominantly synanthropic and demonstrate adaptability to high altitudes (Fremdt and Amendt 2014; Syamsa et al. 2015). However, species composition varies considerably depending on geographic region, bioclimatic zone, and indoor and outdoor distinction (Kumara et al. 2012; Toukairin et al. 2017). In terms of global distribution significance, Pape (1996) compiled a comprehensive dataset of 2510 species of Sarcophagidae, meticulously documenting their presence at the county level, along with specific details regarding states, provinces, or isolated islands.

Liopygia ruficornis stands out as a cosmopolitan species of considerable importance in medical and forensic entomology. It shows a preference for warmer climates and has a wide distribution that includes several Oriental countries, including Malaysia, Thailand, Indonesia, India, Nepal, Japan, and Taiwan. In contrast, a closely related species, *Liopygia crassipalpis*, has a more restricted distribution in temperate countries such as Japan, Korea, China, and Taiwan. Notably, *L. crassipalpis* frequently inhabits urban environments and is commonly found near large cities (Kano et al. 1967).

Parasarcophaga dux (Thomson), a synanthropic flesh fly of considerable medical importance, has a worldwide distribution (Pape 1996). Its distribution is extensive and includes large areas within the Oriental region, encompassing areas such as Japan, India, China, and Southeast Asia. This species is frequently observed due to its diverse dietary preferences, comprising human excrement, animal wounds, and decomposed carcasses. On the other hand, *P. albiceps* is commonly found throughout Japan and inhabits a variety of environments, including seashores, mountainous regions, and residential areas (Kano et al. 1967).

The genus *Boettcherisca* is widely distributed across the Oriental, Palearctic, Australasian, and Oceanic regions (Nandi 2002b). Within this genus, *B. peregrina* emerges as a synanthropic species of notable medical importance. Its distribution ranges from tropical to temperate countries, and it is frequently observed in latrine environments in Japan (Kurahashi and Kano 1984). This species plays a role as a mechanical vector in the transmission of gastroenteric pathogens and has also been associated with cases of human myiasis in Japan (Kamimura and Arakawa 1986; Matsuzaki and Yamazaki 1987; Fan 1992; Chigusa et al. 2005; Miura et al. 2005).

Table 3.3 provides an overview of the geographic distribution of several medically important fly species belonging to the families Calliphoridae, Muscidae, and

Table 3.3 Geographical distribution of medically important fly species from families Calliphoridae, Muscidae, and Sarcophagidae in Asia

Species	Geographical zone	JP	KR	CN	TW	IN	NP	LK	BD	PK	MY	TH	ID	PH	CA	LA	VN	MM
A. grahami	Subtropical—temperate	•	•	•	•	–	–	–	–	•	–	–	–	–	–	–	•	–
C. nigribarbis	Temperate	•	•	•	•	–	–	–	–	–	–	–	–	–	–	–	–	–
C. vicina	Cosmopolitan (temperate)	•	•	•	–	•	•	–	–	•	–	•	–	–	–	–	•	–
C. vomitoria	Subtropical—temperate	•	•	•	•	•	•	–	–	•	–	•	–	•	–	–	–	–
C. albiceps	Temperate	•	–	–	–	•	–	–	–	•	–	–	–	•	–	–	–	–
C. bezziana	Tropical—subtropical	–	–	•	•	•	–	•	•	•	•	•	•	•	•	•	•	•
C. megacephala	Cosmopolitan (worldwide)	•	•	•	•	•	•	•	•	•	•	•	•	•	•	•	•	•
C. pinguis	Tropical—subtropical	•	•	•	•	•	–	•	–	–	•	•	•	•	–	•	•	–
C. rufifacies	Cosmopolitan (tropical—temperate)	•	•	•	•	•	•	•	•	•	•	•	•	•	•	•	•	•
H. ligurriens	Tropical—subtropical	•	•	•	•	•	•	•	•	•	•	•	•	•	–	•	•	–
L. caesar	Temperate	•	•	•	–	–	–	–	–	–	–	–	–	–	–	–	–	–
L. cuprina	Cosmopolitan (warmer zone)	•	•	•	•	•	•	–	•	•	•	•	•	•	•	•	•	•
L. illustris	Temperate	•	•	•	–	•	–	–	–	–	–	–	–	–	–	–	–	•
L. porphyrina	Tropical—subtropical	•	•	•	•	•	•	•	•	•	•	•	•	•	–	•	•	–

Species	Geographical zone	JP	KR	CN	TW	IN	NP	LK	BD	PK	MY	TH	ID	PH	CA	LA	VN	MM
L. sericata	Subtropical—temperate	•	•	•	•	•	–	•	–	•	–	–	–	•	–	–	•	•
M. domestica	Cosmopolitan (worldwide)	•	•	•	•	•	•	•	•	•	•	•	•	•	•	•	•	•
M. conducens	Tropical—temperate	•	•	•	•	•	•	•	•	•	•	•	•	•	–	–	•	•
M. sorbens	Tropical—temperate	•	•	•	•	•	–	•	•	–	•	•	•	–	–	–	–	–
H. spinigera	Tropical—subtropical	•	•	•	•	•	–	•	–	–	•	•	•	•	–	–	•	•
S. calcitrans	Cosmopolitan	•	•	•	•	•	–	–	–	–	•	•	•	–	–	–	–	–
S. nudiseta	Tropical—subtropical	•	–	•	•	•	–	•	–	–	•	•	•	–	–	–	–	–
B. peregrina	Cosmopolitan (worldwide)	•	•	•	•	•	•	•	•	–	•	•	•	•	•	–	•	•
L. ruficornis	Cosmopolitan (warmer zone)	•	•	•	•	•	•	•	•	•	•	•	•	•	•	–	–	•
L. crassipalpis	Subtropical—temperate	•	•	•	•	–	–	–	–	–	–	–	–	–	–	–	–	–
P. albiceps	Tropical—subtropical	•	•	•	•	•	•	–	–	•	•	•	•	•	–	–	•	–
P. dux	Tropical—subtropical	•	•	•	•	•	•	•	•	•	•	•	•	•	•	–	–	–
S. princeps	Tropical—subtropical	–	–	•	•	•	•	•	–	•	•	•	•	–	•	•	•	•

JP Japan, *KR* Korea, *CN* China, *TW* Taiwan, *IN* India, *NP* Nepal, *LK* Sri Lanka, *BD* Bangladesh, *PK* Pakistan, *MY* Malaysia, *TH* Thailand, *ID* Indonesia, *PH* Philippines, *CA* Cambodia, *LA* Laos, *VN* Vietnam, *MM* Myanmar

• Presence, – Not presence

Sarcophagidae within the Asian region. It helps to identify regions where these flies are of particular concern due to their medical, ecological, or agricultural importance. This information can be used for disease surveillance, pest control strategies, and ecological studies.

3.5 Ecology

Numerous fly species within the families Calliphoridae, Muscidae, and Sarcophagidae exhibit beneficial ecological roles, despite their typical association with food waste, animal excrement, garbage, and carrion, which categorizes them as filth flies. They play important roles in the ecosystem as decomposers, predators, parasitoids, pollinators, and prey, interacting with other organisms in the same environment. In particular, various calliphorid, muscid and sarcophagid flies are instrumental in the decomposition of vertebrate carcasses, including human cadavers. Female flies are attracted to the odor of decaying materials, which stimulates them to oviposit their eggs or larvae on the substrates. The larvae actively participate in the decomposition of decaying tissues while concurrently retaining essential nutrients. Because of their abundance, they play a key role in the nutrient recycling processes within the ecosystem. The absence of these flies impedes the efficient recycling of remains and results in the ecological accumulation of waste. This behavior underscores their paramount importance in forensic entomology, particularly concerning the estimation of postmortem intervals during investigative procedures.

3.5.1 Pollinator

Flies exhibit diverse feeding behaviors that include not only carrion consumption but also nectar ingestion for nutrition, with male flies particularly engaged in this behavior. In certain regions, such as Kyushu and Okinawa, *C. megacephala* plays an important role as a valuable mango pollinator within greenhouses (Alqarni et al. 2017). The decline in mango pollinator populations due to excessive and inappropriate use of pesticides to control mango pests has inadvertently led to the elimination of non-target pollinators, including honeybees (*Apis* spp.) and fruit flies, *Bactrocera dorsalis* (Hendel). As a result, *Chrysomya* species, especially *C. megacephala*, has assumed a critical role as mango pollinators in Japan, Taiwan, Australia, Malaysia, and the Philippines (Anderson et al. 1982; Hu et al. 1995; Fajardo et al. 2008; Sung et al. 2006; Huda et al. 2015).

Historically, honeybees have played a dominant role as the primary pollinators for mango trees. However, the attractiveness of mango flowers to honeybees is relatively lower than that of other flowering plants. To address this problem, extensive breeding of *C. megacephala* has been carried out on farms, carefully synchronizing their emergence with the flowering period of mango trees. The large population of these flies that forage for nectar from mango flowers makes them remarkably effective as pollinators for mango trees. In addition, *C. rufifacies* is also employed as a mango pollinator (Alqarni et al. 2017). In Taiwan, the neighboring country to Japan, *C. megacephala*, *C. pinguis*, and *M. domestica* are adopted as mango pollinators (Sung et al. 2006). In India, Sharma et al. (1998) reported an observed increase in pollination within a mango orchard attributed to *Lucilia* sp. and *Sarcophaga* sp. Similarly, a study conducted in Pakistan showed that *Calliphora* species significantly improved both the quality and quantity of mango fruit (Saeed et al. 2016).

In Japan, the blow fly species *L. sericata* has emerged as a substitute pollinator for honeybees in crops such as strawberries, melons, tomatoes, egg plants and other fruits (Hanada et al. 2016). Due to its smaller size, *L. sericata* effectively transfers pollen from flower to flower. The Japan Maggot Company, a medical venture company affiliated with Okayama University's School of Medicine, has named this species "Bee Fly" (green bottles) and taken on the responsibility of supplying them to farmers across Japan, provided in pupal form (Nara Prefecture Agricultural Research and Development Center 2019). In 2011, the company underwent a transformation from being solely a larval supplier for maggot therapy to expanding its services to include both the medical and agricultural sectors. Similarly, in the United States of America, blue bottle flies are commercially available, which has facilitated their application as pollinators for various vegetable crops, including onions, carrots, radishes, cauliflower, broccoli, lettuce, buckwheat, sunflowers, garlic, peppers, and canola. In addition, these flies have been found to significantly contribute to improved pepper and tomato production within greenhouse environments (Forked Tree Ranch 2018). In Australia, the key role of several calliphorid species, namely, *Calliphora augur* (Fabricius), *C. stygia* (Fabricius), and *L. cuprina*, in the successful pollination of avocado crops has been well documented (The New Zealand Institute for Plant and Food Research Ltd 2017).

In the natural environment, a rich diversity of fly species plays a crucial role in flower pollination. Li et al. (2014) conducted a study in China and found that *C. megacephala*, *C. rufifacies*, *C. nigripes*, *Isomyia isomyia* (Séguy), *Pierretia* sp., and *Hemipyrellia* sp. were among the most frequent visitors to the flowers of *Bridelia stipularis* and *Cleistanthus sumatranus*, contributing significantly to flower pollination. Notably, the pollination of Rafflesiaceae, which has the largest flower in the world, is primarily carried out by carrion flies, particularly those of the family Calliphoridae. Fly species such as *C. chani* Kurahashi, *C. defixa* (Walker), *C. megacephala*, *C. pinguis*, *C. rufifacies*, *C. villeneuvi* Patton, *H. ligurriens* (Wiedemann), *Hypopygiopsis tumrasvini* Kurahashi, *H. violacea* (Macquart), *Lucilia papuensis* Macquart, and *L. porphyrina* (Walker), serve as essential pollinators for *Rafflesia catleyi*, *Rafflesia kerrii*, *Rafflesia patma*, *Rafflesia pricei*, *Rhizanthes zipperlii*, and *Sapria himalayana* (Meijer 1985; Beaman et al. 1988; Bänziger 1991, 1996, 2004; Wee et al. 2018). On the other hand, flies of the families Sarcophagidae and Muscidae are thought to be visitors to *Rafflesia* flowers, and they are not considered direct pollinators. *Boettcherisca peregrina*, *Neomyia claripenni* (Malloch), and *Atherigona* species have been observed visiting *R. catleyi*, *R. kerrii*, *R. patma*, and

Rhizanthes zipperlii (Bänziger 1991, 2004; Hidayati et al. 2000). However, Bänziger and Hansen (1997) found that flesh flies are the exclusive pollinators of *Sapria ram* (Rafflesiaceae) and *Bulbophyllum putidum* (Orchidaceae). In their study, *Parasarcophaga scopariiformis* (Senior-White) was identified as the pollinator of *Bulbophyllum putidum*. A similar observation was reported by van der Niet et al. (2011), where *Satyrium pumilum* (Orchidaceae) selectively attracted flesh flies. Another study by Bänziger and Pape (2004) emphasized the more important role of sarcophagid species compared to calliphorids in *Rafflesia* pollination and their greater efficiency in *Sapria ram* pollination. Species recorded in these pollination interactions include *Boettcherisca krathonmai* Pape & Bänziger, *B. nathani* Lopes, *Lioproctia pattoni* (Lopes), *L. saprianovae* (Pape & Bänziger), *Parasarcophaga albiceps*, *P. yunnanensis* Fan, *Robineauella coei* Rohdendorf, *Rosellea notabilis* (Kano & Lopes), and *Sarcorohdendorfia seniorwhitei* (Ho) (Pape and Bänziger 2000; Bänziger and Pape 2004).

Besides *Rafflesia* and orchids, the pitcher plants *Nepenthes* are also pollinated by flies (Kato 1993; Clarke 2001; Handayani 2017). In Malaysia, flies such as *Nepenthomyia malayana* Kurahashi & Beaver, *Shinonagaella urceolla* (Shinonaga & Beaver), and *Wilhelmina nepenthicola* Schmitz & Villeneuve have been observed around and in the *Nepenthes*. These flies are thought to be the pollinators of *Nepenthes*, as both male and female flies are frequently found around the vicinity of the plants and even within the pitcher. The larvae of these flies inhabit the phytotelmata of *Nepenthes* and are referred to as inquilines. This characteristic distinguishes these species from other calliphorid and sarcophagid flies, whose larvae are typically terrestrial. The larvae of these flies develop in the pitcher and obtain nutrients from the *Nepenthes* plant.

The importance of flies as pollinators in natural ecosystems, particularly carrion flies, is often underestimated and overlooked, mainly due to their primary association with medical and forensic entomology. However, their substantial contributions to agriculture and flowering plant biodiversity cannot be overlooked. Despite its importance, fly pollination remains understudied. The fascinating interaction between flowers and flies deserves attention. Flowers provide food and shelter for flies, using sunlight to provide insects with warmth and specific locations for mating. Nectar serves as a source of nutrition and energy for flies, while pollen provides essential protein for their reproduction. Many fly species that play a role in pollination have yet to be fully identified and understood, creating a significant knowledge gap in this area. Considering the challenges posed by global warming and the marked decline in traditional pollinator species, research into alternative pollinators, such as flies, which are more resilient to higher temperatures, maybe a promising avenue of investigation.

3.5.2 Parasitoid and Kleptoparasite

Carrion flies exhibit notably diverse life histories, encompassing parasitic behaviors such as parasitoids, kleptoparasites, and myiasis, as well as scavenging activities across a wide range of substrates. The scavenging behaviors of flies on various substrates categorize them into coprobiodotic (feed on feces), necrobiodotic (feed on vertebrate cadavers or remains of invertebrates) and amphibiodotic (feed on both feces and cadavers) (Bänziger and Pape 2004).

In ecology, sarcophagid flies are recognized for their parasitic behavior, with particular emphasis on the subfamilies Miltogrammatinae and Sarcophaginae. For instance, *Baranovisca banksi* (Senior-White) (=*Sarcophaga litsingeri*) acts as a parasitoid of *Argiope catenulata* spider eggs (Shinonaga and Barrion 1980); *Blaesoxipha rufipes* (Macquart) preys on grasshoppers (Pape 1994; Miura and Ohgushi 2010; Watanabe and Hasegawa 2014); *Miltogramma angustifrons* (Townsend) functions as a parasitoid of leafcutter bees, *Megachile* spp. (Kurahashi 1970); *Parasarcophaga dux* parasitizes Chinese scorpions, *Mesobuthus martensii* (Shi et al. 2015); *Sarcorohdendorfia megafilosia* Pape, McKillup & McKillup and *S. meiofilosia* Pape, McKillup & McKillup act as parasitoids of marine snails (McKillup et al. 2000; Pape et al. 2000); *Sarcorohdendorfia antilope* (Böttcher) serves as a parasitoid of the moth, *Attacus atlas* (Kano et al. 1967); and *Sarcorohdendorfia seniorwhitei* is a parasitoid of Lepidoptera (Kano et al. 1967).

The genera of Miltogrammatinae, namely, *Amobia*, *Macronychia*, *Metopia*, and *Senotainia*, have gained recognition for their kleptoparasitic tendencies toward bees and solitary wasps (Piwczyński et al. 2017; Mulieri et al. 2021). Kleptoparasitism is a form of parasitism in which one species pilfers food collected and stored by another species. *Metopia*, in particular, exhibits a strong association with solitary aculeate wasps and bees. They actively linger and hover around the nests of Hymenoptera, exploiting their food resources (Pape 1986, 1987). In a study conducted by Schwendinger and Pape (2000), they proposed a hypothesis indicating that *Metopia sinensis* Pape may function as a predator of *Liphistius lahu* spiders from Thailand rather than as a kleptoparasitoid.

3.5.3 Predation and Cannibalism

Dipteran flies also function as predators of various organisms. The Calliphoridae family is notably involved in preying on frog egg masses (Boruah and Das 2020). These flies lay their eggs and larvae on the terrestrial eggs of anurans (Rhacophoridae) and subsequently consume the egg clutches for their development. The fly species implicated in this predation include *Caiusa borneoensis* Rognes, *C. coomani* Séguy, *C. karrakerae* Rognes, *C. kurahashii* Rognes, *C. pooae* Rognes, *C. violacea* Séguy, and various *Lucilia* spp. (Yorke 1983; Lin et al. 2000; Lue and Lin 2000; Karraker 2013; Rognes 2015; Huang and Shiao 2016). Boruah and Das (2020) documented a case of larval predation by Sarcophagidae on Terai tree frogs, *Polypedates taeniatus*. However, they were unable to identify the genus and species of the flies involved. Predation by Sarcophagidae on other eggs, such as turtle and lizard eggs, has been documented (Lopes 1982; Trauth and Mullen 1990; Pape and Arribas 1999), but such behavior has not been observed in the Oriental region.

The hairy maggot species *Chrysomya albiceps* (Wiedemann) and *C. rufifacies* exhibit both predaceous and cannibalistic behaviors at the larval stage (Khole 1979; Wells and Kurahashi 1997; Faria and Godoy 2001; Faria et al. 2004; Shiao and Yeh 2008). These larvae prey on other larval species of smaller sizes that coexist on carrion. Simultaneously, they may display cannibalistic behavior, consuming conspecific individuals when the natural food source becomes depleted. The predaceous behavior of *C. albiceps* and *C. rufifacies* could impact the interpretation of fauna recovered from a cadaver in forensic cases, potentially leading to inaccurate conclusions in investigations where the presence of another larval species is obscured by *C. albiceps* or *C. rufifacies* during predation. In Sarcophagidae, a certain level of cannibalistic tendency has been observed in *L. ruficornis*, but only at the third instar larval stage (Khole 1979).

The Muscidae family encompasses various species within the subfamily Coenosiinae, which have gained recognition for their predatory nature. Adult *Lispe* flies are acknowledged predators of small aquatic insects, targeting both the larval and pupal stages, including nematoceran species such as anopheline and chironomid (Werner and Pont 2006; Xue and Zhang 2014). The larvae of *Lispe* inhabit aquatic and semiaquatic habitats, where they also function as predators (Fan 1992). In their natural habitat, the larvae of *Limnophora* demonstrate predatory behavior toward black fly larvae (family Simuliidae), known vectors for filarioid worms and blood protozoans (Merritt and Wotton 1988; Pont et al. 2011). Furthermore, certain *Atherigona* species are acknowledged as agricultural pests in Southeast Asian countries, causing damage to rice, corn, wheat, and durian crops (Waterhouse 1993).

3.5.4 Waste Decomposer

House flies, particularly *Musca domestica*, function as natural decomposers within the ecosystem. They exhibit coprophagous behavior, utilizing animal manure to complete their life cycle. As a result, house flies are employed for waste "recycling" in animal farming operations (Čičková et al. 2015). Adult house flies are permitted to lay eggs in livestock manure, and the emerging larvae contribute to the decomposition process by consuming organic matter. These resulting larvae serve as a valuable and nutrient-rich food source for chickens, shrimp, and fish; alternatively, they can be transformed into organic fertilizer. Scientific studies have substantiated that house fly larvae represent a highly nutritious protein source conducive to the growth of chickens (Hwangbo et al. 2009). Furthermore, incorporating larval meals into chicken diets has been shown to reduce the reliance on and cost of commercial feeds (Dillak et al. 2019). House flies are effectively harnessed in the management of chicken, pig, and cattle manure, effectively mitigating waste accumulation (Hwangbo et al. 2009; Čičková et al. 2012; Wang et al. 2013; Zhu et al. 2015). Larvae play an important role in the biodegradation of manure, which subsequently undergoes conversion into protein-rich animal feeds and organic fertilizers. In Japan, the enterprise MUSCA Inc. has been at the forefront of sustainable insect technology, spearheading the development of a 100% biomass recycling system that utilizes *Musca domestica* to convert manure and waste into organic animal feed and plant fertilizer (Musca Solution 2019).

3.6 Mechanical Transmission of Pathogens by Flies—Focus on the Highly Pathogenic Avian Influenza (HPAI) and Blow Flies in Japan

Flies are medically important insects for public health and are known to carry more than 130 pathogens, including gastrointestinal pathogenic bacteria, protozoa, and helminths (Khamesipour et al. 2018). The method of transmission is not biological, as is often the case with blood-sucking insects such as mosquitoes and ticks, but involves the mechanical transmission of the pathogens that are taken up. In developing countries, where latrines are not widely available, dysentery bacilli are often detected in fecal-feeding house flies (Qureshi et al. 2014; George et al. 2015). In Japan, a nationwide outbreak of food poisoning caused by enterohemorrhagic *Escherichia coli* O157 (hereafter referred to as O157) occurred in 1996. In a nationwide survey conducted afterward, O157 was detected and isolated from collected house flies, strongly suggesting the role played by house flies in the outbreak (Iwasa et al. 1999; Kobayashi et al. 1999).

In December 2003, an outbreak at a poultry farm in Ato Cho, Yamaguchi Prefecture, triggered Japan's first highly pathogenic avian influenza (HPAI) outbreak in 79 years (Infectious Diseases Surveillance Center (IDSC) 2004). By March 2004, a small number of bantam chickens had died in Kokonoe Town, Oita Prefecture, and a large number of layer chickens had died in Tamba Town, Kyoto Prefecture, causing major changes in the livestock industry and consumer society. The HPAI outbreaks have occurred almost annually since 2003 in Japan. In particular, the 2022–2023 season recorded the highest number of outbreaks in history.

This H5N1 subtype of the influenza virus is often endemic to waterfowl and poultry in Asian countries, causing significant economic losses. Most avian influenza viruses are not infectious to humans; however, some subtypes, such as the H5N1 subtype, are directly infectious and cause fatal symptoms, as was the case in Hong Kong in 1997 (six deaths) (Subbarao et al. 1998; Yuen et al. 1998). Since 2003, the H5N1 subtype has caused 865 cases across 17 countries, including 455 deaths. Since 2013, the H7N9 subtype has been reported in 1568 cases, mainly in China, with at least 616 confirmed deaths. Because the HPAI virus causes high mortality in avian species, Japan has repeatedly culled all poultry on outbreak farms whenever there has been an outbreak of HPAI. In this chapter, we focus on HPAI infection in flies.

3.6.1 *The 2004 HPAI Outbreak in Kyoto Prefecture*

3.6.1.1 Blow Fly Collection in the HPAI Outbreak Area

In March 2004, an outbreak of HPAI occurred at a farm in Tamba Town, Kyoto Prefecture, where many chickens were infected with the H5N1 subtype virus and died. The virus is known to be shed in large quantities in chicken secretions and feces, and it was estimated that virus production in the entire poultry house was enormous; more than 600 Self-Defense Forces (SDF) personnel were mobilized, and 225,000 chickens were buried in an adjacent vacant lot after being killed over approximately four days. A second poultry farm, located approximately 4 km from the outbreak farm, was confirmed to be infected with the H5N1 subtype. Although spring in Japan is cold, some fly species are present. However, no studies have reported flies' possible role in transmitting the H5N1 influenza virus. Therefore, an entomological survey was conducted in March 2004 to investigate the possibility of transmitting H5N1 HPAI viruses using flies collected from an infected poultry farm in Tamba Town for virus detection and isolation.

Blow flies were collected on March 10–11, 2004, at six locations within a range of 500 m to 2 km around the first outbreak (farm A) (Sawabe et al. 2006). A sunny place protected from strong winds was selected, and rotten fish bait was placed on the ground. A total of 926 flies were collected within a 2.3 km radius of poultry farm A in Tamba Town (Fig. 3.2), representing eight fly species, with >80% of the flies identified as either *C. nigribarbis* or *A. grahami* (Fig. 3.3).

3.6.1.2 Detection of HPAI Virus from Field Captured Flies

Influenza A virus matrix protein (M) and hemagglutinin (HA) genes were detected in the intestinal organs, crops, and guts of *C. nigribarbis* and *A. grahami* using reverse transcription-polymerase chain reaction (RT-PCR). The prevalence of the H5 subtype virus (20–30%) was higher in flies of both species collected 600–700 m from poultry farm A and lower (10%) in flies collected >2 km from poultry farm A. We found that nearly 5% of *C. nigribarbis* collected around the affected areas contained infectious H5N1 viruses. Using RT-PCR with HA and M gene primers, 44 of the 180 (24.4%) blow flies examined were identified as being virus gene positive. Influenza viruses were isolated from embryonated chicken eggs in the intestinal organs of two of the 10 blow flies (20%). The viral M, HA, and NA genes were amplified by PCR with universal primers and full-length sequences of 991, 1707, and 1362 bp, respectively. These sequences were highly similar to those of strains isolated from chickens (A/chicken/Kyoto/3/2004) and crows (A/crows/

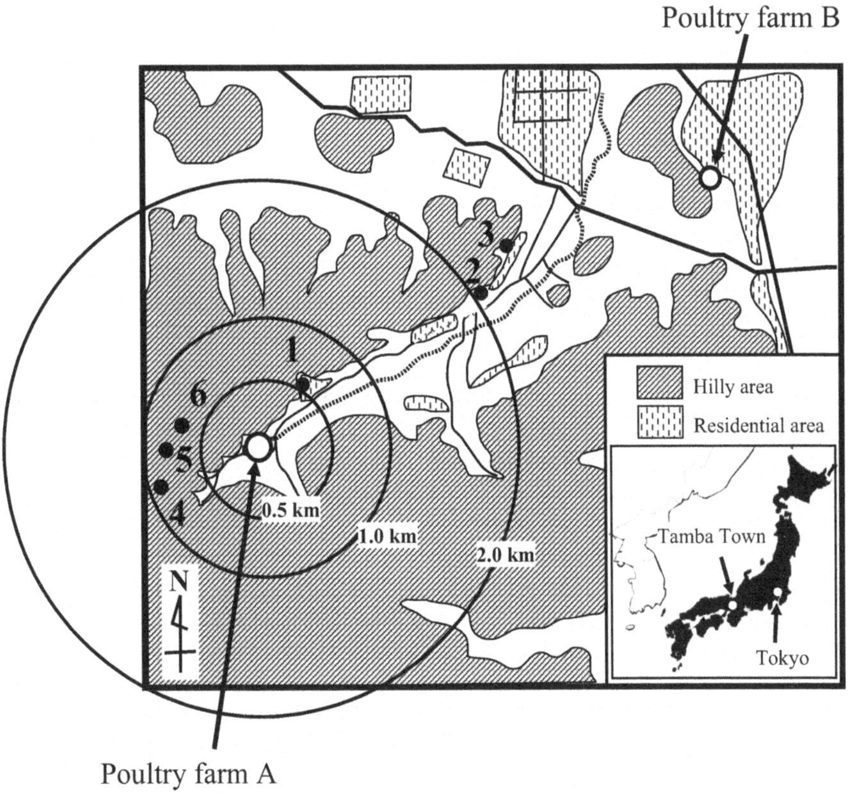

Fig. 3.2 Localities in Tamba Town from which flies used in this study were obtained. Adapted from Sawabe et al. (2006)

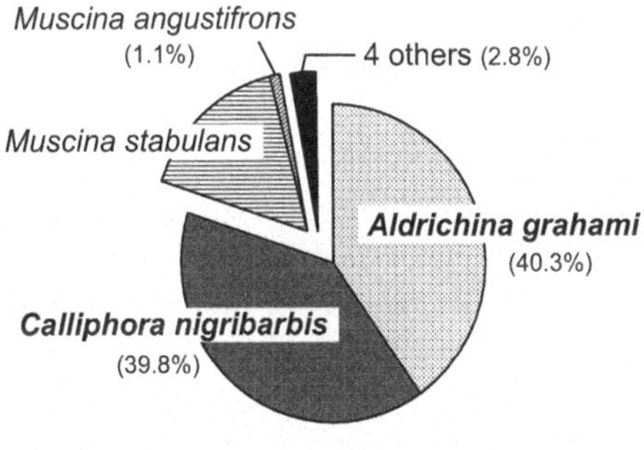

Fig. 3.3 Fly density and location of sample collection sites in Tamba Town, Kyoto, in 2004. Adapted from Sawabe et al. (2006)

Kyoto/53/2004) during the 2004 outbreaks in Kyoto, with >99.9% identity for all three genes. The virus from *C. nigribarbis* (A/blow fly/Kyoto/93/2004) was characterized as an H5N1 subtype of the influenza A virus based on neuraminidase gene (NA) sequences. In addition, the HA1-HA2 connecting peptide sequence in the HA gene segment was RERRRKKR↓G. Finally, the virus isolated from *C. nigribarbis* was characterized as a highly pathogenic H5N1 subtype of the influenza A virus (Sawabe et al. 2009).

3.6.1.3 Experimental Infection of Blow Flies with the H5N1 Virus

To investigate whether the H5N1 virus could survive in the blow fly *C. nigribarbis*, we monitored the titer of the infectious virus in flies after they were exposed to H5N1 avian influenza A virus (A/duck/Hyogo/35/2001). Fifty female blow flies (Kyoto strain *C. nigribarbis*), approximately 14 days old, were placed into a 20 cm^3 fly cage for 3 h at 20 °C with a piece of cotton impregnated with 10^8 EID_{50}/ml allantoic fluid from H5N1 virus (A/duck/Hyogo/35/2001)-infected eggs diluted with MEM diluents. After 3 h of virus exposure, the blow flies were individually reared at 20 °C or 10 °C until testing. Incubation at 10 °C was chosen because the average temperature around Tamba Town was 3.6 °C in February and 6.4 °C in March, with average daytime highs of 11.1 °C in February and 13.1 °C in March (Japan Meteorological Agency (JMA) 2004).

Crops and intestines dissected from flies at various time points after viral exposure were used for virus isolation and titration. The virus was isolated from fly crops and intestines up to 24 h post-exposure and from the feces and vomit matter of one of three blow flies at 48 h post-exposure (Table 3.4). At 14 days post-exposure, no virus was isolated from any blow flies at 20 °C or 10 °C. The H5N1 viral gene was detected in blow flies up to 14 d post-exposure, although no viable virus was detected 48 h post-exposure. Almost all virus-positive samples had a titer of infectious H5N1 virus ranging from 0.5 to 4.63 $TCID_{50}$, but 10 virus-positive samples had no detectable titer (<0.50 $TCID_{50}$). The viable titer of the H5N1 virus in the cotton placed in the fly cages up to 48 h post-exposure ranged from 4.50 to 5.00 $TCID_{50}$ (Sawabe et al. 2009).

3.6.1.4 Blow Flies Are a Possible Transporter of Avian Influenza Virus

Two species of the blow fly, *C. nigribarbis* and *A. graham*, are categorized as larger-sized fly species (Fig. 3.4a, b), particularly in comparison to the house fly *M. domestica*, which has a body size of 5–8 mm (Fig. 3.4c). The body length of female *C. nigribarbis* was 11–15 mm, approximately 1.5 times larger than that of female *A. grahami* (8–13 mm). The crop capacity of female *C. nigribarbis* (average = 23 ml) is approximately five times greater than that of female *A. grahami* (average = 4.4 ml) (Sawabe et al. 2006). The consumption rates of both *C. nigribarbis* and *A. grahami* may have been high because of their large body sizes. Viral genes are found more often in *C. nigribarbis* than in *A. grahami* (Sawabe et al. 2006). Stable flies, *Muscina stabulans* (Falle'n) and *M. angustifrons* (Loew), collected at the same collection sites and at the same time as the fly surveillance in Kyoto, showed much smaller body sizes than *C. nigribarbis* and *A. grahami*, and no virus was detected in these smaller flies.

Table 3.4 Diagnostic results of H5N1 influenza A viruses from crop, intestine, and feces and vomit matters of a blow fly after experimental exposure to the virus solution

		Crop			Intestine			Feces and vomit matters			Virus-exposed cotton		
Postexposure	Replicate	VI	$TCID_{50}$	PCRs	VI	$TCID_{50}$	PCRs	VI	$TCID_{50}$	PCRs	VI	$TCID_{50}$[a]	PCRs
20 °C 3 h	1	+	3.50	+	+	3.50	+				NT	5.00	+
	2	+	3.33	+	+	4.00	+						
	3	+	1.83	+	+	0.50	+						
	4	+	4.60	+	+	3.50	+						
	5	+	2.60	+	+	3.00	+						
	6	+	1.67	+	+	1.67	+						
	7	+	<0.50	+	+	< 0.50	+						
	8	+	4.63	+	+	3.50	+						
	9	+	3.00	+	+	3.63	+						
	10	+	<0.50	+	+	< 0.50	+						
6 h	1	+	<0.50	+	+	3.50	+	-	NT	+	NT	4.50	+
	2	+	<0.50	+	+	1.50	+	-	NT	+			
	3	-	<0.50	+	-	<0.50	+	-	NT	+			
9 h	1	+	1.00	+	+	1.00	+	-	NT	+	NT	5.60	+
	2	+	<0.50	+	+	<0.50	+	-	NT	+			
	3	-	<0.50	+	-	<0.50	+	-	NT	+			
24 h	1	-	<0.50	+	-	<0.50	+	-	NT	+	NT	4.67	+
	2	-	<0.50	+	-	<0.50	+	-	NT	+			
	3	+	1.67	+	+	<0.50	+	-	NT	+			
48 h	1	-	NT	+	-	NT	+	+	<0.50	+	NT	4.83	+
	2	-	NT	+	-	NT	+	-	NT	+			
	3	-	NT	+	-	NT	+	-	NT	+			
14 d	1	-	NT	+	-	NT	+	-	NT	+	NT	<0.50	+

(continued)

Table 3.4 (continued)

Postexposure	Replicate	Crop			Intestine			Feces and vomit matters			Virus-exposed cotton		
		VI	$TCID_{50}$	PCRs	VI	$TCID_{50}$	PCRs	VI	$TCID_{50}$	PCRs	VI	$TCID_{50}$[a]	PCRs
	2	-	NT	+	-	NT	+	-	NT	+			
	3	-	NT	+	-	NT	+	-	NT	+			
10 °C 14 d	1	-	NT	+	-	NT	+	-	NT	+	NT	<0.50	+
	2	-	NT	+	-	NT	+	-	NT	+			
	3	-	NT	+	-	NT	+	-	NT	+			

VI virus isolation using embryonated chicken eggs, $TCID_{50}$ virus titers ($\log_{10}$ $TCID_{50}$/0.05 ml) were calculated by the inoculation onto MOCK cells, *PCRs* RT-PCR performed with specific primers for the HA and M genes and followed by a nested-PCR with primers for the HA gene, *NT* not tested

Adapted from Sawabe et al. (2009)

[a]The average of two times of the virus titration

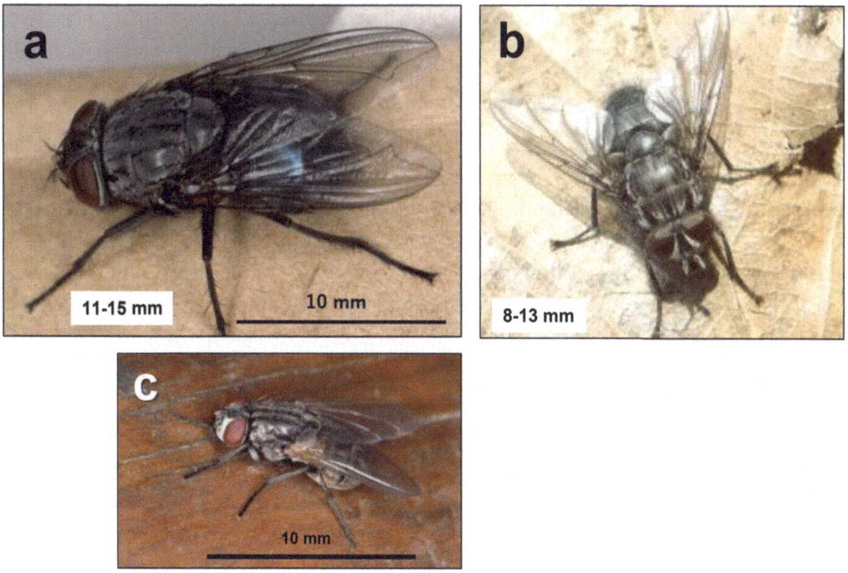

Fig. 3.4 Comparison of body size of three fly species. (**a**) *Calliphola nigribarbis*, (**b**) *Aldrichina grahami*, and (**c**) *Musca domestica*

Blow flies prefer to lick animal carcasses and droppings. If food for blow flies is contaminated with pathogens, the flies may ingest a significant number of pathogens. One possible mechanism for the mechanical transmission of pathogens by blow flies is via regurgitation and feces from food sources (Greenberg 1973; Crosskey and Lane 1993). The effectiveness of mechanical transmission through regurgitation may depend on the viability and titer of pathogens in the fly's body. Accumulated droppings at poultry farms are good breeding sites for blow flies. If flies reproduce at a poultry farm, they have many opportunities for contact with viruses in the feces of infected chickens and/or their dead bodies.

Calliphora nigribarbis has a characteristic temperate zone life cycle. For example, in Japan, they migrate and reproduce more actively during the winter and spring months (Kurahashi et al. 1994; Kurahashi 1979). It is well known that *C. nigribarbis* has an excellent flight capacity and high dispersal ability. They have been identified by weatherships at stations located in the Pacific Ocean and East China Sea, 300–450 km from Kyushu Island, Japan (Kurahashi 1991). It has also been suggested that the number of flies found in autumn in the Kyushu district increases owing to their transoceanic migration (Kurahashi and Suenaga 1997). Female blow flies can survive for approximately one year in Japan (Kurahashi 1991) compared to the house fly *Musca domestica* L., which has a mean longevity of 34.2 days (Rockstein 1957). The longevity and high dispersal ability of blow flies may also result in the wide dispersion of the viruses.

Mark-release-recapture experiments conducted in Tamba Town in 2005 suggested that *C. nigribarbis* could generally migrate up to 2–3 km in 24 h (Tsuda et al.

2009). The distance between the two poultry farms in Kyoto Prefecture, where two H5N1 virus outbreaks occurred in 2004, is approximately 4 km. In fact, 10% of all *C. nigribarbis* flies collected at a site intermediate between the affected farms expressed H5N1 viral genes (Sawabe et al. 2006). Viable titers of the H5N1 influenza virus, but not viral replication, were detected for up to 24 h in the crops and intestines of virus-exposed *C. nigribarbis* (Sawabe et al. 2009). An infectious virus in blow flies for 24 h could have strong implications for virus dispersion because blow flies, with their excellent flight capacity, could transport the H5N1 virus over significant distances. In addition, the H5N1 virus was isolated from the feces and vomit matter of blow flies 48 h post-exposure, but virus titers in flies at 48 h were lower than those of the virus-containing cotton used in these experiments. This suggests that the viability of the influenza virus decreases steadily in the blow fly crop and intestine, although some infectious viruses remain for longer than 24 h. Therefore, *C. nigribarbis* could transport the H5N1 virus to poultry farms 2–3 km apart.

Based on our experience with these cases, we recommend the investigation of fly movement to monitor infectious disease outbreaks associated with massive outbreaks of flies in the areas affected by the Great East Japan Earthquake that occurred in March 2011 (Hayashi and Sawabe 2016).

3.6.2 The 2022–2023 HPAI Outbreaks in Kagoshima Prefecture and Throughout Japan

The 2022–2023 HPAI outbreak is the most severe outbreak to date. Although 5.8 million poultry were killed in 2004, more than 17.4 million poultry were killed in this season. Damage was also inflicted on laying hens, causing eggs' market price to skyrocket (Ministry of Agriculture, Forestry and Fisheries (MAFF) 2023). The HPAI outbreak not only caused tremendous damage to the poultry industry but also brought about serious problems, such as reduced biodiversity in avian species. HPAI viruses have thus far been isolated from 324 birds of 27 species, including four species, *Falco peregrinus*, *Anser fabalis serrirostris*, *Grus vipio*, and *Grus monacha*, which are listed as endangered species in Japan that have died due to HPAI virus infections (Ministry of the Environment, Government of Japan (MOE) 2023). In addition, *Platalea minor* has also been reported in Taiwan and Hong Kong (Liu et al. 2018; Cano-Alonso et al. 2023). It is feared that the HPAI virus not only infects humans but is also a factor in the decline of biodiversity in avian species. The possibility of avian viruses mutating into human pandemic viruses has been highlighted, as mentioned above. There is a fear that the prolonged and serious nature of the epidemic not only infects humans but also contributes to a lack of attention to animal welfare at killing sites, mental health problems of killing workers, and other factors.

3.6.3 Prevention of HPAI Outbreaks

Unfortunately, no measures were taken with flies in mind when the 2004 avian influenza pandemic broke out. This was because the routes and methods of infection were not fully understood. At the time, the website of the Ministry of Agriculture, Forestry, and Fisheries (MAFF) described the route of entry of the virus as follows: (1) chickens infected with the virus were introduced from overseas; (2) there was general use of equipment, vehicles, and egg cases contaminated with the virus; (3) the virus was brought in through people's clothes, hands, boots, etc.; (4) wild birds were allowed to freely come and go; and (5) in poultry houses and outdoor poultry farms, where wild birds can enter and exit, there was a possibility that infected wild birds may have introduced the virus. Accordingly, each municipality, in cooperation with the Pest Control Association, sprayed slaked lime on vehicles entering and leaving the area and disinfected the protective clothing and boots of workers and employees. In addition, each poultry farm took care to keep the areas around the feed tanks and inside the poultry houses clean and thoroughly disposed of chicken manure. Similar measures are being implemented. However, since flies are carriers of various pathogens and may be involved in serious infectious diseases, it is necessary to implement control measures against avian influenza. However, environmental controls alone are not always sufficient, and insecticide control is often the primary means of control. For example, insect growth regulators (IGRs) have been used to control larvae in poultry litter. In addition, in some poultry houses, it is effective to use bait (food poisoning) or residual insecticide treatments (residual spraying of paint) on walls where flies often perch, which have a residual insecticidal effect.

Although the development of resistance of house flies to various insecticides has long been a serious problem in Japan, it has been reported that house flies are still highly susceptible to pyrethroids such as fenitrothion and permethrin (Komagata et al. 2007). Pyrethroid insecticides were effective in controlling blow flies in the areas affected by the Great East Japan Earthquake that occurred after the Great East Japan Earthquake in March 2011. This has resulted in massive outbreaks of blow flies, which are mainly members of the blow fly family. Although fenitrothion and etofenprox emulsions were less effective in house flies, the density of house flies was reduced by 80–90% in the 24 h after application (Tabaru et al. 2012).

3.6.4 Conclusion

It is suggested that blow flies are likely candidates for the mechanical transmission of the HPAI virus because of their ecological and physiological characteristics. Blow flies have been recognized as important vectors for the mechanical transmission of several serious infectious diseases, such as poxvirus (Docherty et al. 1991), rabbit hemorrhagic disease (Asgari et al. 1989), and paratuberculosis (Fischer et al.

2004). Recently, it has been reported that the H5N1 viral gene was detected in house flies (Sievert et al. 2006) and engorged mosquitoes (Barbazan et al. 2008). We suggest that mechanical transmission by flies may also be involved in outbreaks of infectious diseases other than HPAI. Although there are high densities of a variety of fly species in Southeast Asia during all seasons, their ability to transmit viruses has not yet been evaluated. The prevalence of H5N1 avian influenza remains a public health concern for both birds and humans. Therefore, field and laboratory studies on the mechanical transmission of pathogens by flies are important for controlling H5N1 avian influenza outbreaks, at least in Southeast Asian countries.

Recently, H5N1 virus surveillance conducted in Indonesia suggested that pigs are at risk of infection during outbreaks and that pigs can serve as intermediate hosts in which this avian virus can adapt to mammals (Nidom et al. 2010). They also found evidence of pig-to-pig transmission of this virus without significant influenza-like signs. The transmission mechanism of this virus is becoming increasingly complicated and serious. As introduced in the previous section, blow flies prefer to lick carcasses and droppings of chickens and pigs. Furthermore, we can assume that flies can access pigpens more easily than poultry farms. This finding from Indonesia (Nidom et al. 2010) strongly suggests that it is important to pay attention to pigpen and poultry farms within 2–3 km from where viable H5N1 viruses are transmitted by blow flies. Finally, we inferred the role of blow flies in transporting HPAI viruses from the field (mainly bodies of water) into poultry farms, as shown in Fig. 3.5.

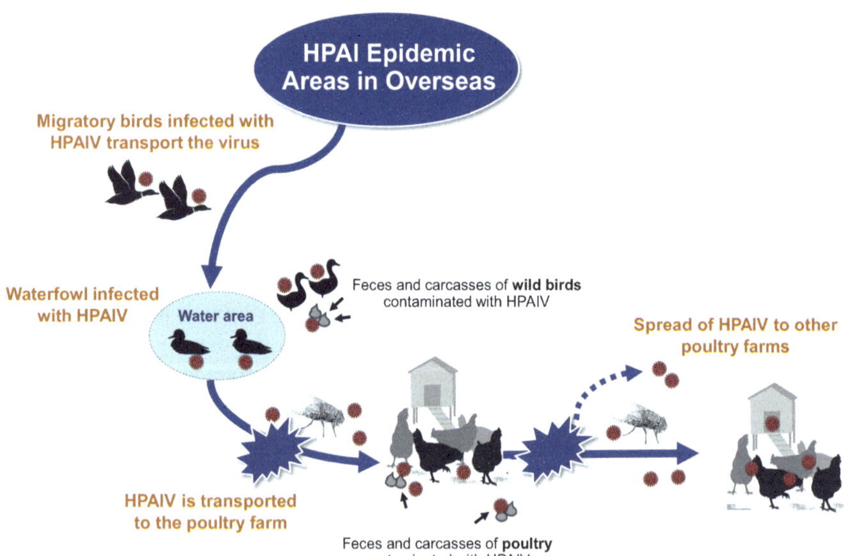

Fig. 3.5 Estimation of the route of transport of the HPAI virus from the field to the poultry farms, and the role of the blow fly. *HPAIV* highly pathogenic influenza virus

3.7 Situation of House Flies in Asia

Synanthropic flies are a species that demonstrate a pronounced affinity for human habitats and have adapted to human activities. As potential pathogens carriers, they can transmit pathogens to humans through their daily interactions. Their diverse dietary preferences include food, waste, feces, blood, wounds, and other fluids found in contaminated environments, which may serve as reservoirs for the pathogens or parasites they ingest. A prominent example of a synanthropic species within the family Muscidae is *Musca domestica.* This species has achieved global domestication status and is known as a cosmopolitan pest, affecting both domesticated animals and humans, raising concerns about hygiene and public health. Increased tension occurs when these flies become overpopulated in certain locations, including landfills, dumping sites, and animal housing facilities such as poultry farms, swine houses, cattle feedlots, and dairies. Waste in these areas creates a favorable habitat for the flies. Adult flies tend to congregate and breed near waste, while larvae feed on the waste material as the source of nourishment.

In the investigation carried out by Junqueira et al. (2017), a comprehensive identification of bacterial taxa was performed, yielding 351 taxa in *M. domestica.* These flies play an important role as mechanical vectors of diseases, contributing to nuisance, irritation, and potential contamination of food and beverages. Notably, the bacteria detected include *Escherichia coli* and *Salmonella* spp., which are representative of the gut flora found in both humans and animals, known to cause food poisoning and gastrointestinal infections. Furthermore, these flies have the potential to act as mechanical vectors for other pathogens, such as *Shigella* spp., Newcastle disease virus, and turkey coronavirus (Levine and Levine 1991; Calibeo-Hayes et al. 2003; Watson et al. 2007; Chaiwong et al. 2014; Lindeberg et al. 2018).

3.7.1 Musca domestica *in Japan*

In Japan, *M. domestica* is hypothesized to play a role in the transmission of *E. coli* O157:H7, which was responsible for an epidemic of enterohemorrhagic colitis during the 1996 outbreak (Kobayashi et al. 1999). Their capacity to enhance the body's survival and multiplication of bacteria elevates public health concerns due to the potential spread of *E. coli* O157:H7 and other pathogens (Sasaki et al. 2000). Moreover, *M. domestica* was found to harbor the highly pathogenic avian influenza virus H5N1 (HPAIV H5N1) on the Island of Java (Tyasasmaya et al. 2016).

The large outbreak of *M. domestica* has elicited considerable attention, particularly in cases where these flies interact with both animals and humans. Such fly outbreaks may also arise following natural disasters, such as earthquakes and tsunamis, where food remnants and carcasses create favorable breeding sites for flies (Srinivasan et al. 2008; Kanno et al. 2016). Over the past few decades, insecticides have been used extensively to manage and control fly populations. Several insecticides have been used, including allethrin, beta-cypermethrin, DDT, dimethoate, deltamethrin, diazinon, dichlorvos, fenitrothion, fenthion, malathion, permethrin, phthalthrin, propetamphos, propoxur, prothifos, pyrethroids, resmethrin, and tetramethrin (Yasutomi 1975; Yasutomi and Shudo 1978; Yasutomi et al. 1988; Bong and Zairi 2010; Wang et al. 2019). Initially, this approach demonstrated efficacy and effectiveness. However, due to the extensive reliance on insecticides, the development of insecticide resistance in *M. domestica* has rendered this strategy ineffective.

In Japan, using insecticide sprays to control the house fly populations in landfills has led to the emergence of insecticide resistance in *M. domestica*. Extensive research on insecticide resistance has been conducted since 1966 at various landfills, including Tokyo Bay (the first Yumenoshima), Landfill No. 15 (No. 2 Yumenoshima), Central Breakwater Landfill (No. 3 Yumenoshima), and the Hokko Landfill in Osaka Bay (Yasutomi 1973, 1975; Yasutomi and Shudo 1978; Imai 1984; Yasutomi et al. 1988). Furthermore, instances of pyrethroid and permethrin resistance in *M. domestica* have been documented in hog farms in the Mashiko area and a poultry house in Tôjô Chô, Hiba-gun, Hiroshima (Motoyama 1984; Yasutomi and Tomioka 2000).

3.7.2 **Musca domestica** *in Malaysia*

In the Cameron Highlands of Malaysia, the persistent prevalence of house flies, *M. domestica*, has raised ongoing concerns (Wharton et al. 1962; Hooi et al. 1970; Singh 1973b; Barrow et al. 2009). Substantial populations of these flies have been consistently observed in various farms and plantations across the region. Their primary breeding grounds are associated with organic compost and fertilizers used in the cultivation of vegetables and flowers on terraced farms. Over time, efforts to address this issue have yielded limited improvement.

Numerous investigations have substantiated the emergence and prevalence of insecticide-resistant strains within the population of *M. domestica* in the Cameron Highlands (Wharton et al. 1962; Singh 1973a; Yoke and Sudderuddin 1975). The *M. domestica* population in the highlands has displayed heightened resistance to insecticides compared to their lowland counterparts (Wharton et al. 1962). This was further corroborated by Hayashi et al. (1978), demonstrating that the *M. domestica* strain originating from the Cameron Highlands exhibited the most pronounced insecticide resistance relative to strains from Ampang Village, Kuala Lumpur, Port Klang, and Tapah. The insecticide resistance of *M. domestica* in the Cameron Highlands could be exacerbated by the application of insecticides intended for the management of other agricultural pests, such as the diamondback or cabbage moth (*Plutella xylostella*) and the black cutworm (*Agrotis ipsilon*) (Mazlan and Mumford 2005).

3.7.3 **Musca domestica** *and Its Impact on Animal Farming in Asia*

The phenomenon of fly nuisance in animal farming areas constitutes a pervasive and significant challenge of global proportions. This issue extends across multiple sectors of animal agriculture, including poultry, cattle, sheep, and other livestock enterprises. The presence of flies in the proximity of animal farming operations presents a complex challenge with profound implications for animal health, welfare, economic productivity, and environmental integrity.

Poultry farming in India poses a significant house fly problem, both in terms of public nuisance and health risks to the surrounding population. The abundance of poultry manure, spilled feed, broken eggs, and carcasses provide favorable breeding grounds for fly populations on the farms. The escalation of fly infestations and their impact on public health has become a major concern, not only locally but also globally, wherever poultry farming is established. The house fly menace has emerged as a noteworthy challenge within the vicinity, leading to disturbances and adversities for nearby residents (Dogra and Aggarwal 2010; Ponnudurai et al. 2015). A study conducted by Dogra and Aggarwal (2010) in Haryana, northern India, demonstrated a positive correlation between high fly density in poultry farms and increased infectious morbidity. To effectively address this problem, the enforcement of laws and regulations governing poultry management practices and insecticide use needs to be strengthened. Proper poultry manure management has emerged as a key strategy to prevent fly infestation. The implementation of dry manure management, which makes it unattractive for flies to breed, has proven to be an effective approach to fly population control. Emphasis on sustainable fly control management, such as "recycling" poultry manure and waste, should be encouraged by local authorities and government agencies. Both environmental and economic benefits can be achieved by converting poultry manure and waste into value-added products such as animal feed, organic fertilizer, and biomass energy (Seal et al. 2015; Ayub et al. 2022).

3.7.4 Development of Insecticide Resistance

In Malaysia, a comprehensive investigation was conducted by Ong et al. (2016) to evaluate the diverse levels of resistance and susceptibility to insecticides observed in *M. domestica* populations from a poultry farm located in Ayer Tawar, Perak. Their findings revealed a state of moderate resistance in the field strain of house flies to the insecticide lambda-cyhalothrin. Likewise, numerous investigations in China have revealed the presence of insecticide resistance in field strains of *M. domestica* (Huang et al. 2015; Li et al. 2018; Wang et al. 2019). In China, the Centers for Disease Control and Prevention (CDC) is responsible for the annual monitoring of insecticide resistance in house flies.

The development of insecticide resistance in insects is closely linked to mutations in the knockdown resistance (*kdr*) allele, which is located in the voltage-sensitive sodium channel (VSSC) gene on chromosome III. These mutations result in reduced nervous system sensitivity to insecticides, including DDT and pyrethroids. Extensive research efforts have been directed at identifying and correlating these mutations with resistance. Globally, numerous point mutations have been documented in *M. domestica*, including *kdr* (L1014F), *kdr-his* (L1014H), *superkdr* (M918T + L1014F), type N (D600N + M918T + L1014F), and 1B (T929I + L1014F) (Williamson et al. 1993; Rinkevich et al. 2012; Sun et al. 2017). These distinct *kdr* mutations exhibit diverse resistance levels and are distributed across varying geographic regions within the house fly population. The L1014F mutation in *M. domestica* has been shown to be significantly correlated with pyrethroid resistance in North America, Europe, and Asia (Roca-Acevedo et al. 2023). Among the detected alleles, *kdr-his* (L1014H) showed the highest level of pyrethroid resistance, followed by *kdr* (L1014F), type N (D600N + M918T + L1014F), *superkdr* (M918T + L1014F), and 1B (T929I + L1014F) (Kasai et al. 2017). The emergence of insecticide resistance in various insect species has become a serious concern, leading to challenges in controlling pest and vector populations.

3.7.5 Alternative Approaches to Fly Control

Despite the successful development and testing of novel insecticides for fly population control, the exclusive reliance on these chemical agents as a fly management strategy is no longer considered sustainable. Imai (1985) proposed an alternative approach that involves a combination of covering the landfill surface with a moderate amount of soil and daily monitoring of fly density. Polyoxyethylene lauryl ether (PEL) is applied to the landfill surface when the fly density exceeds the predetermined control threshold. If PEL does not adequately control the fly population, resorting to insecticide application is required. PEL is recognized as a surfactant with proven efficacy as a non-insecticidal control measure against house flies (Imai and Takamizawa 1979). These measures are aimed at minimizing the overall use and dosage of insecticides.

The adverse effects resulting from the improper and excessive application of insecticides have led to several environmental consequences, including the development of insecticide resistance, unintended harm to beneficial insects, and ecosystem disruption. As a result, the search for alternative approaches that provide safer control methods has become imperative. One of these strategies is biological control, which offers a more environmentally friendly option to replace the conventional use of insecticides to manage the *M. domestica* populations. Parasitoids belonging to the genera *Sphalangia* and *Muscidifurax* (Hymenoptera: Pteromalidae) can be mass-produced and strategically released into the target population (Capinera 2008b). The encouragement of parasitoid establishment for sustainable management is strongly advocated to achieve effective control of the house fly population.

The sterile insect technique (SIT), in which male insects are irradiated with X-ray or gamma rays to induce sterility before their release into the field, is another alternative control strategy. The target population is suppressed by mating sterilized males with wild females, resulting in the production of non-viable offspring. Two successful cases of employing SIT to eradicate fly populations include the New World screwworm fly, *Cochliomyia hominivorax* (Coquerel) (family Calliphoridae), in North America (Knipling 1960), and the tsetse fly, *Glossina austeni* (family Glossinidae), on the island of Unguja (Vreysen et al. 2000).

References

Ahmad NW, Ismail A, Jeffery J, Ibrahim S, Abdul-Hadi A, Ibrahim MN, Chin HC, Lim LH (2009) Aural myiasis in a neonate in peninsular Malaysia. Parasit Vectors 2:63

Alqarni AS, Ahmed KN, Hannan MA, Ghose GK, Munshi JL (2017) Flies for the pollination of greenhouse mango (*Mangifera indica* L., Anacardiaceae) in the subtropical Iriomote Island, Japan. J Asiat Soc Banglad Sci 43(1):135–141

Anderson DL, Sedgley M, Short JRT et al (1982) Insect pollination of mango in northern Australia *Mangifera indica*, includes *Apis mellifera*. Aust J Agric Res 33(3):541–548

Animal Health Australia (AHA) (2017) Old world screw-worm fly: in a diagnostic manual, 3rd edn. Animal Health Australia, p 84

Asgari S, Hardy JRE, Sinclair RG, Cooke BD (1989) Field evidence for mechanical transmission of rabbit haemorrhagic disease virus (RHDV) by flies (Diptera: Calliphoridae) among wild rabbits in Australia. Virus Res 54(2):123–132

Aubertin D (1933) Revision of the genus *Lucilia* R.-D. (Diptera, Calliphoridae). Zool J Linnean Soc 38(260):389–463

Ayub Y, Mehmood A, Ren J, Lee CKM (2022) Sustainable recycling of poultry litter to value-added products in developing countries of South Asia. J Clean Prod 357:132029

Badenhorst R, Villet MH (2018) The uses of *Chrysomya megacephala* (Fabricius, 1794) (Diptera: Calliphoridae) in forensic entomology. Forensic Sci Res 3(1):2–15

Baharuddin O, Ting SH, Zulkifli I (1995) A case of intestinal myiasis in a baby caused by flesh fly (Diptera: Sarcophagidae). Trop Biomed 12:95

Baldacchino F, Muenworn V, Desquesnes M, Desoli F, Charoenviriyaphap T, Duvallet G (2013) Transmission of pathogens by *Stomoxys* flies (Diptera; Muscidae): a review. Parasite 20:26

Bänziger H (1991) Stench and fragrance: unique pollination lure of Thailand's largest flower, *Rafflesia kerrii* Meijer. Nat Hist Bull Siam Soc 39:19–52

Bänziger H (1996) Ecological, morphological, and taxonomic studies on Thailand's fifth species of Rafflesiaceae: *Rhizanthes zippelii* (Blume) Spach. Nat Hist Bull Siam Soc 43:337–365

Bänziger H (2004) Studies on hitherto unknown fruits and seeds of some Rafflesiaceae, and a method to manually pollinate their flowers for research and conservation. Linz Biol Beitr 36(2):1175–1198

Bänziger H, Hansen B (1997) Unmasking the real identity of *Sapria poilanei* Gagnepain emend., and description of *Sapria ram* sp. n. (Rafflesiaceae). Nat Hist Bull Siam Soc 45:149–170

Bänziger H, Pape T (2004) Flowers, faeces and cadavers: Natural feeding and laying habits of flesh flies in Thailand (Diptera: Sarcophagidae, *Sarcophaga* spp.). J Nat Hist 38(13):1677–1694

Barbazan P, Thitithanyanont A, Misse D, Dubot A, Bosc P, Luangsri N, Gonzalez JP, Kittayapong P (2008) Detection of H5N1 avian influenza virus from mosquitoes collected in an infected poultry farm in Thailand. Vector Borne Zoonotic Dis 8(1):105–109

Barrow CJ, Chang NW, Tarmiji M (2009) Issues and challenges of sustainable agriculture in the Cameron Highlands. Malays J Environ Manage 10(2):89–114

Baumgartner DL (1988) Review of myiasis (Insecta: Diptera: Calliphoridae, Sarcophagidae) of Nearctic wildlife. J Wildl Rehab 7:3–46

Bavani MM, Seyyedzadeh SJ, Radi E, Malekian A, Hoseinpour M, Gholizadeh S (2022) Sterilization of *Lucilia sericata* (Diptera: Calliphoridae) eggs for maggot debridement therapy. J Med Entomol 59(3):1076–1080

Bazaliński D, Kózka M, Karnas M, Więch P (2019) Effectiveness of chronic wound debridement with the use of larvae of *Lucilia sericata*. J Clin Med 8(11):1845

Beaman RS, Decker PJ, Beaman JH (1988) Pollination of Rafflesia (Rafflesiaceae). Am J Bot 75(8):1148–1162

Berenji F, Hosseini-Farash BR, Marvi-Moghadam N (2015) A case of secondary ophthalmomyiasis caused by *Chrysomya bezziana* (Diptera: Calliphoridae). J Arthropod Borne Dis 9(1):125–130

Bernhardt V, Finkelmeier F, Verhoff MA, Amendt J (2019) Myiasis in humans–a global case report evaluation and literature analysis. Parasitol Res 118(2):389–397

Bharti M, Kurahashi H (2010) *Lucilia calviceps* Bezzi, new record from India (Diptera: Calliphoridae), with a revised key to Indian species. Halteres 1(2):29–30

Bhatt AP, Jayakrishnan A (2000) Oral myiasis: a case report. Int J Paediatr Dent 10(1):67–70

Bong LJ, Zairi J (2010) Temporal fluctuations of insecticides resistance in *Musca domestica* Linn (Diptera: Muscidae) in Malaysia. Trop Biomed 27(2):317–325

Boruah B, Das A (2020) Predation on an egg mass of the Terai treefrog, *Polypedates taeniatus* (Anura: Rhacophoridae), by flesh fly larvae (Diptera: Sarcophagidae). IRCF Reptil Amphib 27(2):228–230

Calibeo-Hayes D, Denning SS, Stringham SM, Guy JS, Smith LG, Watson DW (2003) Mechanical transmission of turkey coronavirus by domestic houseflies (*Musca domestica* Linnaeaus). Avian Dis 47(1):149–153

Cano-Alonso LS, Grace MK, Yu YT, Chan S (2023) Reversing the decline in a threatened species: the black-faced spoonbill *Platalea minor*. Diversity 15(2):217

Capinera JL (2008a) Australian sheep blowfly, *Lucilia cuprina* Wiedemann (Diptera: Calliphoridae). In: Capinera JL (ed) Encyclopedia of entomology. Springer, Heidelberg, pp 335–338

Capinera JL (2008b) House fly, *Musca domestica* L. (Diptera: Muscidae). In: Capinera JL (ed) Encyclopedia of entomology. Springer, Heidelberg, pp 1877–1880

Chaiwong T, Srivoramas T, Sukontason K, Sanford MR, Moophayak K, Sukontason KL (2012) Survey of the synanthropic flies associated with human habitations in Ubon Ratchathani province of northeast Thailand. J Parasit Res 2012:613132

Chaiwong T, Srivoramas T, Sueabsamran P, Sukontason K, Sanford MR, Sukontason KL (2014) The blow fly, *Chrysomya megacephala*, and the house fly, *Musca domestica*, as mechanical vectors of pathogenic bacteria in Northeast Thailand. Trop Biomed 31(2):336–346

Chan JCM, Lee JSW, Dai DK, Woo J (2005) Unusual cases of human myiasis due to old world screw-worm fly acquired indoors in Hong Kong. Trans R Soc Trop Med Hyg 99(12):914–918

Chemonges-Nielsen S (2003) *Chrysomya bezziana* in pet dogs in Hong Kong: a potential threat to Australia. Aust Vet J 81(4):202–205

Chigusa Y, Kirinoki M, Yokoi H, Matsuda H, Okada K, Yanadori A, Yamazaki S (1996) Two cases of wound myiasis due to *Lucilia sericata* and *L. illustris* (Diptera: Calliphoridae). Med Entomol Zool 47(1):73–76

Chigusa Y, Kawai S, Kirinoki M, Matsuda H, Morita K (1997) A case of myiasis due to *Sarcophaga melanura* (Diptera: Sarcophagidae) in a patient suffering from pontine infarction. Med Entomol Zool 48(2):141–143

Chigusa Y, Otake H, Matsuda H (1998) Pre-condition to alimentary tract myiasis on a healthy subject with *Sarcophaga ruficornis* (Diptera: Sarcophagidae) in Oriental Mindoro, the Philippines. Med Entomol Zool 49(4):345–349

Chigusa Y, Shinonaga S, Kawaguchi T, Kawaguchi N, Kirinoki M, Matsuda H (2002) Cutaneous myasis caused by *Lucilia sericata* (Diptera: Calliphoridae) on skin cancer of the cheek. Med Entomol Zool 53(2):89–94

Chigusa Y, Kirinoki M, Matsuda H (2005) Nosocomial myiasis due to *Sarcophaga peregrina* in an intensive care unit (ICU) in Japan. Med Entomol Zool 56(4):355–358

Chigusa Y, Kawakami K, Shimada M, Kurahashi H, Matsuda H (2006) Hospital-acquired oral myiasis due to *Boettcherisca septentrionalis* (Diptera: Sarcophagidae) in Shimane Prefecture, Japan. Med Entomol Zool 57(2):139–143

Choe S, Lee D, Park H, Jeon HK, Kim H, Kang JH, Jee CH, Eom KS (2016) Canine wound myiasis caused by *Lucilia sericata* (Diptera: Calliphoridae) in Korea. Korean J Parasitol 54(5):667–671

Čičková H, Pastor B, Kozánek M, Martínez-Sánchez A, Rojo S, Takáč P (2012) Biodegradation of pig manure by the housefly, *Musca domestica*: a viable ecological strategy for pig manure management. PLoS One 7(3):e32798

Čičková H, Newton GL, Lacy RC, Kozánek M (2015) The use of fly larvae for organic waste treatment. Waste Manag 35(2015):68–80

Clarke C (2001) Nepenthes of Sumatra and Peninsular Malaysia. Natural History Publications (Borneo), Kota Kinabalu, p 326

Crosskey RW, Lane RP (1993) House-flies, blow-flies and their allies (Calyptrate: Diptera). In: Lane RP, Crosskey RW (eds) Medical insects and arachnids. Chapman & Hall, London, pp 403–428

Dasgupta B (1962) On the myiasis of the Indian toad *Bufo melanostictus*. Parasitology 52(1–2):63–66

Demirel-Kaya F, Orkun Ö, Çakmak A, İnkaya AÇ, Öcal M, Erguven S (2016) A case of extensive wound myiasis caused by *Lucilia sericata* (Diptera: Calliphoridae) in a patient with maxillary sinus squamous cell carcinoma, in Turkey. J Arthropod Borne Dis 10(2):267–270

Dik B, Uslu U, Işik N (2012) Myiasis in animals and human beings in Turkey. Kafkas Univ Vet Fak Derg 18(1):37–42

Dik B, Ilhan C, Ceylan O, Uzunlu EO (2018) The first case of traumatic myiasis caused by *Musca domestica* in a dog in Konya, Turkey. Turk J Vet Anim Sci 42(5):492–495

Dillak SYFG, Suryatni NPF, Handayani HT, Temu ST, Nastiti HP, Osa DB, Ginting R, Yunilas, Henuk YL (2019) The effect of fed maggot meal as a supplement in the commercial diets on the performance of finisher broiler chickens. IOP Conf Ser Earth Environ Sci 260:012056

Docherty DE, Long RI, Flickinger EL, Locke LN (1991) Isolation of poxvirus from debilitating cutaneous lesions on four immature grackles (*Quiscalus* sp.). Avian Dis 35(1):244–247

Dogra SS, Mahajan VK (2010) Oral myiasis caused by *Musca domestica* larvae in a child. Int J Pediatric Otorhinolaryngol Extra 5(3):105–107

Dogra V, Aggarwal AK (2010) Association of poultry farms with housefly and morbidity: a comparative study from Raipur Rani, Haryana. Indian J Community Med 35(4):473–477

Emerson PM, Bailey RL, Mahdi OS, Walraven GE, Lindsay SW (2000) Transmission ecology of the fly *Musca sorbens*, a putative vector of trachoma. Trans R Soc Trop Med Hyg 94(1):28–32

Fajardo AC, Medina JR, Opina OS, Cervancia CR (2008) Insect pollinators and floral visitors of mango. Philippine Agric Sci 91(92):372–382

Fan ZD (1965) Key to the common flies of China, 1st edn. Science Press, Beijing, p 330. (in Chinese)

Fan ZD (1992) Key to the common flies of China, 2nd edn. Academia Sinica, Beijing, p 992. (in Chinese)

Fang F, Chang Q, Sheng Z, Zhang Y, Yin Z, Guillot J (2019) *Chrysomya bezziana*: a case report in a dog from Southern China and review of the Chinese literature. Parasitol Res 118:3237–3240

Faramarzi A, Rasekhi A, Kalantari M, Hatam G (2009) *Chrysomya bezziana* as a causative agent of human myiasis in Fars Province, Southern Iran. Iran J Arthropod Borne Dis 3(1):60–63

Faria LDB, Godoy WA (2001) Prey choice by facultative predator larvae of *Chrysomya albiceps* (Diptera: Calliphoridae). Mem Inst Oswaldo Cruz 96(6):875–878

Faria LDB, Trinca LA, Godoy WAC (2004) Cannibalistic behavior and functional response in *Chrysomya albiceps* (Diptera: Calliphoridae). J Insect Behav 17(2):251–261

Ferraz ACP, Proença B, Gadelha BQ, Faria LM, Barbalho MGM, Aguiar-Coelho VM, Lessa CSS (2010) First record of human myiasis caused by association of the species *Chrysomya*

megacephala (Diptera: Calliphoridae), *Sarcophaga* (*Liopygia*) *ruficornis* (Diptera: Sarcophagidae), and *Musca domestica* (Diptera: Muscidae). J Med Entomol 47(3):487–490
Fischer OA, Matlova L, Dvorska L, Svastova P, Bartl J, Weston RT, Pavlik I (2004) Blowflies *Calliphora vicina* and *Lucilia sericata* as passive vectors of *Mycobacterium avium* subsp. *avium*, *M.a. paratuberculosis* and *M.a. horminissuis*. Med Vet Entomol 18(2):116–122
Forked Tree Ranch (2018) Blue bottle flies pollinators. [Accessed 15 Jan 2023] https://forkedtreeranch.com/blue-bottle-fly-polinators/
Fremdt H, Amendt J (2014) Species composition of forensically important blow flies (Diptera: Calliphoridae) and flesh flies (Diptera: Sarcophagidae) through space and time. Forensic Sci Int 236:1–9
George CM, Ahmed S, Talukder KA, Azmi IJ, Perin J, Sack RB, Sack DA, Stine OC, Oldja L, Shahnaij M, Chakraborty S, Parvin T, Bhuyian SI, Bouwer E, Zhang X, Hasan TN, Luna SJ, Akter F, Faruque ASG (2015) *Shigella* infections in household contacts of pediatric Shigellosis patients in rural Bangladesh. Emerg Infect Dis 21(11):2006–2011
Godhi S, Goyal S, Pandit M (2008) Oral myiasis: a case report. J Maxillofac Oral Surg 7(2):292–293
Gomez-Puerta LA, Cribillero NG, Silva W, Ayala P (2021) Cloacal myiasis by *Lucilia* spp. (Diptera: Calliphoridae) in a rooster (*Gallus gallus domesticus*) and two Harris's hawks (*Parabuteo unicinctus*). Parasitol Int 83:102363
Greenberg B (1973) Flies and disease: Vol. II. Biology and disease transmission. In: Princeton legacy library. Princeton University Press, Princeton, p 464
Grzywacz A, Hall MJR, Pape T, Szpila K (2017) Muscidae (Diptera) of forensic importance — an identification key to third instar larvae of the western Palaearctic region and a catalogue of the muscid carrion community. Int J Legal Med 131:855–866
Gunasegaran N, Seah VQH, Ang SY, Aloweni F, Goh WT, Liew AYJ, Tan WX, Tay HT, Chong TT (2022) Maggot debridement therapy in the tropics – preliminary outcomes from a tertiary hospital. J Tissue Viability 31(3):544–551
Hall MJR, Wall R (1995) Myiasis of humans and domestic animals. Adv Parasitol 35:257–334
Han HS, Sharma R, Jeffery J, Noli C (2017) *Chrysomya bezziana* (Diptera: Calliphoridae) infestation: Case report of three dogs in Malaysia treated with spinosad/milbemycin. Vet Dermatol 28(2):239–e62
Hanada A, Yoshida Y, Sato T, Goto T, Yasuba K, Tanaka Y (2016) Utilization of *Phaenicia sericata* Meig. (green blow fly) as an alternative pollinator to honeybee. Hort Res 15(2):161–169
Handayani T (2017) Flower morphology, floral development and insect visitors to flowers of *Nepenthes mirabilis*. Biodiversitas 18(4):1624–1631
Hashimoto A, Eto H, Takezaki S (1987) Eccrine poroepithelioma with myiasis. Pract Dermatol 9:549–552. (in Japanese)
Hatsushika R, Hyo Y, Okino T (1988) A case study of otomyiasis caused by *Parasarcophaga similis* (Meade, 1876) (Diptera: Sarcophagidae). Kawasaki Med J 14(2):83–89
Hatsushika R, Ushirogawa H, Iwanaga Y (2002) A case study of human otomyiasis caused by *Parasarcophaga similis* (Meade, 1876) (Diptera: Sarcophagidae) found in Hiroshima, Japan. Kawasaki Med J 28(1):33–41
Hayashi T, Sawabe K (2016) Fly emergence surveys in the processing yard for disaster debris discharged by tsunami of 2011 the Great East Japan Earthquake in Iwate Prefecture (2012). Med Entomol Zool 67(2):101–104
Hayashi T, Shinonaga S, Kano R, Hooi C, Singh I (1978) Resistant level of houseflies to six kinds of synthetic insecticides in Malaysia. Bull Tokyo Med Dent Univ 25(2):83–86
Hennig W (1964) Familie 63b: Muscidae. [Part, Lieferung 248]. In: Lindner E (ed) Die Fliegen der Palaearktischen Region. Schweizerbart, pp 961–1054. (in Deutsch)
Hidayati SN, Meijer W, Baskin JM, Walck JL (2000) A contribution to the life history of the rare Indonesian holoparasite *Rafflesia patma* (Rafflesiaceae). Biotropica 32(3):408–414
Hogsette JA, Ruff JP (1985) Stable fly (Diptera: Muscidae) migration in Northwest Florida. Environ Entomol 14(2):170–175

Hooi C, Keiding J, Singh I, Sta M (1970) Susceptibility studies on the housefly, *Musca domestica* L. Southeast Asian J Trop Med Public Health 1:304

Hori K (1953) Morphological studies on muscoid flies of medical importance in Japan, IV. On eleven species of Sarcophaginae (Dipt. Sarcophagidae) from Japan. Sci Rep Kanazawa Univ 2:75–85

Hori K (1954a) Morphological studies on muscoid flies of medical importance in Japan, VI. Descriptions of three new species of the genus *Sarcophaga* (Dipt Sarcophagidae) from Japan. Jpn J Sanit Zool 4(Suppl):296–299

Hori K (1954b) Morphological studies on muscoid flies of medical importance in Japan, VII. Descriptions of six new species of subfamily *Sarcophaginae* (Diptera, Sarcophagidae) from Japan. Sci Rep Kanazawa Univ 2:43–50

Hori K (1955) Taxonomical studies on muscoid flies in Japan, VIII. Description of two new species of genus *Sarcophaga* (Dipt., Sarcophagidae) and *Strongyloneura prasina* Bigot (Dipt., Calliphoridae) from Japan. Sci Rep Kanazawa Univ 3:247–251

Hori K (1966) Taxonomy of house fly. Heredity 20(7):4–7. (in Japanese)

Hu T, Len CH, Lee BS (1995) The laboratory rearing and radiation effects of gamma ray on the pupae of *Chrysomya megacephala* (Fabricius). Chin J Entomol 15(2):103–111. (in Chinese with English abstract)

Huang YZ, Shiao SF (2016) Morphological redescription and population fluctuation of frog flies, *Caiusa violacea* Séguy (Diptera: Calliphoridae), in Taiwan. Formos Entomol 36(2):66–73. (In Chinese with English summary)

Huang ZG, Xian J, Hu J, Ye G, Yang DM, Liu Y, Zeng H (2015) Resistance of *Musca domestica* to commonly used insecticides in China. Chin J Hyg Insect Equip 21(8):306–308. (in Chinese with English summary)

Huda AN, Salmah MR, Hassan AA, Hamdan A, Razak MN (2015) Pollination services of mango flower pollinators. J Insect Sci 15(1):113

Hwangbo J, Hong EC, Jang A, Kang HK, Oh JS, Kim BW, Park BS (2009) Utilization of house fly-maggots, a feed supplement in the production of broiler chickens. J Environ Biol 30(4):609–614

Igari K, Toyofuku T, Uchiyama H, Koizumi S, Yonekura K, Kudo T, Jibiki M, Sugano N, Inoue Y (2013) Maggot debridement therapy for peripheral arterial disease. Ann Vasc Dis 6(2):145–149

Imai C (1984) Development and loss of insecticide resistance in the houseflies, *Musca domestica*, at a waste disposal site in relation to insecticide application. Jpn J Sanit Zool 35(3):225–231

Imai C (1985) A new method to control houseflies, *Musca domestica*, at waste disposal sites. Res Popul Ecol 27:111–123

Imai C, Takamizawa K (1979) Insecticidal effect of polyoxyethyleue lauryl ether on the housefly, *Musca domestica*. Jpn J Sanit Zool 30(3):211–216

Infectious Diseases Surveillance Center (IDSC) (2004) Avian influenza prevailing in Asian and other countries, as of October 2004. IASR 25(11):293–294

Ishii K, Yao T, Kori K, Eguchi Y, Arashiro T, Hirata Y (1988) A case of diabetic gangrene with myiasis. Diabetes J 16:164–166. (in Japanese)

Ito S, Koshimizu D (1955) A case of intestinal myiasis. J Tokyo Women's Cooll 25:133–136. (in Japanese)

Iwasa M, Makino SI, Asakura H, Kobori H, Morimoto Y (1999) Detection of Escherichia coli O157:H7 from Musca domestica (Diptera: Muscidae) at a cattle farm in Japan. J Med Entomol 36(1):108–112

Jang M, Ryu SM, Kwon SC, Ha JO, Kim YH, Kim DH, Jung SM, Lee SI, Sohn WM, Cha HJ, Ock M (2013) A case of oral myiasis caused by *Lucilia sericata* (Diptera: Calliphoridae) in Korea. Korean J Parasitol 51(1):119–123

Japan Meteorological Agency (JMA) (2004) Weather, climate and earthquake information, 2004, Sonobe District data. [Accessed 1 Oct 2023] http://www.data.jma.go.jp/obd/stats/etrn/index.php

Junqueira ACM, Ratan A, Acerbi E, Moses DID, Premkrishnan BNP, Costea PI, Linz B, Purbojati RW, Paulo DF, Gaultier NE, Subramanian P, Hasan NA, Colwell RR, Bork P, Azeredo-Espin

AML, Bryant DA, Schuster SC (2017) The microbiomes of blowflies and houseflies as bacterial transmission reservoirs. Sci Rep 7(1):16324

Kakinuma S, Kurahashi H (2010a) A record of *Angiometopa cicadina* (Diptera, Sarcophagidae) from Yamaguchi Pref., Honshu, Japan. Hana Abu 30:3

Kakinuma S, Kurahashi H (2010b) New records of *Kurahashiodes suenagai* (Diptera, Sarcophagidae) from Kumamoto Pref., Kyushu and Chiba Pref., Honshu, Japan. Hana Abu 30:1

Kakinuma S, Kurahashi H (2012a) A new record of *Myorhina tsintaoensis* (Ye, 1965) (Diptera, Sarcophagidae) from Honshu, Japan. Hana Abu 33:8

Kakinuma S, Kurahashi H (2012b) The first record of *Pterella melanura* Meigen) (Diptera, Sarcophagidae) from Japan. Hana Abu 33:7

Kamimura K, Arakawa R (1986) A case report on nasal myiasis due to the fleshfly (*Boettcherisca peregrina*). Jpn J Sanit Zool 37(2):163–164

Kandi V, Lal SK, Akhila S, Sandhya K, Simar H, Pranuthi M, Kumar MV, Anand K, Rao SD (2013) Persistent pediatric gastro-intestinal myiasis: a case report of fly larval infestation with *Musca domestica* with review of literature. J Global Infect Dis 5(3):114–117

Kaneko K, Amano K, Kubota K, Hosokawa A (1968) A new human otomyiasis caused by larvae of *Parasarcopahag harpax*. Jpn J Sanit Zool 19(4):248–252. (in Japanese)

Kanno K, Kawabata T, Ishikawa Y, Katayama J, Hirao M, Tabaru Y, Watanabe M, Kumon K, Tsuji H, Tanaka K, Saito K, Yoshimura T, Forschler B, Lewis V, Lee CY (2016) Area-wide fly management in the tsunami-affected zones of Tohoku region, Japan. Int Pest Control 58(1):26–33

Kano R, Kurahashi H (1999) Discovery of the female of *Heteronychia nipponensis* Shinonaga & Matsudaira (Diptera, Sarcophagidae). Bull Natn Sci Mus, Tokyo, Ser A 25(1):73–77

Kano R, Kurahashi H (2000) Two new and one newly recorded species of flesh flies from the northern Vietnam (Diptera, Sarcophagidae). Bull Natn Sci Mus Tokyo Ser A 26(2):43–50

Kano R, Shinonaga S (1964) Notes on flies of medical importance in Japan (Part XXII). Descriptions of three new species of sarcophagid flies. Jpn J Sanit Zool 15(4):213–219

Kano R, Kaneko K, Kawashima K, So N (1962) On some cases of myiasis. Med Entomol Zool 13(1):96–97. (in Japanese with English summary)

Kano R, Field G, Shinonaga S (1967) Fauna Japonica – Sarcophagidae (Insecta: Diptera). Biogeograph Soc Japan, pp 12+168, 41pls

Kano R, Thinh TH, Kurahashi H (1999) The flesh flies (Diptera, Sarcophagidae) from the northern part of Vietnam. Bull Natn Sci Mus Tokyo Ser A 25(2):129–141

Karraker NE (2013) Shading mediates the interaction between an amphibian and a predatory fly. Herpetologica 69(3):257–264

Kasai S, Sun H, Scott JG (2017) Diversity of knockdown resistance alleles in a single house fly population facilitates adaptation to pyrethroid insecticides. Insect Mol Biol 26(1):13–24

Kato M (1993) Floral biology of *Nepenthes gracilis* (Nepenthaceae) in Sumatra. Am J Bot 80(8):924–927

Kavitha R, Nazni WA, Tan TC, Lee HL, Azirun MS (2013) Review of forensically important entomological specimens collected from human cadavers in Malaysia (2005–2010). J Forensic Leg Med 20(5):480–482

Khamesipour F, Lnkarani KB, Honarvar B, Kwenti TE (2018) A systematic review of human pathogens carried by the housefly (Musca domestica L.). BMC Public Health 18(1):1049

Khan AR, Huq F (1978) Disease agents carried by flies in Dacca city. Bangladesh Med Res Counc Bull 4(2):86–93

Khan MAJ, Khan RJ (1987) Hematoma of scalp in a baby caused by the common green bottle – *Lucilia sericata* (Meigen) (Diptera: Calliphoridae) in Karachi, Pakistan. Med Entomol Zool 38(2):103–105

Khole V (1979) Some observations on cannibalism in larval blowflies (Calliphoridae: Diptera). Curr Sci 48(8):366–368

Khoso FN, Wong SK, Chia SL, Lau WH (2015) Assessment of non-biting synanthropic flies associated with fresh markets. J Entomol Zool Stud 3(1):13–20

Kim J, Choi H, Han K, Shin D (2011) The comparison between maggot therapy and conventional management in the diabetic foot patients. J Korean Wound Manag Soc 7:14–17

Kim SW, Kang TJ (2008) Clinical application of maggots in chronic difficult wounds. J Korean Soc Plast Reconstr Surg 35(4):400–406

Knipling EF (1960) The eradication of the screw-worm fly. Sci Am 203:54–61

Kobayashi M, Sasaki T, Saito N, Tamura K, Suzuki K, Watanabe H, Agui N (1999) Houseflies: Not simple mechanical vectors of enterohemorrhagic *Escherichia coli* O157:H7. Am J Trop Med Hyg 61(4):625–629

Kobayashi S (1915) Fly larva found in the human intestine. Tokyo Med J 1904:16–18. (in Japanese)

Komagata O, Kasai S, Tsuda Y, Kobayashi M, Tomita T (2007) Insecticide susceptibility of the blow fly, Calliphora nigribarbis Vollenhoven, collected in Yamaguchi Prefecture, Japan. Med Entomol Zool 57(3):205–209

Kumara TK, Disney RHL, Abu Hassan A, Flores M, Tan SH, Mohamed Z, CheSalmah MR, Bhupinder S (2012) Occurrence of oriental flies associated with indoor and outdoor human remains in the tropical climate of north Malaysia. J Vector Ecol 37(1):62–68

Kurahashi H (1964) Studies on the calypterate muscoid flies from Japan I. Revision of the genera *Calliphora*, *Aldrichina* and *Triceratopyga* (Diptera, Calliphoridae). Kontyu 32(2):226–232

Kurahashi H (1966) Studies on the calypterate muscoid flies from Japan V. Revision of the tribe Luciliini (Diptera, Calliphoridae). Sci Rep Kanazawa Univ 11(2):105–124

Kurahashi H (1970) Studies on the calypterate muscoid flies from Japan VII. Revision of the subfamily Miltogramminae (Diptera, Sarcophagidae). Kontyu 38(2):93–116

Kurahashi H (1971) The tribe Calliphorini from Australian and Oriental regions II. *Calliphora*-group (Diptera: Calliphoridae). Pacific Insects 13(1):141–204

Kurahashi H (1972) Studies on the calypterate muscoid flies from Japan IX. Subfamily Macronychiinae (Diptera, Sarcophagidae). Kontyu 40(3):173–180

Kurahashi H (1973) Studies on the calypterate muscoid flies from Japan X. Genus *Synorbitomyia* (Diptera, Sarcophagidae). New Entomol 22(12):17–20

Kurahashi H (1975) Studies on the calypterate muscoid flies from Japan XI. Subfamily Agriinae (Diptera, Sarcophagidae). Kontyu 43(2):202–213

Kurahashi H (1979) Breeding of flies. Insectarium 16:56–61

Kurahashi H (1987a) Blow flies of medical importance in Taiwan (Diptera: Calliphoridae). 1. Tribes Calliphorini and Luciliini. Sieboldia Acta Biol Suppl. 47–59.

Kurahashi H (1987b) New record of *Melinda okazakii* Kano from Okinawa – Honto Island (Diptera: Calliphoridae). Tokkuribachi 53:2–3

Kurahashi H (1988) A new species of *Lucilia* (Diptera, Calliphoridae) from Sulawesi. Kontyu 56(1):144–147

Kurahashi H (1989) Notes on the Bangladesh calliphorid flies of medical importance (Insecta: Diptera). Jpn J Sanit Zool 40(Suppl):97–111

Kurahashi H (1991) The calyptrate muscoid flies collected on weather ships located at the ocean weather stations. Jpn J Sanit Zool 42(1):53–55

Kurahashi H (1994) A new species of *Wohlfahrtiodes* (Diptera, Sarcophagidae) from Kyushu, Japan. Kontyu 62(2):237–241

Kurahashi H (2001) The blow flies recorded from Sri Lanka, with descriptions of two new species (Diptera, Calliphoridae). Jpn J Syst Entomol 7(2):241–254

Kurahashi H (2002) Key to the Sarcophagidae of the Oriental region. In: Greenberg B, Kunich JC (eds) Entomology and the law – flies as forensic indicators. Cambridge University Press, Cambridge, pp 138–142

Kurahashi H (2004) Studies on the calypterate muscoid flies from Japan XII. One new and two newly recorded species of sarcophagid flies (Diptera: Sarcophagidae). Med Entomol Zool 55(4):295–301

Kurahashi H (2005) Studies on the calypterate muscoid flies from Japan XIII. Redescription of *Myorhina asahinai* (Kano)(Diptera: Sarcophagidae). Med Entomol Zool 56(2):73–78

Kurahashi H (2010a) A list of Japanese blow flies (Diptera: Calliphoridae). Proc Res Gro Pestol 7:66–77
Kurahashi H (2010b) Two new species of *Goniophyto* from Honshu, Japan (Diptera: Sarcophagidae). Med Entomol Zool 61(4):375–382
Kurahashi H (2014a) Calliphoridae (Diptera Part 2 Brachycera Schizophora). Cat Insects Jpn 8:807–816
Kurahashi H (2014b) Sarcophagidae (Diptera Part 2 Brachycera Schizophora). Cat Insects Jpn 8:817–831
Kurahashi H, Afzal M (2002) The blow flies recorded from Pakistan, with the description of one new species (Diptera: Calliphoridae). Med Entomol Zool 53(2):213–230
Kurahashi H, Bunchu N (2011) The blow flies recorded from Thailand, with the description of a new species of *Isomyia* WALKER (Diptera, Calliphoridae). Jpn J Syst Entomol 17(2):237–278
Kurahashi H, Chaiwong T (2007) Two new species of *Protomiltogramma* from Thailand (Diptera, Sarcophagidae). Jpn J Syst Entomol 13:173–178
Kurahashi H, Chaiwong T (2013) Keys to the flesh flies of Thailand, with description of a new species of *Robineauella* Enderlein (Diptera: Sarcophagidae). Med Entomol Zool 64(2):83–101
Kurahashi H, Chowanadisai L (2001) Blow flies (Insecta: Diptera: Calliphoridae) from Indochina. Spec Diver 6(3):185–242
Kurahashi H, Harusawa K (2009) *Metopia argentata* Macquart, 1850 (Diptera: Sarcophagidae) newly recorded from Honshu, Japan. hana abu 28:1–2
Kurahashi H, Harusawa K (2012). *Amobia auriceps* (Baranov, 1935) newly recorded from Japan (Diptera: Sarcophagidae)) hana abu 33:1–2
Kurahashi H, Hoshino K (2009) *Parasarcophaga hirtipes* (Wiedemann, 1830) (Diptera: Sarcophagidae) newly recorded from Tsushima I., Japan. hana abu 28:4–6
Kurahashi H, Kakinuma S (2015) Keys to the flesh flies of Japan, with the description of a new genus and species from Honshu (Diptera: Sarcophagidae). Med Entomol Zool 66(4):167–200
Kurahashi H, Kano R (1984) Phylogeny and geographical distribution of the genus *Boettcherisca* Rohdendorf (Diptera: Sarcophagidae*)*. Jpn J Med Sci Biol 37(1):27–34
Kurahashi H, Kano R (1995) A new Luciliine blow fly (Diptera, Calliphoridae) from Taiwan. Spec Bull Jpn Soc Coleopterol 4:487–490
Kurahashi H, Leh MU (2007) The flies from Sarawak, East Malaysia (Diptera: Muscidae, Calliphoridae, Sarcophagidae and Tachinidae). Med Entomol Zool 58(4):261–273
Kurahashi H, Leh MU (2008) Two new species of *Protomiltogramma* from Malaysia (Diptera, Sarcophagidae). Jpn J Syst Entomol 14(1):1–6
Kurahashi H, Leh MU (2009) The blow flies from Sarawak, East Malaysia (Diptera: Calliphoridae), with practical keys and a checklist. Sarawak Mus J 66(87):277–303
Kurahashi H, Magpayo FR (1987) Two new species of the genus *Chrysomya* from Wallacea (Diptera, Calliphoridae). Kontyu 55(1):71–79
Kurahashi H, Magpayo FR (2000) Blow flies (Insecta: Diptera: Calliphoridae) of the Philippines. Raffles Bull Zool 9(Suppl):1–78
Kurahashi H, Pape T (1996) The first endemic *Macronychia* from the Oriental region (Diptera: Sarcophagidae). Kontyu 64(2):267–274
Kurahashi H, Park SH (1972) A new species of the genus *Onesia* from Korea (Diptera: Calliphoridae). Kontyu 40(1):22–24
Kurahashi H, Salazar NP (1977) A new species of *Chrysomya* (Diptera, Calliphoridae) from the Philippines. Kalikasan, Philipp. J Biol 6(3):251–156
Kurahashi H, Suenaga O (1994) A new species of *Goniophyto* from Kyushu, Japan (Diptera: Sarcophagidae). Jpn J Sanit Zool 45(3):227–231
Kurahashi H, Suenaga O (1997) Witnessing hundreds of *Calliphora nigribarbis* in migratory flight and landing in Nagasaki, Western Japan. Med Entomol Zool 48(1):55–58
Kurahashi H, Sukontason KL (2004) A new species of *Sinonipponia* (Diptera: Sarcophagidae) from Thailand. Med Entomol Zool 55(3):191–193

Kurahashi H, Tan SH (2009) The sarcophagid flies from Peninsular Malaysia (Diptera: Sarcophagidae). Med Entomol Zool 60(4):283–296

Kurahashi H, Tan SH (2012a) Two new species of the genus *Phallosphaera* Rohdendorf, 1938 from Malaysia (Diptera, Sarcophagidae). Jpn J Syst Entomol 18(2):319–325

Kurahashi H, Tan SH (2012b) A new genus and species of the flesh flies from Malaysia (Diptera, Sarcophagidae). Med Entomol Zool 63(4):307–312

Kurahashi H, Thapa VK (1994) Notes on the Nepalese calliphorid flies (Insecta: Diptera). Jpn J Sanit Zool 45(Suppl):179–252

Kurahashi H, Tumrasvin W (1979) Six new species of calliphorid flies from Thailand (Diptera: Calliphoridae). Jpn J Sanit Zool 30(4):297–304

Kurahashi H, Kawai S, Shudo C, Wada Y (1994) The life history of *Calliphora nigribarbis* Vollenhoven in Mt. HachijoFuji, Hachijo Island. Jpn J Sanit Zool 45(4):327–332

Kurahashi H, Benjaphong N, Omar B (1997) Blow flies (Insecta: Diptera: Calliphoridae) of Malaysia and Singapore. Raffles Bull Zool 5(Suppl):1–88

Kurahashi H, Yang CH, Kakinuma S (2017) Keys to the flesh flies of Taiwan, with the description of the second species of Papesarcophaga from Tainan (Diptera: Sarcophagidae). Med Entomol Zool 68(1):31–38

Kurahashi H, Tan SH, Leh MU (2021) Keys to the flesh flies of Sarawak, East Malaysia, with description of two genera and eight new species (Diptera: Sarcophagidae). Med Entomol Zool 72(3):119–175

Lan YS, Zhong TW (2001) Case reports of myiasis in cattle. Fujian J Anim Husb Vet Med 5:23. (in Chinese)

Lata J, Kapila BK, Aggarwal P (1996) Oral myiasis – a case report. Int J Oral Maxillofac Surg 25(6):455–456

Lee HL (1989) Recovery of forensically important entomological specimens from human cadavers in Malaysia – an update. Malays J Pathol 11:33–36

Lee HL (1996a) Forensically important fly maggots recovered from human cadavers in Malaysia. Trop Biomed 13:93–94

Lee HL (1996b) Recovery of forensically important insect larvae from human cadavers in Malaysia (1993–1996). Malays J Pathol 18(2):125–127

Lee HL, Krishnasamy M, Abdullah AG, Jeffery J (2004) Review of forensically important entomological specimens in the period of 1972–2002. Trop Biomed 21(2):69–75

Lee HL, Krishnasamy M, Jeffrey J (2005) A case of human nasopharyngeal myiasis caused by *Chrysomya bezziana* Villeneuve, 1914 (Diptera: Calliphoridae) in Malaysia. Trop Biomed 22(1):87–88

Levine OS, Levine MM (1991) Houseflies (*Musca domestica*) as mechanical vectors of Shigellosis. Rev Infect Dis 13(4):688–696

Li Q, Huang J, Yuan J (2018) Status and preliminary mechanism of resistance to insecticides in a field strain of housefly (*Musca domestica*, L). Rev Bras Entomol 62(4):311–314

Li Y, Luo S, Zhang D (2014) Fly pollination and duodichogamy in *Bridelia stipularis* and *Cleistanthus sumatranus* (Phyllanthaceae). Plant Species Biol 29(3):85–92

Lin SS, Kuo CH, Lue KY (2000) Oviposition behaviour and host selection of the frogfly, *Caiusa coomani* (Diptera: Calliphoridae). Chin J Entomol 20:281–292. (in Chinese with English summary)

Lindeberg YL, Egedal K, Hossain ZZ, Phelps M, Tulsiani S, Farhana I, Begum A, Jensen PKM (2018) Can *Escherichia coli* fly? The role of flies as transmitters of *E. coli* to food in an urban slum in Bangladesh. Trop Med Int Health 23(1):2–9

Liu YP, Lee DH, Chen LH, Lin YJ, Li WC, Hu SC, Chen YP, Swayne DE, Lee MS (2018) Detection of reassortant H5N6 clade 2.3.4.4 highly pathogenic avian influenza virus in a black-faced spoonbill (*Platalea minor*) found dead, Taiwan, 2017. Infect Genet Evol 62:275–278

Lopes HS (1961) A contribution to the knowledge of the genus *Boettcherisca* Rohdendorf, 1937 (Diptera, Sarcophagidae). Mem Inst Oswaldo Cruz 59(1):69–82

Lopes HS (1982) On *Eumacronychia sternalis* Allen (Diptera, Sarcophagidae) with larvae living on eggs and hatchilings of the east Pacific green turtle. Rev Bras Biol 42(2):425–429
Lopes HS, Kano R (1979) Notes on *Sarcorohdendorfia* with key of the species (Diptera, Srcophagidae). Rev Bras Biol 39(3):657–670
Lue KY, Lin SS (2000) Investigation of foam nests (Rhacophoridae) infested by frogflies (Diptera) in Taiwan. Chin J Entomol 20(4):267–280. (in Chinese with English summary)
Markus MB (1980) Flies as natural transport hosts of *Sarcocystis* and other *Coccidia*. J Parasitol 66(2):361–362
Matsuzaki S, Yamazaki G (1987) A case report on oral myiasis due to the fleshfly *Boettcherisca peregrina*. Jpn J Sanit Zool 38:155
Mazlan N, Mumford J (2005) Insecticide use in cabbage pest management in the Cameron Highlands, Malaysia. Crop Prot 24(1):31–39
Mckillup SC, Mckillup RV, Pape T (2000) Flies that are parasitoids of a marine snail: the larviposition and life cycles of *Sarcophaga megafilosia* and *Sarcophaga meiofilosia*. Hydrobiologia 439:141–149
Meijer W (1985) Saving the world's largest flower. Natl Geogr 168(1):136–140
Merritt RW, Wotton RS (1988) The life history and behavior of *Limnophora riparia* (Diptera: Muscidae), a predator of larval black flies. J North Am Benthol Soc 7(1):1–12
Ministry of Agriculture, Forestry and Fisheries (MAFF) (2023) Information on Avian Influenza. [Accessed 17 Jan 2024] https://www.maff.go.jp/j/syouan/douei/tori/#3 (in Japanese)
Ministry of the Environment, Government of Japan (MOE) (2023) Information on highly pathogenic avian influenza. [Accessed 17 Jan 2024] https://www.env.go.jp/nature/dobutsu/bird_flu/ (in Japanese)
Mircheraghi SF, Mircheraghi SF, Ramezani Awal Riabi H, Parsapour A (2016) Nasal nosocomial myiasis infection caused by *Chrysomya bezziana* (Diptera: Calliphoridae) following the septicemia: a case report. Iran J Parasitol 11(2):284–289
Mitsui H, Kawabata T, Ugaki S (2005) Successful treatment of severe ischemic foot ulcer with sterile maggots: the first case in this country. Jpn J Vasc Surg 14:653–657
Miura K, Ohgushi T (2010) A new host record for *Balaesoxipha rufipes* (Diptera: Sarcophagidae) on *Atractomopha lata* (Orthoptera: Pyrgomorphidae) from Japan. Entomol News 121(2):213–214
Miura M, Hayasaka S, Yamada T, Hayasaka Y, Kamimura K (2005) Ophthalmomyiasis caused by larvae of *Boettcherisca peregrina*. Jpn J Ophthalmol 49(2):177–179
Miyamoto K, Shinonaga S, Iwasa M, Takahashi T, Nunomura A (1996) Two cases of facultative myiasis due to genus *Sarcophaga*: case reports. Med Entomol Zool 47(2):179–181
Monzon RB, Sanchez AR, Tadiaman BM, Najos OA, Valencia EG, de Rueda RR et al (1991) A comparison of the role of *Musca domestica* (Linnaeus) and *Chrysomya megacephala* (Fabricius) as mechanical vectors of helminthic parasites in a typical slum area of Metropolitan Manila. Southeast Asian J Trop Med Public Health 22(2):222–228
Moon RD (2019) Muscid flies (Muscidae). In: Mullen GR, Durden LA (eds) Medical veterinary entomology, 3rd edn. Academic Press, London, pp 345–368
Moophayak K, Klong-Klaew T, Sukontason K, Kurahashi H, Tomberlin JK, Sukontason KL (2014) Species composition of carrion blow flies in northern Thailand: altitude appraisal. Rev Inst Med Trop São Paulo 56(2):179–182
Moradi-Asl E, Ghalehbin BM, Akbarzadeh K, Mohammad-Shahi J, Afshin H (2021) Wound myiasis caused by *Lucilia cuprina* (Diptera: Calliphoridae) in Ardabil province, northwest of Iran. Punjab Univ J Zool 36(1):21–23
Moribayashi A, Wells JD, Kurahashi H (1999) Chilling period for effective survival of diapausing pupae of the flesh fly, *Boettcherisca peregrina*, collected from Tokyo. Med Entomol Zool 50(2):129–135
Moribayashi A, Shudo C, Kurahashi H (2001) Latitudinal variation in the incidence of pupal diapause in Asian and Oceanian populations of the flesh fly, *Boettcherisca peregrina* (Diptera: Sarcophagidae). Med Entomol Zool 52(4):263–268

Moribayashi A, Hayashi T, Taylor D, Kurahashi H, Kobayashi M (2008) Different responses to photoperiod in non-diapausing colonies of the flesh fly, *Boettcherisca peregrina*. Physiol Entomol 33:31–36

Motoyama N (1984) Pyrethroid resistance in a Japanese colony of the housefly. J Pestic Sci 9:523–526

Mulieri PR, Patitucci LD, Dufek MI (2021) Occurrence of kleptoparasitic sarcophagid flies (Diptera: Sarcophagidae: Miltogramminae) in natural and human modified sites in Southern South American Andean landscapes. Rev Soc Entomol Argent 80(3):1–19

Musca Solution (2019) MUSCA. [Accessed 28 Mar 2023] https://musca.info/en/solution/

Nair HKR, Nazni WA, Ismail AA, Alabed AA, Soffian A (2021) Maggot debridement therapy in the treatment of diabetic foot ulcers. Wounds Asia 4(1):30–36

Nandi B, Sinha SK (2004) On a small collection of muscid flies (Diptera: Muscidae) of Sundarbans Biosphere Reserve, India. Rec Zool Surv India 102(Part 1–2):11–26

Nandi BC (2002a) Blow flies (Diptera: Calliphoridae) of west Bengal, India with a note on their biodiversity. Rec Zool Surv India 100(Part 1–2):117–129

Nandi BC (2002b) Fauna of India and the adjacent countries – Diptera (Volume X) Sarcophagidae. Director (ed) Zoological Survey of India, Kolkata 10:608

Nara Prefecture Agricultural Research and Development Center (2019) "Bee fly" user manual – a novel pollinator for strawberries. [Accessed 1 Feb 2024] https://www.naro.go.jp/publicity_report/publication/files/warc_BeeFly_usage_manual_190320a1.pdf (in Japanese)

Nasiri S, Ershadi S, Abdollahimajd F, Asadi E (2015) Wound and conjunctival myiasis caused by *Lucilia sericata*: a case report. Arch Clin Infect Dis 10(3):e27060

Nazni WA, Jeffery J, Lee HL, Lailatul AM, Chew WK, Heo CC, Sadiyah I, Khairul AM, Heah SK, Mohd HH (2011) Nosocomial nasal myiasis in an intensive care unit. Malays J Pathol 33(1):53–56

Nazni WA, Eugène M, Azahari AH, Shahar MK, Zainah S, Vthylingam I, Lee HL (2013) The potential of house fly, *Musca domestica* (L.) in the mechanical transmission of influenza A subtype H1N1 virus under laboratory conditions. J Gen Mol Virol 5(2):22–28

Nezakati E, Hasani MH, Zolfaghari P, Rashidan M, Sohrabi MB (2020) Effects of *Lucilia sericata* maggot therapy in chronic wound treatment: a randomized clinical trial. Chronic Wound Care Manag Res 7:11–17

Ngoen-klan R, Moophayak K, Klong-klaew T, Irvine KN, Sukontason KL, Prangkio C, Somboon P, Sukontason K (2011) Do climatic and physical factors affect populations of the blow fly *Chrysomya megacephala* and house fly *Musca domestica*? Parasitol Res 109(5):1279–1292

Nidom CA, Takano R, Yamada S, Sakai-Tagawa Y, Daulay S, Aswadi D, Suzuki T, Suzuki Y, Shinya K, Iwatsuki-Horimoto K, Muramoto Y, Kawaoka Y (2010) Influenza a (H5N1) viruses from pigs, Indonesia. Emerg Infect Dis 16(10):1515–1523

Ohashi M, Fukuda R, Abe S, Ikeda S, S. (2012) A case of cutaneous flies caused by the cynomolgus. Jpn J Dermatol 122:122. (in Japanese)

Omar B, Marwi MA, Ahmad A, Zuha RM, Jeffery J (2003) Synanthropic index of flies (Diptera: Muscidae and Calliphoridae) collected at several locations in Kuala Lumpur and Gombak, Malaysia. Trop Biomed 20(1):77–82

Ong SQ, Ahmad H, Jaal Z, Rus AC (2016) Comparative effectiveness of insecticides for use against the house fly (Diptera: Muscidae): determination of resistance levels on a Malaysian poultry farm. J Econ Entomol 109(1):352–359

Otake S, Dee SA, Rossow KD, Moon RD, Trincado C, Pijoan C (2003) Evaluation of mechanical transmission of porcine reproductive and respiratory syndrome virus by houseflies, *Musca domestica* (Linnaeus). Vet Rec 152(3):73–76

Otake S, Dee SA, Moon RD, Rossow KD, Trincado C, Pijoan C (2004) Studies on the carriage and transmission of porcine reproductive and respiratory syndrome virus by individual houseflies (*Musca domestica*). Vet Rec 154(3):80–85

Pape T (1986) Afrotropical species of *Metopia* (Insecta, Diptera, Sarcophagidae). Steenstrupia 124(4):73–84

Pape T (1987) Revision of neotropical *Metopia* Meigen (Diptera: Sarcophagidae). Syst Entomol 12(1):81–101

Pape T (1994) The world *Blaesoxipha* Loew, 1861 (Diptera: Sarcophagidae). Entomol Scand 45(Suppl):1–247

Pape T (1996) Catalogue of the Sarcophagidae of the world (Insecta: Diptera). Associated Publishers, Gainesville, p 558

Pape T, Arribas OJ (1999) *Sarcophaga protuberans* Pandellé – An Old World predator of lizard eggs (Diptera: Sarcophagidae; Reptilia: Lacertidae). Studia Dipterologica 6(1):73–78

Pape T, Bänziger H (2000) Two new species of *Sarcophaga* (Diptera: Sarcophagidae) among pollinators of newly discovered *Sapria ram* (Rafflesiaceae). Raffles Bull Zool 48:201–208

Pape T, Kurahashi H (1995) Taxonomy, systematics and distribution of *Senotainia navigatris* (Meijere) (Diptera: Sarcophagidae, Miltogrammatinae). Kontyu 63(3):485–492

Pape T, Kurahashi H (2000) A new species of *Sarcophaga* (*Lioproctia*) from Timor, Indonesia (Diptera: Sarcophagidae). Entomol Sci 3(3):519–522

Pape T, Kurahashi H (2004) Three new species of *Sarcophaga* (Insecta: Diptera: Sarcophagidae) from Timor, with a lectotype designation for *Myophora peronii*. Species Divers 9(2):165–185

Pape T, Mckillup SC, Mckillup RV (2000) Two new species of *Sarcophaga* Meigen subgenus *Sarcorohdendorfia* Baranov (Diptera: Sarcophagidae), parasitoids of *Littoraria filosa* (Sowerby) (Gastropoda: Littorinidae). Aust J Entomol 39(4):236–240

Parwani RN, Patidar KA, Parwani SR, Wanjari SP (2014) Exuberant oral myiasis caused by *Musca domestica* (Housefly). J Global Infect Dis 6(1):35–38

Paul AG, Ahmad NW, Lee HL, Ariff AM, Saranum M, Naicker AS et al (2009) Maggot debridement therapy with *Lucilia cuprina*: A comparison with conventional debridement in diabetic foot ulcers. Int Wound J 6(1):39–46

Pechter EA, Sherman RA (1983) Maggot therapy: the surgical metamorphosis. Plast Reconstr Surg 72(4):567–570

Pezzi M, Krčmar S, Mendicino F, Carlomagno F, Bonelli D, Scapoli C, Chicca M, Leis M, Bonacci T (2022) *Lucilia sericata* (Diptera: Calliphoridae) as agent of myiasis in a goose in Italy and a review of myiasis by this species in birds. Insects 13(6):542

Pitkin A, Deen J, Otake S, Moon R, Dee S (2009) Further assessment of houseflies (*Musca domestica*) as vectors for the mechanical transport and transmission of porcine reproductive and respiratory syndrome virus under field conditions. Can J Vet Res 73(2):91–96

Piwczyński M, Pape T, Deja-Sikora E, Sikora M, Akbarzadeh K, Szpila K (2017) Molecular phylogeny of miltogramminae (Diptera: Sarcophagidae): implications for classification, systematics and evolution of larval feeding strategies. Mol Phylogenet Evol 116:49–60

Ponnudurai G, Rani N, Velusamy R (2015) Integrated fly management for control of house fly (Diptera: *Musca domestica*) in layer chicken farms: a review. Indian J Anim Sci 85(5):439–445

Pont AC (1980) Family muscidae. In: Crosskey RW (ed) A catalog of the Diptera of the afrotropical region. British Museum (Natural History), London, pp 721–761

Pont AC, Vikhrev N, Werner D (2011) The hunter-flies of *Armenia*. I. Some species of the genus *Limnophora* Robineau-Desvoidy, with the description of a new species. Zool Middle East 52(2011):89–103

Qureshi T, Saeed A, Usmanghani K (2014) Report: prevalence of shigellosis in three different areas of Karachi. Pak J Pharm Sci 27(2):381–388

Rahoma A, Latif B (2010) Human foot myiasis in Malaysia with a review of the literature. J Trop Med Parasitol 33(1):41–43

Rinkevich FD, Hedtke SM, Leichter CA, Harris SA, Su C, Brady SG, Taskin V, Qiu X, Scott JG (2012) Multiple origins of *kdr*-type resistance in the house fly, *Musca domestica*. PLoS One 7(12):e52761

Robinson A, Bickford-Smith J, Abdurahman Shafi O, Aga MA, Shuka G, Debela D, Hordafa G, Alemayehu W, Sarah V, Last A, MacLeod D, Burton MJ, Logan JG (2021) Towards an odour-baited trap to control *Musca sorbens*, the putative vector of trachoma. Sci Rep 11(1):14209

Robinson W (2005) Urban insects and arachnids: a handbook of urban entomology. Cambridge University Press, Cambridge, p 472

Roca-Acevedo G, Boscaro I, Toloza AC (2023) Global pattern of *kdr*-type alleles in *Musca domestica* (L.). Curr Trop Med Rep 10(1):1–10

Rockstein M (1957) Longevity of male and female house flies. J Gerontol 12(3):253–256

Rognes K (2015) Revision of the frog fly genus *Caiusa* Surcouf, 1920 (Diptera, Calliphoridae), with a note on the identity of *Plinthomyia emimelania* Rondani, 1875. Zootaxa 3952(1):1–80

Rohdendorf BB (1937) Fauna U.S.S.R. Ins., Dipt. Sarcophagidae Part 1. Zool Inst Acad Sci USSR 19(1):500

Saeed S, Naqqash MN, Jaleel W, Saeed Q, Ghouri F (2016) The effect of blow flies (Diptera: Calliphoridae) on the size and weight of mangos (*Mangifera indica* L.). PeerJ 4:e2076

Sasaki T, Kobayashi M, Agui N (2000) Epidemiological potential of excretion and regurgitation by *Musca domestica* (Diptera: Muscidae) in the dissemination of *Escherichia coli* O157: H7 to food. J Med Entomol 37(6):945–949

Sawabe K, Hoshino K, Isawa H, Sasaki T, Hayashi T, Tsuda Y, Kurahashi H, Tanabayashi K, Hotta A, Saito T, Yamada A, Kobayashi M (2006) Detection and isolation of highly pathogenic H5N1 avian influenza A viruses from blow flies collected in the vicinity of an infected poultry farm in Kyoto, Japan, 2004. Am J Trop Med Hyg 75(2):327–332

Sawabe K, Tanabayashi K, Hotta A, Hoshino K, Isawa H, Sasaki T, Yamada A, Kurahashi H, Shudo C, Kobayashi M (2009) Survival of Avian H5N1 Influenza A viruses in *Calliphora nigribarbis* (Diptera: Calliphoridae). J Med Entomol 46(4):852–855

Sawabe K, Hoshino K, Isawa H, Sasaki T, Kim KS, Hayashi T, Tsuda Y, Kurahashi H, Kobayashi M (2011) Blow flies were one of the possible candidates for transmission of highly pathogenic H5N1 avian influenza virus during the 2004 outbreaks in Japan. Influenza Res Treat 2011:652652

Schwendinger PJ, Pape T (2000) *Metopia sinensis* (Diptera, Sarcophagidae), an unusual predator of *Liphistius* (Araneae, Mesothelae) in northern Thailand. J Arachnol 28(3):353–356

Seal A, Bera R, Mukhopadhyay K, Mukherjee S (2015) Recycling of poultry litter through Novcom composting method: a case study from Howrah Krishi Vigyan Kendra, West Bengal, India. J Pharm Sci Innov 4(3):176–179

Senior-White RA, Aubertin D, Smart J (1940) Diptera volume VI – family Calliphoridae. The Fauna of British India, including the remainder of the Oriental region. Taylor and Francis Ltd., London, p 288

Sharma J, Mamatha GP, Acharya R (2008) Primary oral myiasis: a case report. Med Oral Patol Oral Cir Bucal 13(11):E714–E716

Sharma S, Abbas SR, Shukiam RP, Sharma S (1998) An easy and quick method of breeding flies for pollination of mango blossoms. Insect Environ 4:76–77

Sherman RA, Wyle FA, Vulpe M (1995) Maggot debridement therapy for treating pressure ulcers in spinal cord injury patients. J Spinal Cord Med 18(2):71–74

Shi CM, Xue-Shu Z, De-Xing Z (2015) Parasitoidism of the *Sarcophaga dux* (Diptera: Sarcophagidae) on the *Mesobuthus martensii* (Scorpiones: Buthidae) and its implications. Ann Entomol Soc Am 108(6):978–985

Shiao SF, Yeh TC (2008) Larval competition of *Chrysomya megacephala* and *Chrysomya rufifacies* (Diptera: Calliphoridae): behavior and ecological studies of two blow fly species of forensic significance. J Med Entomol 45(4):785–799

Shinonaga S (2003) A monograph of the Muscidae of Japan. Tokai University Press, Hiratsuka, p 347

Shinonaga S (2014) Muscidae (Diptera Part 2 Brachycera Schizophora). Cat Insects Jpn 8:785–806

Shinonaga S, Barrion AT (1980) A new species of sarcophagid fly parasitic in the egg of the spider, *Argiope catenulata* (Dolescall) in the Philippines. Kontyu 48:537–539

Shinonaga S, Kano R (1971) Fauna Japonica, Muscidae (Insecta: Diptera), vol 1. Keigaku Publishing Co. Ltd, Tokyo, p 244

Sievert K, Alvarez R, Cortada R, Valks M (2006) House flies carrying avian influenza virus (AIV). Int Pest Control 48(3):114–116

Singh A, Singh D (2016) A study on the incidence of myiasis among dairy animals in the State of Punjab, India. J Agric Vet Sci 9(1):30–34
Singh I (1973a) The housefly (*Musca domestica* L.) problem in Cameron Highlands (West Malaysia) with particular reference to methods for its control. In: Proceeding of First SEAMEO—TROPMED Workshop: vector control in Southeast Asia, Singapore, p 226
Singh I (1973b) Evaluation of insecticides against four strains of the housefly *Musca domestica* L. (Diptera: Muscidae) from West Malaysia. Southeast Asia J Trop Med Public Health 4(4):554–559
Singh I, Kurahashi H, Kano R (1979) A preliminary key to the common calliphorid flies of Peninsular Malaysia (Insecta: Diptera). Bull Tokyo Med Dent Univ 26(1):5–24
Sinha SK (2012) Myiasis in domestic animals: new records of calyptrate Diptera. J Parasit Dis 36(2):277–279
Spradbery JP, Sands DPA, Tozer RS (2019) Distribution and pest status of the old world screwworm fly, *Chrysomya bezziana* (Diptera: Calliphoridae), in Papua New Guinea and the threat to Australian livestock and wildlife. Austral Entomol 58(4):907–918
Sreevatsa, Malaviya GN, Husain S, Girdhar A, Bhat HR, Girdhar BK (1990) Preliminary observations on myiasis in leprosy patients. Lepr Rev 61(4):375–378
Srinivasan R, Jambulingam P, Gunasekaran K, Boopathidoss PS (2008) Tolerance of house fly, *Musca domestica* L. (Diptera: Muscidae) to dichlorvos (76% EC) an insecticide used for fly control in the tsunami-hit coastal villages of southern India. Acta Trop 105(2):187–190
Stevens J, Wall R (1996) Species, sub-species and hybrid populations of the blowflies *Lucilia cuprina* and *Lucilia sericata* (Diptera: Calliphoridae). Proc Biol Sci 263(1375):1335–1341
Stevens J, Wall R (1997) The evolution of ectoparasitism in the genus *Lucilia* (Diptera: Calliphoridae). Int J Parasitol 27(1):51–59
Subbarao K, Klimov A, Katz J, Regnery H, Lim W, Hall H, Perdue M, Swayne D, Bender C, Huang J, Hemphill M, Rowe T, Shaw M, Xu X, Fukuda K, Cox N (1998) Characterization of an avian influenza A (H5N1) virus isolated from a child with a fatal respiratory illness. Science 279(5349):393–396
Sucharit S, Kerdpibule V, Tumrasvin W, Deesin T, Nakorn CN (1981) Myiasis of the vagina of a comatose women caused by *Parasarcophaga ruficornis* Fabricius. J Med Assoc Thail 64(11):580–583
Sueyoshi M (2015) Review of the human myiasis in Japan. Med Entomol Zool 66(3):91–120
Sugiyama E, Kano R (1984) Systematics of the Sarcophaginae of the Oriental region based on the comparative morphology of the male genitalia (Diptera, Sarcophagidae). Jpn J Sanit Zool 35(4):343–356
Sugiyama E, Kurahashi H (1988) A new flesh fly *Sarcophaga dumoga* from Sulawesi, Indonesia (Diptera, Sarcophagidae). Jpn J Sanit Zool 29(1):49–51
Sugiyama E, Shinonaga S, Kano R (1990) Sarcophagine flies from Malaysia and Singapore with the description of six new species (Diptera, Sarcophagidae). Jpn J Sanit Zool 41(2):81–91
Sukontason K, Boonchu M, Khantawa B, Sukontason K, Piangjai S, Choochote W (2000) *Musca domestica* as a mechanical carrier of bacteria in Chiang Mai, North Thailand. J Vector Ecol 25(1):114–117
Sukontason K, Narongchai P, Kanchai C, Vichairat K, Sribanditmongkol P, Bhoopat T et al (2007) Forensic entomology cases in Thailand: a review of cases from 2000 to 2006. Parasitol Res 101(5):1417–1423
Sukontason KL, Narongchai P, Sripakdee D, Boonchu N, Chaiwong T, Ngern-Klun R, Piangjai S, Sukontason K (2005) First report of human myiasis caused by *Chrysomya megacephala* and *Chrysomya rufifacies* (Diptera: Calliphoridae) in Thailand, and its implication in forensic entomology. J Med Entomol 42(4):702–704
Sulaiman S, Sohadi AR, Yunus H, Iberahim R (1988) The role of some cyclorrhaphan flies as carriers of human helminths in Malaysia. Med Vet Entomol 2(1):1–6
Sun H, Kasai S, Scott JG (2017) Two novel house fly *Vssc* mutations, D600N and T929I, give rise to new insecticide resistance alleles. Pestic Biochem Physiol 143:116–121

Sung IH, Lin MY, Chang CH, Cheng AS, Chen WS, Ho KK (2006) Pollinators and their behaviors on mango flowers in southern Taiwan. Form Entomol 26:161–170

Syamsa RA, Omar B, Zuha RM, Faridah MN, Swarhib MS, Hidayatulfathi O, Shahrom AW (2015) Forensic entomology of high-rise buildings in Malaysia: three case reports. Trop Biomed 32(2):291–299

Szostakowska B, Kruminis-Lozowska W, Racewicz M, Knigh R, Tamang L, Myjak P, Graczyk TK (2004) *Cryptosporidium parvum* and *Giardia lamblia* recovered from feral filth flies. Appl Environ Microbiol 70(6):3742–3744

Tabaru Y, Kanno K, Kawanbata T, Ishikawa Y, Tanaka K, Hirao M, Kumon K, Watanabe M (2012) Outbreak of blow flies in the Tsunami affected areas and their control. Med Entomol Zool 63(1):71–83

Tan SH, Kurahashi H, Mohamed Z (2010) A new record of *Iranihindia martellata* (Senior-White, 1924) (Diptera: Sarcophagidae) from Peninsular Malaysia and female identification using both morphology and DNA-based approaches. Trop Med 27(2):287–293

Tan SW, Yap KL, Lee HL (1997) Mechanical transport of rotavirus by the legs and wings of *Musca domestica* (Diptera: Muscidae). J Med Entomol 34(5):527–531

The New Zealand Institute for Plant & Food Research Ltd (2017) Optimising pollination of macadamia and avocado in Australia (Project Number MT13060). Horticulture Innovation Australia. [Accessed 28 Jan 2023] https://www.horticulture.com.au/globalassets/laserfiche/assets/project-reports/mt13060/mt13060-final-report-complete.pdf

Thomas S, Jones M, Shutler S, Jones S (1996) Using larvae in modern wound management. J Wound Care 5(2):60–69

Thomas V, Smith KG, Dear JP (1980) Dermal myiasis in man caused by *Sarcophaga krameri* (Boettcher) (Diptera: Sarcophagidae) in Malaysia. Southeast Asian J Trop Med Public Health 11:140

Toukairin Y, Arai T, Hoshi T, Oliva Trejo JA, Nogami M (2017) The geographical distribution of fly larvae on corpses in Saitama Prefecture in Japan during the summer season. Legal Med 24:75–77

Trauth SE, Mullen GR (1990) Additional observations on sarcophagid fly infestations of *Sceloporus undulatus* (Sauria: Iguanidae) egg clutches in Arkansas. Southwest Nat 35(1):97–98

Tsuda Y, Hayashi T, Higa Y, Hoshino K, Kasai S, Tomita T, Kurahashi H, Kobayashi M (2009) Dispersal of a blow fly, *Calliphora nigribarbis*, in relation to the dissemination of highly pathogenic avian influenza virus. Jpn J Infect Dis 62(4):294–297

Tumrasvin W, Kano R (1979) Studies on medically important flies in Thailand. VI. Report on 48 species of sarcophagid flies, including the taxonomic keys (Diptera: Sarcophagidae). Bull Tokyo Med Dent Univ 26(2):149–179

Tumrasvin W, Kurahashi H, Kano R (1976) Studies on medically important flies in Thailand. I. Discovery of *Calliphora* species first in Thailand (Diptera: Calliphoridae). Bull Tokyo Med Dent Univ 23(4):211–216

Tumrasvin W, Kurahashi H, Kano R (1977) Studies on medically important flies in Thailand. II. Record of four species of *Lucilia* Robineau-Desvoidy (Diptera: Calliphoridae). Bull Tokyo Med Dent Univ 24(l):l–8

Tumrasvin W, Kurahashi H, Kano R (1979) Studies on the medically important flies in Thailand. VII. Report on 42 species of Calliphorid flies, including the taxonomic keys (Diptera: Calliphoridae). Bull Tokyo Med Dent Univ 26(4):243–272

Tyasasmaya T, Wuryastuty H, Wasito W, Sievert K (2016) Avian influenza virus H5N1 remained exist in gastrointestinal tracts of house flies 24 hours post-infection. J Veteriner 17(2):205–210

Urabe S, Kurahashi H, Inokuchi G, Chiba F, Motomura A, Hoshioka Y, Torimitsu S, Yamaguchi R, Tsuneya S, Iwase H (2022) Carrion flies (Insecta: Diptera) found on human cadavers in Chiba prefecture, Honshu, Japan, with the first record of *Fannia prisca* from a human corpse. J Forensic Sci 67(6):2469–2478

van der Niet T, Hansen DM, Johnson SD (2011) Carrion mimicry in a South African orchid: Flowers attract a narrow subset of the fly assemblage on animal carcasses. Ann Bot 107(6):981–992

Verves YG (2005) A catalogue of Oriental Calliphoridae (Diptera). Int J Dipterol Res 16(4):233–310
Villeneuve J (1914) Etude sur quelques types de myodaires supérieurs. Revue Zoologique et Botanique Africaine 3:429–441. (in French)
Vreysen MJ, Saleh KM, Ali M, Abdulla AM, Zhu ZR, Juma KG, Dyck VA, Msangi AR, Mkonyi PA, Feldmann HU (2000) *Glossina austeni* (Diptera: Glossinidae) eradicated on the island of Unguja, Zanzibar, using the sterile insect technique. J Econ Entomol 93(1):123–135
Wadhwa V, Kharbanda P, Rai S, Uppal B (2006) Urogenital myiasis due to *Chrysomya bezziana*. Indian J Med Microbiol 24(1):70–71
Wall R, Shearer D (1997) Arthropod ectoparasites of veterinary importance. In: Veterinary entomology. Chapman & Hall, p 439
Wallace GD (1971) Experimental transmission of *Toxoplasma gondii* by filth-flies. Am J Trop Med Hyg 20(3):411–413
Wanaratana S, Amonsin A, Chaisingh A, Panyim S, Sasipreeyajan J, Pakpinyo S (2013) Experimental assessment of houseflies as vectors in avian influenza subtype H5N1 transmission in chickens. Avian Dis 57(2):266–272
Wang H, Zhang Z, Czapar GF, Winkler MKH, Zheng J (2013) A full-scale house fly (Diptera: Muscidae) larvae bioconversion system for value-added swine manure reduction. Waste Manag Res 31(2):223–231
Wang JN, Hou J, Wu YY, Guo S, Liu QM, Li TQ, Gong ZY (2019) Resistance of house fly, *Musca domestica* L. (Diptera: Muscidae), to five insecticides in Zhejiang Province, China: the situation in 2017. Can J Infect Dis Med Microbiol 2019:4851914
Wang SY, Wang JN, Lv DC, Diao YP, Zhang Z (2010) Clinical research on the biodebridement effect of maggot therapy for treatment of chronically infected lesions. Orthop Surg 2(3):201–206
Wardhana AH, Cameron MM, Ready PD, Hall MJR (2018) Epidemiology of traumatic myiasis due to *Chrysomya bezziana* in Indonesia. Jurnal Ilmu Ternak dan Veteriner 23(1):45–60
Watanabe S, Hasegawa E (2014) A new host record for *Blaesoxipha rufipes* (Diptera: Sarcophagidae) in Japan. Entomol News 124(2):139–140
Waterhouse DF (1993) The major arthropod pests and weeds of agriculture in Southeast Asia: distribution, importance and origin. Australian Centre for International Agricultural Research, Canberra, p 141
Waterhouse DF, Paramonov SJ (1950) The status of the two species of *Lucila* (Diptera: Calliphoridae) attacking sheep in Australia. Aust J Sci Res 3:310–336
Watson DW, Nino EL, Rochon K, Denning S, Smith L, Guy JS (2007) Experimental evaluation of *Musca domestica* (Diptera: Muscidae) as a vector of Newcastle disease virus. J Med Entomol 44(4):666–671
Wee SL, Tan SB, Jürgens A (2018) Pollinator specialization in the enigmatic *Rafflesia cantleyi*: a true carrion flower with species-specific and sex-biased blow fly pollinators. Phytochemistry 153:120–128
Wells JD, Kurahashi H (1996) A new species of *Chrysomya* (Diptera: Calliphoridae) from Sulawesi, Indonesia, with a key to the Oriental, Australasian and Oceanian species. Med Entomol Zool 47(2):131–138
Wells JD, Kurahashi H (1997) *Chrysomya megacephala* (Fabr.) is more resistant to attack by *Chrysomya rufifacies* (Marcquart) in laboratory arena than is *Cochliomyia macellaria* (Fabr.) (Diptera: Calliphoridae). Pan Pac Entomol 73(1):16–20
Werner D, Pont A (2006) New results on diptera predator in the blackfly plaque areas of central Europe and the caucasus. Acta Entomol Serb 4(Suppl):131–140
Wharton RH, Seow CL, Ganapathipillai A, Jabaratnam G (1962) House fly populations and their dispersion in Malaya with particular reference to the fly problem in the Cameron Highlands. Med J Malaysia 17(2):115–131

Wilasrusmee C, Marjareonrungrung M, Eamkong S, Attia J, Poprom N, Jirasisrithum S, Thakkinstian A (2014) Maggot therapy for chronic ulcer: a retrospective cohort and a meta-analysis. Asian J Surg 37(3):138–147

Williamson MS, Denholm I, Bell CA, Devonshire AL (1993) Knockdown resistance (*kdr*) to DDT and pyrethroid insecticides maps to a sodium channel gene locus in the housefly (*Musca domestica*). MGG Mol Gen Genet 240(1):17–22

Xue W, Zhang D (2014) A review of the genus *Lispe* Latreille (Diptera: Muscidae) from China, with descriptions of new species. Orient Insects 39(1):117–139

Yaghoobi R, Bagherani N (2009) *Chrysomya bezziana* infestation in a neglected squamous cell carcinoma on the face. Indian J Dermatol Venereol Leprol 75(1):81–82

Yan WL, Yang CH, Tan SH, Pai CY, Li KK, Chung CC (2023) The study of forensically important insects recovered from human corpses in Taiwan. Insects 14(4):346

Yang ST, Kurahashi H, Shiao SF (2014) Keys to the blow flies of Taiwan, with a checklist of recorded species and the description of a new species of Paradichosia Senior-White (Diptera, Calliphoridae). Zookeys 434:57–109

Yasutomi K (1973) Insecticide resistance in the houseflies of New Yumenoshima, the 15th dumping-island of Tokyo. Jpn J Sanit Zool 23(3):255–256

Yasutomi K (1975) Insecticide resistance in the houseflies of the third Yumenoshima, a new dumping-island of Tokyo. Med Entomol Zool 26(4):257–258

Yasutomi K, Shudo C (1978) Insecticide resistance in the houseflies of the third Yumenoshima, a new dumping-island of Tokyo (II). Med Entomol Zool 29(2):205–208

Yasutomi K, Tomioka Y (2000) Insecticide resistance of housefly, *Musca domestica* L. (Diptera: Muscidae) in poultries, with special reference just in resistance levels. Med Entomol Zool 51(3):187–194

Yasutomi K, Mihara M, Shudo C (1988) Succession of insecticide resistance in the houseflies of the third Yumenoshima, a dumping-island of Tokyo, 1978–1986. Med Entomol Zool 39(1):43–47

Yeung JC, Chung CF, Lai JS (2010) Orbital myiasis complicating squamous cell carcinoma of eyelid. Hong Kong Med J 16(1):63–65

Yoke OP, Sudderuddin KI (1975) Toxicological studies of four insecticides against *Musca domestica* L. Southeast Asian J Trop Med Public Health 6(4):525–531

Yoneda Y, Shinonaga S, Kumashiro H, Fukuma T (1998) Eleven cases of accidental human myiasis since 1990. Med Entomol Zool 49(1):51–56

Yorke C (1983) Survival of embryos and larvae of the frog Polypedates leucomystax in Malaysia. J Herpetol 17:235–241

Yuen KY, Chan PKS, Peiris M, Tsang DNC, Que TL, Shortridge KF, Cheung PT, To WK, Ho ET, Sung R, Cheng AF (1998) Clinical features and rapid viral diagnosis of human disease associated with avian influenza A H5N1 virus. Lancet 351(9101):467–471

Zhou X, Kambalame DM, Zhou S, Guo X, Xia D, Yang Y et al (2019) Human *Chrysomya bezziana* myiasis: a systematic review. PLoS Negl Trop Dis 13(10):e0007391

Zhu FX, Yao YL, Wang SJ, Du RG, Wang WP, Chen XY, Hong CL, Qi B, Xue ZY, Yang HQ (2015) Housefly maggot-treated composting as sustainable option for pig manure management. Waste Manag 35:62–67

Zumpt F (1956) Calliphoridae (Diptera Cyclorrhapha) part I: Calliphorini and Chrysomyiini. Exploration du Parc national Albert. Mission G.F. de Witte (1933–1935). Institut des Parcs nationaux du Congo belge 87:1–200.

Zumpt F (1965) Myiasis in man and animals in the old world: a textbook for physicians, veterinarians and zoologists. Butterworths, London, p 267

Zumpt F (1973) The Stomoxyine biting flies of the world (Diptera, Muscidae): taxonomy, biology, economic importance and control measures. Gustav Fischer Verlag, Stuttgart, p 175

Chapter 4
Other Medically Important Vectors

Tohru Yanase, Yasushi Otsuka, Kandai Doi, Yuichiro Tabaru, Suha Kenan Arserim, Hitoshi Sasaki, Yusuf Özbel, Seray Töz, Takashi Ueda, Naotoshi Tsuji, Yasuyuki Amoh, Chizu Sanjoba, Hüseyin Çetin, and Kyoko Hayashida

T. Yanase
Kagoshima Research Station, National Institute of Animal Health (NARO), Kagoshima, Japan
e-mail: tyanase@affrc.go.jp

Y. Otsuka
International Center for Island Studies, Kagoshima University, Kagoshima, Japan
e-mail: yotsuka@cpi.kagoshima-u.ac.jp

K. Doi
Forestry and Forest Products Research Institute, Tsukuba, Japan
e-mail: kandaidoi@affrc.go.jp

Y. Tabaru
Japan Pest Control Association, Tokyo, Japan

S. K. Arserim
Celal Bayar University, Manisa, Türkiye

H. Sasaki
The Hokkaido University Museum, Sapporo, Japan

Y. Özbel · S. Töz
Department of Parasitology, Faculty of Medicine, Ege University, İzmir, Türkiye
e-mail: yusuf.ozbel@ege.edu.tr; seray.ozensoy.toz@ege.edu.tr

T. Ueda · Y. Amoh
Department of Dermatology, Kitasato University School of Medicine, Sagamihara, Japan

N. Tsuji
Department of Parasitology and Tropical Medicine, Kitasato University School of Medicine, Sagamihara, Japan
e-mail: tsujin@med.kitasato-u.ac.jp

C. Sanjoba (✉)
Graduate School of Agricultural and Life Sciences, The University of Tokyo, Tokyo, Japan
e-mail: asanjoba@g.ecc.u-tokyo.ac.jp

H. Çetin
Department of Biology, Faculty of Science, Akdeniz University, Antalya, Türkiye
e-mail: hcetin@akdeniz.edu.tr

K. Hayashida
International Institute for Zoonosis Control, Hokkaido University, Sapporo, Japan
e-mail: kyouko-h@czc.hokudai.ac.jp

K. Sawabe et al. (eds.), *Medical Entomology in Asia*, Entomology Monographs,
https://doi.org/10.1007/978-981-97-4406-0_4

Abstract In Japan and other Asian countries, in addition to mosquitoes, ticks, and flies, there are other important vector arthropods such as biting midge, black flies, chigger mites, horseflies, sand flies, lice, fleas, and so on. This chapter introduces the taxonomy, behavior, and ecology of these vector arthropods, as well as the pathogens they carry, focusing on arthropods that give a minor impression but cause serious medical infections. This chapter also covers the kissing bug and tsetse fly, which does not inhabit Japan and Asia but is an infectious disease vector of global importance.

Keywords Asia flea · Black fly · Biting midge · Chigger mite (Tsutsugamushi) · Cockroach · Stable fly · Human louse · Horsefly · Kissing bug · Phlebotomine sand fly · Scabies · Tsetse fly

4.1 Introduction

Arthropods are taxonomically composed of 11 classes, but they are mainly blood-sucking insects and ticks that cause infectious diseases in humans and animals. They transmit various pathogens such as viruses, bacteria, and parasites and become vectors or intermediate hosts for the subsequent infection. Vector-borne diseases account for more than 17% of all infectious diseases, causing more than 700,000 deaths annually (WHO 2020). In Asian countries, biting midge, chigger mites, fleas, horseflies, lice, mites, sand flies, etc., are medically important vector arthropods besides mosquitoes, ticks, and flies. For instance, the sand fly is the vector of leishmaniasis, affecting more than one million people worldwide yearly. Lice, which spends its entire life on the body surface of its host animals, still plagues people today. Tsutsugamushi disease is widespread in Asia and Southeast Asia. It used to be endemic in Japan, but now it has spread throughout the country except in some areas. Furthermore, they can transmit pathogens not only to humans but also to other mammals, which requires adequate attention, especially in managing the health of livestock.

4.2 Biting Midge

Tohru Yanase

Ceratopogonidae (Diptera), also known as the "biting midge," is rich in diversity and includes over 6000 recognized species divided into 112 genera (Borkent and Dominiak 2020). Of them, members in four genera, *Leptoconops*, *Forcipomyia*, *Culicoides,* and *Austroconops*, are known to feed on vertebrates. Except for *Austroconops*, the three genera are distributed in Asia and have affected the lives of the people. Their biting habitats have often afflicted humans and livestock animals, causing dermatitis and transmitting pathogens. Huge outbreaks of biting midges have often occurred in limited areas and disturbed outdoor activities, such as leisure, tourism, fishery, and forestry. Many viruses, protozoans, and filariae are transmitted by biting midges, and some have impacted human and livestock animal health.

4.2.1 Biology

Most ceratopogonids do not feed on vertebrates. Many species of the Ceratopogonidae attack other insects. In the case of hematophagous species, females usually require vertebrate blood to develop their eggs, but males do not feed on blood. The host preference of hematophagous midges is highly variable in mammals, birds, lizards, and frogs. It was reported that some *Culicoides* species feed on mud skippers, and others take vertebrate blood from engorged anopheline mosquitoes (Wirth and Hubert 1989). Midge behavior and abundance depend on climate, such as temperature and precipitation. In temperate and subarctic zones, adult midges disappear during winter, and almost all species overwinter as larvae. Although ceratopogonids use various habitats, their immatures generally require aquatic or semiaquatic habitats. Breeding sites of *Culicoides* species include intact dung pats, mud at the soil–water interface, and highly organic soil substrates (Purse et al. 2015). Immatures of *Leptoconops* are usually found in wet or damp sand near fresh waters and marine beaches (Borkent 2004). The larval habitat of *Forcipomyia* is mosses and algae growing on wood or soil.

Biting midges are not able to fly long distances, and thus, their habitats are usually close to their host animals. Airstreams often bring biting midges from their usual habitats to a distant place. The wind dispersal of the *Culicoides* vector is associated with the spread of pathogens transmitted by them (Elbers et al. 2015).

4.2.2 Public Health Importance

Biting midges often cause discomfort to humans through their persistent biting in various regions of the world. Biting by midges induces intense pruritus and swelling in sensitive individuals. In Japan and neighboring regions, their outbreaks have been problematic.

Leptoconops nipponensis nipponensis is distributed along coastal areas in Japan and prefers to feed on humans. Its immature habitat is sands near beaches. The most seriously affected area might be the Yumigahama Peninsula in Tottori Prefecture, located in the western part of Honshu and faces the Japanese Sea (Kurimasa et al. 2015). The peak of midge activity is observed in summer. Its subspecies, *Leptoconops nipponensis oshimaensis* (Fig. 4.1a), is distributed in the Amami-Oshima Island and its surrounding islands (Takaoka and Hayashi 1977) and is a serious nuisance in particular areas of the islands (Matayoshi et al. 1985). It was recently found that

Fig. 4.1 Females of biting midges. (**a**) *Leptoconops nipponensis oshimaensis* (photograph by Y. Otsuka) and (**b**) *Culicoides circumscriptus*

Leptoconops sp., considered a synonym of *L. taiwanensis*, relates to the biting nuisance to humans after long-term harm to residents and tourists at Kume Island in Okinawa Prefecture (Azama et al. 2014; Kimura et al. 2016).

Culicoides species also provide serious problems to humans by biting. *Culicoides circumscriptus* (Fig. 4.1b) is widely distributed in Asia, Europe, and North Africa, and its immatures inhabit wet, muddy sites in saline or brackish coastal regions (Wirth and Hubert 1989). Its biting behavior to humans restricts outdoor activities during its occurrence. Plagues of *C. circumscriptus* have often been reported around brackish lakes, such as Nakaumi facing the Yumigahama Peninsula (Inagaki 1959) and Lake Saloma in Hokkaido (Tabaru et al. 1973). *Culicoides nunomemoguri* and *C. sinanoensis* are annoying pests in mountainous regions of northern Japan (Kitaoka 1980; Natsuaki 2020). It was reported that *C. arakawae*, *C. aterinervis*, *C. dubius*, *C. homotomus*, *C. matsuzawai*, *C. nipponensis*, and *C. sanguisuga* also bite humans in Japan (Tokunaga 1950; Takeda and Mukai 1954; Nishijima and Ono 1964; Maruyama and Takagi 1977).

Restricted species of *Forcipomyia* spp. feed on vertebrates, but they sometimes become nasty pests for humans. *Forcipomyia taiwana* infests most of Taiwan and is one of the most annoying blood-sucking pests there (Chuang et al. 2000). The prevalence of this species has not been recorded in Japan. However, *F. anabaenae,* another blood-sucking pest in Taiwan, was found in Amami-Oshima (Kitaoka and Suzuki 1985).

Because biting midges are tiny and lesser-known insects, their biting might be overlooked. The actual number of cases bitten by biting midges might have been much higher than the reported cases.

Several viruses and filarial nematodes that affect humans are known to be transmitted by biting midges. Oropouche fever is an acute febrile illness caused by a *Culicoides*-borne orthobunyavirus, Oropouche virus (Sakkas et al. 2018). The disease is widely distributed in Central and South America and has considerable social and economic impacts. Several other arboviruses involved in severe clinical cases were isolated from *Culicoides* spp., but their role in transmitting these viruses remains uncertain. *Culicoides*, *Leptoconops,* and *Forcipomyia* species transmit filarial nematodes that cause an infestation in humans, called mansonellosis, through tropical and subtropical regions in Africa and Central and South America (Ta-Tang et al. 2018).

Fortunately, no serious human disease is currently known to be transmitted by biting midges in Asia. However, it cannot be denied that biting midges potentially have vector capacities for emerging and reemerging pathogens if they penetrate Asia. The presence of *Leishmania* DNA in field-collected *Culicoides* spp. has been confirmed in several countries (Slama et al. 2014; Rebêlo et al. 2016; Ríos-Tostado et al. 2021). Recently, *Leishmania martiniquensis*, a causative agent of autochthonous leishmaniasis, was detected from *C. mahasarakhamense* in Thailand (Sunantaraporn et al. 2021). Virus strains isolated from *Culicoides* spp. in Yunnan Province, China, were identified as Banna virus, associated with human patients suffering encephalitis and fever (Song et al. 2017; Duan et al. 2022). Continuous efforts to check human pathogens in biting midges should be conducted to assess their roles as vectors in Asia.

4.2.3 Veterinary Importance

Biting midges may provide a serious nuisance to livestock animals, although it is difficult to quantify their damage. It is well known that *Culicoides* bites cause allergic dermatitis in horses (Schaffartzik et al. 2012). Salivary gland proteins from *Culicoides* species were identified as allergen sources. It was reported that cattle and sheep exposed to *Culicoides* bites also exhibit an allergic skin reaction (Yeruham et al. 1993). Dermatitis usually occurs from spring to autumn when *Culicoides* adults are active. Although seasonal dermatitis in horses widely occurs throughout the world, no detailed study has been conducted in Asia to the best of our knowledge.

Culicoides species play a crucial role as vectors of viruses, blood protozoans, and filarial nematodes that affect livestock animals (pathogens of veterinary importance in Japan are listed in Table 4.1). Several *Culicoides*-borne pathogens severely affect livestock production in various regions throughout the world. In contrast, many endemic diseases associated with *Culicoides* transmission might have been neglected or not well known.

Three important diseases caused by *Culicoides*-borne orbiviruses are listed as notifiable diseases by the World Organisation of Animal Health (WOAH) because of their significant economic impacts on livestock production and the global trade of livestock products. Bluetongue is an infectious hemorrhagic disease in domestic ruminants and is caused by an orbivirus named bluetongue virus (BTV), transmitted by the *Culicoides* species (Maclachlan et al. 2015). Its mortality rate is significantly higher in sheep than in cattle and goats. Another *Culicoides*-borne orbivirus, epizootic hemorrhagic disease virus (EHDV), produces similar clinical signs in cattle (Savini et al. 2011). Serious manifestations caused by EHDV are often observed in infected white-tailed deer in North America. BTV and EHDV are found all over the

Table 4.1 Pathogens of veterinary importance transmitted by biting midges in Japan

Taxon	Vector species	Affected vertebrate hosts	Major manifestations
Viruses			
Orbivirus			
Bluetongue	*C. asiana*, *C. tainanus*	Cattle, sheep	Fever, anorexia, dysphagia, facial edema, ulceration of the oral mucosa, excessive salivation
Epizootic hemorrhagic disease	*C. oxystoma*, *C. lungchiensis*, *C. punctatus*	Cattle	Fever, anorexia, dysphagia, foamy salivation, stillbirths, abortion
Chuzan	*C. oxystoma*	Cattle	Congenital malformations
D'Aguilar	*C. oxystoma*, *C. sumatrae*	Cattle	Congenital malformations
Orthobunyavirus			
Akabane	*C. oxystoma*	Cattle, sheep, goats	Abortion, stillbirth, premature birth, congenital malformations, encephalomyelitis
Aino	*C. oxystoma*, *C. punctatus*	Cattle	Abortion, stillbirth, premature birth, congenital malformations
Peaton	*C. jacobsoni*	Cattle	Abortion, congenital malformations
Shamonda	*C. tainanus*	Cattle	Congenital malformations
Sathuperi	*C. oxystoma*	Cattle	Congenital malformations
Ephemerovirus			
Bovine ephemeral fever	Unknown	Cattle, water buffalos	High fever, limping, astasia
Protozoa			
Leucocytozoon caulleryi	*C. arakawae*	Chickens	Anemia, hemorrhagic, reduction of egg production, weight loss
Filarial nematodes			
Onchocerca cervicalis	Unknown	Horses	Pruritus, dermatitis, hair loss
Onchocerca gutturosa	Unknown	Cattle	Pruritus, dermatitis
Onchocerca gibsoni	Unknown	Cattle	Subcutaneous nodules

world, except a part of the colder portion and island areas. Although BTV circulation has been recognized in various East Asian countries, outbreaks of bluetongue have been reported only in Japan and China (Yanase et al. 2020). A strain of EHDV, named Ibaraki virus, causes severe clinical symptoms in cattle, characterized as deglutitive difficulty and called Ibaraki disease. A large epidemic of Ibaraki disease occurred in 1959 and 1960 in Japan, and over 4000 cattle were lost during the epidemic. Outbreaks of Ibaraki disease were also confirmed in South Korea and Taiwan. In addition, other EHDV strains were suggestively involved in clinical signs resembling Ibaraki disease and in epizootic abortions. African horse sickness is a fatal equine disease caused by the African horse sickness virus (AHSV), which is grouped into orbiviruses (Dennis et al. 2019). African horse sickness is endemic in sub-Saharan Africa but is not prevalent outside of the margin, except for several occasional spreads. In 2020, horses showing typical symptoms of the disease were suddenly found in Thailand, and then AHSV spread and affected domestic horses throughout the country (Bunpapong et al. 2021). This outbreak reminds us of its potential threat to other Asian countries, including Japan.

Several *Culicoides*-borne orthobunyaviruses can cross the ruminant placenta and damage their fetuses, manifesting as abortion, stillbirth, and congenital malformations (Yanase et al. 2020). Of them, the Akabane virus (AKAV) has repeatedly caused large- or medium-scale epidemics with the abovementioned symptoms in cattle in Japan (Forman et al. 2008). Approximately 42,000 calves were affected during the largest epidemic between 1972 and 1975. AKAV is known to be distributed in Africa, the Middle East, Asia, and Australia, and outbreaks have been reported throughout its expanded regions (Sick et al. 2019). Recently, AKAV has also been associated with outbreaks of bovine encephalomyelitis in Japan and Korea (Kono et al. 2008; Oem et al. 2012; Yanase et al. 2018). Outbreaks of abortion and congenital malformations in cattle were also caused by the Aino virus in Japan (Tsuda et al. 2004; Forman et al. 2008). Three other orthobunyaviruses, Peaton, Shamonda, and Sathuperi, are suggestively involved in teratogenic lesions of calves (Yanase et al. 2012; Hirashima et al. 2017; Matsumori et al. 2018). Other than orthobunyaviruses, two *Culicoides*-borne orbiviruses, Chuzan and D'Aguilar viruses, are associated with bovine congenital malformations with hydranencephaly in Japan (Goto et al. 1988; Ohashi et al. 2004).

In Japan, most of the abovementioned viruses have been isolated from *C. oxystoma* more frequently than other *Culicoides* species, suggesting that the distribution and activity of *C. oxystoma* could be high-risk factors for virus spread (Yanase et al. 2005; Kato et al. 2016). However, AKAV occasionally spread in northern Japan, where *C. oxystoma* was not distributed (Yanase et al. 2019a). Oral susceptibility of several other *Culicoides* species to AKAV was confirmed under laboratory conditions (Yanase et al. 2019b). BTV strains were isolated from *C. asiana* and *C. tainanus* but not from *C. oxystoma* in Japan (Shirafuji et al. 2012).

Previous epidemiological and virological investigations indicated that *Culicoides*-borne viruses have not been established in Japan (Forman et al. 2008; Yanase et al. 2018). The infected midges are probably brought by seasonal winds from neighboring regions every summer and may contribute to initiating viral spread in Japan.

Bovine ephemeral fever (BEF) is caused by an ephemerovirus, bovine ephemeral fever virus (BEFV). It is an acute febrile illness of cattle and buffalos, which often results in a significant decrease in milk production in dairy cattle and loss of condition in beef cattle (Lee 2019). Outbreaks of bovine ephemeral fever have repeatedly occurred and caused significant economic loss in Japan and other Asian countries. BEFV strains were isolated from both *Culicoides* biting midges and mosquitoes collected in the field, but their transmission of BEFV to vertebrate hosts has not been proven under laboratory conditions (Stokes et al. 2020). The crucial vector species contributing to the spread of BEFV have not been identified worldwide.

Many avian blood protozoans are transmitted by *Culicoides* biting midges. Although most of them cause little or no apparent symptoms in their avian hosts, a few are harmful to poultry production. Infection of *Leucocytozoon caulleryi* in chickens induces leucocytozoonosis characterized by anemia, hemorrhagia, reduction of egg production, weight loss, and sometimes death (Bermudez 2008). *Leucocytozoon caulleryi is* principally transmitted by *C. arakawae* (Akiba 1960) and is common in various Asian countries with overlapping rage of the vector (Morii et al. 1981; Li et al. 2003; Lee et al. 2016).

Onchocerca species are parasitic filarial nematodes that cause several clinical symptoms in livestock animals. In Japan, three *Onchocerca* species transmitted by *Culicoides* biting midges have been found thus far (Takaoka 1999). Of them, *O. cervicalis* is associated with dermatitis in horses. *O. gutturosa* develops various skin lesions called 'Wahi' in cattle. *O. gibsoni* in cattle forms subcutaneous nodules encapsulating adult worms. These *Onchocerca* species probably spread in Japan and neighboring regions, but no detailed information, including vector identification, has been obtained.

4.2.4 Control and Prevention

Chemical insecticides and repellents have been used to reduce the impacts caused by biting midges. Several trials with chemical insecticides were successful in Japan. For example, the population of *C. circumscriptus* was largely reduced after insecticide treatment to immature habitats in a seashore and a brackish lake (Tabaru et al. 1973; Tanaka et al. 1987). Insect growth regulators were also effective against larvae of *C. circumscriptus* under laboratory conditions (Takahashi et al. 1985). However, because the immature habitats of biting midges are rich in variety and expand vast areas, larviciding might be ineffective for midge control in most other cases. In addition, the impact of chemical insecticides on non-target organisms and food pollution is not acceptable due to increasing environmental and health concerns. Notably, *C. arakawae* and *C. oxystoma*, which are principal vectors of pathogens of veterinary importance, use rice paddies for their immature growth (Yanase et al. 2013). Adultcidal chemicals knock down biting midges in laboratory-based studies and have been partially used in the field to reduce their successful blood-feeding and survival rates (Harrup et al. 2016). In Japan, insecticides are often used for *C. arakawae* in poultry houses to protect chickens from *L. caulleryi* infection (Sakai 2007). Repellents, such as N,N-Diethyl-meta-toluamide (DEET), have been frequently selected for humans and horses to avoid biting by hematophagous midges. However, it is necessary to consider cost-effectiveness if these chemicals are applied to livestock animals.

Physical barriers, such as facilities enclosed by fine meshes, may reduce contact with biting midges. WOAH recommends the barriers to protect horses from the potential transmission of AHSV (OIE 2021). Vaccinations prevent arboviral diseases without vector controls. Commercial vaccines against six arboviruses (Akabane, Aino, Peaton, Chuzan, Ibaraki, and BEF viruses) were developed and are being used for cattle in Japan.

4.2.5 Conclusion

There are many research gaps about the biology of biting midges distributed in Japan and its neighboring countries. In particular, the terrible lack of information on the immature habitat of biting midges makes their management difficult. Taxonomy of biting midges, even though the most studied genera, *Culicoides*, might have to be rearranged. Molecular taxonomic tools, such as DNA barcoding, should be extensively introduced for this group. Although many human and animal pathogens are probably transmitted by biting midges, crucial vector species have been identified for only a few. In addition, wildlife-origin pathogens transmitted by biting midges could possibly cause emerging zoonoses. Biting midges may be neglectable for many people, but continuous and advanced research is essential to reduce their current harm and potential threats in the future.

4.3 Black Fly

Yasushi Otsuka

Black flies belong to the family Simuliidae of the order Diptera. More than 2300 species of black flies have been recorded globally (Adler 2022). Of them, species in eight genera, *Helodon, Levitinia, Prosimulium, Twinnia, Metacnephia, Simulium, Stegopterna* and *Sulcicnephia*, are distributed in Asia. Still, in the Oriental region, only species of the genus *Simulium* are found. Some of these species feed on the blood of humans and animals and transmit filarial nematodes, protozoa, and viruses. The blood-sucking habits of female black flies are responsible for considerable deleterious effects on humans and their economic welfare. Reduction of tourism, death of domesticated birds and mammals and transmission of diseases are medical and socioeconomic impacts associated with black flies.

4.3.1 Biology

Black flies have four developmental stages: egg, larva, pupa and adult. The larval stage usually consists of seven instars (some species are different) and is closely associated with clean and running waters in streams and rivers, where their immature stages develop, and the female adults return for oviposition. Mating occurs soon after adults emerge from pupae. The female and male adults take nectars of flowers as a source of carbohydrates, but only females take blood for egg development, which requires 2–4 days. However, some species are known to be autogenous, enabling egg maturation without blood meals, e.g., *Simulium aureohirtum*, of which females can develop the first batch of eggs without blood meals (Takaoka and Noda 1979). Females with mature eggs in their abdomen are called gravid, and gravid females return to lotic environments to seek sites suitable for oviposition. The gravid females lay eggs in a gelatinous mass. It is approximately 0.6 mm long. Black fly larvae are sausage-shaped organisms with a well-sclerotized head capsule that bears a pair of labral fans. The thorax bears a single ventral proleg, and the last abdominal segment bears a posterior proleg that serves as an attachment organ. The larvae have glands that produce a silken protein to hold the abdomen onto the substrate. The labral fans curve back upstream and serve to catch small bits of food, which are then transferred to the mouth. Food consists of finely divided plant material and small invertebrates. Black flies are one of the principal processors of plant debris in streams. The final-instar larva spins a variously shaped silken cocoon in which to pupate. The cocoons vary in shape. The pupa is recognized by its enclosure in a cocoon, which is attached to the surface of the substrate, and gills arise from both sides of the thorax. Black flies are pool feeders (not capillary feeders like a mosquito). Females cut and injure the skin with the serrated mandibles and imbibe the blood oozing from the wound (blood is not taken directly from a capillary as in the case of mosquitoes). This type of feeding enables black flies to ingest the skin-borne microfilariae of *Onchocerca* parasites.

4.3.2 Public Health Importance

The importance of black flies to humans revolves primarily around the harmful habits of blood-seeking females. The problem of biting and discomfort caused by black flies has had a serious impact on most outdoor activities, including agriculture, forestry, industrial development, military exercises, mining, and tourism. The actual financial losses due to grievous and nuisance problems in different sectors of the economy are significant (Adler et al. 2004; Sariözkan et al. 2014).

In Japan, approximately 20 species of black flies in four genera, *Prosimulium, Twinnia, Stegopterna and Simulium,* are known to suck human blood (Bentinck 1955; Ogata 1955; Ogata et al. 1956; Ono 1982). Among the species, *Simulium arakawae, Simulium nacojapi* and *Simulium japonicum* (Fig. 4.2a) are known to be

Fig. 4.2 Females of black flies. (**a**) *Simulium japonicum* and (**b**) *Simulium bidentatum*

intensely man-sucking in certain areas. In Kashiwazaki, Niigata Prefecture, *S. arakawae* is found in large numbers and has caused considerable blood-sucking damage to residents (Hirabayashi et al. 2018). In Kyoto, the damage caused by *S. nacojapi* was so severe that in the 1950s, lindane emulsion was used to control the larvae of the black fly in natural streams (Shogaki and Yoshida 1958). *Simulium japonicum* causes severe damage in some areas, especially in Hokkaido and some islands of the Ryukyu Archipelago. Nakano-shima Island in Kagoshima Prefecture has long suffered from blood-sucking damage caused by *S. japonicum.* Since the 1980s, the larvae of black flies have been exterminated by using insecticides in the river. Initially, they used Abbate, but now they use microbial control agent (*Bacillus thuringiensis*). Since the use of insecticides, the number of daylilies has decreased, but there are still many, mainly in May, making life difficult for people (Otsuka 2020). In Thailand, seven black fly species of the genus *Simulium* are regarded as man-biting species (Choochote et al. 2005; Pramual et al. 2016; Takaoka et al. 2017). Among them, two species, *Simulium nigrogilvum* and *Simulium nodosum*, are considered pests and are only found in some areas in northern Thailand. In the northern part of India, some black fly species of the genus *Simulium* have been identified (Lewis 1974; Datta 1992). Of them, *Simulium indicum* and *S. nodosum* are considered pests. In particular, *S. indicum* is widely distributed at the southern foot of the Himalayas, from Afghanistan in the west through Pakistan, India, Nepal, Bangladesh and Burma to southern China in the east (Adler 2022). In Malaysia, swarming and biting activities of black flies are hardly observed, but *Simulium aziruni* was found to be attracted to and landing on human skin (Izwan-Anas et al. 2021).

Skin dermatitis as a result of a hypersensitivity reaction to biting black flies is characterized by chronic eruptions of pruritic papules, vesicles, and erythematous wheals. It can result in severe skin complications where the man-biting species of black flies are available. The clinical manifestations and severity of dermatitis caused by biting black flies are serious matters of concern in hypersensitive persons (Borah et al. 2012). Characterization of the IgE-binding proteins in the salivary glands of *S. nigrogilvum*, a black fly pest species in northern Thailand, detected nine proteins involved in allergic reactions (Hempolchom et al. 2019).

The most significant human disease transmitted by black flies is human onchocerciasis (also called river blindness in endemic areas in Africa) caused by *Onchocerca volvulus,* a species of filamentous fungus found in Africa and Central and South America. In the early 1970s, when the World Health Organization (WHO) and other organizations initiated a control program in the Volta River basin of West Africa, it was estimated that approximately 18 million people were infected with the disease, which was a major health and socioeconomic burden for the residents of the endemic areas (Adler et al. 2004).

Since there were no previously known important black fly-borne infectious diseases in Asia, including Japan, few studies have been conducted on black flies as disease vectors. In 1987, a human infection by the animal parasite *Onchocerca* was found in Oita, Japan (Beaver et al. 1989; Hashimoto et al. 1990). This was the first case of zoonotic onchocerciasis in Japan and Asia. Before the first case was reported in Japan, zoonotic onchocerciasis had been reported only in Canada (Ali-Khan 1977), Switzerland (Siegenthaler and Gubler 1965), Ukraine (Azarova et al. 1965), and the United States (Beaver et al. 1974) and was considered to be a rare disease, although it is widespread worldwide. In Japan, a second case was found in Oita in 1995 (Takaoka et al. 1996), followed by four cases in Oita (one of which was diagnosed in a hospital in Fukuoka Prefecture; the location of infection was not specified), three in Hiroshima, and one each in Shimane, Shiga, Fukushima, and Chiba, for a total of 13 cases (Takaoka et al. 2001, 2004, 2005; Uni et al. 2010, 2015b; Fukuda et al. 2011, 2015, 2019; Ishizawa et al. 2015; Okazaki et al. 2022). Initially, the causal species was unknown but was later identified as *Onchocerca japonica*, a parasite of wild boars (Takaoka et al. 2001; Uni et al. 2001). Approximately 20 cases of zoonotic onchocerciasis have been reported outside Japan since then, most of which were caused by *Onchocerca lupi*, which infects dogs and cats, and one case in Iran caused by *O. lupi* was reported from Asia (Mowlavi et al. 2014; Grácio et al. 2015; Cambra-Pellejà et al. 2020). There is also a case from Kuwait caused by *Onchocerca* sp. (Hira et al. 2008).

Since the discovery of the first case of zoonotic onchocerciasis in Japan, efforts have been made to identify the vector species. Black flies and biting midges (*Culicoides* spp.) are considered possible candidate vectors, and natural infection of filarial larvae has been found in black flies (Takaoka et al. 1989, 1992; Takaoka and Bain 1990; Takaoka et al. 1995). Fukuda et al. (2010a, b) dissected adult female black flies collected in Oita and were the first to show that anthropophilic *Simulium bidentatum* (Fig. 4.2b) is the natural vector of wild boar parasites, *O. japonica*, based on the morphological characteristics of the third-stage larvae detected and the results of DNA analysis. Five other anthropophilic blackfly species, *S. arakawae*, *S. japonicum*, *Simulium oitanum*, *Simulium tanegashimaense,* and *Simulium yamatoense*, have been revealed as putative vectors of *O. japonica* through intrathoracic inoculation experiments (Fukuda et al. 2008). In the United States, the DNA of *O. lupi* has been found in *Simulium tribulatum* and implicates *S. tribulatum* as a putative vector for *O. lupi* in southern California (Hassan et al. 2015).

4.3.3 Veterinary Importance

The veterinary importance of black flies is attributed to biting and swarming. Persistent attacks by black flies can cause unruly host behavior, weight loss, decreased egg and milk production, malnutrition in young animals, dermatitis and epidermal necrosis, abortions, and possibly stress-related diseases such as pneumonia, which can lead to enormous economic losses (Adler et al. 2004). Black fly-borne infections can also cause problems, such as filarial nematodes, protozoa, and possibly some viruses being transmitted to animals.

Onchocerca species are known in Japan as parasites of cattle, horses, and other domestic animals. In connection with the etiology of a bovine skin disease called "kose" or "wahi," *Onchocerca gutturosa* was reported from cattle in Tottori Prefecture, Japan, and was identified from adult and microfilarial morphology (Sato et al. 1954). A year before this report, Niimi and Kono (1953) reported the presence of two types of microfilariae (referred to as X- and Y-types) in the skin of cattle. The tail of the X-type is strongly coiled, while that of the Y-type is not. Additionally, the body length of the X-type is shorter than that of the Y-type. (0.110–0.140 mm vs. 0.195–0.210 mm). Subsequently, the similarity of *Onchocerca lienalis* to *O. gutturosa* became an issue in Europe and the United States, and *O. lienalis* was distinguished as an independent species (Bain et al. 1978; Eberhard 1979), but this issue was not discussed in Japan. Incidentally, in Europe and the United States, the microfilaria of both species is Y-type.

As mentioned above, to identify vectors of zoonotic onchocerciasis in Japan, black flies collected from cattle sheds in Oita were dissected to investigate the larvae of filaria that black flies carried. Takaoka and Bain (1990) found two types of microfilariae (X- and Y-types) and three types of third stage larvae (infective larvae) (Type I, II, III). Type X-type microfilariae were found in both the neck and abdomen of the skin of cattle in Oita, and experimental infection of adult *S. bidentatum* that had hatched from pupae demonstrated that they develop into Type I third instar larvae (Takaoka 1990). Adults (*Onchocerca* sp.) of Type I larvae have not been identified, but *S. bidentatum* and *S. arakawae* are known to transmit them. In Iwate, *Simulium daisense* and *S. oitanum* also transmitted them. In Thailand, type I larvae were found in *S. nigrogilvum* (Saeung et al. 2020).

Y-type larvae developed into Types II and III; Type III was identified as *O. lienalis*, and Type II was tentatively identified as *O. gutturosa* (Takaoka 1997). In addition to *O. japonica*, which is the causative species of zoonotic filariasis parasites of wild boars, *Onchocerca takaokai* is also known as wild boars (Uni et al. 2015a). Using the same method as for *O. japonica*, *S. bidentaum* was found to be a vector of *O. takaokai* (Fukuda et al. 2008). Fukuda et al. (2008) used microfilariae of *O. japonica* and *O. takaokai* obtained from wild boar skin to experimentally infect blackfly adults, which resembled each other in size and morphology and were difficult to distinguish from Type II previously reported by Takaoka and Bain (1990). It is possible that the third-stage larvae (Type II) found in the *S. bidentatum* collected in cattle sheds in Oita were not *O. gutturosa* but *O. japonica* or *O. takaokai* in wild boars. Considering these findings, it is difficult to conclude that *S. bidentatum* is the vector of *O. gutturosa*, and further investigation is needed. Outside of Asia, *O. gutturosa* has been shown to be transmitted by *Simulium* spp. and biting midges *Culicoides* spp. in infection experiments (Mikhailyuk 1967; Bain 1979; El-Sinnary and Hussein 1980; Dohnal et al. 1990). In addition to *S. arakawae* in Oita, *Simulium kyushuense* in Kumamoto and *S. daisense* in Iwate were added as vectors of *O. lienalis*.

As a wild boar parasitic filaria, *Onchocerca dewittei* is known in Malaysia, but the vector is unknown (Bain et al. 1977).

Among Japanese cattle, *Onchocerca gibsoni* has been reported in cattle in Tokyo (Itagaki 1954). This species is commonly found in cattle in Asia and Australia, but there are no subsequent records in Japan, and the vector is unknown in Japan. It has been reported that biting midges of *Culilioides* spp. were transmitted in Malaysia (Buckley 1938).

Although the vector of *Onchocerca suzukii* in Japanese serow is still unknown, Iikawa et al. (2009) and Sato et al. (2009b) detected the specific *COI* gene sequence of *O. suzukii* in *S. japonicum* and *Prosimulium* sp., suggesting that these two species were the vectors of *O. suzukii*. Iikawa et al. (2009) and Sato et al. (2009b) also detected *Onchocerca skrjabini*-specific gene sequences from these two species and *Twinnia japonensis*.

The association between the incidence of teat papillomatosis in cattle and the number of *S. oitanum* in Japan suggests that *S. oitanum* is a species that plays an important role in the transmission of teat papillomatosis in cattle (Iwasa and Shido 2016).

Blood-parasitic protozoa (*Leucocytozoon lavati*) have been found to infect Japanese rock ptarmigans, one of the special monuments of Japan that inhabit the Japanese Alps, suggesting the three *Prosimulium* species, *S. japonicum* and *Simulium uchidai,* as vectors (Sato et al. 2009a; Imura et al. 2010). In Thailand, at least two species (*Simulium asakoae* and *Simulium chumpornense*) are possible vectors of avian parasites of the genera *Leucocytozoon* and *Trypanosoma* (Jumpato et al. 2019; Thaijarern et al. 2019; Pramual et al. 2020).

4.3.4 Conclusion

Since human onchocerciasis does not exist in Asia, there has not been much research on black fly-borne diseases in Asia, including Japan. However, since the first case of zoonotic onchocerciasis in Japan in 1989, studies have focused on onchocerciasis vectors. The results of many studies have gradually clarified the role of black flies as vectors in Japan. In recent years, research on black flies has progressed remarkably in Southeast Asia, and filariasis and protozoa transmitted by black flies have been reported. It is hoped that further research in Asia outside of Japan will provide a comprehensive view of the role of black flies as vectors in Asia.

4.4 Chigger Mite (Tsutsugamushi)

Kandai Doi

Chigger mites, also known as "harvest mites" or "chiggers," are a group of mites in the families Trombidiculidae and Leewuwenhoekiidae (there are still many discussions in the present day) distributed worldwide. Approximately 3000 species of chigger mites have been confirmed to date. In Japan, 139 species are known to be found (Takada and Takahashi 2019).

The larval mite (approximately 0.2 mm) bites the rodent and makes a stylostome. The tube penetrates the epidermis with its saliva to feed on tissue fluid of mammals (mainly rodents), birds, and reptiles (*Vatacarus ipoides*, infesting the lung of sea snakes and *Doloisia syntonic*, infesting the nasal cavity of bats, and *Laurentella* spp. infesting the nasal cavity of rodents are the exceptions), and sometimes Gecarcinids, terrestrial crab, (*Eutrombicula* (*Siseca*) *hematocheiri* infests 57% of terrestrial crabs) (Takada and Takahashi 2019). Engorged larvae drop off from the host, become protonymphs (pupal form), and develop into deutonymphs after molting. Deutonymph and other older developmental stages, Tritonymph and adult stage (approximately 1 mm), are free-living, unlike the parasitic larva. They usually feed on the eggs of insects (e.g., mosquitoes, dragonflies, springtails, etc.). For reproduction, males deposit stalked spermatophores in the substrate, and females take spermatophores into genital valves to germinate eggs. The deposited eggs clustered. Thus, the larvae were often found in small spots within a 50 cm diameter circle (Takada 1990; Traub and Wisseman Jr 1974).

4.4.1 Medical Importance of Chigger Mites

Among the species of chigger mites in the world, medically important species, which were confirmed to function as vectors of *Orientia tsutsugamushi* (or *Rickettsia tsutsugamushi*; hereafter "Ot"), the pathogen of tsutsugamushi disease or scrub typhus, is limited to several species of the genus *Leptotrombidium* (Fig. 4.3). *Leptotrombidium akamushi*, *L. fuji*, *L. pallidum*, *L. palpale*, *L. scutellare*, and *L. tosa* are vectors of tsutsugamushi disease in Japan, and *L. delicense* and *L. fletcheri* are vectors in the Southeast Asian region. *Leptotrombidium delicense* is also found in Okinawa, Japan (Bonell et al. 2017; Takada and Takahashi 2019; Lv et al. 2021). However, Takada and Takahashi (2019) discussed that *L. delicense* in Japan was probably introduced with imported soil from Taiwan for sugarcane plantation during the first half of the twentieth century.

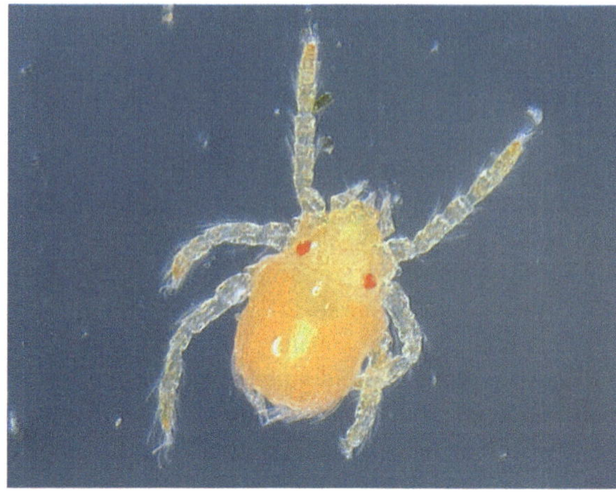

Fig. 4.3 Larva of *Leptotrombidium* sp.

4.4.2 History of Tsutsugamushi Disease

The history of tsutsugamushi disease can be traced back to the book written by Shizhen Li, a Chinese herbalist of the Ming Dynasty. He wrote that tiny bugs in the mountain sometimes bite people and cause severe fever. "Tsutsugamushi" means "disastrous bug" in Japanese and was described in a book written during the eighteenth century as a monster, who lived deep in the mountains and sucked the living blood of people at night, causing severe fever. In endemic regions, local people call chigger mites, as "Aka-mushi" means a red bug, or "Kedani" means a small mite. In those endemic regions, stone pillars and shrines dedicated to the deity, called "Akamushi Daimyojin" or "Kedani Myojin," were built and have been handed down to the present day (Kobayashi 2016).

After World War II, soldiers from the United States stationed in Japan were infected with tsutsugamushi disease in the area surrounding Mt. Fuji. Similar endemic diseases were found in many regions of Japan, such as on Izu Islands called "Shichito fever" means "seven island fever", "Hatsuka fever" means "twenty-day fever" in Chiba Prefecture, and "Hoppan" means "red skin rush" in Kochi Prefecture. These diseases have now been proven to be caused by Ot. Tsutsugamushi disease is also known to have two types: the classical type emerges during the summer season in the northern part of Japan and is caused by *L. akamushi,* and the new type emerges between autumn and early spring in other parts of the region. This seasonal pattern is known to be caused by the climatic characteristics that affect the phenology of vector chigger mites (Takada 1990).

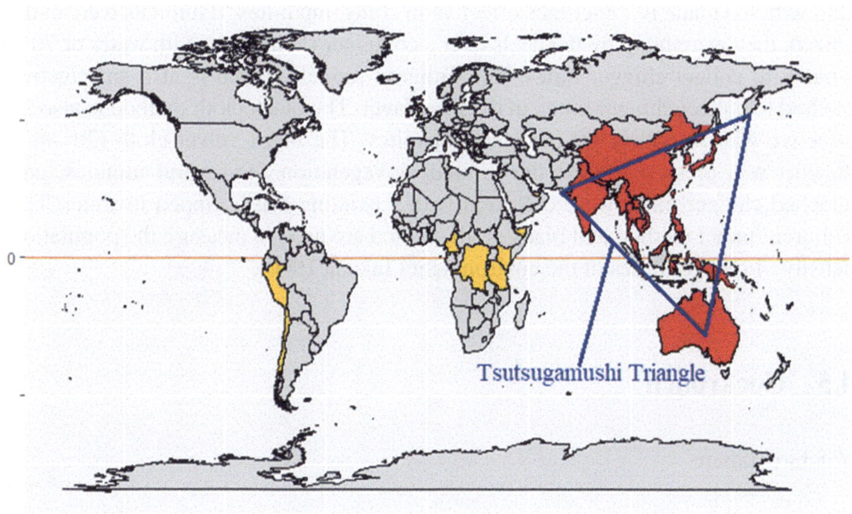

Fig. 4.4 Map of emergence of tsutsugamushi disease. The red region is the classically known endemic region of tsutsugamushi disease in the Tsutsugamushi Triangle (blue triangle). Yellow regions were where tsutsugamushi disease recently emerged. Adapted from Bonell et al. (2017), Xu et al. (2017), and Jiang and Richards (2018)

4.4.3 Ecology of Tsutsugamushi Disease

The relationship between chigger mites and Ot is also remarkably mysterious. The Ot can be transmitted from infected mites to host animals but not from the infected host to mites. The Ot in mites is only passed to other individuals of mites through transovarial infection. This seems to indicate that the population of Ot declines rapidly, but a previous study suggested that Ot causes the feminization of the next generation of infected chigger mites (Ponnusamy et al. 2018).

For a long time, the emergence of tsutsugamushi disease was known to be limited to the area called the "Tsutsugamushi triangle," the area of the eastern rim of the Pacific Ocean (Fig. 4.4). However, recent studies have reported the emergence of tsutsugamushi diseases in South America, Africa, and the Arabian Peninsula (Bonell et al. 2017; Jiang and Richards 2018; Xu et al. 2017). Even though medical treatment is effective in curing this disease, the expansion of endemic areas and imported cases requires vigilance.

4.4.4 Collecting of Chigger Mites

There are several standard methods of collecting chigger mites. Collecting from animals is probably the most well-known method. Chigger mites are often found in the ear canal of mammals. Forceps can pick up the mites directly, or scraping the skin with the blade is sometimes effective in collecting mites. If animals were euthanized, they were hung by the heels over a container (Petri dish) with water or 70% ethanol to collect chigger mites. The Tullgren funnel method is also an effective method to collect chigger mites in the litter layer. The black cloth method is also an effective way to collect off-host chigger mites. The black velvet cloth (20 cm x 20 cm) was pressed against the ground or vegetation for several minutes, and attached chigger mites were collected with a painting brush dipped in water. The Tullgren funnel method and black cloth method are used to measure the population density of chigger mites in the environment (Takada 1990).

4.5 Cockroach

Yuichiro Tabaru

The ancient type of cockroach trapped in a fossil is estimated to have been in the Carboniferous period (approximately 350 million years ago). The cockroach is classified as the order Blattodea, family Blattidae, and has recorded more than 4400 species in the world (Cornwell 1968). It belongs to incomplete metamorphosis insects. Some domesticated cockroaches are widely distributed in urban areas in tropical, subtropical, and temperate zones in Asia. However, the majority of these organisms inhabit rainforests and act as decomposers of organic matter. Even in Japan, 55 species have been recorded. Still, the majority are distributed in forests on the Ryukyu Islands and Ogasawara Islands (Asahina 1991), and recently Komatsu and Itou (2023) recorded 64 species and 7 sub-species from Japan. In temperate zones, *Blattella germanica*, *Periplaneta americana*, *P. brunnea*, and *P. australasiae* are distributed only in air-conditioned environments and are difficult to adapt to in residents in Japan. The smoky-brown cockroach *P. fuliginosa* is widely distributed among Japanese residents. The cockroach is recognized as a nuisance pest as well as an industrial pest. Cockroach control is the most important for pest control operators in developed countries. They are the most prejudiced creatures in residents, restaurants, hotels, hospitals, and food processing plants. German cockroach control is facing difficulty due to insecticide resistance in restaurants, hotels, and food processing plants.

4.5.1 Medical Importance

In developed countries, biological transmission of infectious diseases carried by cockroaches has not been proven. Roth and Willis (1957) reported that cockroaches could transmit viruses, bacteriocins, protozoa, and parasitic diseases. Still, it was the background of an era of unsatisfactory environmental health, and sewerage systems were not available. When cockroaches inhabit unsanitary environments, they

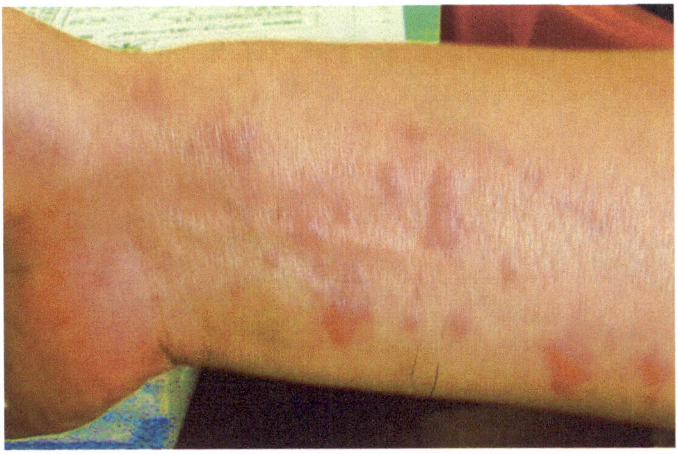

Fig. 4.5 Dermal allergy symptoms appeared on the author's forearm after handling German cockroaches in a laboratory

might carry unexpected pathogens. It is clear that cockroaches cause allergies as well as house dust. Powdered feces and exuvia (left old skin at molting) cause allergies and asthma in humans due to the existence of allergens (Fig. 4.5). Hitomi (1957) reported for the first time in an academic journal on the German cockroach outbreak in ocean ships. After cockroach infestation in hospitals, coal mining in Nagasaki and Hokkaido was reported in the 1950s, cockroaches might transmit polio, which was spreading around Japan at that time. Cockroach phobia or vehement hatred of cockroaches is common among housewives in Japan.

4.5.2 Nuisance Pest

The cockroach is omnivorous and one of the most obnoxious insects because it is very common in resident's kitchens. The German cockroach is the most serious industrial pest. Food processing companies and restaurants are very nervous about the existence of cockroaches because food contamination, food container damage, and package pollution by cockroaches are serious (Fig. 4.6). If the carcass or fragments of cockroaches are found in their products by consumers, the fact may be reported in the media. Then, the producer will suffer from irrevocable damage. We can see numerous cockroaches crowded on religious offerings in the streets in some Asian countries. It is a wonder why Japanese housewives hesitated seriously about cockroaches. Author made a questionnaire against 100 housewives about why they were disgusted with cockroaches; the answers were the following: bud smell, gathering on foods, haunting the kitchen, quick movement, flying toward me, difficult to catch, etc. (Tabaru 2015). The smoky brown cockroach is very common among residents who introduce sticky cockroach traps, which are sold at the top in drug stores

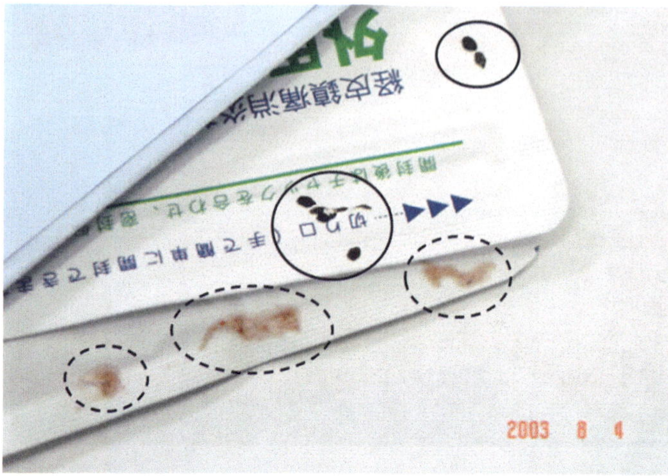

Fig. 4.6 Fecal contamination on the poultices medicines packages by the smoky-brown cockroach. Two different colors, brown (solid circles) and black (dashed circles), cause different tablet colors that they ingested

and DIY during the warm season in Japan. There are many TV CMs on cockroach control items in the summer season.

4.5.3 Main Species

4.5.3.1 German Cockroach (*Blattella germanica*)

Blattella germanica is the smallest cockroach in buildings; the female is 12 mm, and the male is 11 mm long. The color is light brown, and they have four wings but cannot fly. Females keep the egg capsule until hatching. In temperate regions such as Japan, the cockroach habitat is maintained only in air-conditioned buildings, maintaining a high breeding rate but not in residents, and it is the most important cockroach habitat for the pest control operator (PCO). Most colonies develop insecticide resistance. A similar species to the German cockroach, *B. nipponica*, inhabits outdoors in Japan, but sometimes they invade rural factories. The pronotum is different in the two species (Fig. 4.7).

4.5.3.2 Smoky Brown Cockroaches (*Periplaneta fuliginosa*)

Periplaneta fuliginosa is glossy black: the adult is 25–30 mm. They can fly toward the light at night. Femail sticks egg capsule on wood panels. The first instar nymph is black with a white band in the abdomen. The cockroaches live in residential

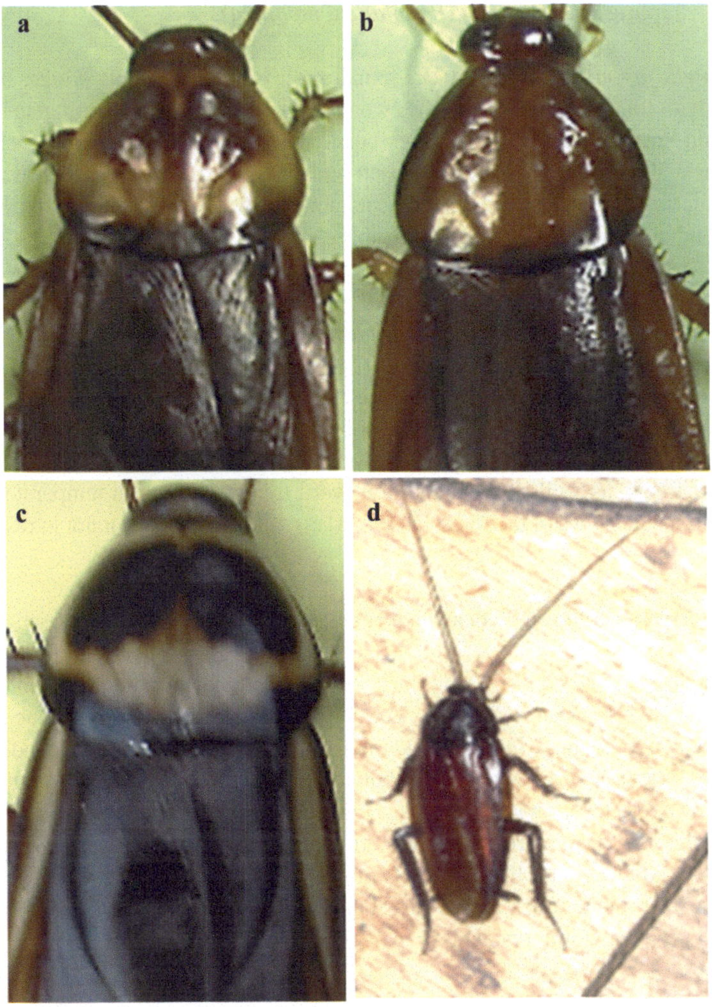

Fig. 4.7 Difference of pronotum. (**a**) *Periplaneta amaricana*, (**b**) *P. brunnea*, (**c**) *P. australasiae*, and (**d**) *P. fuliginosa*

houses and high-rise buildings, and animals are burning from metropolitan areas to southern Japan. The cockroach is very active. Feces of this species have a high viscosity that causes package pollution (Hirao et al. 2006). In the resident, the cockroach survives in a dormant stage during the winter season, followed by repeat regeneration from March through October.

4.5.3.3 Japanese Cockroach (*Periplaneta japonica*)

Periplaneta japonica is light black: female is 20 mm with short wings; male is 25 mm with long wings. Cockroaches were very common in northern Japan before. They can survive under fallen leaves covered with snow in winter (Tabaru and Kobayashi 1971). Recently the cockroach replaced *P. fuliginosa* due to difficulty in overwintering in the residents because inhabitants wake up late under light. This species inhabits outdoor areas, namely, in tree hollows in parks.

4.5.3.4 American Cockroaches (*Periplaneta americana*)

Periplaneta americana is 30–40 mm long and yellowish spots in the pronotum and is widely distributed in the tropical and subtropical regions of Asia. The cockroach inhabits not only residents but also outdoor environments, such as manholes and altars in streets, as well as trash and boulevard tree roadsides. In temperate countries, this species lives in factories, supermarkets, and manholes but not in residents.

4.5.3.5 Brown Brunnea Cockroach (*Periplaneta brunnea*)

Periplaneta brunnea is 30–37 mm, lacks a yellowish mark in the pronotum, but is similar to the American cockroach. The living area is almost the same as that of American cockroaches. In Japan, adapting areas are limited to air-conditioned buildings, manholes, and zoo gardens but not to residents. The activity is dull compared to that of the previous species. *Periplaneta australis* resembles a previous species; the yellowish marks in the pronotum are clearer than those in the American cockroach, and there are light yellowish marks on the front wings. This species is common in tropical zones and is distributed only in air-conditioned buildings in southern districts, not in residents (Fig. 4.8).

4.5.3.6 Brown-Banded Cockroach (*Spella lomgipalpa*)

Spella lomgipalpa is a small cockroach that is 12–13 mm long. The female cockroach has short wings, similar to the German cockroach but darker in color. Recently, this species has been expanding in distribution in tropical and subtropical areas. This species inhabits a dry environment compared with *B. germanica*.

4.5.4 Behavior of the Important Species

Cockroaches are nocturnal. They hide in cracks and clevis during the daytime. The young German cockroach gathers in a narrow crevice via an aggregation pheromone. Cockroaches are omnivorous. Tsuji (1966) reported on the attractive and

Fig. 4.8 Difference of pronotun in *B. germanica* (**a**) and *B. nipponia* (**b**). Two lines are almost strate in *B. germanica* whereas *B. nipponia* are bent in under

feeding stimulative effect of fatty acids on cockroaches. Methyl myristate was the most attractive for several kinds of cockroaches.

German cockroaches don't invade residents in Japan (except Okinawa) because it is hard to get through the winter. However, this species is most popular in air-conditioned environments in Japan, such as restaurants, hotels, food-processing plants, and hospitals. Kosone et al. (1995) reported the movement of *B. germanica* by the catch-and-release method in a restaurant. The cockroach did not move to a remote location from the released point. The difference in insecticide susceptibility levels in the 13 restaurants on the same floor of multitenant buildings in Tokyo, *B. germanica* didn't seem to move frequently inter-restaurants. They repeated breeding, staying in each kitchen of the restaurants (Watabe and Tabaru 2010).

Periplaneta fuliginosa is very popular in mainland Japan. Its habitat consists of residents, high-rise apartments, supermarkets, and animal burns from spring to late autumn. The cockroach also inhabits outdoors, such as tree caves, roadside trees, under vending machines, and compost storages.

Periplaneta japonica was common in residents of northern Japan by the 1950s. The cockroach can overwinter in piled fire trees in the snowy area of Japan (Tabaru and Kobayashi 1971). However, this species has been very rare recently in residents, and the cockroach is common in outdoor environments, such as hollows in large tree hollows in parks. *B. japonica* has been translated to an outdoor environment by residents because inhabitants stay up late at night under light, so the cockroach cannot enter overwintering without recognizing short-day conditions. Recently, wandering *P. amaricana* was observed in the center of Tokyo (Nakano 2023). This species develops in a manhole. Resistance to thirst is very weak in *B. germanica,* and they cannot survive for weeks without water.

4.5.5 *Control*

Cockroach control is the most important issue for pest control operators. Restaurants, hotels, and food-processing companies have contracted with PCO to control cockroaches. After the introduction of the Act on Maintenance of Sanitation in Buildings was established in 1970, cockroach control was scheduled by the management of buildings. In addition to the act, the Japan Pest Control Association (JPCA) launched the original standard to control pests in industries (IPM manifesto), including cockroaches, in 2002. In this standard, the investigation stipulated a more important issue for pests in industrial companies. In Japan, the bait poison hydramethylnon, fipronil, or imidacloprid is used, and the organophosphorus compound propetamphos MC has been successfully used for residual treatment. Recently, broflanilide was launched in the market as a cockroach control insecticide. Pyrethroid compounds are not recommended for cockroach control due to repellent activities. If they use pyrethroid compounds, the cockroaches will spread over a wide area. Insecticide resistance is a serious problem in urban environments against *B. germanica.* Sticky traps are commonly used for cockroach density inspection and control evaluation in fields. However, we should pay attention to sticky trap avoidance (Tabaru et al. 2011). When the sticky traps are placed for long days, the cockroaches might be reluctant to cross the traps, introducing inaccurate information (Fig. 4.9).

4.6 Flea

Suha Kenan Arserim

Fleas are small (1–8 mm), wingless, laterally compressed bodies and obligate blood-sucking ectoparasitic insects of mammalian and avian hosts. Ninety-four percent of known species of fleas occur in mammals, and the rest occur in birds (Bitam et al. 2010). Fleas are distributed worldwide, including in Antarctica, and are the most common ectoparasites of domesticated animals such as cats and dogs. In the wild, fleas have been identified in rodents, bats, and opossums, which could be a potential route for parasite transmission to humans. Fleas impact human health in several ways, including biting pests, intermediate hosts, and vectors of human pathogens. The most notable flea-borne human diseases include plague, murine typhus, tungiasis, cat scratch disease and flea allergy dermatitis.

4.6.1 *Morphology*

Adult fleas have numerous bristles, shiny, varying yellowish brown to almost black bodies with modified hind legs and metathorax for jumping (Fig. 4.10). Different species have varying numbers and placements of setae, the presence of frontal

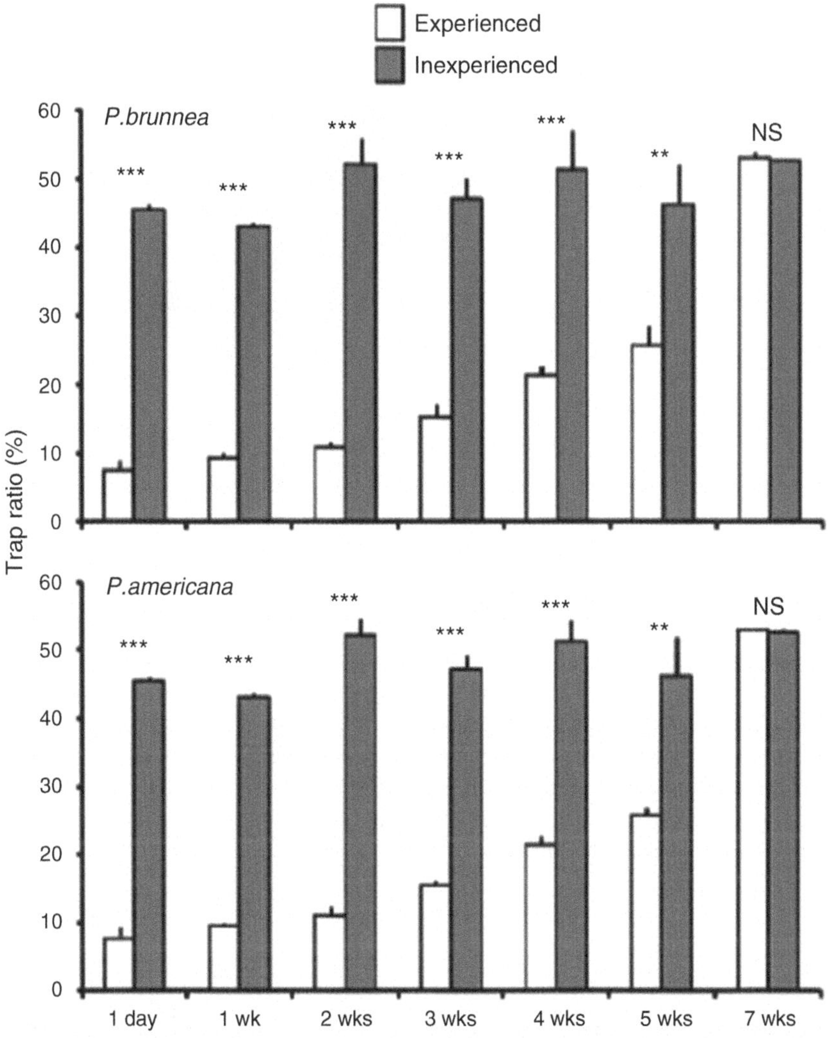

Fig. 4.9 Sticky trap avoidance in two kinds of cockroaches; the trapping ratio was quite different between experienced ones and inexperienced ones. Experienced cockroaches survived the group for 10 days without trapping whereas inexperienced ones were transferred from breeding containers. Asterisks indicate a significant difference between the experienced and inexperienced groups (***$p < 0.001$, **$p < 0.01$, ns: $p > 0.05$; t-test)

tubercles on the head and the appearance of mouthparts. Unlike other insects, fleas do not have compound eyes; instead, they have ocelli-like eyes. In some fleas, the eyes are either vestigial or absent, as in burrowing fleas (Gage 2005). The mouthparts are specialized for piercing and blood sucking. The abdomen is divided into pregenital segments (I–VII) and genital segments (VIII–X) (Durden and Hinkle

Fig. 4.10 General flea morphology

2019). In males, the genital segments are more modified compared to females and are useful in taxonomical identification, mostly in separating species with identical females. These modified male genital structures help to grasp the female during copulation. In females, the most important female reproductive structure is the spermatheca, whose structure (the head and tail) varies greatly among different taxa (Gage 2005). Another important taxonomical structure is the leg, which bears setae and bristles whose positions are significant in separating different taxa. While some fleas can jump to a maximum of approximately 20–30 cm, some flies have lost the ability to jump especially on certain birds, bats and slow-moving mammals (Gage 2005; Durden and Hinkle 2019).

4.6.1.1 Life Cycle

Fleas are holometabolous insects that have complete metamorphosis, with four major life stages: egg, larva, pupae and adult. On average, a female flea deposits several eggs under optimal conditions on the host itself or the host's nest or burrow. The eggs usually hatch after approximately 5 days into worm-like larvae with mouthparts to help them feed on organic matter (Durden and Hinkle 2019). However, in some species, larvae need to feed on blood, which adults supply through ejected feces with remnants of digested blood (Bitam et al. 2010; Durden and Hinkle 2019). After 2 or 3 weeks, the larvae molt twice and produce silk cocoons, where they molt into pupa. The pupa requires a stimulus (vibration) to emerge from the cocoon as adults (Hastriter and Whiting 2009; Bitam et al. 2010). Upon emergence, adults can copulate and deposit eggs after sufficient nourishment. Temperature and humidity affect the time it takes fleas to go from egg to adult. The life cycle length also varies with the type of host, with burrowing hosts being infested with fleas with an extended life cycle.

4.6.1.2 History in Japan

Fleas and flea-borne diseases in Japan can be traced back to August 1899 when the first case of plague, a bacterial disease transmitted by fleas, was detected in Yokohama (Eldridge 1900). Since then, flea-borne diseases have become a public health problem in the country (Tsukahara 2002). For instance, the human flea, *Pulex irritans*, was a common ectoparasite detected on humans, dogs, livestock, and wild animals until the 1970s and then reemerged in 1983 when the last case was reported. Since then, new cases of human flea infestation have been recently reported in Hokkaido, Japan, in 2015 (Takahashi et al. 2016b). Fleas have also been detected in wild animals such as mongoose, bats, and rodents (Ishibashi et al. 2009; Takahashi et al. 2016a).

4.6.2 Taxonomy

Fleas are arthropods and insects belonging to the order Siphonaptera, with approximately 2500 species placed in 18 families (Durden and Hinkle 2019). Evidence suggests that the first flea was identified in mammals 125 or more million years ago and that avian fleas evolved from mammalian fleas (Durden and Hinkle 2019). The 18 families are: Pulicidae, Tungidae, Malacopsyllidae, Rhopalopsyllidae, Vermipsyllidae, Coptopsyllidae, Ancistropsyllidae, Lycopsyllidae, Pygiopsyllidae, Stivaliidae, Hystrichopsyllidae, Chimaeropsyllidae, Macropsyllidae, Stephanocircidae, Ceratophyllidae, Leptopsyllidae, Ischnopsyllidae, and Xiphiopsyllidae. Species belonging to Pulicidae, Tungidae, Leptopsyllidae, Vermipsyllidae and Ceratophyllidae are of medical and veterinary importance (Durden and Hinkle 2019). The taxonomy of fleas is primarily based on morphological structures of the head, thorax, and abdomen (Bitam et al. 2010). Molecular techniques are currently being used to identify ungrouped species and their phylogenetic relationships (Yssouf et al. 2014).

4.6.3 Distribution

Although fleas are found all over the globe except Antarctica, some families are restricted in their distribution. Members of Ceratophyllidae (rodents and bird fleas) are predominantly found in the Holarctic region, with few genera and species identified in the Southern Hemisphere, and are cosmopolitan (Bitam et al. 2010). Most species in this family are parasites of small rodents, while a few are associated with birds. Significantly, members of this family are the only fleas found in Antarctica.

Species belonging to the family Pulicidae (common fleas) are cosmopolitan and found on every continent except Antarctica. They have a very broad range of hosts, including carnivores, insectivores, ungulates, hyraxes, bats and birds. Some important fleas belonging to this family include *Pulex irritans* (house or human flea), *Xenopsylla cheopis* (oriental or tropical rat flea), *Ctenocephalides canis* (dog flea) and *Ctenocephalides felis* (cat flea). *Xenopsylla cheopis* is commonly found in rat-infested regions of Africa and central and southern Asia. The dog flea and cat flea are closely related in appearance and biology, and both have nearly cosmopolitan distributions (Bitam et al. 2010). *Ctenocephalides felis* is the most common dog and cat flea, and its wide distribution is believed to result from increased human movement and domestication of dogs and cats. A survey of parasitic fleas on Asian mongoose on Okinawa Island, Japan, in 2009 reported 370 ectoparasites belonging to *C. felis* (Ishibashi et al. 2009). In addition, a wide distribution of fleas on cats and dogs was reported on the island. Seven species of the family Pulicidae in neighboring Russia, such as *P. irritans*, *C. felis*, *C. canis,* and *X. cheopis*, have been reported on mammals and birds in the Great Caucasus.

Family Tungidae includes species found in the Americas, tropical Africa, and Eastern Asia (Durden and Hinkle 2019). The hosts of these fleas include humans, domestic livestock, rodents, and insectivores. The most medically important species is *Tunga penetrans*, commonly known as chigoe or jigger, which causes tungiasis.

Hystrichopsyllidae (rodent fleas) are distributed in the Nearctic, Palearctic, Neotropics and Australia. They predominantly hosted on rodents and insectivores. Members of the family Ischnopsyllidae (bat fleas) are cosmopolitan. Leptopsyllidae family members have been reported in the Palearctic, Nearctic, and Oriental regions, and few Australian species have been identified. Distributions of species belonging to both Chimaeropsyllidae and Xiphiopsyllidae are restricted in Ethiopia. The Vermipsyllidae family consists of Holarctic fleas that are hosted by carnivores and ungulates. The Stephanocircidae family is primarily Neotropical, and two species have been reported in Australia. Their major hosts are rodents and a few species of marsupials. Species of the Coptopsyllidae family are Palearctic and primarily hosted by rodents such as gerbils. Ctenophthalmidae are primarily Holarctic, with some species found in the Southern Hemisphere. Major hosts of these fleas include rodents, marsupials, and mustelids. Rhopalopsyllidae family are distributed in the Neotropics, Southern Nearctic, and Oceanic, where they primarily parasitize rodents and oceanic seabirds. Species of the Pygiopsyllidae family are found in Ethiopia. Oriental, Australian, and Neotropical regions, where their hosts are rodents, marsupials, insectivores, monotremes, and birds. The Ancistropsyllidae family is found in the Oriental region, and its major hosts are ungulates. Members belonging to the Malacopsyllidae family are restricted to the Neotropics, and their primary host is the armadillos. Species of the families Hystrichopsyllidae, Ischnopsyllidae, Leptopsyllidae, Ceratophyllidae, Coptopsyllidae, and Vermipsyllidae have been identified in Japan, Korea, and Russia.

4.6.4 Ecology

Fleas spend their whole life in their host's dwellings, where male and female adult blood feed on their hosts. On the other hand, the immature stages are free-living organisms feeding on organic matter and blood remnants in adult feces, and some are predators and scavengers of small arthropods and dead adults in the nest (Durden and Hinkle 2019). Most adults are not host or nest-specific; they feed on available hosts while waiting for their primary. Microclimate and availability of food and shelter for larvae are important in determining a suitable host or nest. Although fleas can jump, their dispersal from one habitat to another primarily depends on the mobility of their hosts. Poor sanitation is a risk factor for flea infestation.

4.6.5 Flea-Borne Diseases

Plague is one of the most common flea-borne diseases, best known for the Black Death in the fourteenth century, accounting for 25 million deaths (Durden and Hinkle 2019). Japan and its Asian counterparts, such as China, Thailand, India, Mongolia, Vietnam, Indonesia, and Kazakhstan, have reported cases of plague outbreaks (Bitam et al. 2010). It is caused by *Yersinia pestis*, a gram-negative bacterium that is transmitted to humans via flea bites. The bacteria can also infect humans if they contact infected material or animals or inhale infectious respiratory droplets. Human plague continues to exist around the world via transmission foci involving infected fleas and rodents. *Xenopsylla cheopis* is the world's primary vector of Y. pestis. Other confirmed vectors of the bacterium include *X. braziliensis* (Africa, India, and South America), *Nosopsyllus fasciatus* (almost worldwide), *X. astia* (Indonesia and Southeast Asia), *Oropsylla montana* (USA), *X. vexabilis* (Pacific islands), *C. felis* (Africa), *C. canis* (Africa), *Echidnophaga gallinacea* (USA), *E. oschanini* (Asiatic USSR and Mongolia) and *P. irritans* (Africa) (Bitam et al. 2010). However, they are relatively poor vectors of plague compared to *X. cheopis*. Antibiotics are used to treat *Y. pestis* infections.

Fleas are vectors of rickettsial diseases such as murine typhus and flea-borne spotted fever. Murine typhus is a flea-borne disease caused by *Rickettsia typhi* that is commonly transmitted to humans by a bite of infected *X. cheopis;* however, other flea species have been incriminated as vectors. The disease clinically presents as high fever, severe headaches, chills, nausea, myalgia and, in extreme cases, rashes. Murine typhus is a neglected disease in Asia (Parola et al. 2003), even though the *R. typhi* bacterium continues to be isolated from *C. felis* fleas (Reeves et al. 2015). Treatments with tetracycline, doxycycline and fluoroquinolone have proven effective. Flea-borne fever, caused by *Rickettsia felis*, occurs worldwide, including in Japan and neighboring countries such as Thailand, Indonesia and South Korea (Bitam et al. 2010). *Rickettsia felis* is transmitted by a variety of species of fleas. The disease has similar clinical manifestations as murine typhus with other signs, such as photophobia, vomiting, abdominal pain, diarrhea, and black crusted skin lesions.

Fleas are important vectors of *Bartonella* sp., a gram-negative bacterium that is the causative agent of Bartonelloses such as cat scratch diseases, bacillary angiomatosis, bacteremia and endocarditis. Cat scratch is a worldwide zoonotic disease caused by the bacterium B. henselae, mainly transmitted by the cat flea *C. flea* (Tsukahara 2002). However, there are reported cases of disease transmission by *C. canis*, a dog flea, in Japan (Bitam et al. 2010). Human transmissions are also possible via a cat bite or scratch. Infected people show clinical signs such as fever, malaise, anorexia, weight loss, fatigue, headaches, and lymphadenopathy. The disease can be fatal in immunocompromised patients. In Hokkaido Prefecture, Japan, Kabeya et al. (2011) detected *B. grahamii* and *B. taylorii* in *Ctenophthalmus congener truncus* and *Neopsylla sasai* fleas. *Bartonella grahamii* is a causative agent of neuroretinitis in humans, while *B. taylorii* causes endocarditis.

Other notable flea-borne diseases include Tungiasis (*Tunga penetrans*) and Tularemia (*Francisella tularensis*). Tungiasis is a tropical skin disease that occurs in Central and South America and tropical regions of Africa (Durden and Hinkle 2019). Tularemia is a bacterial zoonotic disease transmitted to humans and animals via exposure to infected vectors or consumption of infected food or water (Durden and Hinkle 2019). Although ticks are the primary vectors of the disease, fleas play a limited role in disease transmission in the USSR, United States, and Scandinavia.

4.6.6 Discovery of Novel Pathogens

The Discovery of novel pathogens involves rigorous scientific steps to prevent incorrect linkage between causal agents and diseases. Any incorrect linkage may result in inappropriate and potentially fatal treatment recommendations. According to Lipkin (2013), it is important to first find evidence of exposure to the pathogen in one or more infected individuals. For vector-borne diseases, evidence may include isolation of the pathogen in both the vector (Victoria and Yermia Sem 2021) and the infected person(s). Molecular techniques such as PCR, DNA sequencing, and mass spectrometry are available to detect pathogen nucleic acids or proteins. Immunological methods and microscopy may be used to identify specific components of a pathogen. Second, a probable causal relationship between the pathogen and clinical signs must be established. Evidence of causal relation may include the presence of microorganism's nucleic acids, proteins, or microorganism-specific antibodies in or near infected host cells. Third, using Koch's postulate can prove that the pathogen is the causative agent of the disease. Finally, in the case of vector-borne pathogens, it is important to demonstrate successful transmission of the pathogen by the vector (Tokarz and Lipkin 2020).

4.7 Horsefly

Hitoshi Sasaki

4.7.1 Classification

Tabanids form a large family named Tabanidae, which includes approximately 4400 species and subspecies in 156 genera worldwide (Morita et al. 2016). Approximately 100 species in nine genera have been recorded from Japan (Sasaki and Watanabe 2016). They are divided into three groups, namely, deer flies (*Crysops*), cleg (*Haematopota*), and horseflies (*Tabanus*, *Atylotus*, *Hybomitra,* and so on). Tabanids inhabit every continent except Antarctica and have colonized even remote islands such as the Galápagos.

4.7.2 Morphological Characteristics

Although adult tabanids show variation among species in general appearance, this family is easily recognized as a family of order Diptera. They are generally large, stout-bodied flies ranging from 6 (deer fly) to 30 mm (horsefly). Their bodies are generally brown, black, or gray, and they have pale-colored pattern areas. The head is large and much broader than the length. The compound eyes are particularly well developed and differentiated in size between sexes, with the males being holoptic and the females dichoptic. The eyes are often colored brilliantly and spotted in deer flies, with zigzag bands in cleg and unicolorous or horizontally banded in horseflies. The flagella of the antennae are elongated and annulated, and the number of segments varies from four to eight. The mouthparts are adapted for blood-sucking (pool feeding), both cutting the skin and licking the blood. The wing venation is well-developed, and all wing veins are prominent. Veins on r4 and r5 diverge (shaped like a "Y") to enclose the wing tip. Wings in deer flies have a dark costal region and a single, broad, transverse dark band. In cleg, wings have rosette patterns, and those in horseflies are usually clear but sometimes may be dark or banded. The legs are stout and end in three pads because the empodium is pad-like, like the pulvilli.

4.7.3 Bionomics

Eggs are usually laid in a large mass containing 200–1000 eggs. Eggs of deer flies are often laid in a single layer, and those of horseflies and cleg are stratified into three to four and two to three layers, respectively (Chvála et al. 1972). They are

Fig. 4.11 Natural enemies of Tabanidae. (**a**) Egg mass and their parasitic wasp and (**b**) robber fly preying *T. trigonus*

usually laid on vegetation hanging over suitable larval breeding sites but are sometimes laid directly on rocks beside pools and streams. Eggs are initially creamy white, and with age, they become dark gray to black and hatch in four or more days, depending on air temperature. Larvae of many species breed in mud or damp soil, but some breed in water, dry soil, mosses, and rot holes in trees. Their breeding sites include pasture, meadows, forest floors, pools, streams, boggy depressions, salt marshes, and occasionally beaches. The larvae spend 6–13 instars (but this is variable even in the same species) and need several months (up to several years) to mature. The larvae of most species are actively carnivorous, and many are cannibalistic. Larvae of some species feed on any living flesh that is available, including humans in paddy fields (Otsuru and Ogawa 1959). Pupation occurs close to or within the dried parts of the larvae breeding medium. The pupal stage generally lasts 1–3 weeks, and adults live for 2–4 weeks. Eggs are parasitized by parasitic wasps (Scelionidae and Trichogrammatidae) (Fig. 4.11a), and larvae are parasitized by Bombyliidae and Tachinidae (Diptera). Larvae of another or the same species are the main predators of larvae, and large Odonata nymphs, birds and mice also prey upon the larvae. Predators of adults include birds, spiders, dragonflies, wasps, and robber flies (Fig. 4.11b).

4.7.4 Medical and Veterinary Importance

Females of many species of tabanids suck blood from humans and animals to develop their ovaries. According to blood-sucking, tabanids transmit various pathogens to their hosts. Almost all the methods of transmission of pathogens are mechanical except for nematodes, such as loiasis by deer flies (*Crysops*) in West Africa and elaeophorosis by horseflies (*Hybomitra* and *Tabanus*) in southern Europe and North America (Table 4.2). Among the pathogens transmitted by tabanids, *Trypanosoma evansi*, a monomorphic Trypanosoma (24 μm long) with a free flagellum, is one of

Table 4.2 Disease agents transmitted by Tabanidae

Disease agent	Geographic occurrence	Transmission
Viruses		
Equine infectious anemia	Worldwide	Mechanical
Bovine leukemia	Worldwide	Mechanical
Classical swine fever	Worldwide, eradicated from	Mechanical
	North America, Australia,	
	New Zealand, South Africa	
Bacteria/rickettsia		
Anaplasma marginale	Worldwide (Tropic, Subtropic)	Mechanical
Francisella tularensis	Japan, North America, Russia	Mechanical
Bacillus anthracis	Worldwide	Mechanical
Protozoa		
Besnoitia besnoiti	Asia, Africa, Southern Europe,	Mechanical
	South America	
Trypanosoma evansi	Asia, India, Russia, North	Mechanical
	Africa, South America	
T. vivax	Africa, South America	Mechanical
Filarial nematodes		
Loa loa	Central Africa	Biological
Elaeophora schneideri	Southern Europe, North America	Biological

the most serious and causes fever, weakness and lethargy in horses, donkeys, mules, cattle, buffalo, deer, camels, llamas, dogs, and cats. Infected animals show weight loss and anemia. The mortality of this disease in horses is nearly 100%, but it is only slightly pathogenic to sheep, goats, and pigs. In cattle and buffalo, it is usually a stable enzootic condition with low mortality but is subjected to occasional severe outbreaks, such as that in Timur-Indonesia, which brought 25.7 billion IDR (1.64 million USD) economic loss during the outbreak between 2010 and 2016 (Dewi et al. 2020). This disease is called "surra" and is seen throughout southern Asia, including India, China, Mongolia, parts of Russia, Bhutan, Nepal, Myanmar, Laos, Vietnam, Cambodia, Thailand, Malaysia, the Philippines, and Indonesia (Table 4.3). It is also present in the Middle East, South America, and northern Africa. Its presence was suspected in Papua New Guinea but not confirmed, and it is thus far absent from Australia (Whitehouse 2022).

The main vectors of *T. evansi* in southeastern Asia are *Tabanus rubidus* and *T. striatus* (Fig. 4.12a, b). We studied the profile of internal transcribed spacer 2 in the ribosomal DNA of *T. rubidus* collected in Thailand and Myanmar and constructed a phylogenetic tree (Fig. 4.13) (Tawatsin et al. 2017).

Bovine leukemia virus (BLV) is also known to be a serious pathogen causing enzootic bovine leukosis (EBL) in cattle (Sasaki 2021). It is a tumor (cancer) of lymphoid cells, and symptoms begin with emaciation, lethargy, anorexia, decreased milk production, diarrhea, and constipation. Then, swollen lymph nodes throughout the body and bulging eyes (cancer eyes) occur. This virus also infects buffalo, house

Table 4.3 Present reports of *Trypanosoma evansi* from Asian countries

	Cattle	Buffalo	Deer	Camel	Donkey	Dog	Horse	Goat	Mule	Pig	Sheep	Human
China	○											
India	○	○		○	○	○	○	○	○	○	○	○
Indonesia	○	○					○					
Malaysia	○	○	○			○	○					
Pakistan		○		○		○	○					
Sri Lanka						○						
Thailand	○	○	○			○	○		○			
Vietnam		○										○

Remade based upon Aregawi et al. (2019)

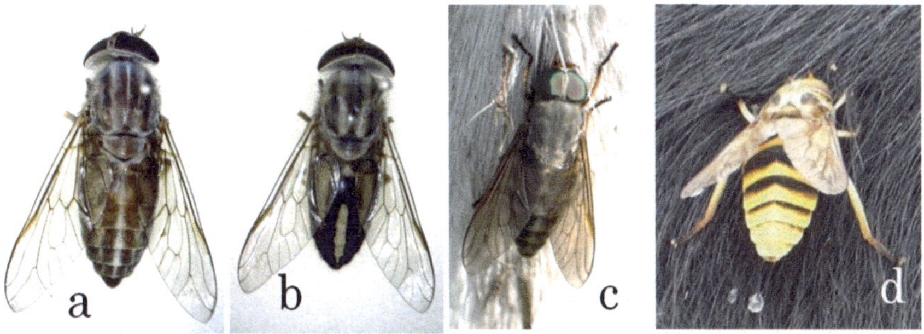

Fig. 4.12 Vector Tabanidae. (**a**) *Tabanus rubidus*, (**b**) *T. striatus*, (**c**) *T. nipponicus*, and (**d**) *T. chrysurus*

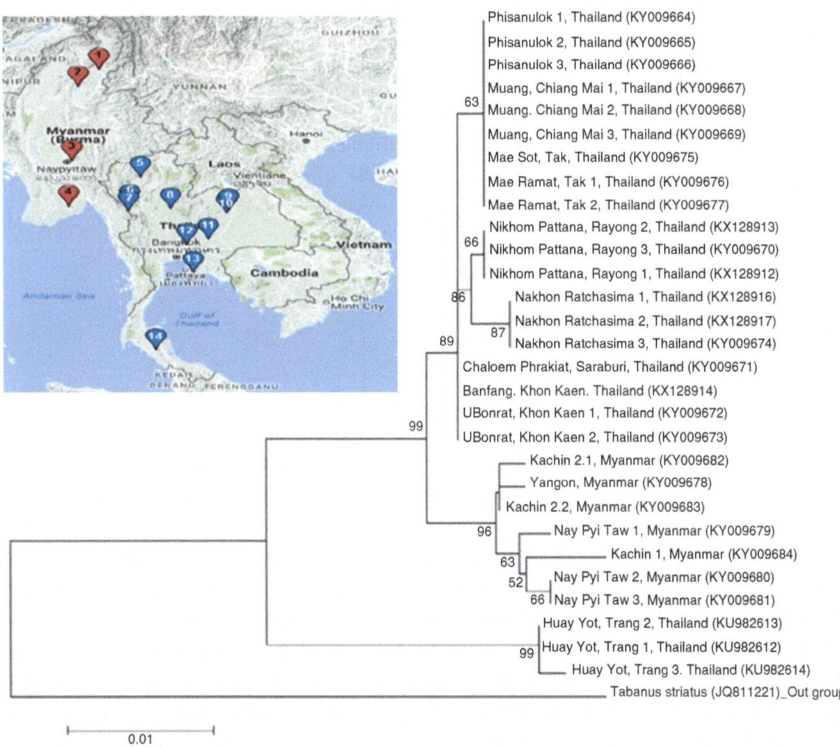

Fig. 4.13 The 1000 bootstrapped NJ phylogenetic tree under the K2P model based on ITS2 region sequence obtained in this study and using the sequences of *T. striatus* as an outgroup (Tawatsin et al. 2017). Sampling locations in Myanmar (1,2: Kachin, 3: Nay Pyi Taw, 4: Yangon) and Thailand (5: Chiang Mai, 6,7: Tak, 8: Phitsanulok, 9,10: Khon Kaen, 11: Nakhon Ratchasima, 12: Saraburi, 13: Rayong, 14: Trang)

mice, rabbits, and sheep. Although this disease spread widely worldwide, present records in Asia are reported only from Kazakhstan, Sri Lanka, South Korea, and Japan. The main vectors of BLV in northern Japan are *Tabanus nipponicus* and *T. chrysurus* (Fig. 4.12c, d).

4.7.5 Control

During the larval stage, tabanids prey on small animals such as springtails and earthworms, which can result in cannibalism. Therefore, they live sparsely in a wide range of soils. This is the main reason why it is difficult to control larvae of tabanids, and the current situation is that the measures against them must be aimed exclusively at adults. As a countermeasure against adults, traps such as Nzi traps (Mihok 2002) are used for the purpose of reducing the number of tabanids in livestock by catching and killing them and reducing damage. In addition to the use of traps, attempts are being made to prevent blood-sucking by them by applying chemicals to livestock that are safe for livestock and humans. Finally, I make a point that all measures to control tabanids must be kind to all the environment, livestock, and humans.

4.8 Human Lice

Yusuf Özbel and Seray Töz

Lice are obligate blood-feeding parasitic arthropods found only on humans and taxonomically divided into two genera: *Pediculus* (fam. Pediculidae) and *Pthirus* (fam. Pthiridae). Two *Pediculus* and one *Pthirus* species have parasitic behavior on humans by occupying different habitats on the body. Pediculidae and Pthiridae belong to the suborder Anoplura, order Phthiraptera, class Insecta, and phylum Arthropoda (Ko and Elston 2004). *Pediculus capitis* is called head lice, and *Pediculus corporis* is called body lice. They generally have similar morphology, but there are slight differences to distinguish. Head lice are smaller than body lice, and some other features such as the length of tibial and tarsal segments of the second pair of legs can be used especially for identification when they are found simultaneously in the same host. These two species are genetically distinct, do not interbreed, and always remain in their ecological environment on the body (Heukelbach and Feldmier 2010).

The head and body lice are 1–2 mm and 2–4 mm long, respectively. Their width is around 0.33 mm and is similar to sesame seeds. They are dorsoventrally flattened, wingless, and elongated (Fig. 4.14). They have hidden mouthparts, short antennae, eyes, stylets, and vestigial labial palps on the head. They have three pairs of clawed legs ending with grasps to grasp the hair. The louse inserts its mouthparts into the

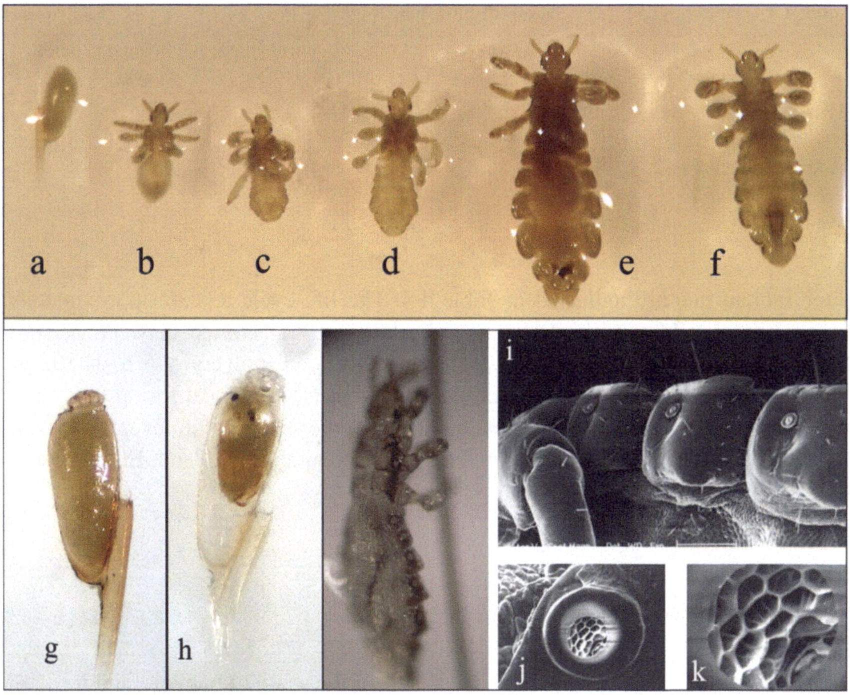

Fig. 4.14 (**a–f**) The life stages of head louse, (**g**) alive louse egg, (**h**) nymph ready to hatch in egg, (**i–k**) stigmas on the side of female louse. (Photos: Özbel. Y., original)

scalp and injects saliva for vasodilatory effects. The head louse moves along the hairs rapidly (23 cm/min) but usually stays close to the scalp. *Pthirus pubis* (pubic louse, crab louse) is approximately 0.8–1.2 mm long and almost as wide as it is long (square shape). This morphology allows it to grasp wide space in pubic hairs. The legs of the pubic louse are clawed, except for the first pair, which are shortened and vestigial (Ko and Elston 2004).

Lice eggs called nits are cemented to human hair; therefore, removing them is more difficult than other things, such as dandruff and gel drops. *Pthirus* eggs can be distinguished from *Pediculus* eggs because of their locations or relatively more acute angle on the hair (Ko and Elston 2004).

4.8.1 Biology and Life Cycle

Lice usually lays nits within several millimeters of the scalp. A female head louse can lay 50–150 eggs during an average lifespan of 16 days while a female body louse lays 270–300 eggs in its average lifespan of 18 days. A female pubic louse can live about 14 days and produce about 25 eggs. The life cycle periods and stages of

Table 4.4 The biological specifications of three human lice

	Adult lifespan (days)	Eggs (n)	Egg to nymph (days)	Nymph to adult (days)
Head lice (Pediculus capitis)	16	50–150	7	>7
Body lice (Pediculus corporis)	18	270–300	8–10	>14
Pubic lice (Pthirus pubis)	14	25	7	>14

three human lice are indicated in Table 4.4. The life cycle repeats approximately every 3 weeks. Mostly third instars and adults are transmitted between persons rather than first and second instars. They usually feed in every 4–5 h. The body louse can survive in the seams of clothing without a blood meal for up to 3 days, while head lice can survive only several hours on the scalp, but in favorable conditions, they may live for up to 4 days experimentally (Ko and Elston 2004) (Fig. 4.14a–k).

4.8.2 *Head Lice*

Pediculus capitis, head lice, causes infestation of the scalp of humans, and it is still a worldwide public health and social problem because it can infest people of all ages and economic status. Related to developing new medications using new medical tools other than insecticides, the prevalence has decreased in developed countries but still has a presence in all countries and infests hundreds of millions of people globally.

Head lice are mainly childhood infestations affecting children between three and 12 years old, particularly in schools, because of close contact (Ko and Elston 2004). When the infested children carry the head lice to the home, adult people, especially older people who mainly stay at home, can be infected with a few head lice that may be difficult to be aware of. The infestation rate is always higher in girls than boys, probably because long hair causes close and prolonged head-to-head contact among girls (Canyon et al. 2002; Fu et al. 2022).

4.8.2.1 Clinical Manifestations

The head lice infestation called "pediculosis capitis" can be asymptomatic. After the first infestation, the symptoms can develop in 4–6 weeks as an immune system reaction against louse saliva (Frankowski and Weiner 2002). Related to this immune sensitization, papules or wheals can develop because of repeated bite reactions of the lice (Dagrosa and Elston 2017). Itching is the most common reaction besides cervical lymphadenopathy and conjunctivitis. New bites made by recently hatching instars may also cause reactivation of healed papules (Ko and Elston 2004). In most school children, the symptoms are mild to moderate, but heavy and chronic infestations can cause anemia. Pediculosis capitis, despite very common infestation in the communities, mostly causes psychological and sociological problems among children and their parents (Fu et al. 2022).

4.8.2.2 Diagnosis and Treatment

An active infestation is diagnosed by identifying any forms of living lice. The inspection can be done visually or by using a lice comb. Visual inspection can be difficult because head lice avoid light and can crawl very quickly. The combing is much easier and increases the chances of finding live lice. It can be used by teachers, parents, etc., in addition to health personnel.

The presence of nits alone is not adequate for the diagnosis of active pediculosis capitis. However, nits at a distance of 0.5–1.0 cm from the scalp indicate an active infestation, and nits more than 5 cm from the scalp indicate past infestation if adult lice are not found (Williams et al. 2001). Differential identification from other things that can be found on the hair and resemble nits is very important. Hair casts (pseudonits), especially, may resemble nits stuck to hair shafts. In contrast to nits, hair casts are freely movable along the hair shaft, and this can be tested by fingers. Nits can also be confused with debris on the hair shaft left by hair spray and accumulated flakes of seborrheic dermatitis. Dermoscopic or microscopic examination also allows easy distinction of the hair cast from real nits (Ko and Elston 2004). This is also a good tool for differentiating eggs containing nymphs from the empty cases of hatched lice. Wet combing always increases the possibility of finding living lice showing active infestation (Jahnke et al. 2009).

All adults, nymphs, and eggs need to be eradicated for successful treatment. Lice and nit removal combs used for the diagnosis are also a mechanical treatment option. Combs having a small distance (0.2–0.3 mm) between the teeth can serve as a treatment tool before or after topical therapies are applied because of the physical removal of all life stages. Actually, wet combing (every 3–4 days) alone has also been promoted as an adequate therapy (Larsen 2010). Shaving parasitized hair on the scalp also eradicates head lice, but it is not usually acceptable for cosmetic and aesthetic reasons, especially for girls.

Shampoos containing pyrethroids can be applied topically and are considered the first line of therapy (Dagrosa and Elston 2017). Pyrethroids interfere with sodium transport in the louse, causing depolarization of neuromembranes and then respiratory paralysis. The efficacy of pyrethrins derived from the chrysanthemum plant is increased when combined with piperonyl butoxide, an insecticide synergist, but pyrethroid resistance is gradually becoming an important problem worldwide (Elston 2005). Besides treatments using pediculicides with the neurotoxic mode of action, the products with a physical mode of action are a new treatment approach to pediculosis because of the developments of resistance less likely. Silicone-based fluids/gels (dimethicone) occluding the respiratory system of the louse physically have been shown to have advantages according to pyrethroids (Burgess et al. 2013) and can also kill viable eggs if the fluidity is arranged suitable. Although this therapy is still expensive compared to insecticide-based therapies that can be neurotoxic for humans, they will be preferred in time due to not developing resistance and good safety profile. In all kinds of therapy, the re-treatment after 7–10 days is advisable to ensure that all stages of lice have been killed (Dagrosa and Elston 2017).

4.8.2.3 Epidemiology

Head lice infestation is a global problem and is seen in almost all countries, not depending on the economic conditions. The studies related to the distribution or epidemiology of this infestation were usually carried out in local areas among school children using different observing and/or combing techniques in different countries. For this reason, the prevalence rates vary greatly depending on the competence and diligence of the researchers carrying out the survey. It is difficult to find adult lice with visual inspection because of the low number of adult lice on the head and the small number of eggs associated with a new or recent infestation, which are not easily found. A WHO report published in 1997 shows that the infestation rate is high in Asian countries. The prevalence mostly among school children was given as 75–100% in Bangladesh, 17–60% in India, 10–41% in Malaysia, 25–40% in Pakistan, 31% in Japan, 44–73% in Korea and 21–67% in Taiwan (Gratz 1997).

The main transmission route of the head lice is head-to-head contact. Fomite-to-head contact also occurred, but it is relatively unimportant. The control of head lice infestation should focus on the head and not only the infested person but also those who have close contact at home, school, or work.

4.8.3 Body Lice

Pediculus corporis, known as body lice, is slightly bigger than head lice and has the same life cycle. They require blood meals 4 or 5 times per day and cannot survive more than two days without feeding. Its prevalence is higher among populations with poor hygiene conditions such as homeless people and persons living in refugee camps or lack of regular bathing (Nyers and Elston 2020). Transmission usually occurs via physical contact or sharing the infected person's personal items (Feldmeier and Heukelbach 2009).

4.8.3.1 Clinical Manifestations

The body lice infestation called "pediculosis corporis" is diagnosed less than pediculosis capitis but is more related to the transmission of different pathogens to humans. Although they have a similar morphology to head lice, body lice live in the host's clothing, localizing the seams and temporarily migrating to the host's skin for blood meals. Host reactions against their bites are similar to head lice. The infested person presents generalized pruritus, erythematous, and copper-colored maculae; with chronic infestation the skin may be thickened and hyperpigmented in a generalized distribution. Early infestation can cause misdiagnosis, such as atopic dermatitis, contact dermatitis, irritant contact dermatitis, or a drug reaction (Ko and Elston 2004).

4.8.3.2 Diagnosis and Treatment

To diagnose body lice, the seams of clothing need to be checked, especially in areas with higher body temperature, like the waist. The adults, nymphs, and viable nits can usually be found within inner clothing seams, and dermoscopy can be a useful tool for increasing the sensitivity of the diagnosis (Nyers and Elston 2020). The clothing can also be stained with blood or feces of lice, which can indicate the presence of body lice in the person (Ko and Elston 2004).

Treatment of body lice is primarily environmental, requiring all bedding and clothing materials that may harbor lice or nits to be disposed of or laundered and dried for 30 min at 65 °C. Ironing the clothes and underwear, especially at the seams, is also effective. In some cases, permethrin spray (0.5%) can treat clothing and reduce the reinfection rate (Mikhail et al. 2007).

4.8.3.3 Epidemiology

Pediculosis corporis is more common among adults. Some people can be infested with *Sarcoptes scabiei* and *P. corporis* simultaneously, especially in unfavorable and unhygienic conditions such as refugee camps and homeless life, etc. The outbreaks caused by *P. corporis* usually occur during natural disasters, and the prevalence of this condition reaches up to 90%. The infestation rate could be as high as 11–22% among homeless people in public shelters. Head and body lice co-infestation is noted at 4.3% among homeless people in San Fransico, while it is 59% among street children in Katmandu and Nepal (Badiaga and Brouqui 2012). There are occasional reports of infestations in Asia (Nepal, India, and China), the Middle East, America, and eastern African countries (Gratz 1997).

4.8.4 Public Lice

Pubic louse commonly infests pubic hairs and its infestation is classified as a sexually transmitted disease; therefore, it is more common among adults than in children. It also resides in perineal hair, hair of the legs, abdomen, chest, and back. It can be seen in the eyelashes, beard, axillary hair, and scalp (Ko and Elston 2004;

Galiczynski and Elston 2008). Approximately 2% of adults are infested worldwide (Dholakia et al. 2014).

4.8.4.1 Clinical Manifestations

The primary complaint of infested persons is pruritus. It increases at night, and the patients can feel the pubic lice while itching their hair. In chronic cases, usually, excoriations are present, and these patients can have inguinal lymphadenopathy. Maculae ceruleae with blue spots in the thigh are a specific symptom of pthiriasis pubis. They are thought to be secondary to the anticoagulant activity of louse saliva, and presumably, they are related to deep dermal hemosiderin deposition from the bites (Ko and Elston 2004). Underwears of the patients may be stained by blood and crusts (Galiczynski and Elston 2008). Itching may also cause erythema, small skin ulcerations, and secondary bacterial infections (Dholakia et al. 2014). Eyelash infestations can cause *Pthiriasis palpebrarum* and other ocular symptoms (Mumcuoğlu 2015). Pubic lice in the scalp is called "*Pthiriasis capitis*" and causes pruritic lesions on the nape (Veraldi et al. 2020).

4.8.4.2 Diagnosis and Treatment

Although pubic lice can be seen by the naked eye, magnification or dermoscopy is also useful for certain diagnoses. The presence of lice and nits on pubic or body hairs is enough for diagnosis. If the infestation is chronic, a great number of adults, nymphs, and eggs can be found on body hairs. Seeing small gray to bluish macules (>1 cm) on the lower abdominal wall, buttocks, and upper thighs by eye is also an indicator of the presence of pubic lice. Differential diagnosis from white piedra or trichomycosis pubis is also important (Ko and Elston 2004; Galiczynski and Elston 2008).

If the infestation is located only in the pubic area, shaving is the best mechanical treatment if it is acceptable to the patient. However, if more adults, nymphs, or nits are present on body hairs, insecticide solutions can be used similar to scabies therapy. Other topical lotions killing lice via asphyxiation can also be used as an alternative treatment agent. Because pediculosis pubis is highly contagious, sexual partner(s) of the patients need to be checked for the presence of pubic lice or nits. The infestation can also be acquired by contact with infested bed linens, bath towels, or shared clothes (Galiczynski and Elston 2008). Reinfestation is very common, and clothing and bedding need to be machine-washed in high temperatures.

4.8.5 Diseases Transmitted by Human Lice

Since 1909, the identified human pathogens in the body, head, and pubic lice are listed in Table 4.5. The frequently highlighted pathogen agents for the body lice in the literature are *Rickettsia prowazekii* (agent of epidemic typhus), *Bartonella*

Table 4.5 Pathogens associated with lice

	Head lice	Body lice	Pubic lice
DNA detection	*Borrelia recurrentis*	*Rickettsia prowazekii Borrelia recurrentis*	*Acinetobacter johnsonii*
	Bartonella quintana	*Bartonella quintana*	*Acinetobacter* spp.
	Borrelia theileri	*Yersinia pestis*	*Rickettsia prowazekii*
	Rickettsia prowazekii	*Acinetobacter* spp.	*Bartonella quintana*
	Acinetobacter spp.	*Serratia marcescens*	
	Yersinia pestis	*Coxiella burnetti*	
	Serratia marcescens	*Rickettsia* spp.	
	Coxiella burnetii	*Anaplasma phagocytophilum*	
	Rickettsia aeschlimannii	*Ehrlichia muris*	
	Anaplasma spp.		
	Ehrlichia spp.		
	Moraxellaceae		
	Staphylococcus spp.		
	Streptobacillus moniliformis		
	Haemophilus influenzae		
	Bordetella pertussis		
	Acinetobacter baumannii		
	Acinetobacter johnsonii		
	Acinetobacter variabilis		
Experimental model	*Bartonella quintana*	*Yersinia pestis*	No reported
	Rickettsia prowazekii	*Rickettsia rickettsii*	
		Rickettsia conorii	
		Rickettsia typhi	
		Rickettsia akari	
		Acinetobacter lwoffii	
		Acinetobacter baumannii	
Disease transmission	No reported cases	Trench fever	No reported cases
		Relapsing fever	
		Epidemic typhus	

quintana (agent of trench fever, endocarditis, and bacillary angiomatosis), and *Borrelia recurrentis* (agent of relapsing fever) that can cause illnesses mostly seen in poverty and overcrowded conditions and be fatal up to 40%, especially in epidemic typhus and relapsing fever (Fournier et al. 2002; Feldmeier 2010; Mana et al. 2017; Coates et al. 2020; Fu et al. 2022).

For the listed agents, the vectorial role of body lice has been proved in laboratory conditions, and some of them were also identified from field samples. Multiple pathogens were frequently identified in louse-born epidemics, and *Y. pestis* and *B. quintana* were reported in several cases (Barbieri et al. 2021). However, regarding head lice, there is insufficient evidence for its vectorial capacity and increasing the risk of clinical diseases (Table 4.5) (Feldmeier 2010).

High prevalence of *B. quintana, B. recurrentis,* and *R. prowazekii* transmitted by body lice are associated with people living in unhygienic conditions such as soldiers, refugees, and the homeless (Coates et al. 2020). The authors suggested that head lice could become a vector for *R. prowazekii*, *B. quintana* and *B. recurrentis* with a combination of different epidemiological factors such as the presence of the agent in the environment, high concentration of head lice on the scalp and immunosuppression of the host due to malnutrition, chronic infections, poverty, etc. (Feldmeier 2010).

4.9 Mite (*Scabies mintes*)

Takashi Ueda, Naotoshi Tsuji, and Yasuyuki Amoh

Scabies is an itchy skin rash caused by the tiny burrowing mite *Sarcoptes* and affects an estimated 130 million individuals worldwide at any given time. This estimate is supported by annual global reports of 300 million cases. However, scabies are still an underrecognized cause of illness in many developing countries and are overlooked as a major underlying cause of high rates of bacterial skin infections and their complications. Scabies control programs, which are directly affected by social stigma, cannot be achieved in resource-poor countries. Therefore, international cooperation is critical for novel, robust, and workable measures against scabies in the expanding framework for controlling neglected tropical diseases (Romani et al. 2015).

4.9.1 Medical Importance

Scabies is a contagious disease that is easily transmitted through close skin contact with an infested individual but is easily treated. Accurate diagnostics and prompt treatment prevent community spread. However, scabies are commonly overlooked; thus, affected individuals may not receive appropriate treatment. In developing countries, outbreaks in health institutions and vulnerable communities result in a significant economic burden. Scabies was formally designated as a neglected tropical disease by The World Health Organization (WHO) in 2017 in recognition of the lack of awareness among healthcare professionals and health organizations in addition to a global consensus not yet being reached on control guidelines and strategies. The increasing incidence of scabies points to the need for a prompt and collaborative response from healthcare professionals, governmental institutions, and nongovernmental organizations, particularly in the settings of widespread population migration and poor living standards. Outbreaks of scabies are prevented by education programs for whole populations and accessible healthcare. Accurate national data and proper disease reporting are crucial for developing strategies worldwide that not only prevent but also control the community spread of scabies. Scabies was recently included as part of the WHO roadmap for neglected tropical diseases 2021–2030, with the aim of ending neglect to attain Sustainable Development Goals (Arlian and Morgan 2017).

4.9.1.1 Control in Japan

The annual estimated number of scabies patients in Japan ranges between 80,000 and 150,000, and the disease continues to be transmitted from person to person mainly through sexual activity and among nursing home residents. However, various measures to achieve a rapid and accurate diagnosis, appropriate treatment, and cure are being taken to control the spread of scabies. Dermatologists are now taking the lead in preventing outbreaks and raising awareness, and scabies treatment guidelines have been developed (Executive Committee of Guideline for the Diagnosis and Treatment of Scabies 2017).

4.9.2 Causative Agent

4.9.2.1 Classification

The causative agent of scabies, a common and contagious ectoparasitic infection, is *Sarcoptes scabiei* var. *hominis*, a mite specific to humans. *Sarcoptes scabiei* infects nonhuman domestic animals and wild animals and causes scabies of various lesions. All *S. scabiei* have been phylogenetically determined to be infected by the variety of *Sarcoptes scabiei* var. *hominis*. Although mutations are morphologically indistinguishable, clear differences have been detected in physiological and genetic characteristics. Phylogenetic studies showed that the original host of *S. scabiei* was humans. The successive domestication of wild animals, beginning with dogs 12,000 years ago, increased the frequency of contact between humans and animals, resulting in human parasitism. Var. *hominis* adapting to and becoming parasitic on animals and, ultimately, the birth of various varieties, including var. *canis* on dogs, var. *suis* on swine, var. *ovis* on sheep, and var. *bovis* on bovines. In addition, *S. scabiei* was suggested to infect wild animals that strayed into villages, giving rise to variants such as var. *vulpes* on foxes (Niedringhaus et al. 2019).

Scabies in domestic and wild animals are referred to as mange to distinguish them from human scabies. Taxonomically, scabies in animals is mostly caused by *Psoroptes*, *Choripotes*, and *Knemidokoptes*, which are parasitic on birds, and these are collectively called *Psoroptidae* and sometimes psoroptic or choripotic mange.

There are more than 200 species of *Sarcoptes*, most of which are morphologically indistinguishable. Nevertheless, taxonomic analyses revealed that the consistent clustering of geographic points or host specificity is still possible by using microsatellites and mitochondrial DNA as markers; therefore, the utility of ITS-2 as a gene in *S. scabie*i phylogenetic analyses is now uncertain. The Sarcoptes-World Molecular Network was newly established to develop better detection methods for Sarcoptes and provide a centralized site for integrating and comparing phylogenetic data. Although a formal consensus on taxonomy based on morphological and genetic features has not yet been reached, *S. scabiei* variants appear to be of the same genetically diverse species.

4.9.2.2 Morphology

Sarcoptes scabiei have oval, ventrally flattened, and dorsally convex tortoise-like bodies (idiosoma), mouthparts, and four pairs of short stubby ambulatory legs, the terminal segments of which have claws. The dorsal idiosoma has cone- and spine-shaped lateral and dorsal setae, multiple cuticular spines, and coarse, transversely ridged cuticular striations. The gnathosoma (capitalum) consists of short, stout chelicerae and pedipalps. In females, the anal opening is in the posterior/dorsal region, and the nipple-like papilla of the bursa copulatrix is located anterior to the anal opening (Fig. 4.15a).

The traditional classification of variant forms of *S. scabiei* is based on the host species and the lack of pronounced clinical disease in taxonomically distinct hosts. However, few morphological differences have been detected among mites examined from different host species.

4.9.2.3 Life History

Sarcoptes scabiei undergoes five stages in its life cycle: egg, larva, protonymph, tritonymph, and adult scabies mite. Scabies mites infect a host by creating tunnels and burrowing into the stratum corneum of the epidermis of the skin, which is partly achieved by the cutting mouthpiece and hooks on the legs (Fig. 4.15b). All life stages of *S. scabiei* penetrated the host skin.

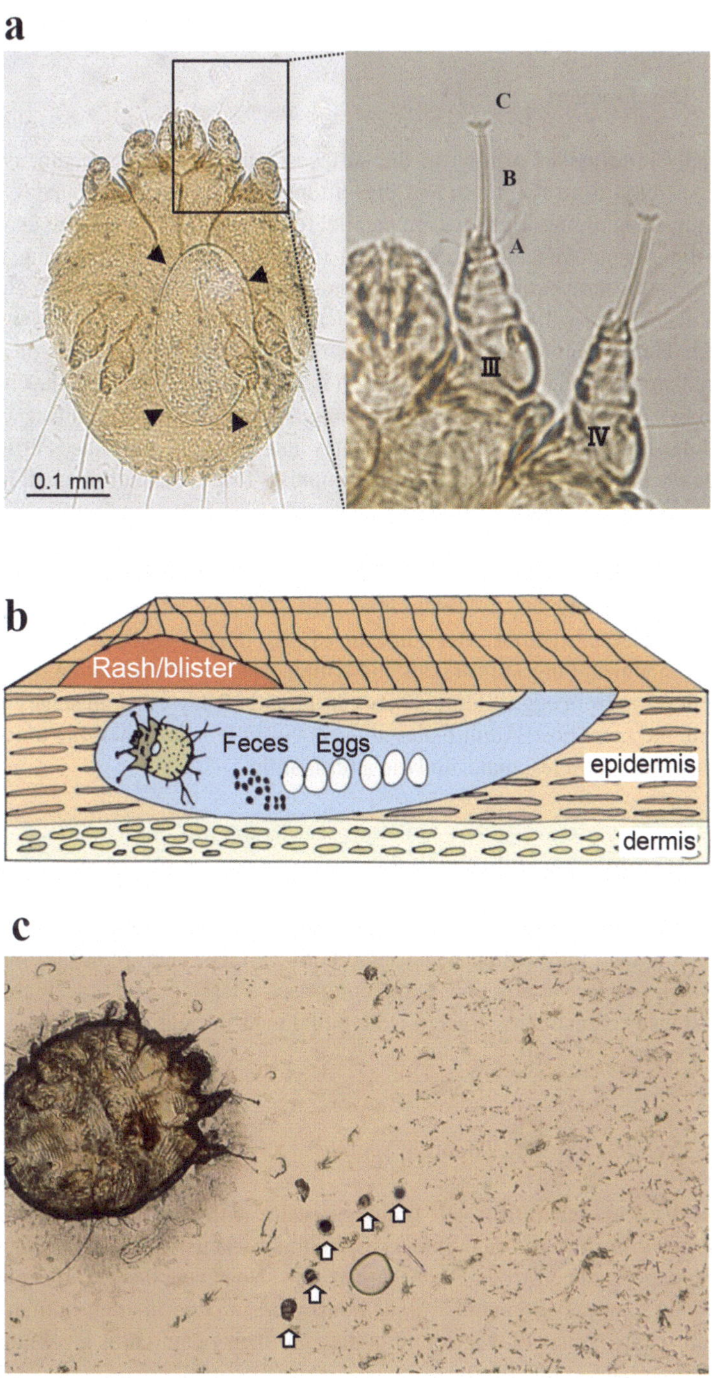

Fig. 4.15 (**a**) *Sarcoptes* female (ventral side). The body shape is almost spherical, with wrinkles on the body surface and many spines. An egg (arrowheads) is enclosed in the body (*left*). Legs III and IV show a tarsus with claws (**a**) and a stalked empodium (**b**) that terminates in a pad (**c**) (*right*). (**b**) Schematic diagram of a scabies burrow. Pads on short legs facilitate walking in tunnels. Once female mites perforate the stratum corneum, they dig tunnels almost horizontally when they reach just under the dermis, excreting and laying eggs. Secretions in the burrow are the cause of itching. (**c**) Skin scraping specimen of *Sarcoptes scabies* var. *hominis*. Arrows indicate scabies fecal pellets

4.9.2.4 Development

The female mite makes a hole on the surface of the stratum corneum using an enzyme secreted from its mouth and digs a tunnel called a scabies burrow, which runs in a horizontal line. Burrows are parallel to the skin surface and are characterized by interconnecting honeycombs with tunnels made by neighboring female mites. *Sarcoptes scabiei* orally digests the decomposed cells of the stratum corneum and inoculates them with intracellular fluid leaking into the tunnel but does not feed on blood. This nonblood-sucking nature is related to *S. scabiei's* lack of vector-borne/transmitting infections, even though it is a typical ectoparasite. Mites become completely submerged in the stratum corneum and are in constant contact with the intracellular fluid; therefore, mite metabolites and antigenic substances efficiently stimulate the host's humoral and cellular immunity. This is the pathological basis of scabies, which is characterized by delayed-type hypersensitivity reactions (Fig. 4.15c).

Each adult female lays a small number of eggs (3–4) daily, with one mite potentially producing more than 50 eggs during the duration of its lifespan, namely, 4–6 weeks. Larvae, nymphs, and adult males also inhabit scabies burrows in the stratum corneum and ingest host cells and lymph. The reported duration for scabies mites to develop from egg to adult is between 12 and 17 days, and all stages may be present on the same host. Adult males ambush virgin females emerging around the openings of hair follicles, mate, and die after mating.

4.9.2.5 Transmission and Infectivity

Adult female mites are almost always present in scabies, burrows, larvae, and nymphs, and male mites often wander the body surface. Therefore, the immature stage, which is superficially active and comprises a large population, is the main agent of transmission of *S. scabiei* to other individuals through skin contact with the body of the host.

4.9.2.6 Pseudoscabies or Zoonotic Scabies

Pseudoscabies are generally a self-limiting disease with a short incubation period and transient clinical signs on human skin. Animal variants of *S. scabiei* mites typically do not reproduce or persist on human skin. However, successful replication with hatching and development was demonstrated in experimental human infections with scabies mites of dog origin, indicating their potential as a zoonotic disease (Niedringhaus et al. 2019).

4.9.3 Topics of Clinical Scabies Research

4.9.3.1 Importance of Scabies Burrows for Clinical Diagnosis

Scabies burrows are scabies-specific eruptions that are found in both common scabies, which is a mild but intensely itchy infestation, and crusted scabies, which is infested with a large number of mites (Fig. 4.16a). They are macropathologically recognized as "linear scales" and distinguished from the scales observed in other

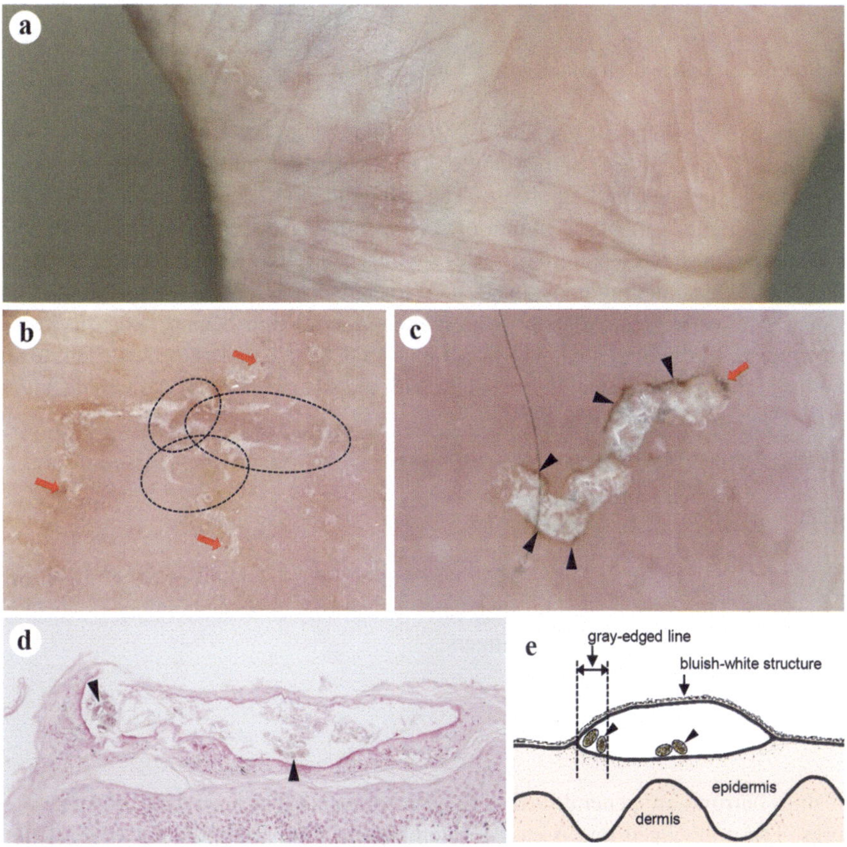

Fig. 4.16 (**a**) Clinical image showing tracks on the skin, known as burrows, on a patient's heel, indicating typical scabies. (**b**) Dermoscopic image showing scabies burrows spreading out in three directions. Each has a blackish-brown triangle at the tip (red arrows) and three wakes (dotted lines) overlapping in the center. (**c**) A black–brown triangle at the tip (red arrow) and a gray-edged line (black arrowheads) of black–brown lines on the outer edge of the tunnel are observed in the scabies burrow with gray-edged lines on dermoscopy. (**d**) Histopathological cross-section of a scabies burrow. Multiple fecal pellets (arrowhead) comprising melanin are present on the left wall of the scabies burrow and in its center. Eggshells (dashed line) are evident above feces in the center. (**e**) Schematic diagram of a scabies burrow with the gray-edged line sign. Feces with abundant amounts of melanin (arrowheads) are detected as the gray-edged line sign or bluish-white structure

skin diseases. To ensure a definitive diagnosis of scabies burrows, scabies tunnels must be suspected in a physical examination and confirmed by dermoscopy, followed by the application of KOH to the scales to detect mite bodies and eggs at a high rate. In seminal studies using human clinical specimens, we recently discovered a new specific finding, the gray-edged line sign, which facilitates the diagnosis of scabies (Ueda et al. 2021).

4.9.3.2 Dermoscopic Images with Scabies Burrows for a Diagnosis of Scabies

Delta Glider Sign with a Black-Gray Tone in Color

Scabies burrows generally appear as white lines with "delta glider" ("gray delta structures," "hang glider," and "triangle"), "jetliner with contrail," "wake," or "noodle pattern" findings on dermoscopy. The delta glider and contrail parts of the sign indicate the gray part of the mouth of scabies mites and the burrow, respectively. The wake sign shows a pattern of scales reminiscent of the wake left on the surface of the water by a moving body. In addition to the blackish-brown triangle at the tip, a nearly transparent circular thoracic abdomen extending behind it or blackish-brown parts of the hind legs may sometimes be detected. A dark brown triangle at the tip of the scabies tunnel is consistent with the advancement of the female adult in the stratum corneum (Fig. 4.16b).

Wake

Wake is the trajectory of water spray that follows a ship or bird moving along the waterfront. This sign is scaling that flares posteriorly in scabies tunnels over the creases of the wrists, palms, and fingers.

Gray-Edged Line

Scabies burrows are generally whitish lines but sometimes have dark brown structures. The gray-edged line is easy to identify and forms on the outer edge of the burrow as it curves. This finding is caused by the melanin pigment in feces sticking to the edges of the burrow (Fig. 4.16c).

Other Signs

The white linear scabies tunnel itself is sometimes called a jetliner with contrail, and together with the black–brown triangle at the tip, is occasionally expressed as the delta wing jet with the contrail sign. A dense cluster of scabies tunnels is sometimes called the noodle pattern and appears as a bluish-white structure inside the tunnels. However, these names are not generally recognized.

Common Sites of These Signs

Delta gliders are commonly found in scabies tunnels; however, wake and gray-edged lines are rare. Wake and gray-edged lines are larger structures than the dark brown triangle at the tip and are often recognized in a physical examination, although the most common sites are completely distinct. The former refers to the palms and soles (surrounding interdigits and wrist joints) at which the stratum corneum is thin, while the latter refers to the extremities, trunk, and genitals (other than the palms and soles). It is often found on the palms and soles because the scales are sloughed off before the wake occurs. The gray-edged line is rarely observed on palmar and plantar surfaces because of the limited amount of melanin present in the keratinocytes at these sites, which is eaten by the female imago. Thus, a low amount is in its feces (Fig. 4.16d, e).

Diagnostic Significance of Scabies Burrows

The search for scabies burrows in a physical examination is facilitated by linear scales with wakes on the palms and soles and those with gray-edged lines on the extremities, trunk, and genitals. The identification of these two findings contributes to the diagnosis of scabies burrows without confirmation by dermoscopy. However, scabies burrows are not detected in all cases of scabies, and a diagnosis by dermoscopy is also limited. Scabies is still an easily misdiagnosed disease because there are types of scabies in which the eruption is not visible on gross lesions (scabies in clean individuals). Nevertheless, a definite diagnosis may be reached at a high rate based on the presence of scabies burrows; therefore, searching for tunnels by dermoscopy is extremely effective when scabies are suspected.

4.10 Sand Fly

Chizu Sanjoba and Yusuf Özbel

Phlebotomine sand flies are small insects with a body length of about 2–3 mm that belong to the order Diptera, the family Psychodidae, and the subfamily Phlebotominae. Sand flies are widely distributed from 50° north latitude to 40° south latitude, including southern Europe, Asia, Africa, Australia, and Central and South America. Sand flies have adapted to all zoogeographic regions found in the rainforests of South America and the Gobi Desert of Mongolia. Furthermore, their habitat is moving northward due to climate change, especially in Europe.

Phlebotomine sand flies have medical importance due to their biological vectorial capacity for transmitting pathogens such as parasitic protozoa, several pathologic viruses, and bacteria affecting humans and animals in many countries. More than 1000 sand fly species have been identified in The Old and New World, and around 98 (42 in The Old World) have a vectorial role.

4.10.1 Biology

Phlebotomine sand flies undergo a complete metamorphosis from egg (Fig. 4.17a) to larva pupa and adult. The number of days required for each growth stage varies depending on the species, but the eggs hatch in about 4–20 days, and the larva grows to the fourth instar. Its growth is relatively slow, and reaching the pupa stage takes at least 3 weeks (Fig. 4.17d). The pupal period is 6–13 days, and the life span of adults can vary from 2 to 6 weeks.

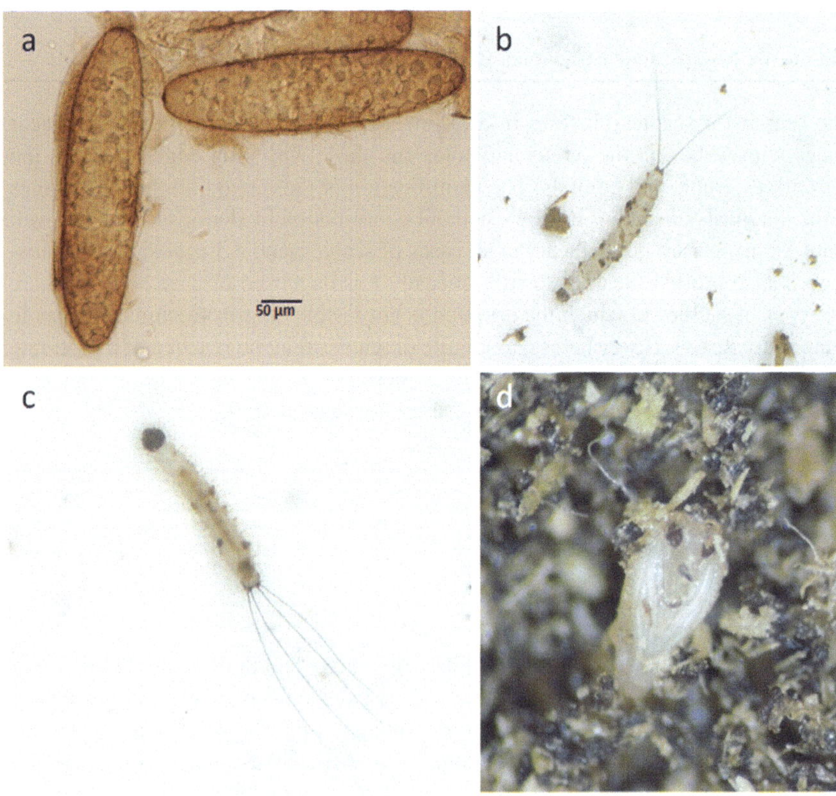

Fig. 4.17 (**a**) Eggs, (**b**) second instar larva, (**c**) fourth instar larva, and (**d**) pupa of Phlebotomine sand fly

First and second instar larvae have two setae on their rear ends (Fig. 4.17b), and third and fourth instar larvae have four setae (Fig. 4.17c). The size of the fourth instar larva is approximately 3–4 mm. The morphological characteristics that distinguish male and female larvae are unknown. The species can be identified by the morphology of eggshells, and around 100 species have been described (Jariyapan et al. 2022). However, an electron microscope is required to observe tiny eggs less than 1 mm in size. Female sandflies lay eggs in terrestrial microhabitats rich in organic matter, such as soil and animal burrows, which serve as nutrients for the larvae. Therefore, it is almost impossible to find tiny eggs in nature. Morphological species identification is usually performed on adult sandflies. When the female's spermatheca, which is necessary for morphological identification on adult sand flies, cannot be observed because the female carries eggs in her abdomen, identification using eggs can be a good option.

Although the behavior of the adult flies of different species is different in terms of living and activity conditions (indoor, outdoor, etc.), most adults are active in the evening and at night, and during the day, they stay in relatively cool and humid areas such as houses, livestock sheds, caves, cracks in the rock.

The mouthparts of males are not suitable for sucking blood; they can only feed with plant juice. Females, besides plant juice, can suck blood, usually at night, to produce eggs. Mating can occur before or after blood-sucking. The females take blood from a subcutaneous pool produced by injury to the blood vessels. Saliva produced by two rounded salivary glands helps the blood-feeding process with vasodilatory and anticlotting components. Some species need blood-feeding, and some need multiple blood meals for egg development.

4.10.2 Classification

Phlebotomine sand flies are grouped in the suborder Nematocera of the order Diptera, family Psychodidae, subfamily Phlebotominae, and they share the family Psychodidae with the non-vector, non-biting moth flies (Subfamily: Psychodinae), often seen around shower drains (Maroli et al. 2013). Among six subfamilies, Psychodinae and Phlebotominae are the most common, and members of Phlebotominae have medical importance. Because the term "sand fly" has also been used for biting midges (Ceratopogonidae) and occasionally blackflies (Simuliidae), it causes misunderstanding. Therefore, it is better to use the term "phlebotomine sand flies".

Approximately 380 species of Phlebotomine sand flies in three genera: *Phlebotomus*, *Sergentomyia*, and *Chinius*, have been reported in The Old World. The genus *Phlebotomus* includes 13 subgenera and is widely distributed in the Mediterranean region, Africa, the Middle East, Central Asia, Asia, and Australia. The genus *Sergentomyia* includes 10 subgenera and, like the genus *Phlebotomus*, is widespread in The Old World, particularly in the tropics. *Chinius* is the only genus distributed mainly in Asia (Alkan et al. 2013).

Approximately 470 species of Phlebotomine sand flies in three genera: *Lutzomyia*, *Warileya*, and *Brumptomyia*, have been reported in The New World. *Lutzomyia* is the largest genus of Phlebotomine sandflies, containing approximately 440 species in 15 genera and 11 groups, and is found primarily in Central and South America. *Warileya* consists of two genera and is found mainly in the Neotropics. Brumptomyia is the only genus widespread in Central and South America and does not feed on human blood.

Sergentomyia squamirostris (Fig. 4.18) is widespread in Japan. The first Phlebotomine sand flies were collected in Japan on June 25, 1916, in Matsuyama City, Ehime Prefecture (one male and one female). Also on July 19, 1916, a male individual was collected from Ogoori in Yamaguchi Prefecture. These three sand flies were reported by Robert Newstead in 1923 as a new species from Japan (Newstead 1923). For a long time, it was thought that only one sand fly species inhabited Japan. However, in 2022, sand flies collected on Iriomote Island, Okinawa Prefecture, were reported as a new species and were named *Sergentomyia iriomotensis* (Fig. 4.19) (Sanjoba and Miyagi 2022).

4.10.3 Public Health Importance

Only females of sand flies suck blood to lay eggs. At that time, they transfer bacteria, viruses, and protozoa to the humans or animals they feed, causing various infectious diseases; a bacterial infection known as Bartonellosis, a viral infection known as sand fly fever, and Leishmaniasis caused by *Leishmania* protozoa.

Fig. 4.18 *Sergentomyia squamirostris*. Female

Fig. 4.19 *Sergentomyia iriomotensis*. Male

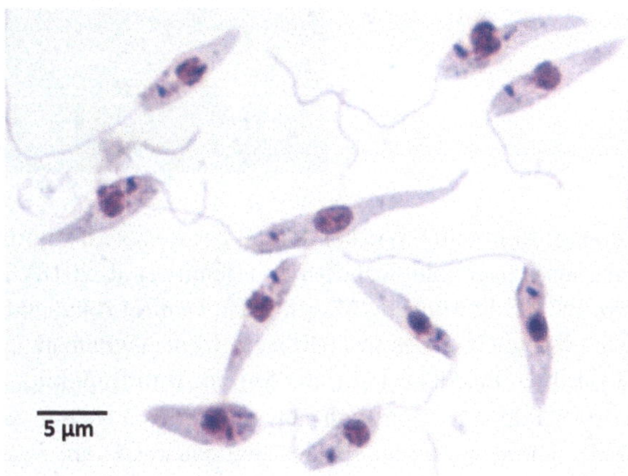

Fig. 4.20 *Leishmania* parasite. Promastigote stage

4.10.3.1 Leishmaniasis

Leishmaniasis caused by the protozoan parasite *Leishmania* (Fig. 4.20), which is transmitted by phlebotomine sand flies, is one of the Neglected Tropical Diseases. Leishmaniases are endemic in more than 98 countries, including tropical, subtropical, and Mediterranean regions, and a total of 350 million people are at risk of infection (Bennis et al. 2018). The genus *Phlebotomus* sandfly in The Old World and *Lutzomyia* sandfly in The New World are mainly responsible for transmitting the *Leishmania* parasites among mammals, and *Sergentomyia* sandflies usually involve the transmission of *Leishmania* parasites to cold-blooded animals such as reptiles.

Leishmaniasis is broadly divided into three clinical types: cutaneous, visceral, and mucocutaneous leishmaniasis. Cutaneous leishmaniasis (CL) causes skin lesions such as single or multiple papules, nodules, and ulcers on exposed areas of the body, such as the hands, feet, and face (Fig. 4.21), which leaves lifelong

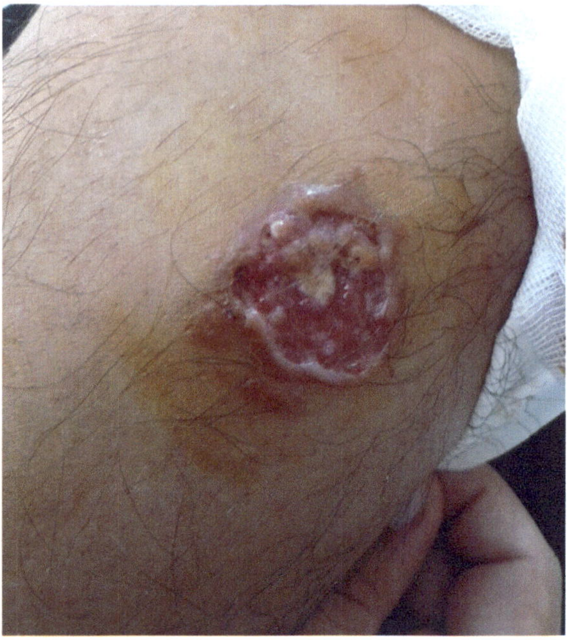

Fig. 4.21 The clinical manifestation (ulcer form) of cutaneous leishmaniasis caused by *Leishmania major*

scarring. Although CL is not fatal, it causes social stigma due to permanent scarring and causes psychological distress (Bennis et al. 2018). Most CL cases are reported from the Mediterranean, Middle East, Central Asia, and South America. More than 85% of new CL cases in 2020 were from Afghanistan, Algeria, Brazil, Colombia, Iraq, Libya, Pakistan, Peru, the Syrian Arab Republic, and Tunisia. It is estimated that 600,000 to one million new cases of CL occur each year worldwide (WHO 2023). The *Leishmania* species that cause CL are *Leishmania major*, *L. tropica*, *L. infantum*, *L. donovani*, *L. aethiopica*, *L. mexicana*, *L. venezuelensis*, *L. amazonensis*, *L. braziliensis*, *L. guyanensis*, *L. lainsoni*, *L. lindenbergi*, *L. naiffi*, *L. panamensis*, *L. peruviana*, *L. shawi*, etc.

The clinical symptoms of visceral leishmaniasis (VL), or Kala-azar, are fever, weight loss, anemia, and enlargement of the spleen and liver. It is fatal in more than 95% of cases without treatment. Most VL cases have been reported from Brazil, East Africa, Bangladesh, and India. An estimated 50,000–90,000 new cases of VL occur each year worldwide, but only 25–45% are reported to WHO (WHO 2023). The main *Leishmania* species that cause VL are *L. donovani* and *L. infantum*. *Phlebotomus argentipes* is the most important vector of *L. donovani* in the Indian subcontinent. The proven/suspected vectors of *L. infantum* in The Old World are *Phlebotomus alexandri*, *P. ariasi*, *P. balcanicus*, *P. halepensis*, *P. kandelakii*, *P. langeroni*, *P. major*, *P. mascittii*, *P. perfiliewi*, *P. perniciosus* and *P. tobbi*. *Leishmania infantum* is the main causative species of not only human leishmaniasis but also dogs and cats.

Mucocutaneous leishmaniasis (MCL) causes ulcers that partially or completely infiltrate the mucous membranes of the nose, mouth, and throat. More than 90% of MCL cases are reported from Bolivia, Brazil, Ethiopia, and Peru (WHO 2023). The *Leishmania* species that cause MCL are mainly *Leishmania braziliensis, L. guyanensis, L. amazonensis, L. panamensis,* and *L. peruviana.*

Bartonellosis

The Gram-negative aerobic bacillus *Bartonella bacilliformis*, formerly known as South American bartonellosis, is transmitted by bites of infected female sandflies (genus *Lutzomyia*) to humans. The disease is divided into two phases: the acute phase, called Oroya fever, and the chronic phase, called Verruga peruana (Peruvian warts).

During the Oroya fever, fever, headache, muscle aches, abdominal pain, and severe anemia may occur. Complications may occur in up to 70% of patients with Oroya fever, including secondary infections and cardiopulmonary complications such as heart failure, pericardial effusion, pulmonary edema, and cardiogenic shock. The Verruga peruana phase presents numerous red-to-purplish papules, nodules, and sores, mainly on the extremities and face (CDC 2023). These skin lesions can last for several years. Bartonellosis is endemic in the Andes Mountains at an altitude of 900–3000 meters in western South America, including Peru, Colombia, and Ecuador, but most cases are from Peru (CDC 2023). Although Bartonellosis does not exist in Asia, a research team, including a Japanese biologist, found that sand flies are vectors of Bartonella bacillus infection (Noguchi et al. 1929).

Sand Fly-Transmitted Viruses

Viruses transmitted by sand flies mainly belong to the genera *Phlebovirus*, *Vesiculovirus*, and *Orbivirus* (Depaquit et al. 2010). Of these, the most important in terms of public health is sandfly fever, also known as pappataci fever or 3-day fever, which is caused by viruses of the genus *Phlebovirus*. Sand fly-borne phleboviruses are transmitted by *Lutzomyia* flies in The New World and by mainly *Phlebotomus* flies in The Old World.

Toscana virus (TOSV) infection, a virus of the genus *Phlebovirus*, may be asymptomatic, but after an incubation period of 3–7 days, it begins as a mild febrile illness that causes symptoms such as headache, fever, nausea, and myalgia, later on, meningitis or encephalitis may develop (Alkan et al. 2013). Severe meningoencephalitis (Baldelli et al. 2004; Kuhn et al. 2005), hearing loss (Pauli et al. 1995), and lethal encephalitis (Bartels et al. 2012) have been reported. Infections caused by Sandfly fever Naples virus (SFNV) and Sandfly fever Sicilian virus (SFSV) cause influenza-like symptoms such as fever, headache, malaise, myalgia, and retroorbital pain, which usually resolve within 1 week, even in severe cases. Infections with SFNV and SFSV are clinically indistinguishable (Alkan et al. 2013).

Sandfly fever is widely distributed in the Mediterranean region, Africa, the Indian subcontinent, the Middle East, and Central Asia; however, SFNV and SFSV cause fever that lasts for 3–5 days and is therefore not given much attention by healthcare workers. It is difficult to prove an accurate prevalence of *Phlebovirus* infections in the Mediterranean area, but it is estimated that at least 250 million people are exposed (Moriconi et al. 2017).

4.11 Stable Fly (*Stomoxys* Spp. (Insecta: Diptera: Muscidae))

Hüseyin Çetin

Many ectoparasite species, such as ticks, fleas, and flies, adversely affect the health of the animals raised in livestock farms, cause stress to the animals (swine, cattle, and horses), and cause economic losses by decreasing cattle growth and milk production (Catangui et al. 1993; Campbell et al. 2001). One of the groups of ectoparasitic flies in the genus *Stomoxys* includes 18 species, most of which are endemic to Africa (Fig. 4.22). The best-known species is *Stomoxys calcitrans* (L.) (Diptera: Muscidae), which is the most widely distributed species.

Stomoxys calcitrans, commonly known as the biting housefly, stable fly, or dog fly, is a well-known fly species on animal farms. Both male and female adults of the stable fly bite warm-blooded farm animals, pets, and humans during the daytime. In the United States, the livestock sector is estimated to have lost hundreds of millions of dollars due to stable flies per year. It is estimated that dairy cows lose an average of 139 liters of milk per year, and cattle and calves lose between 6 and 26 kg of live weight in enterprises with high fixed fly populations (Taylor et al. 2012b).

These flies are very persistent biters, and their bites are extremely painful for their hosts. Field observations in the United Kingdom show that *S. calcitrans*

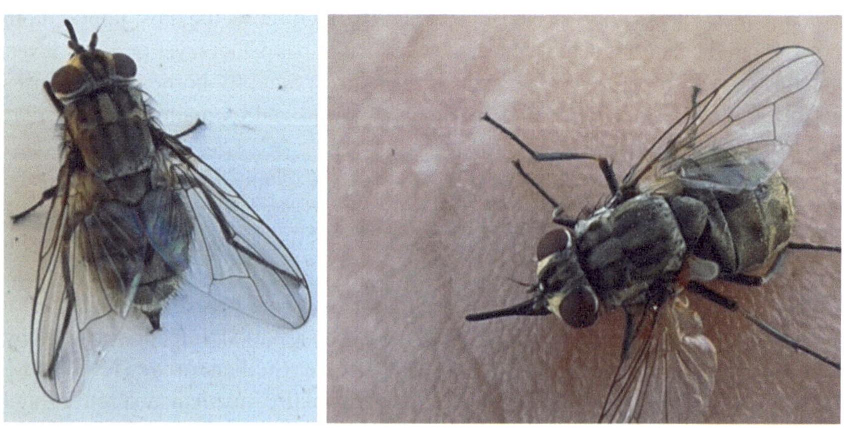

Fig. 4.22 *Stomoxys* (Diptera: Muscidae)

primarily prefers cattle as hosts, followed by horses. It is also fed on domestic animals (cats, dogs, goats, and sheep) and humans (Warnes and Finlayson 1987). They are potential vectors of various animal pathogens (bacteria, helminths, protozoans, viruses) and have a serious negative impact on the tourism industry in some parts of the world (Baldacchino et al. 2013). The stable fly shows diurnal behavior; in summer and hot days, it prefers feeding on its hosts in the early morning and late afternoon, and in cooler months, it prefers the middle of the day (Rochon et al. 2021).

4.11.1 Life Cycle

Stable flies have four stages in their life cycle: egg, larva, pupa, and adult (Fig. 4.23). The average life cycle is usually approximately 20 days at 25 ± 2 °C and 50 ± 10% RH under a 12-h light cycle, but it may change depending on environmental conditions in the field (Salem et al. 2012). Wet straw and manure, spilled feed, silage, and grass-clipping decaying vegetation are the main breeding sites of this insect. In areas where hay is stored for animals to consume during the winter months, old substrates (grass clippings, garden compost, and silage) that have undergone microbial degradation are favorable sites for development (Taylor et al. 2012b).

A female fly can lay approximately 600–800 eggs in her lifetime, 60–150 eggs at a time after each bloodsucking. The eggs of the stable fly are white in color and sausage-shaped, approximately 1 mm long. Eggs hatch within 24 h, and the hatched larva is creamy white, grows to approximately 1–1.2 cm, and goes through three instars (stages). After the larval stage, the maggot turns into a 4.5–6 mm long

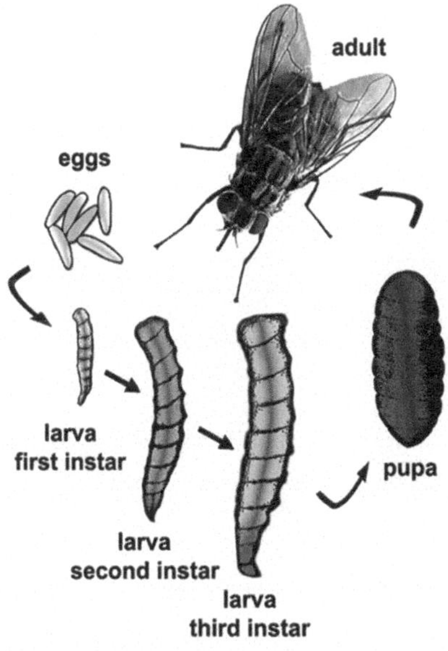

Fig. 4.23 Life cycle of *Stomoxys* species

reddish-brown pupa. Although stable fly adults and housefly adults are similar in size (5–7 mm) and color, adult stable flies have circular spots on their abdomens in a checkerboard pattern, while houseflies have unpatterned abdomens (Fig. 4.24). In addition, houseflies have sponging and sucking mouthparts that do not suck blood, whereas stable flies have developed mouthparts that suck blood. Adult stable flies can live in the field for approximately 10 days.

4.11.2 Control

To control adult flies in and around animal shelters, spray application of residual insecticides to target areas where flies rest or congregate, such as poles, door frames, windows, beams, railings and walls, can be used. Neonicotinoids, synthetic pyrethroids, and organic phosphorus insecticides are used in spray applications. In addition to the application of pesticides in and around animal shelters, some biocidal products can be sprayed on the animal body to provide protection against stable flies for approximately 2 weeks (Schmidt et al. 1976; Morgan and Bailie 1980). Topical repellent products can protect humans and livestock from stable fly bites (Parashar et al. 1993). Although many insecticide groups are successful in controlling adult stable flies, it should not be forgotten that flies can develop resistance to the insecticides used against them in a short period (Pitzer et al. 2010; Reissert-Oppermann et al. 2019; Lorn et al. 2022).

Insect growth regulators (IGRs) have been used to control stable fly larvae. The application of IGRs to adult resting surfaces or larval developmental sites (such as cattle, swine, and poultry manure, soiled animal bedding, rotting vegetable matter, and old rotting hay) in addition to animal bedding inhibits stable fly development (Reissert-Oppermann et al. 2019). Injection into animals or oral administration of IGRs can be highly effective on stable flies. The application of cyromazine at a dose of 0.5 g/m^2 in mid-May reduced the number of stable flies emerging from winter hay-feeding sites by over 95% (Taylor et al. 2012a). Cyromazine and pyriproxyfen

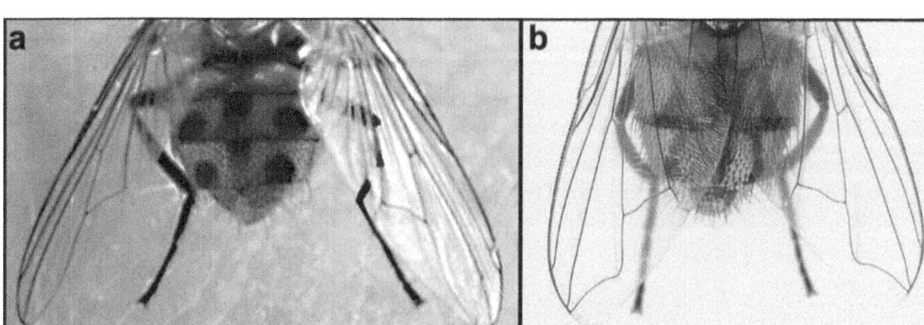

Fig. 4.24 Dorsal view of the abdomen of stable fly (**a**) and house fly (**b**)

at recommended concentrations caused 100% mortality in stabile fly larvae originating from 10 dairy farms in Germany (Reissert-Oppermann et al. 2019).

The success of sticky sheets and strips in controlling flies has been studied by different researchers. Various types and sizes of traps were examined, and large sticky traps and pyramid traps reduced stable fly populations in dairy calf greenhouse facilities and farms (Black and Krafsur 1985; Kaufman et al. 2005; Ose and Hogsette 2014). Sticky traps can also be used to monitor local stable fly populations.

4.12 Kissing Bug, Triatominae

Yuichiro Tabaru

The vectors of *Trypanosoma cruzi*, a protozoan pathogen of Chagas disease, are insects classified into the order Hemiptera, family Reduviidae, and subfamily Triatominae. The vectors are called kissing bugs or assassinate bugs, and many species are potential vectors of Chagas disease. Most of the kissing bugs are distributed in Central and South America (Lent and Wygodzinsky 1979). Chagas disease is one of the 17 neglected tropical diseases designated by the WHO. Vector control and house improvement have been successively conducted in several South American countries (Schofield 1994) and Central America (Tabaru et al. 1998; Ponce 2007).

One species classified as *Triatoma rubrofasciata* has been migrating from South America into Asian countries and Pacific countries since the 1920s. This species might be a potential vector of Chagas disease because many positive persons have migrated from Latin American countries to Asian countries.

Most kissing bugs inhabit rural houses constructed with mud walls and thatched roofs. Chagas disease is a chronic heart disease without vaccine and medicine. Vector control is the most effective method for preventing Chagas disease and improving the house. In Asian countries, including Japan, there are some groups of Triatominae, but they are predators of insects.

4.12.1 Chagas Disease

Chagas disease was first reported by a Brazilian surgeon, Carlos Chagas, in 1909 for chronic heart disease. This disease is mainly transmitted by kissing bugs, which are widely distributed in Central America and South America. Most bugs inhabit rural houses constructed with mud walls and thatched roofs.

Blood of humans and wild animals, including domestic animals and chickens, are sources of nutrition for bugs. These animals are carriers of the protozoan pathogen *Tr. cruzi*. Approximately 80% of Chagas disease is transmitted by bugs; others are blood transfusion, vertical transmission by the positive mother to her baby, and contaminated foods. The vectorial transmission occurs through the mucosa, mouth, nose, eyes, and wounds by contact with the infested dropping of the bugs but not through blood-sucking.

In 2000, the number of Chagas disease patients was estimated to be 1.8 million, approximately 4.9% of the total population of Central America, and 60,000 were annual new patients (Schofield 2000).

The acute phase lasts approximately 2 weeks after transmission, and sometimes the eyes are swollen at this moment, so-called Romaña symptoms (Fig. 4.25a). After the acute phase, the disease moves to the chronic phase without any symptoms. Ten to 20 years after transmission, patients are fatal due to heart storks.

In the acute periods, some anti-parasitic medicines are available. Medical treatments should be performed under strict supervision by a physician to avoid side effects. Benznidazole (BZN) or nifurtimox (NFX) is recommended as an equivalent to treat *Tr. cruzi* infection.

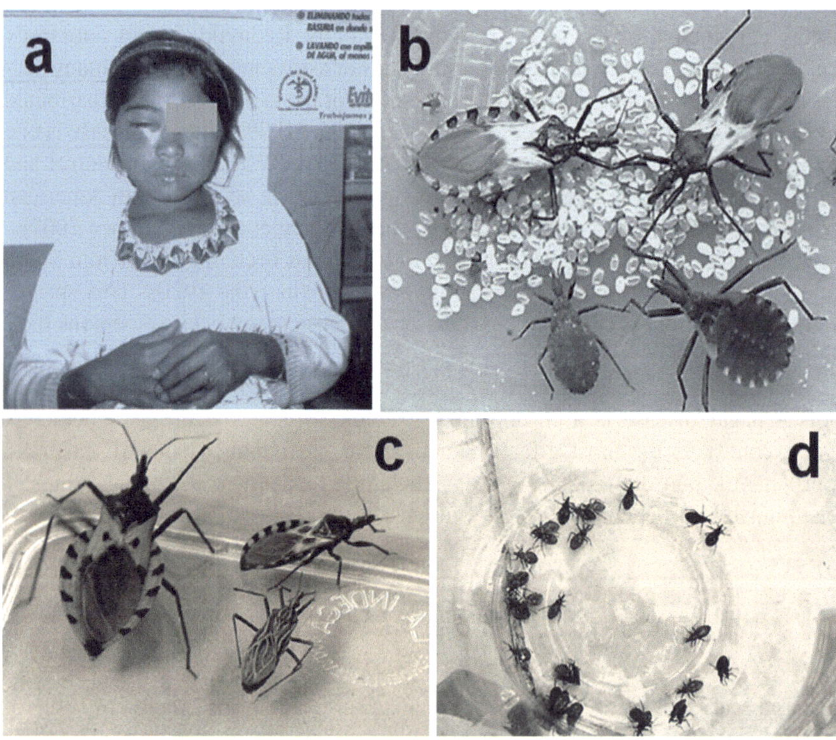

Fig. 4.25 Acute stage of Chagas disease. (**a**) Ronaña in the right eye, (**b**) *T. dimidiata*, female (upper left), male (upper right), nymph (under two bugs), and eggs (white is after hatch, translucent is still alive), (**c**) *T. dimidiata* (left), *T. nitida* (upper right), and *R. prolixus* (lower right), (**d**) *T. rubrofasciata*

4.12.1.1 Triatominae and Their Biology

More than 110 species of kissing bugs have been reported in Latin American countries (Lent and Wygodzinsky 1979). *Triatoma infestans, T. sordita, T. brasiliensis,* and *Panstrongylus megistus* are common vectors in South America. However, in Central America, such as Guatemala, El Salvador, Honduras, Nicaragua, and Costa Rica, only two species are important: *T. dimidiata* and *Rhodnius prolixus* (Schofield 1994: Tabaru et al. 1999b). The former is an indigenous species in Central America, and the latter one migrated from South America. This species inhabits stork nests in Venezuela but only inhabits human residents in Central America (Fig. 4.25b, c).

Triatoma dimidiata inhabits the cracks and crevices on the mud wall house. The indoor distribution was focused on the bedside walls in the sleeping rooms, which was confirmed after the demolition of rural houses (Monroy et al. 1998; Tabaru et al. 1999a). The bugs' feces on the walls are good evidence of bug inhabitation. *R. prolixus* inhabits roof materials, and the bug adheres eggs to fiber. The majority of the bugs are distributed just over the beds.

The blood resources of the bugs revealed that humans were the most abundant, followed by chickens and opossums in Guatemala (Sasaki et al. 2003). Kissing bugs are incompletely metamorphosed, so nymphs and adults live in the same places and suck the blood of animals. *Tripanosoma cruzi* was highly positive after the second instar of the bugs; the mean positive rates were 19.1% (*T. dimidiata*) and 20.6% (*R. prolixus*) in nationwide surveillance in Guatemala (Tabaru et al. 1999b: Monroy et al. 2003). The villages that inhabited *T. dimidiata* were high (29.2%) in the vectors but low in the residents (8.9%); however, villages with *R. prolixus* were low in the vectors (13.2%) and high in the humans (38.3%) in Guatemala (Paz-Bailey et al. 2002). This phenomenon might be introduced by direct mouth contact with infested feces of *R. prolixus* while sleeping.

On *T. dimidiata,* some colonies were collected in the deep limestone sinus, and tropical jungles in Guatemala were sylvatic strains that might not invade human dwellings (Monroy et al. 2003).

4.12.1.2 Migrated Kissing Bug in Asia

Triatoma rubrofasciata (Fig. 4.25d) has migrated into Asian countries and Pacific countries, such as Andaman Islands, Myanmar, Cambodia, China, Taiwan, India, Indonesia, Malaysia, Okinawa, the Philippines, Seychelles, Singapore, Sri Lanka, Thailand, and Vietnam, since the 1920s (Lent and Wygodzinsky 1979). This species might be a potential vector of Chagas disease because there is a large number of migrated positive carriers from Latin American countries. In Hanoi, Vietnam, this species was distributed in villages close to Hanoi city. Many bugs were collected in accumulated firewood in front of dwellings. The bugs might suck blood from rats and chickens. This species was very active compared with *T. dimidiata.*

People who want to travel to rural areas in Latin America must pay attention to contaminated materials with *Tr. cruzi* and blood transfusions.

4.12.1.3 Vector Control in Central America

A field vector control investigation against two main species was carried out in Guatemalan villages in 1997. *Triatoma dimidiata* was decreased after spraying pyrethroid insecticides onto walls and roofs, and *R. prolixus* was very sensitive to the insecticides and was dramatically controlled after spraying to the attic side of the roof (Tabaru et al. 1998). We must be careful in controlling the evaluation that bug collection by man/hour depends on the experience of collectors; the mean recovery rate in six houses was only 8.8% after demolition (Monroy et al. 1998).

In October 1997, the Central American Initiative for Chagas Disease Control (IPCA) was launched. The objectives of this subregional Initiative are (1) the elimination of *R. prolixus* in Central America; (2) the reduction of the domiciliary infestation of *T. dimidiata*; and (3) the elimination of the transfusion transmission of *Tr. cruzi*.

After introducing the Japan International Cooperation Agency (JICA)'s vector control project in Guatemala in 2000, the two species of kissing bugs dramatically decreased in the control area. Five years later, *R. prolixus* was eliminated from Guatemala, El Salvador, Honduras, and Nicaragua (Ponce 2007). Community surveillance for vectors has functioned in each of Central countries since 2007; residents are requested to bring the bugs to the local health center when they collect them, and if the bugs are identified as the vector triatomine, spray insecticides in the houses.

4.13 Tsetse Fly, *Glossinidae*

Kyoko Hayashida

Tsetse flies are hematophagous flies with brown color in the family *Glossinidae*, which resembles other common flies such as stable flies in Family *Muscidae;* however, they are characterized by their prominent proboscis, completely overlayed scissor-like wings over the abdomen at the resting position (Figs. 4.26 and 4.27). The tsetse flies infest vast sub-Saharan Africa, covering over nine million km^2 in 37 countries (FAO 2019). The spatial distribution of tsetse flies mainly depends on the temperature and humidity, as well as the rich animal density and species-specific preferable land cover. Due to their strict habitat preferences, tsetse flies have not been reported in other areas apart from Africa, although some species have been reported in Saudi Arabia in the past (Elsen et al. 1991). During the blood meal, they carry unicellular flagellated parasite African trypanosomes, the causative agents of African animal trypanosomiasis (AAT) and human African trypanosomiasis (HAT). The presence of tsetse flies and AAT/HAT has posed a serious threat to people in endemic areas and has been a major constraint on socio-economic development. As no effective vaccine has yet been developed, animal/human case detection, treatment, and tsetse fly control are all indispensable to combat AAT and HAT.

Fig. 4.26 *Glossina morsitans centralis* viewed from a vertical surface

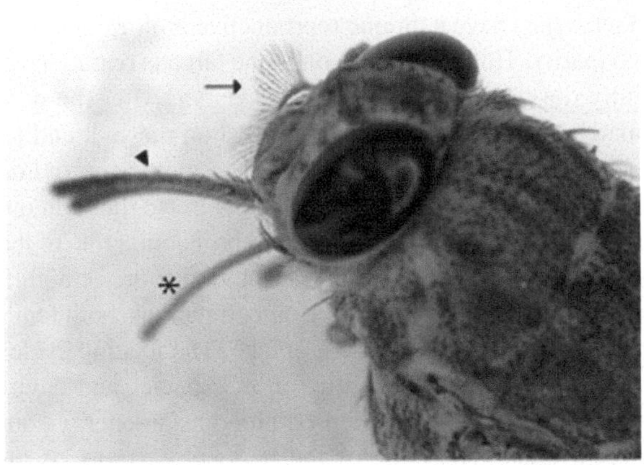

Fig. 4.27 Lateral view of *G. m. centralis*. Long forwarded proboscis (*) with two maxillary pulps (arrowhead). The elongated third segment of the antenna (arrow) has a branched fringe of hair (arista)

4.13.1 Taxonomy

There are 21 species and several unclassified species belonging to the genus *Glossina*, a sole member of the family *Glossinidae* (Leak 1998). *Glossina* are classified into three groups based primarily on morphological features of the adult genitalia (Jordan 1993) and their distribution and ecological habitats: morsitans ("savannah" group), fusca ("forest" group), or palpalis ("riverine" group). The morsitans group is mainly distributed in arid savanna woodland areas in the central and eastern parts of Africa. Among them, *G. morsitans morsitans* and *G. pallidipes* are important transmitting vectors of *Trypanosoma brucei rhodesiense* responsible for the acute form of HAT. Blood meal analysis indicated that they preferably feed on Suidae (warthog, bushpig) and ruminants (cattle, bushbuck, and buffalo) as well as humans, suggesting that these animals are important natural reservoirs of rhodesiense HAT (Clausen et al. 1998). The palpalis group is distributed waterside in western and central Africa. *Glossina palpalis palpalis* is one of the principal vectors of *T. b. gambiense* and is responsible for the chronic form of HAT. While the distribution of the morsitans group is often confined to the conserved area where the wildlife population is dense, *G. p. palpalis* adapts more to the urban settlement, which partially explains its anthroponotic transmission nature of gambiense HAT (Courtin et al. 2005).

The fusca group is distributed in forest vegetation in western and central Africa. The group is genetically ancestral to the other two groups, and the body size is relatively large (11–14 mm) compared to the other two groups (6–11 mm length). Although the fusca group is not considered an important vector of HAT, it is an important vector of AAT.

4.13.2 Life Cycle

Tsetse flies have a unique reproductive biology characterized by their adenotrophic viviparity. This means that only one larva is reared in the uterus and nourished until fully developed into a third-instar larva. After the development cycle of approximately 9 days, the larva is deposited on the soil, and pupation occurs within a few hours of deposition. Adult tsetse flies emerge approximately 30 days later. With this unique mode of reproduction, female tsetse flies can only produce ten or fewer offspring in their lifetime. Thus, the reproduction rate of the tsetse fly is relatively slow, making the control strategy adaptation more straightforward and effective compared to other vectors. Unlike other dipterans, both females and males rely solely on vertebrate blood for nutrient uptake. The feeding cycle is approximately 2–4 days, although it varies depending on the species and environmental conditions. Since tsetse flies exclusively feed on blood as an energy source, the density of animals/host availability is one of the important factors of their distribution. With more expansive urbanization, tsetse flies are currently confined to protected areas such as national parks and forest reserves.

4.13.3 Animal and Human African Trypanosomiasis

"Tsetse belt" is a vast tsetse-infested area covering most of the sub-Saharan Africa region where livestock farming is challenging due to the presence of the disease. Exotic cattle breeds are more susceptible than local breeds, and wildlife is usually asymptomatic and acts as a natural reservoir of parasites. The direct economic loss from the AAT is estimated to be over US$1.0 billion, hindering agricultural development and poverty alleviation (FAO 2019). Moreover, two trypanosome species, *T. b. rhodesisense* in eastern and southern Africa and *T. b. gambiense* in western Africa, cause the fatal disease HAT, also known as sleeping sickness, which threatens human health. In 2012, the WHO set up the 2030 goal of zero transmission of gambiense HAT, which accounts for over 98% of HAT cases, and eliminating rhodesiense HAT as a public health problem. Gambiense HAT is assumed to circulate anthroponotically among humans. Thus, active/passive case detection and treatment have contributed to the dramatic reduction of cases. The recent introduction of nifurtimox-eflornithine combination therapy (NECT), as well as the newly developed oral therapeutic fexinidazole, was an important breakthrough for the successful treatment of patients (Lindner et al. 2020). For rhodesiense HAT, since wild animals and domestic animals are the reservoirs of the disease, its control is more complex than that of gambiense HAT. Nevertheless, in both forms of HAT, vector tsetse control has substantially reduced disease burden. Notably, these multisectoral control efforts have been successful, and the number of new HAT cases is rapidly declining (Franco et al. 2022).

4.13.4 Advances in Tsetse Control

Many approaches have been taken to control the number of tsetse flies. These include the ground and sequential aerial treatment (SAT) of low-dose insecticides (e.g., pyrethroids), which usually target focused areas because of their financial burden and negative environmental effects. The sterile insect technique (SIT) is a powerful, environmentally friendly method that uses irradiation to produce sterile male tsetse flies released by air to reduce the number of offspring. The success story for tsetse elimination using SIT was achieved on the island of Unguja, Zanzibar, where the aerial release of a total of 8.5 million sterile tsetse flies was conducted from 1994 to 1997. In the end, tsetse flies were completely eradicated in 1997, followed by the disappearance of AAT (Vreysen et al. 2014). Since then, many other elimination programs using SIT have been started across Africa, and some are successful. Blue/black-colored stationary traps (Figs. 4.28 and 4.29) or targets with residual insecticide with/without odor attractants (e.g., acetone or octenol) are also widely used at the community level. Insecticide-treated cattle (ITC), in which insecticides are applied to cattle, has also been proven to effectively control the tsetse population.

Fig. 4.28 The epsilon trap, one of the stationary traps for capturing savanna tsetse species

Recent advancements in molecular science may further accelerate the control of tsetse flies. The complete genomes of nine *Glossina* species have recently been characterized (International Glossina Genome Initiative 2014; Attardo et al. 2019). One of the applications of genomics is population genetics analysis for strategically designing control regions by estimating population structure. Additionally, the biological understanding of the olfactory and visual attraction mechanisms will improve trap/bait design. "Paratransgenesis" is a recently proposed unique strategy that uses genetically modified tsetse symbionts that can produce antiparasitic molecules. For its capability of culturing and genetic modification, one of the commensal symbionts, *Sodalis glossinidius, has* been investigated as a target symbiont of paratransgenesis, where this bacterium produces an anti-trypanosome nanobody (Vooght et al. 2014). Although further investigation and risk assessment are still needed, updated tsetse fly ecological/biological information will be beneficial in designing an effective control strategy for tsetse flies and African trypanosomes.

Fig. 4.29 The caught tsetse flies

References

Adler PH (2022) World blackflies (Diptera: Simuliidae): a comprehensive revision of the taxonomic and geographical inventory. 145 pp. http://biomia.sites.clemson.edu/pdfs/blackflyinventory.pdf. Accessed 3 Mar 2023

Adler PH, Currie DC, Wood DM (2004) The black flies (Simuliidae) of North America. Cornell University Press, Ithaca, 941 pp

Akiba K (1960) Studies on the *Leucocytozoon* found in the chicken, in Japan: II. On the transmission of *L. caulleryi* by *Culicoides arakawae*. Jpn J Vet Sci 22:309–317

Ali-Khan Z (1977) Tissue pathology and comparative microanatomy of *Onchocerca* from a resident of Ontario and other enzootic *Onchocerca* species from Canada and the USA. Ann Trop Med Parasitol 71(4):469–482

Alkan C, Bichaud L, Lamballerie X, Aletn B, Gould EA, Charrel RN (2013) Sandfly-borne phleboviruses of Eurasia and Africa: epidemiology, genetic diversity, geographic range, control measures. Antivir Res 100(1):54–74

Aregawi WG, Agga GE, Abdi RD, Philippe Büscher P (2019) Systematic review and meta-analysis on the global distribution, host range, and prevalence of *Trypanosoma evansi*. Parasit Vectors 12(1):67

Arlian LG, Morgan MS (2017) A review of *Sarcoptes scabiei*: past, present and future. Parasit Vectors 10(1):297

Asahina S (1991) Blattaria of Japan. Nakayama-Shoten, Tokyo, 253 pp (in Japanese)

Attardo GM, Abd-Alla AMM, Acosta-Serrano A, Allen JE, Bateta R, Benoit JB, Bourtzis K, Caers J, Caljon G, Christensen MB, Farrow DW, Friedrich M, Hua-Van A, Jennings EC, Larkin DM, Lawson D, Lehane MJ, Lenis VP et al (2019) Comparative genomic analysis of six *Glossina* genomes, vectors of African trypanosomes. Genome Biol 20(1):187

Azama Y, Fukuchi Y, Okano S, Kato T, Takara T, Nakama Y, Arakaki E, Kudaka J (2014) Studies on epidemiology of bite cases, ecology and control methods for biting midge (Ceratopogonidae) in Kume Island, Okinawa Prefecture (4) studies on ecology and control methods for biting midge. Ann Rep Okinawa Prefect Instit Health Environ 48:37–41. (in Japanese)

Azarova NS, Miretskij OY, Sonin MD (1965) The first discovered case of nematode parasitism in the USSR in a human being (genus *Onchocerca* Diesing, 1841). Med Parazitol (Mosk) 34:156–158. (in Russian)

Badiaga S, Brouqui P (2012) Human louse-transmitted infectious diseases. Clin Microbiol Infect 18(4):332–337

Bain O (1979) Transmission d'Onchocerque bovine, *Onchocerca gutturosa*, par *Culicoides*. Ann Parasitol Hum Comp 54:483–488

Bain O, Ramachandran CP, Petter F, Mak JW (1977) Description d'*Onchocerca dewittei* n. sp. (Filarioidea) chez *Sus scrofa* en Malaisie. Ann Parasitol Hum Comp 52:471–479

Bain O, Petit G, Poulain B (1978) Validity of the two species of *Onchocerca lienalis* and *O. gutturosa* from bovines. Ann Parasitol Hum Comp 53(4):421–450

Baldacchino F, Muenworn V, Desquesnes M, Desoli F, Charoenviriyaphap T, Duvallet G (2013) Transmission of pathogens by *Stomoxys* flies (Diptera, Muscidae): a review. Parasite 20:26

Baldelli F, Ciufolini MG, Francisci D, Marchi A, Venturi G, Fiorentini C, Luchetta ML, Bruto L, Pauluzzi S (2004) Unusual presentation of life-threatening Toscana virus meningoencephalitis. Clin Infect Dis 38(4):515–520

Barbieri R, Drancourt M, Raoult D (2021) The role of louse-transmitted diseases in historical plaque pandemics. Lancet Infect Dis 21(2):e17–e25

Bartels S, de Boni L, Kretzschmar HA, Heckmann JG (2012) Lethal encephalitis caused by the Toscana virus in an elderly patient. J Neurol 259(1):175–177

Beaver PC, Horner GS, Bilos JZ (1974) Zoonotic onchocerciasis in a resident of Illinois and observations on the identification of *Onchocerca* species. Am J Trop Med Hyg 23(4):595–607

Beaver PC, Yoshimura H, Takayasu S, Hashimoto H, Little MD (1989) Zoonotic *Onchocerca* in a Japanese child. Am J Trop Med Hyg 40(3):298–300

Bennis I, De Brouwere V, Belrhiti Z, Sahibi H, Boelaert M (2018) Psychosocial burden of localised cutaneous Leishmaniasis: a scoping review. BMC Public Health 18(1):358

Bentinck W (1955) The black flies of Japan and Korea (Diptera: Simuliidae). 406 Medical General Laboratory, U.S. Army, Tokyo, 23 pp

Bermudez AJ (2008) Miscellaneous and sporadic protozoal infections. In: Saif YM, Fadly AM, Glisson JR, McDougald LR, Nolan LK, Swayne DE (eds) Diseases of poultry, 12th edn. Blackwell, Oxford, pp 1105–1117

Bitam I, Dittmar K, Parola P, Whiting MF, Raoult D (2010) Fleas and flea-borne diseases. Int J Infect Dis 14(8):e667–e676

Black WC, Krafsur ES (1985) Use of sticky traps to investigate seasonal trends in the spatial distribution of house flies and stable flies (Diptera: Muscidae). J Med Entomol 22(5):550–557

Bonell A, Lubell Y, Newton PN, Crump JA, Paris DH (2017) Estimating the burden of scrub typhus: a systematic review. PLoS Negl Trop Dis 11(9):e0005838

Borah S, Goswami S, Agarwal M, Rahman I, Deka M, Chattopadhyay P, Singh L (2012) Clinical and histopathological study of *Simulium* (blackfly) dermatitis from North-Eastern India–a report. Int J Dermatol 51(1):63–66

Borkent A (2004) The biting midges, the Ceratopogonidae (Diptera). In: Marquardt WC (ed) Biology of disease vectors, 2nd edn. Elsevier Academic Press, Burlington, pp 113–126

Borkent A, Dominiak P (2020) Catalog of the biting midges of the world (Diptera: Ceratopogonidae). Zootaxa 4787(1). https://doi.org/10.11646/zootaxa.4787.1.1

Buckley JJC (1938) On *Culicoides* as a vector of *Onchocerca gibsoni* (Cleland & Johnston, 1910). J Helminthol 16:121–158

Bunpapong N, Charoenkul K, Nasamran C, Chamsai E, Udom K, Boonyapisitsopa S, Tantilertcharoen R, Kesdangsakonwut S, Techakriengkrai N, Suradhat S, Thanawongnuwech R, Amonsin A (2021) African horse sickness virus serotype 1 on horse farm, Thailand, 2020. Emerg Infect Dis 27(8):2208–2211

Burgess IF, Brunton ER, Burgess NA (2013) Single application of 4% dimethicone liquid gel versus two applications of 1% permethrin creme rinse for treatment of head louse infestation: a randomised controlled trial. BMC Dermatol 13:5

Cambra-Pellejà M, Gandasegui J, Balaña-Fouce R, Muñoz J, Martínez-Valladares M (2020) Zoonotic implications of *Onchocerca* species on human health. Pathogens 9(9):761

Campbell JB, Skoda SR, Berkebile DR, Boxler DJ, Thomas GD, Adams DC, Davis R (2001) Effects of stable flies (Diptera: Muscidae) on weight gains of grazing yearling cattle. J Econ Entomol 94(3):780–783

Canyon DV, Speare R, Muller R (2002) Spatial and kinetic factors for the transfer of head lice (*Pediculus capitis*) between hairs. J Invest Dermatol 119(3):629–631

Catangui MA, Campbell JB, Thomas GD, Boxler DJ (1993) Average daily gains of Brahman-crossbred and English x exotic feeder heifers exposed to low, medium, and high levels of stable flies (Diptera: Muscidae). J Econ Entomol 86(4):1144–1150

Centers for Disease Control and Prevention (CDC) (2023) *Bartonella bacilliformis* infection. https://www.cdc.gov/bartonella/bartonella-bacilliformis/index.html. Accessed 10 Dec 2023

Choochote W, Takaoka H, Fukuda M, Otsuka Y, Aoki C, Eshima N (2005) Seasonal abundance and daily flying activity of black flies (Diptera: Simuliidae) attracted to human baits in Doi Inthanon national park, northern Thailand. Med Entomol Zool 56(4):335–348

Chuang YY, Lin CS, Wang CH, Yeh CC (2000) Distribution and seasonal occurrence of *Forcipomyia taiwana* (Diptera: Ceratopogonidae) in the Nantou area in Taiwan. J Med Entomol 37(2):205–209

Chvála M, Lyneborg L, Moucha J (1972) Horse flies of Europe (Diptera, Tabanidae). Entomological Society of Copenhagen, Copenhagen, 500 pp

Clausen PH, Adeyemi I, Bauer B, Breloeer B, Salchow F, Staak C (1998) Host preferences of tsetse (Diptera: Glossinidae) based on bloodmeal identifications. Med Vet Entomol 12(2):169–180

Coates SJ, Thomas C, Chosidow O, Engelman D, Chang AY (2020) Ectoparasites: pediculosis and tungiasis. J Am Acad Dermatol 82(3):551–569

Cornwell PB (1968) The cockroach. The Rentokil Library, Hutchinson/London, 391 pp

Courtin F, DuPont S, Zeze DG, Jamonneau V, Sane B, Coulibaly B, Cuny G, Solano P (2005) Human African trypanosomiasis: urban transmission in the focus of Bonon (Ivory Coast). Trop Med Int Health 10(4):340–346

Dagrosa AT, Elston DM (2017) What's eating you? Head lice (Pediculus humanus capitis). Cutis 100(6):389–392

Datta M (1992) An overview of the simuliidae (Diptera) of West Bengal, India. J Bengal Nat Hist Soc New Ser 11:41–62

Dennis SJ, Meyers AE, Hitzeroth II, Rybicki EP (2019) African horse sickness: a review of current understanding and vaccine development. Viruses 11(9):844

Depaquit J, Grandadam M, Fouque F, Andry PE, Peyrefitte C (2010) Arthropod-borne viruses transmitted by Phlebotomine sandflies in Europe: a review. Euro Surveill 15(10):19507

Dewi RS, Damajanti R, Wardhana AH, Mulatsih S, Poetri ON, Steeneveld W, Hogeveen H (2020) The economic losses of Surra outbreak in Sumba Timur, Nusa Tenggara Timur-Indonesia. Trop Anim Sci J 43(1):77–85

Dholakia S, Buckler J, Jeans JP, Pillai A, Eagles N, Dholakia S (2014) Pubic lice: an endangered species? Sex Transm Dis 41(6):388–391

Dohnal J, Blinn J, Wahl G, Schulz-Key H (1990) Distribution of microfilariae of *Onchocerca lienalis* and *Onchocerca gutturosa* in the skin of cattle in Germany and their development in

Simulium ornatum and *Culicoides nubeculosus* following artificial infestation. Vet Parasitol 36(3–4):325–332
Duan Y, Yang Z, Bellis G, Xie J, Li L (2022) Full genome sequencing of three *Sedoreoviridae* viruses isolated from *Culicoides* spp. (Diptera, Ceratopogonidae) in China. Viruses 14(5):971
Durden LA, Hinkle NC (2019) Fleas (Siphonaptera). In: Mullen GR (ed) Medical and veterinary entomology, 3rd edn. Academic Press, Amsterdam, pp 145–169
Eberhard ML (1979) Studies on the *Onchocerca* (Nematoda: Filarioidea) found in cattle in the United States. I. Systematics of *O. gutturosa* and *O. lienalis* with a description of *O. stilesi* sp. n. J Parasitol 65(3):379–388
Elbers AR, Koenraadt CJ, Meiswinkel R (2015) Mosquitoes and *Culicoides* biting midges: vector range and the influence of climate change. Rev Sci Tech 34(1):123–137
Eldridge S (1900) Japan. Plague in Japan from July 1 to September 15, 1900. Public Health Rep (1896–1970) 15(44):2717–2719
Elsen P, Amoudi MA, Leclercq M (1991) The discovery in Saudi Arabia of 2 species of tsetse flies, vectors of human and animal trypanosomiasis. Rev Med Liege 46(4):225–231. (in French)
El-Sinnary K, Hussein HS (1980) *Culicoides kingi* Austen: a vector of *Onchocerca gutturosa* (Neumann, 1910) in the Sudan. Ann Trop Med Parasitol 74(6):655–656
Elston DM (2005) Drugs used in the treatment of pediculosis. J Drugs Dermatol 4(2):207–211
Executive Committee of Guideline for the Diagnosis and Treatment of Scabies (2017) Guideline for the diagnosis and treatment of scabies in Japan (third edition). J Dermatol 44(9):991–1014
Feldmeier H (2010) Lice as vectors of pathogenic microorganisms. In: Heukelbach J (ed) Management and control of head lice infestations. UNI-MED Verlag AG, Bremen, pp 132–135
Feldmeier H, Heukelbach J (2009) Epidermal parasitic skin diseases: a neglected category of poverty-associated plagues. Bull World Health Organ 87(2):152–159
Food and Agriculture Organization of the United Nations (FAO) (2019) The disease programme against African trypanosomosis (PAAT). http://www.fao.org/paat/the-programme/the-disease/en/. Accessed 25 Dec 2023
Forman S, Hungerford N, Yamakawa M, Yanase T, Tsai HJ, Joo YS, Yang DK, Nha JJ (2008) Climate change impacts and risks for animal health in Asia. Rev Sci Tech Off Int Epizoot 27(2):581–597
Fournier PE, Ndihokubwayo JB, Guidran J, Kelly PJ, Raoult D (2002) Human pathogens in body and head lice. Emerg Infect Dis 8(12):1515–1518
Franco JR, Cecchi G, Paone M, Diarra A, Grout L, Kadima Ebeja A et al (2022) The elimination of human African trypanosomiasis: achievements in relation to WHO road map targets for 2020. PLoS Negl Trop Dis 16(1):e0010047
Frankowski BL, Weiner LB (2002) Head lice. Pediatrics 110(3):638–643
Fu YT, Yao C, Deng YP, Elsheikha HM, Shao R, Zhu XQ, Liu GH (2022) Human pediculosis, a global public health problem. Infect Dis Poverty 11(1):58
Fukuda M, Takaoka H, Uni S, Bain O (2008) Infective larvae of five *Onchocerca* species from experimentally infected *Simulium* species in an area of zoonotic onchocerciasis in Japan. Parasite 15(2):111–119
Fukuda M, Otsuka Y, Uni S, Bain O, Takaoka H (2010a) Genetic evidence for the presence of two species of *Onchocerca* from the wild boar in Japan. Parasite 17(1):33–37
Fukuda M, Otsuka Y, Uni S, Bain O, Takaoka H (2010b) Molecular identification of infective larvae of three species of *Onchocerca* found in wild-caught females of *Simulium bidentatum* in Japan. Parasite 17(1):39–45
Fukuda M, Otsuka Y, Uni S, Boda T, Daisaku H, Hasegawa H, Takaoka H, Bain O (2011) Zoonotic onchocerciasis in Hiroshima, Japan, and molecular analysis of a paraffin section of the agent for a reliable identification. Parasite 18(2):185–188
Fukuda M, Uni S, Otsuka Y, Eshita Y, Nakatani J, Ihara K, Yoshikawa Y, Goto M, Fujiwara S, Ramli R, Sofian Azirun M, Takaoka H (2015) New case of zoonotic onchocercosis in northern Kyushu, Japan. Parasitol Int 64:519–521

Fukuda M, Uni S, Igari T, Utsumi Y, Otsuka Y, Nakatani J, Uga S, Hara T, Hasegawa H, Takaoka H (2019) Human case of *Onchocerca dewittei japonica* infection in Fukushima, Northeastern Honshu, Japan. Parasitol Int 72:101943

Gage KL (2005) Fleas, the Siphonaptera. In: Marquardt WC (ed) Biology of disease vectors, 2nd edn. Elsevier Academic Press, Burlington, pp 77–92

Galiczynski EM, Elston DM (2008) What's eating you? Pubic lice (*Pthirus pubis*). Cutis 81(2):109–114

Goto Y, Miura Y, Kono Y (1988) Serologic evidence for the etiologic role of Chuzan virus in an epizootic of congenital abnormalities with hydranencephaly cerebellar hypoplasia syndrome of calves in Japan. Am J Vet Res 49(12):2026–2029

Grácio AJS, Richter J, Komnenou AT, Grácio MA (2015) Onchocerciasis caused by *Onchocerca lupi*: an emerging zoonotic infection. Systematic review. Parasitol Res 114(7):2401–2413

Gratz NG (1997) Human lice – their prevalence, control and resistance to insecticides. A review 1985–1997, WHO/CTD/WHOPES/97.8. World Health Organization, Geneva, 61 pp

Harrup LE, Miranda MA, Carpenter S (2016) Advances in control techniques for *Culicoides* and future prospects. Vet Ital 52(3–4):247–264

Hashimoto H, Murakami I, Fujiwara S, Takayasu S, Takaoka H, Uga S, Akao N, Kondo K, Yoshimura Y (1990) A human case of zoonotic onchocerciasis in Japan. J Dermatol 17(1):52–55

Hassan HK, Bolcen S, Kubofcik J, Nutman TB, Eberhard ML, Middleton K, Wekesa JW, Ruedas G, Nelson KJ, Dubielzig R, Wekesa JW, Ruedas G, Nelson KJ, Dubielzig R, Lombaert MD, Silverman B, Schorling JJ, Adler PH, Unnasch TR, Beeler ES (2015) Isolation of *Onchocerca lupi* in dogs and black flies, California, USA. Emerg Infect Dis 21(5):789–796

Hastriter MW, Whiting MF (2009) Chapter 236 - Siphonaptera: (Fleas). In: Resh VH, Cardé RT (eds) Encyclopedia of insects, 2nd edn. Academic Press, pp 924–928

Hempolchom C, Sookrung N, Srisuka W, Reamtong O, Sakolvaree Y, Chaicumpa W, Dedkhad W, Jariyapan N, Takaoka H, Saeung A (2019) Characterization of IgE-binding proteins in the salivary glands of *Simulium nigrogilvum* (Diptera: Simuliidae). Parasitol Res 118(8):2353–2359

Heukelbach J, Feldmier H (2010) Head lice biology. In: Heukelbach J (ed) Management and control of head lice infestations. UNI-MED Verlag AG, Bremen, pp 24–31

Hira PR, Al-Buloushi A, Khalid N, Iqbal J, Bain O, Eberhard ML (2008) Zoonotic filariasis in the Arabian peninsula: autochthonous onchocerciasis and dirofilariasis. Am J Trop Med Hyg 79(5):739–741

Hirabayashi K, Abe T, Kamisako M, Komata T, Takahashi H (2018) Occurrence of blackflies and plans for their control in the upper reaches of the Ukawa River in Kashiwazaki City, Niigata Prefecture, Japan. 1. Focus on distribution of blackfly larvae. Med Entomol Zool 69(1):7–11. (in Japanese with English summary)

Hirao K, Takahashi C, Mochizuki K, Tabaru Y (2006) Viscosity of feces in different kind of foods in domically cockroach. Pestology 2(1):9–11. (in Japanese with English summary)

Hirashima Y, Kitahara S, Kato T, Shirafuji H, Tanaka S, Yanase T (2017) Congenital malformations of calves infected with Shamonda virus, Southern Japan. Emerg Infect Dis 23(6):993–996

Hitomi M (1957) Cockroaches prevalent in overseas vessels with special reference to *Blattella germanica* L. Jpn J Sanit Zool 8(3):112–118. (in Japanese with English summary)

Iikawa R, Sato Y, Harada S, Matsumoto J, Nogami S, Sakanakura T, Murata K, Takaoka H, Yukawa M (2009) Prevalence and phylogeny of filarial nematode detected from blackflies distributed in the alpine regions of Japan. Jpn J Vet Parasitol 9:64. (in Japanese)

Imura T, Sato Y, Ejiri H, Tamada A, Isawa H, Sawabe K, Omori S, Murata K, Yukawa M (2010) Molecular identification of blood source animals from black flies (Diptera: Simuliidae) collected in the alpine regions of Japan. Parasitol Res 106(2):543–547

Inagaki T (1959) Studies on the ecology and control of the biting midges, *Culicoides circumscriptus* Kieffer. J Yonago Med Ass 10:997–1012. (in Japanese)

International Glossina Genome Initiative (2014) Genome sequence of the tsetse fly (*Glossina morsitans*): vector of African trypanosomiasis. Science 344(6182):380–386

Ishibashi O, Niizuma J, Sudo K, Ogura G, Sunagawa K, Nakada T (2009) Survey of parasitic fleas on small Asian mongoose on Okinawajima Island, Japan. Jpn J Zoo Wildl Med 14(1):67–72. (in Japanese with English abstract)

Ishizawa H, Kato S, Nishimura H, Tanaka R, Fukuda M, Takaoka H, Tsutsumi Y (2015) Zoonotic onchocerciasis. Pathol Int 65(5):271–273

Itagaki S (1954) Review on studies on filariae. Bull Azabu Univ 1:1–10. (in Japanese)

Iwasa M, Shido Y (2016) Participation of dipterous insects and search for vector in outbreak of teat papillomatosis in cattle. Med Entomol Zool 67(2):109. (in Japanese)

Izwan-Anas N, Ya'cob Z, Low VL, Lourdes EY, Ramli R, Bolongan G, Takaoka H (2021) Simulium (Gomphostilbia) aziruni: first record of a black fly (Diptera: Simuliidae) attracted to a human in Malaysia. Acta Trop 218:105904

Jahnke C, Bauer E, Hengge UR, Feldmeier H (2009) Accuracy of diagnosis of pediculosis capitis: visual inspection vs wet combing. Arch Dermatol 145(3):309–313

Jariyapan N, Tippawangkosol P, Sor-Suwan S, Mano C, Yasanga T, Somboon P, Depaquit J, Siriyasatien P (2022) Significance of eggshell morphology as an additional tool to distinguish species of sand flies (Diptera: Psychodidae: Phlebotominae). PLoS One 17(2):e0263268

Jiang J, Richards AL (2018) Scrub typhus: no longer restricted to the Tsutsugamushi triangle. Trop Med Infect Dis 3(1):11

Jordan AM (1993) Tsetse flies (Glossinidae). In: Lane RP, Crosskey RW (eds) Medical insects and arachnids. Chapman and Hall, London, pp 333–388

Jumpato W, Tangkawanit U, Wongpakam K, Pramual P (2019) Molecular detection of *Leucocytozoon* (Apicomplexa: Hemosporida) in black flies (Diptera: Simuliidae) from Thailand. Acta Trop 190:228–234

Kabeya H, Inoue K, Izumi Y, Morita T, Imai S, Maruyama S (2011) *Bartonella* species in wild rodents and fleas from them in Japan. J Vet Med Sci 73(12):1561–1567

Kato T, Shirafuji H, Tanaka S, Sato M, Yamakawa M, Tsuda T, Yanase T (2016) Bovine arboviruses in *Culicoides* biting midges and sentinel cattle in southern Japan from 2003 to 2013. Transbound Emerg Dis 63(3):e160–e172

Kaufman PE, Rutz DA, Frisch S (2005) Large sticky traps for capturing house flies and stable flies in dairy calf greenhouse facilities. J Dairy Sci 88(1):176–181

Kimura G, Kanasugi T, Uezu T, Nakamoto A (2016) Massive occurrence of *Leptoconops* (*Leptoconops*) *taiwanensis* (Diptera: Ceratopogonidae) at Kume Islands, Japan. Med Entomol Zool 67(Suppl):98. (in Japanese)

Kitaoka S (1980) Six new species and males of some species of Japanese *Culicoides* (Diptera: Ceratopogonidae). Natl Inst Aim Health Q (Tokyo) 20(1):11–22

Kitaoka S, Suzuki H (1985) Record of the blood-sucking midge, *Forcipomyia* (*Dacnoforcipomyia*) *anabaenae* Chan et Sunders, 1965 from Amami-oshima, Japan. Jpn J Sanit Zool 36(2):103–105. (in Japanese)

Ko CJ, Elston DM (2004) Pediculosis. J Am Acad Dermatol 50(1):1–14

Kobayashi T (2016) The bug of death. Chuokouron, Tokyo, 268 pp

Komatsu N, Itou F (2023) Cockroach guide book of Japan. Hokkaido University Press, Sapporo, 189 pp (in Japanese)

Kono R, Hirata M, Kaji M, Goto Y, Ikeda S, Yanase T, Kato T, Tanaka S, Tsutsui T, Imada T, Yamakawa M (2008) Bovine epizootic encephalomyelitis caused by Akabane virus in southern Japan. BMC Vet Res 4:20

Kosone K, Itou M, Kanayama A (1995) Movement of German cockroaches, *Blattella germanica* (L.) in a restaurant. Pest Control Res 10:32–36. (in Japanese with English summary)

Kuhn J, Bewermeyer H, Hartmann-Klosterkoetter U, Emmerich P, Schilling S, Valassina M (2005) Toscana virus causing severe meningoencephalitis in an elderly traveler. J Neurol Neurosurg Psychiatry 76:1605–1606

Kurimasa A, Miyano Y, Yakura H, Youdou S, Furukawa H, Iwasaki Y (2015) Studies of questionnaire on epidemiology of bite cases, ecology and control method for biting

midges(Ceratopogonidae)in Yumi-ga-hama Peninsula, Yonago, Tottori Prefecture. J Yonago Med Ass 66(23):19–35. (in Japanese with English abstract)
Larsen K (2010) Mechanical removal of lice and eggs. In: Heukelbach J (ed) Management and control of head lice infestations. UNI-MED Verlag AG, Bremen, pp 54–56
Leak SGA (1998) Classification and anatomy. In: Leak SGA (ed) Tsetse biology and ecology: their role in the epidemiology and control of trypanosomosis. CABI Publishing, Wallingford, pp 7–13
Lee F (2019) Bovine ephemeral fever in Asia: recent status and research gaps. Viruses 11(5):412
Lee HR, Koo BS, Jeon EO, Han MS, Min KC, Lee SB, Bae Y, Mo IP (2016) Pathology and molecular characterization of recent *Leucocytozoon caulleryi* cases in layer flocks. J Biomed Res 30(6):517–524
Lent H, Wygodzinsky P (1979) Revision of the Triatominae (Hemiptera, Reduviidae), and their significance as vectors of Chagas' disease. Bull Am Mus Nat Hist, New York 163(3):125–520
Lewis DJ (1974) Man-biting simuliidae (Diptera) of Northern India. Israeli J Entomol 9:23–53
Li GQ, Hu YL, Xiao SM (2003) The establishment of a diagnostic PCR assay for *Leucocytozoonosis caulleryi*. Chin J Prev Vet Med 25(2):148–151
Lindner AK, Lejon V, Chappuis F, Seixas J, Kazumba L, Barrett MP, Mwamba E, Erphas O, Akl EA, Villanueva G, Bergman H, Simarro P, Kadima Ebeja A, Priotto G, Franco JR (2020) New WHO guidelines for treatment of gambiense human African trypanosomiasis including fexinidazole: substantial changes for clinical practice. Lancet Infect Dis 20(2):e38–e46
Lipkin WI (2013) The changing face of pathogen discovery and surveillance. Nat Rev Microbiol 11(2):133–141
Lorn S, Klakankhai W, Nusen P, Sumarnrote A, Tainchum K (2022) Pyrethroid susceptibility in *Stomoxys calcitrans* and *Stomoxys indicus* (Diptera: Muscidae) collected from cattle farms in Southern Thailand. Insects 13(8):711
Lv Y, Guo X-G, Jin D-C, Fan R, Zhao C-F, Zhang Z-W, Mao K-Y, Song W-Y, Dong W-G, Qian T-J, Yang Z-H (2021) Distribution and host selection of the chigger mite vector of scrub typhus, *Leptotrombidium deliense*, in Southwest China. Int J Acarol 47(3):233–241
Maclachlan NJ, Mayo CE, Daniels PW, Savini G, Zientara S, Gibbs EP (2015) Bluetongue. Rev Sci Tech 34(2):329–340
Mana N, Louni M, Parola P, Bitam I (2017) Human head lice and pubic lice reveal the presence of several Acinetobacter species in Algiers, Algeria. Comp Immunol Microbiol Infect Dis 53:33–39
Maroli M, Feliciangeli MD, Bichaud L, Charrel RN, Gradoni L (2013) Phlebotomine sandflies and the spreading of leishmaniases and other diseases of public health concern. Med Vet Entomol 27(2):123–147
Maruyama K, Takagi M (1977) Results of ecological investigation on *Culicoides nipponensis* TOKUNAGA, 1955 in Meiwa-cho, Mie Prefecture. Jpn J Sanit Zool 28(4):367–371. (in Japanese with English summary)
Matayoshi S, Noda S, Sato A (1985) Ecological study of *Leptoconops* (*Leptoconops*) *nipponensis oshimaensis* (Diptera, Ceratopogonidae) at Katoku, Amami-oshima Island, Japan. Jpn J Sanit Zool 36(3):219–225
Matsumori Y, Aizawa M, Sakai Y, Inoue D, Kodani M, Tsuha O, Beppu A, Hirashima Y, Kono R, Ohtani A, Yanase T, Shirafuji H, Kato T, Tanaka S, Yamakawa M (2018) Congenital abnormalities in calves associated with Peaton virus infection in Japan. J Vet Diagn Invest 30(6):855–861
Mihok S (2002) The development of a multipurpose trap (The Nzi)for tsetse and other biting flies. Bull Entomol Res 92(5):385–403
Mikhail M, Weinberg JM, Smith BL (2007) What's eating you? Body lice (Pediculus humanus var corporis). Cutis 80(5):397–398
Mikhailyuk AP (1967) Study of the biology of the causative agent of cattle onchocerciasis in forest-steppe zone conditions of the Ukrainian SSR. Veterinariya 11:62–67. (in Russian)
Monroy C, Mejia M, Rodas A, Rosales R, Horio M, Tabaru Y (1998) Comparison of indoor searches with whole house demolition collection of vectors of Chagas disease and their indoor distribution. Med Entomol Zool 49(3):195–200

Monroy C, Rodas A, Mejia M, Rosales R, Tabaru Y (2003) Epidemiology of Chagas disease in Guatemala: infection rate of *Triatoma dimidiata, T. nitida* and *Rhodnius prolixus* with *Tripanosoma cruzi* and *T. rangeli*. Mem Inst Ozwaldo Cruz 98(3):305–310

Morgan DW, Bailie HD (1980) A field trial to determine the effect of fly control using permethrin on milk yields in dairy cattle in the UK. Vet Rec 106(6):121–123

Moriconi M, Rugna G, Calzolari M, Bellini R, Albieri A, Angelini P, Cagarelli R, Landini MP, Charrel RN, Varani S (2017) Phlebotomine sand fly-borne pathogens in the Mediterranean Basin: human leishmaniasis and phlebovirus infections. PLoS Negl Trop Dis 11(8):e0005660

Morii T, Shiihara T, Lee YC, Manuel MF, Nakamura K, Iijima T, Hoji K (1981) Seroimmunological and parasitological surveys of *Leucocytozoon caulleryi* infection in chickens in several Asian countries. Int J Parasitol 11(3):187–190

Morita SI, Bayless KM, Yeates DK, Wiegmann BM (2016) Molecular phylogeny of the horse flies: a framework for renewing tabanid taxonomy. Syst Entomol 41(1):56–72

Mowlavi G, Farzbod F, Kheirkhah A, Mobedi I, Bowman DD, Naddaf SR (2014) Human ocular onchocerciasis caused by *Onchocerca lupi* (Spirurida, Onchocercidae) in Iran. J Helminthol 88(2):250–255

Mumcuoğlu KY (2015) Pubic louse (*Pthirus pubis*) infestation of the scalp in a 4-years old infant. Chonnam Med J 37(3):241–243

Nakano K (2023) Investigation of a cockroach habitat in an outdoor urban environment. Seasonal prevalence of American cockroaches at a sidewalk site in the center of Tokyo. Pestology 38(1):1–4. (in Japanese with English summary)

Natsuaki M (2020) Insect-inducing dermatitis. Jpn J Entomol (NS) 23(1):2–11. (in Japanese)

Newstead R (1923) On a new species of *Phlebotomus* from Japan. Ann Trop Med Parasit 17:531–532

Niedringhaus KD, Brown JD, Sweeley KM, Yabsley MJ (2019) A review of sarcoptic mange in North American wildlife. Int J Parasitol Parasites Wildl 9:285–297

Niimi D, Kono I (1953) Etiological studies on the bovine disease 'Kose' in Kagoshima, Japan. Jui Chikusan Shinpo 116:768–772. (in Japanese)

Nishijima Y, Ono H (1964) Identification of blood meals of Heleid biting midges by the precipitin test (Diptera). Jpn J Sanit Zool 15(3):131–135

Noguchi H, Shannon RC, Tilden EB, Tyler JR (1929) Etiology of Oroya fever: XIV. The insect vectors of Carrion's disease. J Exp Med 49(6):993–1008

Nyers ES, Elston DM (2020) What's eating you? Human body lice (Pediculus humanus corporis). Cutis 105(3):118–120

Oem JK, Lee KH, Kim HR, Bae YC, Chung JY, Lee OS, Roh IS (2012) Bovine epizootic encephalomyelitis caused by Akabane virus infection in Korea. J Comp Pathol 147(2–3):101–105

Ogata K (1955) Observation of the animal preference of Japanese blackfly. Jpn J Appl Zool 20(1,2):83–89. (in Japanese)

Ogata K, Sasa M, Suzuki T (1956) Blackflies and their control. DDT Association, Tokyo, 162 pp (in Japanese)

Ohashi S, Matsumori Y, Yanase T, Yamakawa M, Kato T, Tsuda T (2004) Evidence of an antigenic shift among Palyam serogroup orbiviruses. J Clin Microbiol 42(10):4610–4614

Okazaki D, Fukuda M, Hebisawa A, Uni S, Junker K, Suzuki Y, Nakano M, Agatsuma T, Hasegawa H, Yamada M, Nakatani J, Hara T, Martin C, Kimura D, Takaoka H (2022) Zoonotic infection caused by *Onchocerca japonica* (Nematoda: Filarioidea) in a 69-year-old woman in Kanto Region, Eastern Honshu, Japan. Parasitol Int 91:102643

Ono H (1982) Taxonomic study of black flies in Hokkaido, with notes on their veterinary viewpoint (Diptera: Simuliidae). Res Bull Obihiro Univ 12:277–316

Ose GA, Hogsette JA (2014) Spatial distribution, seasonality and trap preference of stable fly, *Stomoxys calcitrans* L. (Diptera: Muscidae), adults on a 12-hectare zoological park. Zoo Biol 33(3):228–233

Otsuka Y (2020) Control of black fly in Nakano-Shima Island. In: Otsuka K, Terada R, Nishimura S (eds) The Tokara Islands: culture, society, industry. Hokuto Shobo Publishing, Tokyo, pp 130–135

Otsuru M, Ogawa S (1959) Observations on the bite of horsefly larvae in the paddy-fields. Jpn J Sanit Zool 10(1):27–37. (in Japanese)

Parashar BD, Gupta GP, Sikder N, Rao KM (1993) Diethylphenylacetamide: a new insect repellent against stable fly, *Stomoxys calcitrans*. Med Vet Entomol 7(3):259–262

Parola P, Miller RS, McDaniel P, Telford SR, Rolain J, Wongsrichanalai C, Raoult D (2003) Emerging rickettsioses of the Thai-Myanmar border. Emerg Infect Dis 9(5):592–595

Pauli C, Schwarz TF, Meyer CG, Jager G (1995) Neurological symptoms after an infection by the sandfly fever virus. Dtsch Med Wochenschr 120(43):1468–1472. (in German with English abstract)

Paz-Bailey G, Monroy C, Rodas A, Rosales R, Tabaru Y, Davles R, Lines J (2002) Incidence of *Tripanosoma cruzi* infection in two Guatemalan communities. Trans R Soc Trop Med Hyg 96(1):48–52

Pitzer JB, Kaufman PE, Tenbroeck SH (2010) Assessing permethrin resistance in the stable fly (Diptera: Muscidae) in Florida by using laboratory selections and field evaluations. J Econ Entomol 103(6):2258–2263

Ponce C (2007) Current situation of Chagas disease in Central America. Mem Inst Ozwaldo Cruz 102(Suppl 1):41–44

Ponnusamy L, Willcox AC, Roe RM, Davidson SA, Linsuwanon P, Schuster AL, Richards AL, Meshnick SR, Apperson CS (2018) Bacterial microbiome of the chigger mite Leptotrombidium imphalum varies by life stage and infection with the scrub typhus pathogen *Orientia tsutsugamushi*. PLoS One 13:e0208327

Pramual P, Thaijarern J, Wongpakam K (2016) DNA barcoding of human-biting black flies (Diptera: Simuliidae) in Thailand. Acta Trop 164:33–40

Pramual P, Thaijarern J, Tangkawanit U, Wongpakam K (2020) Molecular identification of blood meal sources in black flies (Diptera: Simuliidae) suspected as *Leucocytozoon* vectors. Acta Trop 205:105383

Purse BV, Carpenter S, Venter GJ, Bellis G, Mullens BA (2015) Bionomics of temperate and tropical *Culicoides* midges: knowledge gaps and consequences for transmission of *Culicoides*-borne viruses. Annu Rev Entomol 60:373–392

Rebêlo JM, Rodrigues BL, Bandeira MD, Moraes JL, Fonteles RS, Pereira SR (2016) Detection of *Leishmania amazonensis* and *Leishmania braziliensis* in *Culicoides* (Diptera, Ceratopogonidae) in an endemic area of cutaneous leishmaniasis in the Brazilian Amazonia. J Vector Ecol 41(2):303–308

Reeves WK, Durden LA, Iwakami M, Vince KJ, Paul RR (2015) Rickettsial diseases and ectoparasites from military bases in Japan. J Parasitol 101(2):150–155

Reissert-Oppermann S, Bauer B, Steuber S, Clausen PH (2019) Insecticide resistance in stable flies (*Stomoxys calcitrans*) on dairy farms in Germany. Parasitol Res 118(9):2499–2507

Ríos-Tostado JJ, Castillo-Ureta H, Torres-Montoya EH, Torres-Avendaño JI, Olimón-Andalón V, Romero-Higareda CE, Silva-Hidalgo G, Zazueta-Moreno JM (2021) Molecular detection of *Leishmania* (L.) *mexicana* (Kinetoplastida: Trypanostomatidae) DNA in *Culicoides furens* (Diptera: Ceratopogonidae) from an area with autochthonous canine leishmaniasis in northwestern Mexico. Acta Parasitol 66:1055–1058

Rochon K, Hogsette JA, Kaufman PE, Olafson PU, Swiger SL, Taylor DB (2021) Stable fly (Diptera: Muscidae) – biology, management, and research needs. J Integr Pest Manag 12(1):38

Romani L, Whitfeld MJ, Koroivueta J, Kama M, Wand H, Tikoduadua L, Tuicakau M, Koroi A, Andrews R, Kaldor JM, Steer AC (2015) Mass drug administration for scabies control in a population with endemic disease. N Engl J Med 373(24):2305–2313

Roth LM, Willis ER (1957) The medical and veterinary importance of cockroaches. Smithson Misc Collect 134(10):1–147

Saeung A, Srisuka W, Aupalee K, Fukuda M, Otsuka Y, Taai K, Maleewongf W, Takaoka H (2020) Natural infections with larvae of Onchocerca species type I in the human-biting black fly, *Simulium nigrogilvum* (Diptera: Simuliidae), in western Thailand. Acta Trop 204:105344

Sakai T (2007) Current status and countermeasures of chicken leucocytozoonosis in Japan. J Anim Protozoosis 22(1):14–21. (in Japanese with English summary)

Sakkas H, Bozidis P, Franks A, Papadopoulou C (2018) Oropouche fever: a review. Viruses 10(4):175

Salem A, Franc M, Jacquiet P, Bouhsira E, Liénard E (2012) Feeding and breeding aspects of *Stomoxys calcitrans* (Diptera: Muscidae) under laboratory conditions. Parasite 19(4):309–317

Sanjoba C, Miyagi I (2022) A new species of the genus *Sergentomyia* França and Parrot (Diptera: Psychodidae) from Iriomote Island, Ryukyu Archipelago, Japan. Med Entomol Zool 73(3):125–129

Sariözkan S, Inci A, Yildirim A, Düzlü Ö, Gray EW, Adler PH (2014) Economic losses during an outbreak of *Simulium* (Wilhelmia) species (Diptera: Simuliidae) in the Cappadocia region of Turkey. Turk Parazitol Derg 38(2):116–119

Sasaki H (2021) Risk of enzootic bovine leukosis by vectors and preventive measures. Jpn J Anim Hyg 47(3):135–141. (in Japanese)

Sasaki H, Watanabe M (2016) Tabanids of Japan – species and fauna, and bionomic and control. In: Matsuoka H (ed) Progress of medical entomology and zoology 2. Mie University Press, Tsu, pp 155–178. (in Japanese with English summary)

Sasaki H, Rosales R, Tabaru Y (2003) Host feeding profile of *Rhodnius prolixus* and *Triatoma dimidiata* in Guatemala. Med Entomol Zool 54(3):283–289

Sato T, Hayashi S, Tanaka K (1954) Studies on *Onchocerca gutturosa* Neumann, 1910 as a causative parasite of bovine dermatitis 'Wahi' and *Onchocerca cervicalis* Railliet et Henry, 1910 as a causative parasite of equine dermatitis 'Kasen'. Jpn J Parasitol 2:199–206. (in Japanese)

Sato Y, Tamada A, Mochizuki Y, Nakamura S, Okano E, Yoshida C, Ejiri E, Omori S, Yukawa M, Murata K (2009a) Molecular detection of *Leucocytozoon lovati* from probable vectors, black flies (Simuliudae) collected in the alpine regions of Japan. Parasitol Res 104(2):251–255

Sato Y, Yoshida T, Harada S, Matsumoto J, Nogami S, Murata K, Takaoka H, Yukawa M (2009b) Prevalence of filarial DNA in blackflies in alpine regions of Japan. Jpn J Vet Parasitol 8:110. (in Japanese)

Savini G, Afonso A, Mellor P, Aradaib I, Yadin H, Sanaa M, Wilson W, Monaco F, Domingo M (2011) Epizootic heamorragic disease. Res Vet Sci 91(1):1–17

Schaffartzik A, Hamza E, Janda J, Crameri R, Marti E, Rhyner C (2012) Equine insect bite hypersensitivity: what do we know? Vet Immunol Immunopathol 147(3–4):113–126

Schmidt CD, Matter JJ, Meurer JH, Reeves RE, Shelley BK (1976) Evaluation of a synthetic pyrethroid for control of stable flies and horn flies on cattle. J Econ Entomol 69(4):484–486

Schofield CJ (1994) Triatominae: biologícal & control. Eurocommunica Publications, Bognor Regis, 80 pp

Schofield CJ (2000) Challenges of Chagas disease vector control in Central America, WHO/CDS/WHOPES/G CDPP, pp 1–36

Shirafuji H, Yanase T, Kato T, Yamakawa M (2012) Genetic and phylogenetic characterization of genome segments 2 and 6 of bluetongue virus isolates in Japan from 1985 to 2008. J Gen Virol 93(7):1465–1473

Shogaki Y, Yoshida Y (1958) Studies on the black flies of Yase District in Kyoto-city and their control. II. On the control of black fly larvae by lindane emulsions in the natural streams. Jpn J Appl Entomol Zool 2(3):157–170. (in Japanese with English summary)

Sick F, Beer M, Kampen H, Wernike K (2019) *Culicoides* biting midges-underestimated vectors for arboviruses of public health and veterinary importance. Viruses 11(4):376

Siegenthaler R, Gubler R (1965) Paraarticulares Nematodengranulom (einheimische *Onchocerca*). Schweiz Med Wochenschr 95(33):1102–1104. (in Germany)

Slama D, Haouas N, Remadi L, Mezhoud H, Babba H, Chaker E (2014) First detection of *Leishmania infantum* (Kinetoplastida: Trypanosomatidae) in *Culicoides* spp. (Diptera: Ceratopogonidae). Parasit Vectors 7:51

Song S, Li Y, Fu S, Lei W, Guo X, Feng Y, Gao X, Li X, Yang Z, Xu Z, He Y, Wang H, Zhou H, Wang B, Lu X, Liang G (2017) Genome sequencing and phylogenetic analysis of Banna virus (genus *Seadornavirus*, family *Reoviridae*) isolated from *Culicoides*. Sci China Life Sci 60:1372–1382

Stokes JE, Darpel KE, Gubbins S, Carpenter S, Fernández de Marco MDM, Hernández-Triana LM, Fooks AR, Johnson N, Sanders C (2020) Investigation of bovine ephemeral fever virus transmission by putative dipteran vectors under experimental conditions. Parasit Vectors 13:597

Sunantaraporn S, Thepparat A, Phumee A, Sor-Suwan S, Boonserm R, Bellis G, Siriyasatien P (2021) *Culicoides* Latreille (Diptera: Ceratopogonidae) as potential vectors for *Leishmania martiniquensis* and *Trypanosoma* sp. in Northern Thailand. PLoS Negl Trop Dis 15(12):e0010014

Tabaru Y (2015) Eternal challenging to cockroaches. Tokai University Press, Hiratsuka, 286 pp (in Japanese)

Tabaru Y, Kobayashi A (1971) Outdoor hibernation of *Periplanater japonica* in snowy area. Jpn J Sanit Zool 22(2):76–77. (in Japanese with English summary)

Tabaru Y, Ikeda Y, Hasegawa M, Hattori K (1973) Studies on the control of the biting midges in Hokkaido III. On the chemical control and vertical distribution of the larvae of *Culicoides circumscriptus* KIEFFER (Diptera: Ceratopogonidae) in Saroma, shore of Lake Saroma, Hokkaido. Jpn J Sanit Zool 23(3):215–218. (in Japanese with English summary)

Tabaru Y, Monroy C, Rodas A, Mejia M, Rosales R (1998) Chemical control of *Triatoma dimidiata* and *Rhodnius prolixus*, the vector of Chagas disease in Guatemala. Med Entomol Zool 49(2):87–92

Tabaru Y, Monroy C, Rodas A, Mejía M (1999a) Chagas disease vector surveillance in various residents in Guatemala. Med Entomol Zool 50(1):19–25

Tabaru Y, Monroy C, Rodas A, Mejia M, Rosales R (1999b) The geographical distribution of vectors of Chagas disease and population at risk of infection in Guatemala. Med Entomol Zool 50(1):9–17

Tabaru Y, Kamiya S, Watabe Y (2011) Adhesive trap avoidance in German cockroach, *Blattella germanica*, American cockroach, *Periplaneta americana*, and brown cockroach, *P. brunnea* under laboratory conditions. Med Entomol Zool 62(2):101–107

Takada N (1990) A pictorial review of medical acarology. Kinpodo, Kyoto, 216 pp (in Japanese)

Takada N, Takahashi M (2019) Trombiculidae and diseases. In: Takada N (ed) Medical acarology in Japan. Hokuryukan, Tokyo, pp 26–113. (in Japanese)

Takahashi K, Yagi K, Hattori K (1985) The effects of two insect growth regulators on the biting midges, *Culicoides circumscriptus* Kieffer (Diptera: Ceratopogonidae). Jpn J Sanit Zool 36(4):353–355. (in Japanese with English summary)

Takahashi M, Misumi H, Kawai K, Sato M (2016a) The first finding of a bat flea *Myodopsylla trisellis* (Siphonaptera: Ischnopsyllidae) on *Myotis gracilis* (Chiroptera: Vespertilionidae) in Japan. Med Entomol Zool 67(1):29–33

Takahashi K, Takahashi M, Hamasaka K (2016b) Finding of the human flea *Pulex irritans* (Siphonaptera: Pulicidae) in Hokkaido, Japan with human dermatitis caused by this flea. Med Entomol Zool 67(4):233–235

Takaoka H (1990) Development of microfilariae of two bovine *Onchocerca* species in blackflies in Japan. Jpn J Trop Med Hyg 18(2):183–188

Takaoka H (1997) Interest in vectors of bovine *Onchocerca* species in Japan. Bull Oita Vet Assoc 8:3–6. (in Japanese)

Takaoka H (1999) Review on zoonotic *Onchocerca* species and their insect vectors in Japan. Med Entomol Zool 50(1):1–8. (in Japanese with English summary)

Takaoka H, Bain O (1990) Infections of blackflies (Diptera: Simuliidae) with three types of zoonotic *Onchocerca* larvae in Oita, Japan. Jpn J Trop Med Hyg 18(1):1–10

Takaoka H, Hayashi Y (1977) A new species of the genus *Leptoconops* from Amami-oshima, Japan (Ceratopogonidae: Diptera). Jpn J Sanit Zool 28(4):385–388

Takaoka H, Noda S (1979) Autogeny of the black fly *Simulium* (*Eusimulium*) *aureohirtum* (Diptera: Simuliidae). J Med Entomol 15(2):183–184

Takaoka H, Baba M, Bain O (1989) Natural infections of *Simulium bidentatum* (Diptera: Simuliidae) with larvae of *Onchocerca* spp. in relation to a human zoonotic onchocerciasis in Oita, Japan. Jpn J Trop Med Hyg 17(4):279–284

Takaoka H, Aoki C, Hayakawa H (1992) Natural infections of blackflies with larvae of zoonotic *Onchocerca* spp. in Northeast Japan. Jpn J Trop Med Hyg 20(1):1–9

Takaoka H, Aoki C, Bain O, Ogata K, Baba M (1995) Investigation of *Culicoides* (Diptera: Ceratopogonidae) in relation to the transmission of bovine *Onchocerca* and other filariae in Central Kyushu, Japan. Parasite 2(4):367–371

Takaoka H, Bain O, Tajimi S, Kashima K, Nakayama I, Korenaga M, Aoki C, Otsuka Y (1996) Second case of zoonotic *Onchocerca* infection in a resident of Oita in Japan. Parasite 3:179–182

Takaoka H, Bain O, Uni S, Korenaga M, Tada K, Ichikawa H, Otsuka Y, Eshita Y (2001) Human infection with *Onchocerca dewittei japonica*, a parasite from wild boar in Oita, Japan. Parasite 8:261–263

Takaoka H, Bain O, Uni S, Korenaga M, Kozek WJ, Shirasaka C, Aoki C, Otsuka Y, Fukuda M, Eshita Y, Daa T (2004) Zoonotic onchocerciasis caused by a parasite from wild boar in Oita, Japan: a comprehensive analysis of morphological characteristics of the worms for its diagnosis. Parasite 11(3):285–292

Takaoka H, Yanagi T, Daa T, Anzai S, Aoki C, Fukuda M, Uni S, Bain O (2005) An *Onchocerca* species of wild boar found in the subcutaneous nodule of a resident of Oita, Japan. Parasitol Int 54(1):91–93

Takaoka H, Srisuka W, Saeung A (2017) A new human-biting black fly species of *Simulium* (*Simulium*) (Diptera: Simuliidae) from Thailand. J Med Entomol 54(4):945–948

Takeda T, Mukai I (1954) Studies on the *Culicoides*, broken out at the Kashima area (Saga Prefecture) and its skin diseases. Jpn J Sanit Zool 5(3–4):86–92

Tanaka I, Ogata K, Muto A, Kashiwagi H (1987) A case for the control of biting midge, *Culicoides circumscriptus* (Diptera: Ceratopogonidae) around the seashore at Yaizu city. Bull Jpn Env Sanit Cent 14:108–112. (in Japanese with English summary)

Ta-Tang TH, Crainey JL, Post RJ, Luz SLB, Rubio JM (2018) Mansonellosis: current perspectives. Res Rep Trop Med 9:9–24

Tawatsin A, Bhakdeenuan P, Yang C, Amano T, Sasaki H, Thavara U, Sangkitporn S, Chareonviriyaphap T, Siriyasatien P (2017) Profile of the internal transcribed spacer 2 in ribosomal DNA of *Tabanus rubidus* in Thailand and Myanmar; the possibility of using for molecular identification. Vector Biol J 2(2):1–5

Taylor DB, Friesen K, Zhu JJ, Sievert K (2012a) Efficacy of cyromazine to control immature stable flies (Diptera: Muscidae) developing in winter hay feeding sites. J Econ Entomol 105(2):726–731

Taylor DB, Moon RD, Mark DR (2012b) Economic impact of stable flies (Diptera: Muscidae) on dairy and beef cattle production. J Med Entomol 49(1):198–209

Thaijarern J, Tangkawanit U, Wongpakam K, Pramual P (2019) Molecular detection of *Trypanosoma* (kinetoplastida: trypanosomatidae) in black flies (Diptera: simuliidae) from Thailand. Acta Trop 200:105196

Tokarz R, Lipkin WI (2020) Discovery and surveillance of tick-borne pathogens. J Med Entomol 58(4):1525–1535

Tokunaga M (1950) Culicoid flies from Kyushu, Japan (Ceratopogonidae, Diptera). Jpn J Sanit Zool 1(3):64–67. (in Japanese)

Traub R, Wisseman CL Jr (1974) The ecology of chigger-borne Rickettsiosis (scrub typhus). J Med Entomol 11(3):237–303

Tsuda T, Yoshida K, Ohashi S, Yanase T, Sueyoshi M, Kamimura S, Misumi K, Hamana K, Sakamoto H, Yamakawa M (2004) Arthrogryposis, hydranencephaly and cerebellar hypoplasia

syndrome in neonatal calves resulting from intrauterine infection with Aino virus. Vet Res 35(5):531–538
Tsuji H (1966) Attractive and feeding stimulative effect of some fatty acids and related compounds on three species cockroaches. Jpn J Sanit Zool 17(2):89–97
Tsukahara M (2002) Cat scratch disease in Japan. J Infect Chemother 8(4):321–325
Ueda T, Katsura Y, Sasaki A, Minagawa D, Amoh Y, Shirai K (2021) Gray-edged line sign of scabies burrow. J Dermatol 48(2):190–198
Uni S, Bain O, Takaoka H, Miyashita M, Suzuki Y (2001) *Onchocerca dewittei japonica* n. subsp., a common parasite from wild boar in Kyushu Island, Japan. Parasite 8:215–222
Uni S, Boda T, Daisaku K, Ikura Y, Maruyama H, Hasegawa H, Fukuda M, Takaoka H, Bain O (2010) Zoonotic filariasis caused by *Onchocerca dewittei japonica* in a resident of Hiroshima Prefecture, Honshu, Japan. Parasitol Int 59(3):477–480
Uni S, Fukuda M, Agatsuma T, Bain O, Otsuka Y, Nakatani J, Matsubayashi M, Harada M, Omar H, Ramli R, Hashim R, Sofian-Azirun M, Takaoka H (2015a) *Onchocerca takaokai* n. sp. (Nematoda: Filarioidea) in Japanese wild boar (*Sus scrofa leucomystax*): description and molecular identification of intradermal females. Parasitol Int 64(6):493–502
Uni S, Fukuda M, Otsuka Y, Hiramatsu N, Yokobayashi K, Takahashi H, Murata S, Kusatake K, Morita E, Maruyama H, Hasegawa H, Shiwaku K, Ramli R, Sofian-Azirun M, Takaoka H (2015b) New zoonotic cases of *Onchocerca dewittei japonica* (Nematoda: Onchocercidae) in Honshu, Japan. Parasit Vectors 8:59
Veraldi S, Scanni G, Nazzaro G (2020) "Eczema" of the nape: a marker of *Pthiriasis capitis*. Parasitol Int 75:102026
Victoria RD, Yermia Sem M (2021) The metagenomic analysis of potential pathogenic emerging bacteria in fleas. Pak J Biol Sci 24(10):1084–1090
Vooght L, Caljon G, Ridder K, Abbeele J (2014) Delivery of a functional anti-trypanosome Nanobody in different tsetse fly tissues via a bacterial symbiont, *Sodalis glossinidius*. Microb Cell Fact 13:156
Vreysen MJB, Saleh K, Mramba F, Parker A, Feldmann U, Dyck VA, Msangi A, Bouyer J (2014) Sterile insects to enhance agricultural development: the case of sustainable tsetse eradication on Unguja Island, Zanzibar using an area-wide integrated pest management approach. PLoS Negl Trop Dis 8(5):e2857
Warnes ML, Finlayson LH (1987) Effect of host behaviour on host preference in *Stomoxys calcitrans*. Med Vet Entomol 1(1):53–57
Watabe Y, Tabaru Y (2010) Efficacy of organophosphate formulations against German cockroach collected from several restaurants in two multitenant buildings in Tokyo. Med. Entomol. Zool. 61(1):1–8. (in Japanese with English summary)
Whitehouse C (2022) Surra. https://www.cabi.org/isc/datasheet/62413#todistribution. Accessed 16 Feb 2024
Williams LK, Reichert A, MacKenzie WR, Hightower AW, Blake PA (2001) Lice, nits, and school policy. Pediatrics 107(5):1011–1015
Wirth WW, Hubert AA (1989) The *Culicoides* of Southeast Asia (Diptera: Ceratopogonidae). Mem Am Entomol Soc 44:1–508
Word Health Organization (WHO) (2020) Vector-borne diseases. https://www.who.int/news-room/fact-sheets/detail/vector-borne-diseases#Vector-BorneDiseases. Accessed 22 Feb 2024
World Health Organization (WHO) (2023) Leishmaniasis. https://www.who.int/news-room/fact-sheets/detail/leishmaniasis. Accessed 1 Dec 2023
World Organisation for Animal Health (OIE) (2021) African horse sickness: OIE guidelines for the practical control of viral transmission by reducing vector-host contact in the Asian context. OIE, Paris, 20 pp
Xu G, Walker DH, Jupiter D, Melby PC, Arcari CM (2017) A review of the global epidemiology of scrub typhus. PLoS Negl Trop Dis 11(11):e0006062

Yanase T, Kato T, Kubo T, Yoshida K, Ohashi S, Yamakawa M, Miura Y, Tsuda T (2005) Isolation of bovine arboviruses from *Culicoides* biting midges (Diptera: Ceratopogonidae) in Southern Japan: 1985–2002. J Med Entomol 42(1):63–67

Yanase T, Kato T, Aizawa M, Shuto Y, Shirafuji H, Yamakawa M, Tsuda T (2012) Genetic reassortment between Sathuperi and Shamonda viruses of the genus *Orthobunyavirus* in nature: implications for their genetic relationship to Schmallenberg virus. Arch Virol 157(8):1611–1616

Yanase T, Matsumoto Y, Matsumori Y, Aizawa M, Hirata M, Kato T, Shirafuji H, Yamakawa M, Tsuda T, Noda H (2013) Molecular identification of field-collected *Culicoides* larvae in the southern part of Japan. J Med Entomol 50(5):1105–1110

Yanase T, Kato T, Hayama Y, Akiyama M, Itoh N, Horiuchi S, Hirashima Y, Shirafuji H, Yamakawa M, Tanaka S, Tsutsui T (2018) Transition of Akabane virus genogroups and its association with changes in the nature of disease in Japan. Transbound Emerg Dis 65(2):e434–e443

Yanase T, Hayama Y, Shirafuji H, Tsutsui T, Terada Y (2019a) Surveillance of *Culicoides* biting midges in northern Honshu, Japan, during the period of Akabane virus spread. J Vet Med Sci 81(10):1496–1503

Yanase T, Kato T, Hayama Y, Shirafuji H, Yamakawa M, Tanaka S (2019b) Oral susceptibility of Japanese *Culicoides* (Diptera: Ceratopogonidae) species to Akabane virus. J Med Entomol 56(2):533–539

Yanase T, Murota K, Hayama Y (2020) Endemic and emerging arboviruses in domestic ruminants in East Asia. Front Vet Sci 7:168

Yeruham I, Braverman Y, Orgad U (1993) Field observations in Israel on hypersensitivity in cattle, sheep and donkeys caused by *Culicoides*. Aust Vet J 70(9):348–352

Yssouf A, Socolovschi C, Leulmi H, Kernif T, Bitam I, Audoly G, Parola P (2014) Identification of flea species using MALDI-TOF/MS. Comp Immunol Microbiol Infect Dis 37(3):153–157

Chapter 5
Biology of Non-Vector Arthropods

Kimio Hirabayashi, Hironori Sakamoto, Kyoko Sawabe, Tetsuhiko Sasaki, and Masakazu Nishimura

Abstract There has been an increase in the number and types of damage caused by arthropods in recent years. This chapter focuses on blood-sucking, poisonous, and nuisance arthropods. We look at bedbugs as a representative of the blood-sucking arthropods; the redback spider, tropical fire, and Argentine ants, wasps, and honeybees as representatives of poisonous arthropods; and chironomid midges as a representative of nuisance arthropods. This chapter presents basic information (i.e., taxonomy, morphology, life cycle, biology, distribution, etc.) on these five arthropods and their harmful effects (i.e., medical importance, etc.). First, we focus on bedbug blood feeding at night. There has been an increase in bedbug bites in high-class hotels, inns, and public facilities, and existing insecticides have proven ineffective in many cases. Second, there has also been an increase in cases of redback spider bites. Envenoming by this spider is characterized by local, radiating, and regional pain, which may be associated with local and regional diaphoresis, nonspecific systemic features, and, less commonly, other autonomic or neurological effects. Third, imported red fire ants and Argentine ants have recently been designated invasive alien species (IAS) in Japan based on the Invasive Alien Species Act (IASA).

K. Hirabayashi (✉)
Faculty of Textile Science and Technology, Shinshu University, Ueda, Japan
e-mail: kimio@shinshu-u.ac.jp

H. Sakamoto
Biodiversity Division, National Institute for Environmental Studies, Tsukuba, Japan
e-mail: sakamoto.hironori@nies.go.jp

K. Sawabe
Department of Medical Entomology, National Institute of Infectious Diseases, Tokyo, Japan
e-mail: sawabe@niid.go.jp

T. Sasaki
Graduate School of Brain Sciences, Tamagawa University, Tokyo, Japan
e-mail: tsasaki@lab.tamagawa.ac.jp

M. Nishimura
Department of Agri-Production Sciences, College of Agriculture, Tamagawa University, Tokyo, Japan
e-mail: mnishimura@lab.tamagawa.ac.jp

K. Sawabe et al. (eds.), *Medical Entomology in Asia*, Entomology Monographs,
https://doi.org/10.1007/978-981-97-4406-0_5

The damage caused by invasive alien ants is not limited to native biodiversity and ecosystems but also includes direct damage to human health, agriculture, and tourism. Fourth, recent extreme weather events have also led to increased bites by wasps and honeybees. People who have previously experienced hypersensitivity reactions or anaphylaxis to wasps and honeybees' stings should avoid further stings. Desensitization immunotherapy with antigens is an effective treatment to reduce the risk for such patients. It is effective in >80% of adult patients, and its effects last for >10 years. Finally, chironomids, the "non-biting midges," are one of the most important groups of insects worldwide in freshwater aquatic ecosystems. Large populations of midges with densities amounting to several thousand larvae/m^2 have been reported. They are a source of economic burden, nuisance, and health problems. Moreover, many researchers have found that chironomid-related particles are strong allergens, taking the form of an inhalant antigen that causes asthma and other respiratory problems in humans. Large midge population densities create conflict with nearby human settlements.

Keywords Allergen · Anaphylaxis · Bee · Bedbug · Chironomid midges · Hornet · Invasive alien species · Redback spider · Wasp

5.1 Introduction

In recent years, there has been an increase in the number and types of damage caused by arthropods. In this chapter, we focus on blood-sucking, poisonous, and nuisance arthropods. We look at bedbugs as a representative of the blood-sucking arthropods; the redback spider, tropical fire, and Argentine ants, wasps, and honeybees as representatives of poisonous arthropods; and chironomid midges as a representative of nuisance arthropods. This chapter presents basic information (i.e., taxonomy, morphology, life cycle, biology, distribution, etc.) on these five arthropods and their harmful effects (i.e., medical importance, etc.).

First, we focus on bedbug blood feeding at night when people are asleep, so they are often unaware that they have been bitten. There has been an increase in bedbug bites in high-class hotels, inns and public facilities, and existing insecticides have proven ineffective in many cases. According to Jerome and Richard (2009), bedbug bites are caused mainly by two species of insects, *Cimex lectularius* (the common bedbug) and *Cimex hemipterus*, found primarily in the tropics. Bedbugs occur in all regions of the globe (Centers for Disease Control and Prevention (CDC), 2020).

Second, there has also been an increase in cases of redback spider (*Latrodectus hasselti*) bites. Whyte and Anderson (2017) noted in a review that *L. hasselti* is a species of highly venomous spider believed to originate in South Australia or adjacent Western Australian deserts but is now found throughout Australia, Southeast Asia, and New Zealand, with colonies discovered elsewhere outside Australia as well. Bites from this spider occur in the warmer months (Isbister and Fan 2011). According to Isbister (2006), envenoming by this spider is characterized by local, radiating, and regional pain, which may be associated with local and regional diaphoresis, nonspecific systemic features, and less commonly other autonomic or neurological effects. Pain increases over approximately an hour and may radiate proximally to the limb or, less commonly to the trunk. Local erythema is common, and local diaphoresis occurs in approximately one-third of cases. Common nonspecific effects include nausea, lethargy, malaise, and headache. Numerous other systemic effects are reported less commonly. The effects last approximately 1–4 days, with almost all cases resolving within 1 week.

Third, imported red fire ants (tropical fire ants; *Solenopsis geminata*) and Argentine ants (*Linepithema humile*) have recently been designated invasive alien species (IAS) in Japan based on the Invasive Alien Species Act (IASA). The Ministry of the Environment (MOE), Japan (2015, 2022a, b, c, 2023), states that an alien species designated by a Cabinet Order based on the IASA is a species that will cause or is likely to cause adverse impacts on ecosystems, human life, the body, and agriculture. Raising, growing, storing, carrying, importing, and handling these species are regulated. These invasive ants are native to the central part of South America and have invaded the United States and Caribbean islands as stowaways in containers and cargo carried by ships and planes since the 1940s. By the 2000s, they had spread far from their origin to Australia, New Zealand, China and Taiwan. These ants have also invaded Japan from abroad. If these ants become established, they will have various adverse effects, including health damage to humans and pets, damage to electric equipment and infrastructure, damage to agriculture and industries, and finally damage to ecosystems (MOE 2022a).

Fourth, recent extreme weather events have also led to increased bites by wasps and honeybees. Vespidae is a large group comprising nearly 5000 species. They are a diverse, cosmopolitan family of wasps, including nearly all the known eusocial wasps, i.e., *Polistes fuscatus*, *Vespa orientalis*, *Vespula germanica*, and many solitary wasps. The subfamilies Polistinae and Vespinae are composed solely of eusocial species. Polistinae is a subfamily of eusocial wasps belonging to the family Vespidae. They are closely related to more familiar wasps and true hornets of the subfamily Vespinae, containing four tribes. With approximately 1100 species in total, it is the second-most diverse subfamily within Vespidae. While most species are tropical or subtropical, they also include some of the most frequently encountered large wasps in temperate regions. The subfamily Vespinae contains the largest and best-known eusocial wasps, including true hornets, the genus Vespa, and the genera Dolichovespula and Vespula. Collectively, the group can be found on all continents except Antarctica. A bee sting is a wound and pain caused by the stinger of a female bee puncturing skin. Stings differ from insect bites, with the venom of stinging insects having considerable chemical variation. A person's reaction to a sting may vary according to the species. While stinger venom is slightly acidic and causes only mild pain in most people (Hall 2019), allergic reactions may occur in people with allergies to venom components, e.g., anaphylactic shock (Ewan 1998).

Finally, chironomids, the "nonbiting midges" (Diptera: Chironomidae), are one of the most important groups of insects worldwide in freshwater aquatic ecosystems. Reviewed by Cranston (1995a), chironomids can be found on all continents, including Antarctica. They are abundant in terms of the number of species known to exist (an estimated number between 15,000 and 20,000), their relative biomass, or both. They play an important role in the decomposition process. Large populations of midges with densities amounting to several thousand larvae/m^2 have been reported. They are a source of economic burden, nuisance, and health problems. For example, some chironomids breed enormous larvae in lakes, rice paddy rivers, and sewage ditches, causing serious problems for people residing in nearby areas. Moreover, many researchers have found that chironomid-related particles are strong allergens, taking the form of an inhalant antigen that causes asthma and other respiratory problems in humans (Ali 1995; Cranston 1995b). Large midge population densities create conflict with nearby human settlements. This phenomenon is recognized worldwide in Europe, North America, Asia, Africa and Australia.

5.2 Bedbug

Kimio Hirabayashi

5.2.1 Taxonomy

The Cimicidae are a family of small parasitic bugs that feed exclusively on the blood of warm-blooded animals. They are called cimicids or bedbugs. This family includes human bed bugs, bat bugs, chicken bugs, swallow bugs, pigeon bugs, and others for which no common names have been proposed. When Usinger (1966) was writing, 74 species had been recognized, arranged in 22 genera and 6 subfamilies, but the family contains over 100 species now. Bedbug bites are caused primarily by two species, i.e., *Cimex lectularius* (the common bedbug) and *Cimex hemipterus*, found primarily in the tropics (Jerome and Richard 2009).

5.2.2 Morphology

Khan and Rahman (2012) reported the morphology and biology of the bedbug, *C. hemipterus*. The body of the bedbug is oval-shaped and flattened dorso-ventrally (Fig. 5.1). The average length and width of adult bedbugs are 5.5 and 2.5 mm, respectively. The head is short, broad, and pointed at the tip. It has a pair of prominent compound eyes, in front of which a pair of antennae is present. Compound eyes are oval, black, and sessile. The antenna is four-segmented. They have piercing and sucking mouthparts located on the ventral side of the head. These consist of a triangular labrum, a long three-segmented labium reaching the base of the prothorax,

Fig. 5.1 Morphology of bedbugs, *Cimex lectularius* (Usinger 1966). Left, female; right, male; above, a cluster of nymphs and adults

and paired mandibular and maxillary stylets, which are blade-like. The mouth parts are modified for piercing the host's skin and sucking the blood. The mandibular stylets puncture the skin of the host, and the maxillary stylets enter the wound. The saliva runs down into the puncture and prevents the clotting of blood, which is subsequently sucked up. The thorax is three-segmented. Compared to *C. lectularius*, *C. hempiterus* is observed to have a slightly narrower prothorax. Its abdomen is eight-segmented. Slight differences exist between the sexes; females are typically larger than males and contain a more rounded abdomen tip than the pointed tip seen in males.

5.2.3 Life Cycle

Parola and Izri (2020) noted that the common bedbug has five developmental life stages (Fig. 5.2). In each immature life stage, a blood meal must be taken to develop into the next life stage. Moreover, both the male and female bedbugs suck blood. After growing through five instar molts, the bedbug reaches the adult stage. Both male and female adult bedbugs must take regular blood meals to reproduce (Miller and Polanco 2023). The total development process from egg to adult can occur in approximately 37 days at optimal temperatures (>72 ° F). Adult bedbugs have a life span of nearly one year depending on regular access to blood meals and favorable

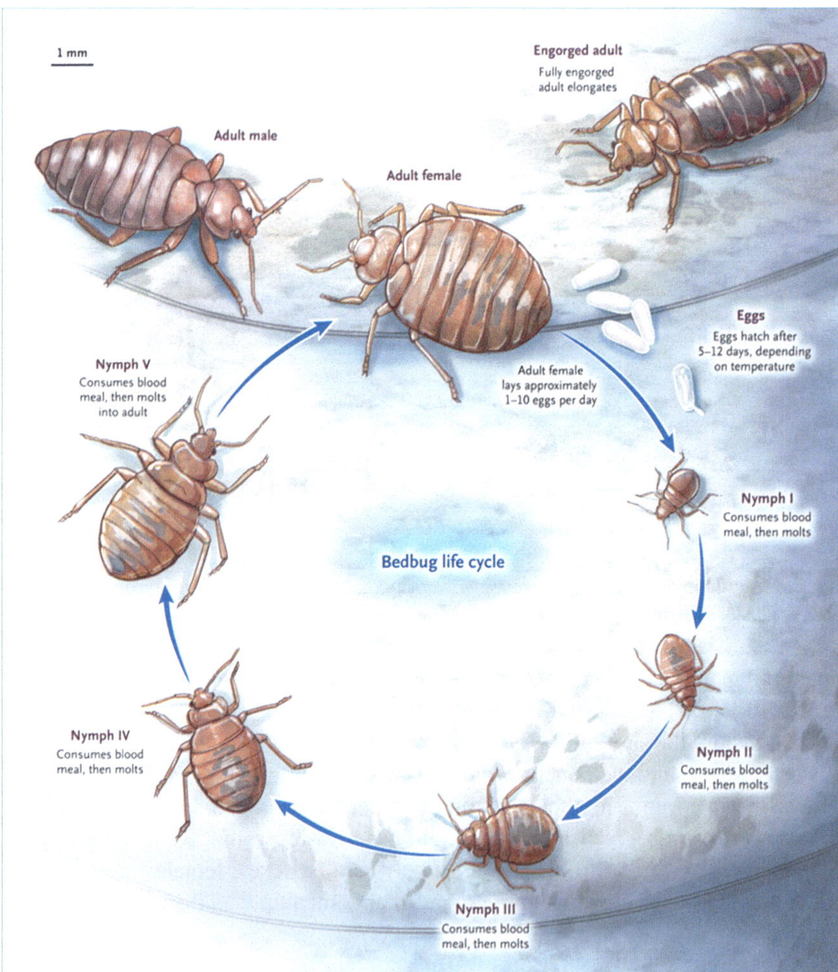

Fig. 5.2 The life cycle of the common bedbug, *C. lectularius*. (Parola and Izri 2020)

temperatures. Compared to *C. lectularius*, *C. hemipterus* goes through slightly longer developmental phases. Adults can live anywhere from 6–12 months, with females typically living longer than males.

5.2.4 Biology

Bedbugs fear light and are generally active in the dark, i.e., nocturnal insects. They hide in small, dark places, such as bedclothes, mattresses, springs, bed frames, cracks, crevices, and wallpaper. They emit an easily recognized, offensive odor caused by an oily secretion produced by special glands (Khan and Rahman 2012). Bedbug blood feeds at night when people are asleep, so they are often unaware that they have been bitten. There has been an increase in cases of bedbug-bites in high-class hotels, inns and public facilities, and existing insecticides have proven ineffective in many cases.

5.2.5 Distribution

Bedbugs occur in all regions of the globe (CDC 2020). Infestations are relatively common and have increased since the 1990s (Jerome and Richard 2009). The exact causes of this increase are unclear; theories include increased human travel, more frequent exchange of second-hand furnishings, a greater focus on controlling other pests, and increasing resistance to pesticides (Kolb et al. 2009). The map shows the distribution of two cosmopolitan species, *C. lectularius* and *C. hemipterus* (Zorrilla-Vaca et al. 2015) (Fig. 5.3). *Cimex hemipterus* is known as the tropical bedbug and primarily resides in tropical climates. However, it has been reported that this species can live in more temperate climates along with the closely related bedbug species *C. lectularius*.

5.2.6 Medical Importance

According to Usinger (1966), bedbugs have been known human parasites for thousands of years. Bedbug bites are painless (they inject saliva containing a variety of components, including anesthetics, vasodilating compounds, and anticoagulants) and may occur on any exposed part of the body. However, bites and lesions commonly occur on exposed areas that are not covered by sheets and blankets, such as the arms, legs, feet, face, and neck (Parola and Izri 2020). Ibrahim et al. (2017) reported that bites begin as pink macules that progress to papules (Fig. 5.4), large plaques, or wheals (hives). Bites can arise minutes or even days after the initial assault. Some papules and plaques may have a central crust or erosion, suggesting a

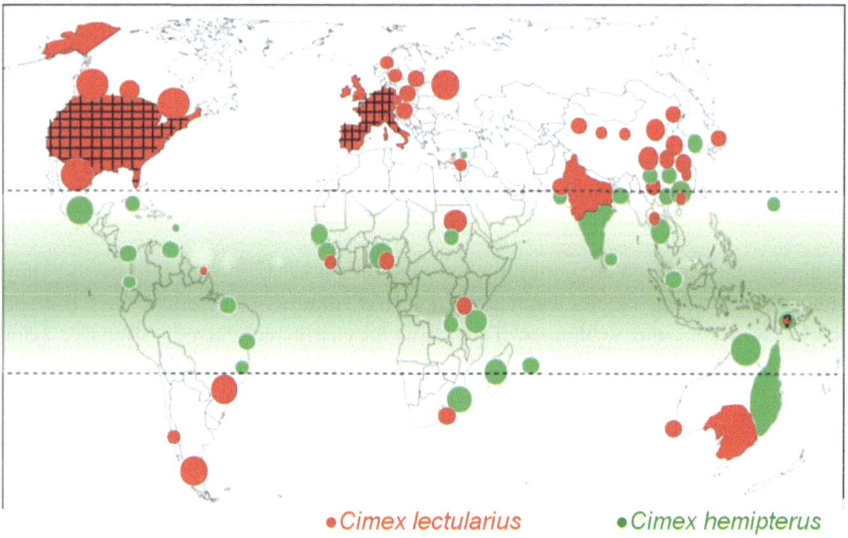

Fig. 5.3 Distribution of bedbugs, *Cimex lectularius,* and *C. hempiterus* in the world. (Zorrilla-Vaca et al. 2015)

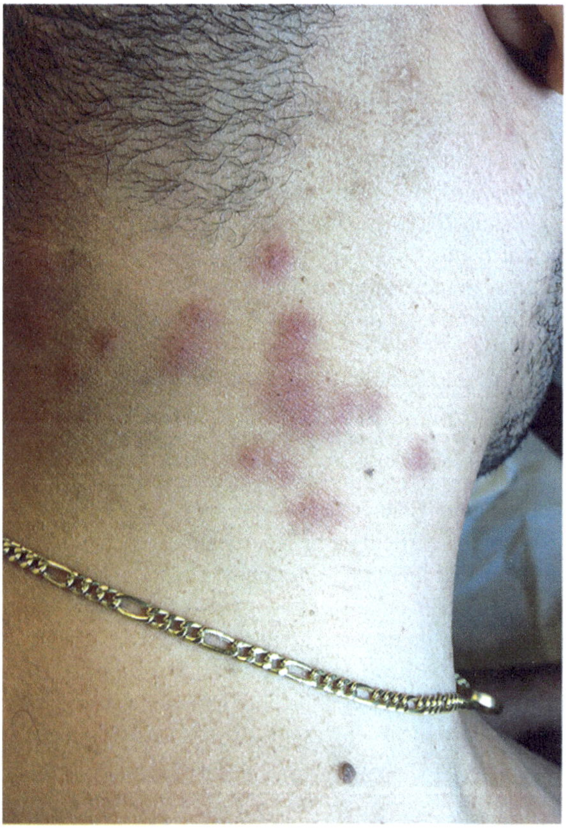

Fig. 5.4 Skin lesions after bedbug bites (Parola and Izri 2020). Shown are erythematous and maculopapular lesions

bite. Bites are typically intensely pruritic, and occasionally, hypersensitive victims can develop bullae, necrotic plaques, or even vasculitis. New papules and plaques form as older ones heal. Some patients may have fever and malaise. Approximately 30% of patients may not develop skin disease from bedbugs, making diagnosis in those individuals impossible. Urticarial reactions and anaphylaxis can also occur (Delaunay et al. 2011). Bedbugs receive essential nutrients through blood meals, which has raised concerns regarding the transmission of blood-borne infections through bites. Bedbugs may be infected with at least 28 different human pathogens but have never been proven capable of transmitting those infections to humans (Kolb et al. 2009). For more information on bedbugs and infectious diseases, see the review articles of Delaunay et al. (2011) and Goddard and deShazo (2009).

5.2.7 *Treatment of Bedbug Bite Reactions*

Goddard and deShazo (2009) reported the treatment of bedbug bite reactions. Treatments of common and complex cutaneous reactions are usually symptomatic and not evidence-based. Table 5.1 shows the reported treatments and outcomes for dermatitis from bedbug bites.

5.2.8 *Expanding*

Jerome and Richard (2009) noted that the dispersal of human-associated bedbugs generally depends on their human hosts for movement from one location to another. This may occur through furniture, clothing, suitcases, used mattresses, or other personal possessions. Bugs may also migrate from one apartment or hotel room to another through holes in walls, water pipes, or gutters.

5.3 Fire Ant and Argentine Ant

Hironori Sakamoto

Species that have expanded their distribution outside their original range because of human activities, whether intentional or unintentional, are called alien species. Among them, many species of ants (Hymenoptera: Formicidae) have expanded their distribution throughout the world through unintentional introduction via sea cargo by their tiny body sizes (Bertelsmeier et al. 2017). Of these, those that cause serious damage to native ecosystems and human activities in invaded areas, such as the red imported fire ant, *Solenopsis invicta* Buren, and the Argentine ant, *Linepithema humile* (Mayr), are especially called invasive alien ants. Ants are

Table 5.1 Reported treatments and outcomes for dermatitis from bedbug bites[a] (Goddard and deShazo, 2009)

Source	No. of patients	Treatment(s)	No. improved	Remarks
Scarupa and Economides (2006)	17[b]	Antihistamines and topical and/or systemic corticosteroids	17	Antihistamines minimize itch in some patients, although appearance and duration of dermatitis were not affected; lesions in most patients were poorly responsive to relatively high-dose oral corticosteroids
Masetti and Bruschi, (2007)	2	Oral antibiotics, oral antihistamines, and topical corticosteroids	2	Remission in 2 week
Sansom et al. (1992)	2	Prednisolone	2	Both patients had resolution of dermatitis over a 2-week period, but only one was treated
Goddard and deShazo, (2008)	1	Diphenhydramine, injection of 80-mg depomedrol; another injection 12 d later	1	Completely resolved in 2 week
Mumcuoglu (2008)	1	Systemic corticosteroids	0	Little improvement in clinical symptoms
Stucki and Ludwig, (2008)	1	Topical corticosteroids	1	Symptoms completely resolved within 2 days
Ter Poorten and Prose, (2005)	1	Oral corticosteroids, oral antihistamines, oral antibiotics, topical permethrin, antibiotic ointments, mid-potency corticosteroid creams	0	"Unsuccessful"
Fletcher et al. (2002)	1	Oral prednisolone and flucloxacillin	1	"Rapid response"
Liebold et al., (2003)	1	100-mg prednisolone and 4-mg dimethindene maleate intravenously; 2-g/d erythromycin for 5 d; topical ointments containing betamethasone	1	Remission in 1 week
Brasch and Schwarz (2006)	1	Topical corticosteroids	1	"Sting" reactions became short-term corticosteroid treatments

[a]No study designs were reported in three treatment regimes

[b]Over the past 7 months, authors have evaluated multiple patients with suspected urticaria, of which at least 17 actually had bedbug bites

taxonomically related to bees and wasps; therefore, some ant species can sting and inject venom into humans through the stinger. The red imported fire ant, the tropical fire ant, *S. geminata* (Fabricius), and the electric ant (little fire ant), *Wasmannia auropunctata* (Roger), are representative human stinging ants among the invasive alien ants that have invaded Asia. In addition, ants, which are social insects, have a high fecundity of queens due to the reproductive division of labor, i.e., the queens specialize in reproduction. The Argentine ant and the browsing ant, *Lepisiota frauenfeldi* (Mayr), do not possess venomous stingers, but they proliferate in numerous numbers in invaded areas, causing serious damage to both native ecosystems and human activities. These typical invasive alien ants are summarized in Fig. 5.5. The damage caused by invasive alien ants is not limited to native biodiversity and ecosystems but also includes direct damage to human health, agriculture, and tourism (Angulo et al. 2022). In fact, ants are the group with the largest number of species on the "World's 100 Most Invasive Alien Species (IAS)"list published by the International Union for Conservation of Nature (IUCN) (Lowe et al. 2000). Under Japan's Invasive Alien Species Act, which was enacted to control damage caused by invasive alien species, fire ants (including the red imported fire ant and the tropical fire ants), the Argentine ant, the electric ant, and the browsing ant are listed as designated invasive alien species, and breeding without permission is regulated. Furthermore, the amended Invasive Alien Species Act (IASA), which will go into effect in April 2023, establishes a new category of "designated invasive alien species requiring urgent action" for fire ants.

Fig. 5.5 Typical invasive alien ant species. (**a**) The red imported fire ant *Solenopsis invicta*. Workers of different body sizes (polymorphism) and a winged new queen (center). (**b**) Argentine ant *Linepithema humile*. Workers swarming over sweetbreads. (**c**) Tropical fire ant *S. geminata*. A new queen collected from the nest in a Japanese port. (**d**) The browsing ant, *Lepisiota frauenfeldi*. Workers attacking hands that touch the nest. (**e**) The electric ant, *Wasmannia auropunctata*. Workers gathering on potato chips placed as attractant baits

5.3.1 Fire Ant

5.3.1.1 General Biology of Fire Ant

"Fire ants" is a general term for the large ant species of the genus *Solenopsis* stinging humans with their venomous stingers that are distributed in the Americas. Although detailed classification differs among taxonomists, according to a recent review, fire ants include 23 species in 4 species groups (Pitts et al. 2018). These fire ants are closely related, and interspecific hybridization has been reported frequently.

In Japan, all fire ant species and their hybrids were listed as designated invasive alien species because they are morphologically very similar and difficult to classify. They form hybrids and are ecologically similar and may cause damage similar to that of invasive alien ants. Furthermore, the amended Invasive Alien Species Act, enforced in 2023, elevated these fire ants to the new, more regulated category, "designated Invasive Alien Species requiring urgent action" (MOE 2022b).

Henceforth, we will describe their representative species, the red imported fire ant *S. invicta*. *Solenopsis invicta* workers show continuous polymorphism in size, ranging from 2.5 mm to 6 mm (Tschinkel 2006) (Fig. 5.6). The body color of the worker is generally reddish from head to thorax, with a black abdomen. *S. invicta* is a typical omnivorous species, preferring both plant and animal food. A mature nest of *S. invicta* contains up to 200,000–400,000 workers, which move through foraging tunnels in the soil. *S. invicta* workers are highly aggressive, actively stinging humans when their nests are attacked.

Fire ant venom is composed mostly (>95%) of an alkaloid called Solenopsin, named after the fire ant genus *Solenopsis*, and a small amount (<1%) of several proteins. Alkaloids are generally found in plant venoms and are rarely found in animals. Solenopsin is a cell-dissolving toxin, and fire ant stings cause pus-filled "pustules" under the skin. Proteins in fire ant venom account for only <1% of the venom, but there have even been reports of anaphylactic shock, an acute allergic reaction that can cause death in humans.

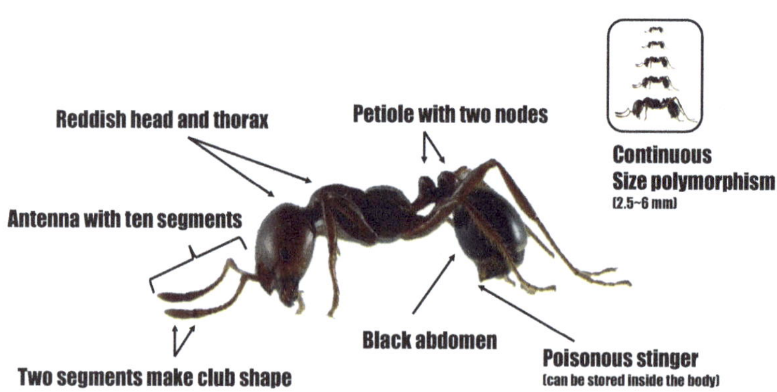

Fig. 5.6 Morphological characters of the red imported fire ant *Solenopsis invicta* worker

One of the important characteristics of fire ants is that both a single queen in one nest (monogeny) and multiple queens (polygeny) can be exhibited within the same species. In monogenetic colonies, the next generation of queens can self-distribute 2–5 km by mating flight, whereas in polygenetic colonies, queens disperse by budding with their workers. These different characteristics are governed by only one gene named *gp-9*. The fact that they could switch from one generation to the next is one of the factors that makes fire ant control difficult.

5.3.1.2 Invasion History and Damage of Fire Ant

The native range of *S. invicta* is the Parana River basin in South America. In the 1930s, they were transported to the port of Mobile, U.S.A., where they were established. In the United States, large-scale pesticide control measures were implemented, but once established, *S. invicta* could not be eradicated. As a result, *S. invicta* continues to cause serious damage in the United States to this day. Since the beginning of the twenty-first century, *S. invicta* has spread from the United States to other Asia-Oceanic countries. Except for New Zealand, where *S. invicta* was first discovered in the early stages of invasion and has been thoroughly eradicated through control measures, *S. invicta* invaded countries such as China, Taiwan, and Australia and has not been eradicated to date. These countries are now being used as new hubs for the spread of fire ants.

5.3.1.3 Fire Ant Invasion in Japan and Newly Developed Control Methods

Due to the serious damage caused in past invaded countries, fire ants and tropical fire ants were listed as the designated invasive alien species in 2005, when the Invasive Alien Species Act was enforced. The first detection of the fire ant in Japan was 12 years later, in 2017, when a sea container landed at Kobe Port. Since then, fire ant invasions into Japan have continued continuously, accumulating to 111 cases in 18 prefectures as of December 2023 (MOE 2023) (Fig. 5.7). Notably, since 2019, large colonies containing approximately 1000 workers and reproductives (queens and males) have been discovered at major international ports such as the Ports in Tokyo, Nagoya, and Osaka (Sakamoto and Goka 2023). These international ports are close to metropolitan areas, and there is strong concern about the invasion of the new queens to metropolitan areas through mating flights (Fig. 5.8). The flight dispersal distance of new queens is at least within 2–5 km (Tschinkel 2006). Furthermore, in October 2022, a large population of fire ants, including more than 70,000 workers with >700 males and >20 females, were confirmed from a container in Fukuyama Port, Hiroshima (MOE 2022c). Thus, the development of control methods for fire ants is an urgent issue.

Under these circumstances, we propose a three-stage approach to fire ant control. The first stage is pre-invasion control, which is preventive control targeting the

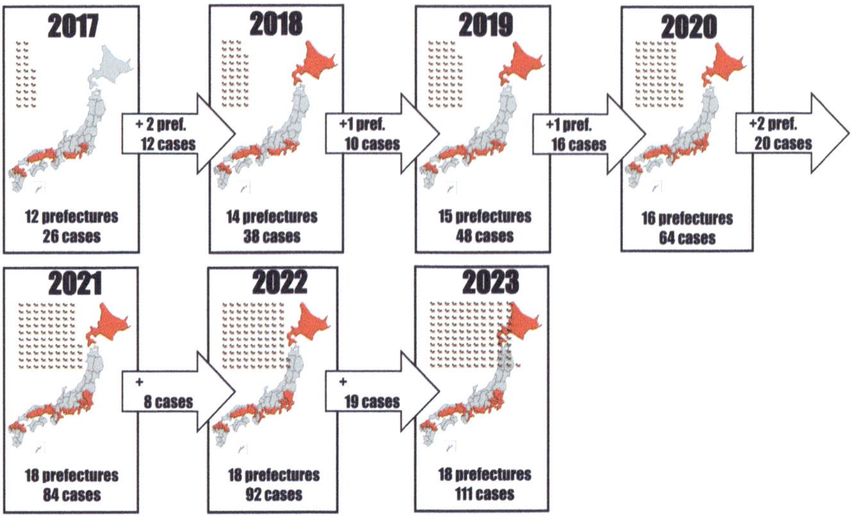

Fig. 5.7 Yearly increase in the number of *S invicta* invasions in Japan (2017–2023). Areas highlighted in red show prefectures where *S. invicta* invasions have been reported. It should be noted that *S. invicta* has not been confirmed as established in Japan.

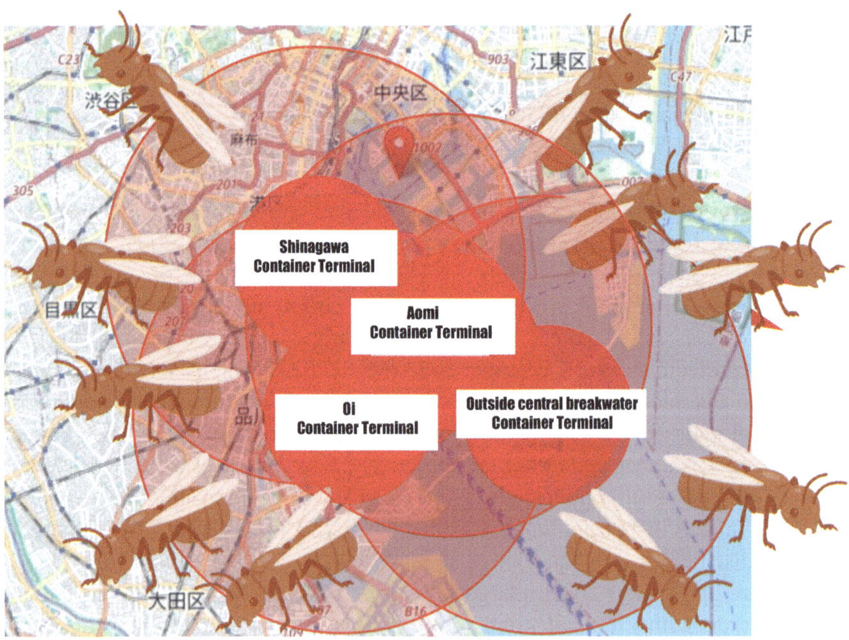

Fig. 5.8 Container terminal where *S. invicta* was found in the Tokyo port and the flight dispersal area in a metropolitan city. The inner circle indicates a 2 km radius area, and the outer circle indicates a 5 km radius area. The map was drawn using OpenStreetMap. https://www.openstreetmap.org/copyright/ja

inside of containers before fire ants arrive in Japan and the container yard before fire ants' nest. To achieve this goal, we are trying to use microencapsulated allyl isothiocyanate (AITC), a volatile component derived from Wasabi, to repel fire ants from sea cargo (Hashimoto et al. 2019, 2020, 2022). The second phase is pre-establishment control, i.e., batch disinfection of fire ants while they are still inside containers. We compared and verified the efficacy of six synthetic pyrethroids against *S. Invicta* (Sakamoto and Goka 2021) and developed a disinfection method for invasive alien ants in sea cargo using effective pyrethroid aerosol agents (Sasaki et al. 2019). The third stage is post-establishment or emergency control after a fire ant nest is discovered on the field. First, for early detection of *S. invicta* invasions, which are difficult to distinguish from other ant species, we developed a genetic identification technique using the loop-mediated isothermal amplification (LAMP) method, a simple gene amplification method for fire ants and other invasive alien ants (Nakajima et al. 2019). Furthermore, we conducted acute (Sakamoto and Goka 2021) and colony-level (Sakamoto and Goka 2023) toxicity tests on ants and found that both fipronil and pyriproxyfen were highly effective in controlling fire ants. As described above, we are developing efficient new methods for all phases.

5.3.2 Argentine Ant

5.3.2.1 General Biology of Argentine Ant

Argentine ant workers are single form, approximately 2.5 mm long, and brown in color (Tatsuki 2014) (Fig. 5.9). The specific name "*humile*" means "humid" in English, and Argentine ants prefer humid environments as on riverbanks and beside waterways. In their native distribution range in South America, Argentine ants live in riverside floodplains. Living in an environment highly prone to disturbance, Argentine ants build temporary nests in crevices in sediments rather than burrows, as most Japanese ants do. Notably, the Argentine ant is a typical polygynous

Fig. 5.9 Morphological characters of the Argentine ant *Linepithema humile* worker

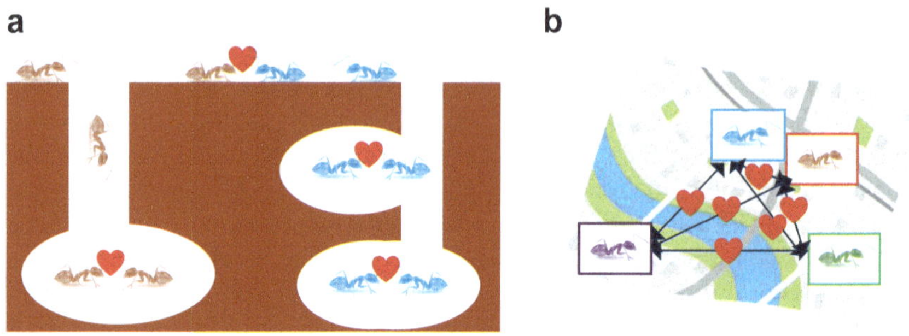

Fig. 5.10 Schematic diagram of a super colony composed of Argentine ants. (**a**) Inter- and intra-colonial Interactions. Ants of different colors indicate that they belong to different colonies. In supercolonies, individuals from different colonies behave in the same friendly way as individuals from the same colony (in other words, there is no intraspecific competition). (**b**) Supercolonies when there is distance between colonies. Ants of different colors indicate that they belong to different colonies. The characteristics of supercolony hold not only when colonies are adjacent to each other, but also when the colonies are widely separated from each other

species. The number of queens in a colony varies with the seasons, but in the summer, when there are the most queens, there is one queen for every 70 workers (Keller et al. 1989). Argentine ants do not engage in nuptial flights like other ants, and after mating, queens proliferate their colonies by colony budding with their workers in tow. Moreover, in invaded areas, Argentine ants form large social structures called supercolony, in which individual ants can move freely between nests physically separated from each other without intraspecific aggression (Pederson et al. 2006) (Fig. 5.10).

5.3.2.2 Invasion History of Argentine Ant

As the name implies, Argentine ants originate from South America. The earliest record of Argentine ants outside their natural distribution range was reported in the 1850s on the Macaronesian archipelagos of Madeira (Wetterer and Wetterer 2006). The Madeiran population is considered to have been introduced unintentionally with shipments from South America. Nearly two centuries later, Argentine ants spread their distribution plan globally, from all continents except Antarctica to the Hawaiian Islands and other oceanic islands. Argentine ants are the most successful pan-global invasive species of invasive alien ants. Argentine ants are highly competitive with other ants and often eliminate other ants in invaded areas. Many severe negative impacts on ecosystem services such as pollination have been reported because of these effects.

5.3.2.3 Invasion of Argentine Ants into Japan and Their Management

Like fire ants, this species was listed as a designated invasive alien species in 2005, when the Invasive Alien Species Act was first enforced. The first record of Argentine ants in Japan was in 1993 in Hiroshima Prefecture. In the approximately 30 years since the first record, Argentine ants have been found to be established in 14 prefectures in Japan, nearly 30% of all Japanese prefectures, including two prefectures (Shizuoka and Tokyo) where eradication has been achieved (NIES 2023) (Fig. 5.11). In 2022, even a critical case of establishment at Osaka (Itami) International Airport was reported. In these invaded areas, the density of Argentine ants is extremely high, which has a strong negative impact on other arthropods represented by ants (Tatsuki 2014). Furthermore, negative impacts on daily life, such as intrusions into machinery and homes, have also been recorded. In Japan, the effectiveness of toxic baits and liquid insecticides based on fipronil as the main active ingredient for controlling Argentine ants has been proven (Inoue et al. 2015; Sakamoto et al. 2019). The eradication of Argentine ants from Shizuoka and Tokyo has been achieved by fipronil-based control methods (NIES 2023). Based on these successful cases, Argentine ant populations established at Osaka International Airport have also been successfully reduced to low densities using toxic baits with fipronil as the main ingredient. The eradication of Argentine ants is a rare and notable achievement in the world. Additionally, statistical models have been developed to confirm the success of eradication in a common, rather than an empirical, manner (Sakamoto et al. 2017).

5.3.3 Other Invasive Alien Ants in Japan and Asia

This section introduces invasive alien ants that require vigilance in Japan and Asia. Among the ants listed as designated invasive alien species, only the electric ant *W. auropunctata* was not found to have invaded Japan until July 2023. The establishment of this species in Taiwan has been reported in 2022 (Hsu et al. 2022), and we have been particularly concerned about the invasion of this species into Japan. The electric ant is a tiny orange ant, approximately 1.5 mm long (Fig. 5.5e). The tiny body size and slow movement of the electric ant make it difficult to detect in the early stages of invasion. This species is highly carnivorous and has negative impacts on invaded areas, from arthropods to humans, due to its poisonous stingers. Moreover, it is difficult to eradicate because the electric ant is a parthenogenetic species.

A polygynous species similar to the Argentine ant, which increases to enormous numbers in invaded areas, is the browsing ant, *L. frauenfeldi* (Fig. 5.5d). In Japan, invasions and establishments of browsing have already been reported in several ports (Ueda and Murakami 2022). We have confirmed that a liquid formulation containing fipronil as the main ingredient is effective for controlling this species, as well as Argentine ants.

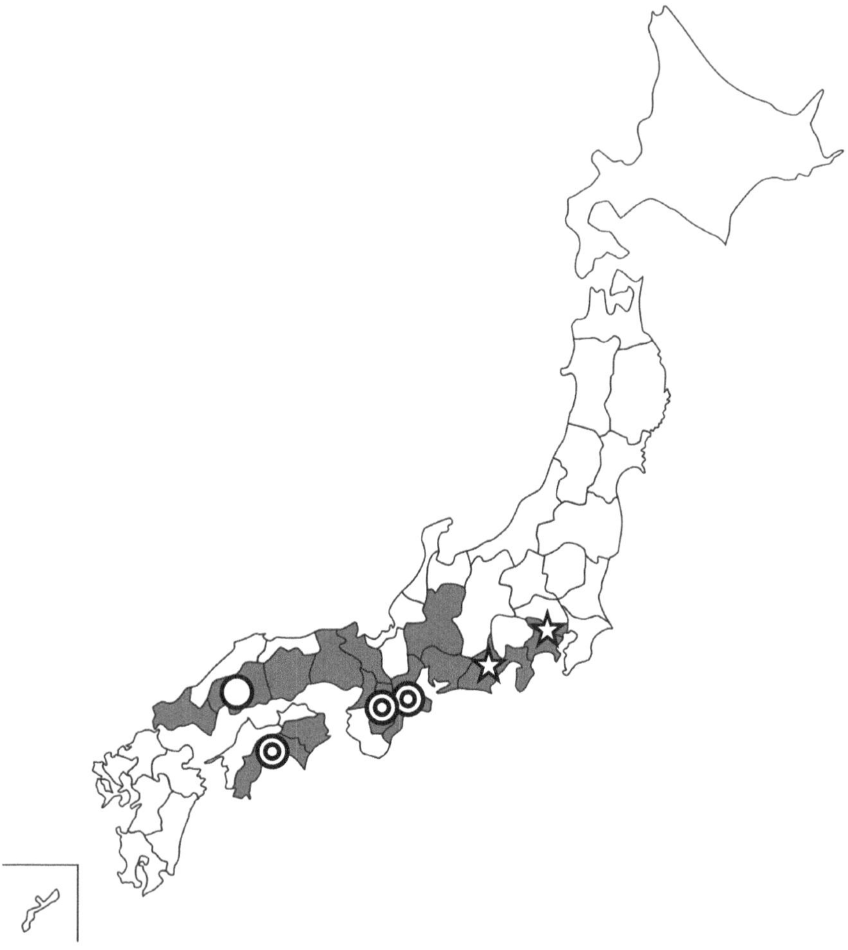

Fig. 5.11 Prefectures where Argentine ant invasions and establishment have been recorded in Japan. The establishment of the Argentine ant was reported in the prefectures marked in black. Prefectures marked with a white circle (Hiroshima Pref.) indicate that this is where Argentine ants were first found in Japan in 1992. Two prefectures placed with a white star indicate that eradication of the Argentine ant has been achieved in the prefectures. Three prefectures with a double circle indicate that the prefectures are new invasion areas where Argentine ants have been found to have established for less than 5 years

The number of invasive alien ant species continues to increase with the globalization of logistics. For example, in the United States, the ant species *Nylanderia fulva* Mayr, called the Rasberry crazy ant or tawny crazy ant, which prefers to invade machinery, is a new serious problem (Wang et al. 2016) (Fig. 5.12). We will need to continue to collect the most recent information on invasive alien ants from various countries and remain vigilant.

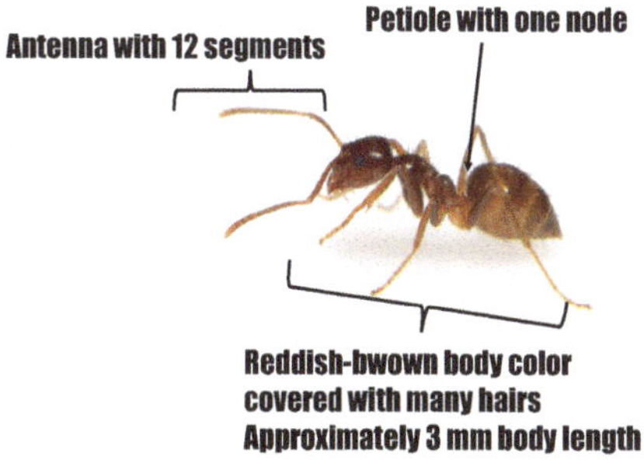

Fig. 5.12 Morphological features of the Rasberry crazy ant *Nylanderia fulva* worker. (This photo was taken by Bentleypkt—Own work, CC BY-SA 4.0, https://commons.wikimedia.org/w/index.php?curid=37887707)

5.4 Venomous Widow Spiders of the Genus *Latrodectus*

Kyoko Sawabe

There are approximately 35,000 species of spiders worldwide (WSC 2023), and approximately 1660 species have been identified in Japan (Ono and Ogata 2018), but not many are venomous enough to affect humans. All of which are carnivores that catch and eat insects and other prey. They have fangs and produce digestive juices from their fangs to digest and absorb their prey outside of their bodies. Some digestive juices may be toxic to humans (Natsuhara 1996). Only a few Japanese native species can be regarded as venomous spiders, including the genus *Latrodectus* and the *Cheiracanthium* spiders. Thus, many of the most dangerous venomous spiders are non-native species, having been introduced to Japan from overseas.

The genus *Latrodectus* Walckenaer widow spiders are classified in the order Araneae family Theridiidae, and 31 species, including undescribed and unknown species, have been recorded from around the world. All of them are venomous. Of these, five widow species have been found in Japan to date (Shimizu et al. 2014). They are *Latrodectus elegans* Thorell, *L. geometricus* C. L. Koch, *L. hasseltii* Thorell, *L. hesperus* Chamberlin & Ivie, and *L. mactans* (Fabricius) (Table 5.2). *Latrodectus elegans* is regarded as a native species in Japan, but the other four were successively introduced into Japan after 1995. Three of these species (*L. geometricus*, *L. hasseltii,* and *L. mactans*) were designated as "Specified Invasive Alien Species (SIAS)" in 2015 under the Invasive Alien Species Law enacted in 2005. Additionally, *L. tredecimguttatus* (Rossi), currently unidentified in Japan, was designated as a SIAS species along with the above three species.

This chapter will focus on *L. hasseltii* (Redback spider, RBS) and will also briefly introduce the biology and distribution of other important widow spiders of the genus *Latrodectus*. (Table 5.2).

Table 5.2 Invasive and native species of the genus *Latrodectus* that may cause noteworthy spider bites in Japan

Species		Distribution	Condition in Japan	
L. elegans Thorell, 1898	Black widow spider	Mainland China, Taiwan, Myanmar, Vietnam, Bhutan, Nepal, Japan	Native	
L. geometricus C. L. Koch, 1841	Brown widow spider	Australia Central America South America The Pacific Islands	Invasive	SIAS[a]
L. hasseltii Thorell, 1870	Redback spider	Australia South east Asia New Zealand Japan	Invasive	SIAS[a]
L. hesperus Chamberlin & Ivie, 1935	Western black widow spider	Israel Western North America	Invasive	
L. mactans (Fabricius, 1775)	Southern black widow spider	Southeastern USA	Invasive	SIAS[a]
L. tredecimguttatus (Rossi, 1790)	Mediterranean black widow spider	from Europe to Central Asia	Unconfirmed	SIAS[a]

[a]SIAS: They were designated as "Specified invasive alien species" by the Ministry of the Environment in 2015

5.4.1 Latrodectus hasseltii *Thorell, 1870 (Redback Spider, RBS)*

5.4.1.1 Taxonomy

Latrodectus hasseltii (RBS), named in 1870 by Thorell, is a small, venomous spider species classified in the family Theridiidae genus *Latrodectus*.

5.4.1.2 Morphology

RBS females are 10–14 mm long, and males are less than half the size of females, approximately 3 mm. It is entirely black, with a rounded abdomen like a sphere and a red pattern in the center of the back (Fig. 5.13a). The ventral surface of the abdomen also has a red pattern. Its Japanese name means "red-backed spider". Their oocysts are nearly spherical, approximately 6–10 mm in diameter, milky white, and contain 50–200 eggs in an oocyst (Fig. 5.13b).

Fig. 5.13 A redback spider (RBS) and its nest. (**a**) RBS female. The arrow indicates a red pattern on its abdomen. (**b**) RBS female and its oocysts. Arrows indicate that RBS abdomen and an oocyst are approximately the same size. (**c**) Nest of the RBS. They nest in the brake area of the bicycle. (**d**) Nest of the RBS. The arrow indicates a nest built in rock crevices

5.4.1.3 Life Cycle

RBS grows at temperatures above 10 °C and takes 28–45 days for males and 45–74 days for females to reach the adult stage at 25 °C. The difference in developmental time between males and females is explained by the fact that females require seven to eight ecdysis to become adults, whereas males become adults after only five ecdysis. The lifespan of female RBSs is estimated to be 2–3 years and the male is 6–7 months (Foster 1984). The number of RBS begins to increase in June, with the largest number found between August and October.

5.4.1.4 Biology

RBS requires 2–3 months of relatively warm summers of 15–25 °C for survival and reproduction (Vink et al. 2010). Because of this, it was generally assumed that RBS was a tropical/subtropical spider and could not overwinter in Japan. However, RBS have been found in Osaka City in November and December (Yoshida et al. 2003). RBS is also considered to be a temperate-zone spider. In addition, the annual mean temperature of 16.8 °C in Adelaide, the capital of South Australia (34.92°S,

138.60°E), where the spider is distributed in large numbers, is not significantly different from 15.4 °C in Osaka City (34.68°N, 135.50°E), where RBSs were first observed establishing in Japan. It was confirmed that RBS can survive in temperatures ranging from −0.5 °C to 46.1 °C (Yoshida et al. 2003). In fact, they have been found in areas with temperatures ranging from below freezing to 41 °C in Australia (Forester 1995). These results suggest that although RBS is a tropical and subtropical species, it is also a temperate spider that can establish itself in temperate regions of Japan (Yoshida et al. 2003). RBS is a spider with excellent survival capabilities that can adapt to a wide range of environments, from tropical to temperate zones.

RBSs inhabit close to human living environments. They nest and breed in shadows and crevices in sunny, warm areas, and insects to feed on, such as inside gutters, gaps in their metal lids, inside drainage pipes on residential lots, at the base of fences, metal fittings for attaching bicycle brake shoes (Fig. 5.13c), and inside blocks in flower beds (Fig. 5.13d)). RBSs are not aggressive but can bite if you touch them unintentionally.

5.4.1.5 Distribution

RBSs originated in Australia but have now been confirmed in New Zealand, various European countries, Southeast Asia, and the United States. In Japan in September 1995, RBS was first discovered in Takaishi City, Osaka Prefecture (Kanazawa et al. 1996; Nishikawa and Kanazawa 1996), in Yokkaichi City, Mie Prefecture the following November, and in Miyakojima City, Okinawa Prefecture that December (Ori et al. 1996) (Fig. 5.14). The distribution of RBS in Japan has since expanded and was first sighted in Metropolitan Tokyo in 2014 (Tokyo Metropolitan Government Bureau of Public Health 2014). RBS has been found in large numbers in or adjacent to port areas, and it is highly likely that they have invaded by attaching themselves to containers and other objects. In Japan, it is thought that its habitat has expanded over a wide area as a result of human transport of nests in cargo, containers, building materials, automobiles, and other items. RBS is now widely established in Japan. It has been shown to be established in a total of 45 prefectures, except for two prefectures, Akita and Aomori, as of July 2020 (NIES 2020). There is concern that its distribution range will continue to expand, and its population will continue to increase.

5.4.1.6 Medical Importance of RBS Bite

In Australia, its origin, the redback spider is known as a venomous spider species. There are 3000 to 5000 bite cases every year (Isbister and Gray 2003). While in Japan, the occurrence of RBS bites in Osaka Prefecture has been well studied, and it was reported that there were three to six cases annually until 2020 (Fig. 5.15). Since the RBS bite cases in Osaka Prefecture and the rest of Japan show almost the

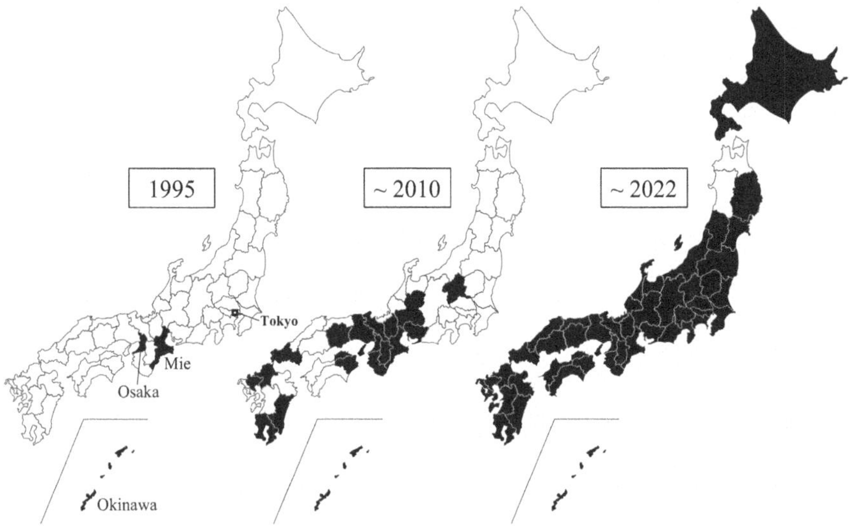

Fig. 5.14 The expanding area where RBSs were discovered in Japan. In 1995, RBSs were found in three prefectures, Okinawa, Mie, and Osaka (black areas). In 2010, the distribution area expanded to 19 prefectures. By 2022, RBS has expanded its distribution area throughout Japan and is found in all but two prefectures

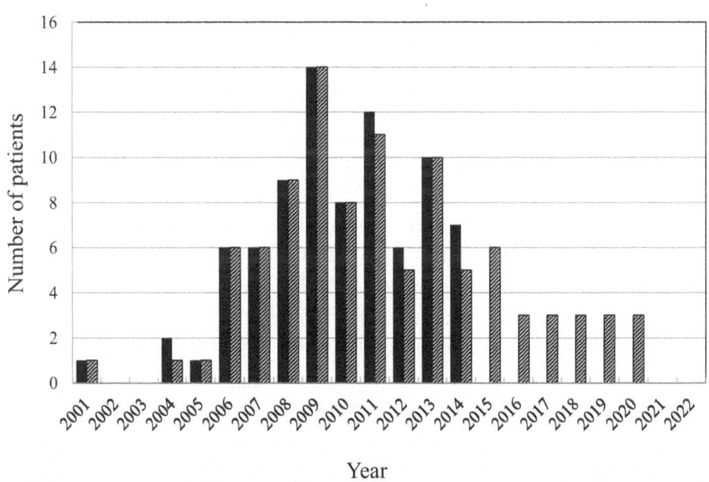

Fig. 5.15 Annual incidence of RBS bite cases. The bite cases that occurred in Osaka Prefecture (shaded bars) show the same trend as that of Japan (closed bars)

same trend, the possibility of RBS bites in Japan in the future is considered high enough.

It is known that free amino acids, enzymes such as hyaluronidase, and α-latrotoxin, a proteinaceous neurotoxin, are commonly found in the venom of widow spiders. The amount of venom extracted from a single female of the redback spider is as low as 0.08–0.120 mg, and the LD_{50} (mg/kg) is 145 for frogs, 0.9 and 1.00 for mice, and 0.075 for Guinea pigs (McCrone 1964; Minton and Ori 1977; Tu 1996). The effects of venom vary greatly depending on the type of animal. The lethal index, calculated by dividing the amount of venom a single spider has by the LD_{50} for mice, is 0.404 for *L. tredecimguttatus*, 0.226 for *L. geometricus*, 0.120 for RBS, 0.106 for *L. mactans* (McCrone 1964). This indicated *L. geometricus* is more dangerous than the RBS bites. However, considering the actual number of people bitten and symptoms, the risk is considered to be *L. mactans* > RBS and *L. elegans* > *L. geometricus* (Nishikawa and Kanazawa 1996). It is suggested that in Japan, attention should be paid not only to RBS but also to bites by other widow spiders.

5.4.1.7 Symptoms

Symptoms of RBS bites are usually mild and localized, such as local pain and erythema. However, fatal cases had been reported before the development of antivenom (Hifumi et al. 2014). The localized symptoms begin with localized pain that increases in intensity during the first 5–60 min after the bite. Over time, the pain spreads to the entire bitten extremity and eventually to the involved lymph node. This may take from 30 min to several hours. Local sweating also occurs, often accompanied by a burning and itching sensation. The onset of systemic symptoms is gradual and often takes more than 12 h. Severe cases occur in children, the elderly, and those with weak constitutions, and the main systemic symptom is pain. Muscle relaxation occurs throughout the body but rarely leads to paralysis. Other major symptoms include nausea, vomiting, fever, insomnia, dizziness, headache, generalized rash, hypertension, diarrhea, hemoptysis, dyspnea, dysuria, severe difficulty opening the mouth, anorexia, eyelid edema, generalized joint pain, tremors, anxiety, photophobia, lacrimation, mental abnormalities, bradycardia or tachycardia, urinary retention following sphincter spasm (IASR 1997).

Hifumi et al. (2014) reported cases of RBS bites in a total of 28 patients (approximately 40% male and 60% female, median age 32 years) during a 5-year study period. RBS bites occurred most frequently on the fingers, followed by the forearm. More than 80% of patients developed local pain and erythema, and 35.7% (10 patients) developed systemic symptoms. Six patients who developed systemic symptoms were given antitoxin, four of whom experienced symptom relief after antitoxin administration. No anaphylaxis occurred in any of the cases. His group found that systemic symptoms occurred in about one-third of cases and that the antitoxin was administered effectively and safely in severe cases.

5.4.1.8 Treatment of RBS Bites

The primary cause of the pain appears to be due to venom called α-latrotoxin (LT), which is demonstrated to be a protein with a molecular weight of 130,000 Da that is found in the spider salivary gland (Hiraoka et al. 1996). In Australia, where the spider species were first identified, there was a risk of developing severe or life-threatening conditions following a spider bite if not treated with equine antivenom. Therefore, an antivenom produced from LT-immunized horses was approved for commercial manufacturing and marketing in 1956 (Nimorakiotakis and Winkel 2004) and has been used clinically for over a half-century. Similarly in Japan, an antivenom is eventually used for the treatment. Therefore, an antivenom produced from LT-immune horses was approved for marketing in 1956 (Nimorakiotakis and Winkel 2004) and has been used clinically worldwide for more than half a century. Similarly in Japan, an antivenom preparation (liquid type, bioCSL, Australia) was eventually used for the treatment. However, import restrictions from Australia have increased the likelihood that antivenom products will not be available after 2014 when nearly all antivenom in Japan will expire unless alternative supply sources can be identified (Hifumi et al. 2015).

In the wake of this incident, a scientific research group of the Ministry of Health, Labor and Welfare, formulated a plan to quickly produce an equivalent domestic product and secure a stockpile for the safety and security of the people of Japan. The previous studies indicated that a comparative study on the characteristics of Australian LT and Japanese one demonstrated these two species are not different physicochemically and immunologically (Nimorakiotakis and Winkel 2004). Therefore, from June to December 2015, more than 450 staff members were mobilized in the Kansai region, and more than 10,000 red spiders were collected. Based on the crude venom purified from these collected RBSs, the first horse antivenom in Japan was manufactured. Laboratory tests have been performed to confirm the quality of the antivenom and it is currently stockpiled for research purposes and emergency treatment (Mori et al. 2017; Matsumura et al. 2018). In addition, it was suggested that therapeutic efficacy could be achieved when the administration of the RBS antivenom in addition in patients with other widow spider bites (by *L. geometricus, L. mactans,* and/or *L. tredecimguttatus*), administration of the RBS antivenin is expected to relieve pain that cannot be relieved by other agents and to improve autonomic symptoms.

5.4.2 Latrodectus geometricus *C. L. Koch, 1841 (Brown Widow Spider)*

Latrodectus geometricus originates from Australia, Central America, South America, and the Pacific Islands. Females are 12–16 mm long and males 6–8 mm long, with body coloration varying from white to black. The spherical abdomen is blackish-brown with reddish-brown spots. Egg sacs are light yellow with numerous projections on the surface, distinguishing them from the white, smooth egg sacs of other spiders of the genus. Only females have α-latrotoxin, which causes severe pain and vomiting when bitten, but they seem to be less aggressive than those of RBS.

In Japan, it was first recorded in Honmoku Wharf in Yokohama City, Kanagawa Prefecture in 1995 (Ono 1995). In the same year, it was discovered in five prefectures: Fukuoka, Okinawa, Osaka, Aichi, and Tokyo. In 2001, it was discovered in Hyogo and Kagoshima prefectures, followed by a series of discoveries in eight more prefectures, and the discovery has been confirmed in 16 Prefectures throughout Japan by 2022. The brown widow spider has already been established in Japan. They are found around buildings and in gutters in urban areas. They inhabit windowsills, protrusions, inside ventilators, behind pipes, and under awnings of man-made buildings. Its habitat is almost the same as that of RBS, but it is considered more resistant to cold conditions than RBS.

5.4.3 Latrodectus mactans *(Fabricius, 1775) (Southern Black Widow Spider)*

Latrodectus mactans was first found in the southeastern United States (east of the Mississippi in the southern United States). It has since expanded its range to the Americas. In recent years, it has become established in the Hawaiian Islands, along with the common high-lipped gopher spider. It is estimated that it may have entered Hawaii with imported agricultural products. Mature females are 8–10 mm in length. The body is jet black with red spots on the back and an hourglass-shaped red pattern on the abdomen. Although this spider is not particularly large, its venom is very potent.

In Japan, immature individuals were first discovered in 1995 on the premises of an agricultural machinery manufacturing company in Shiga Prefecture. It was most likely hiding in an empty box returned from a farmer in Atlanta, USA, to whom the tractor had been exported. The following year, in 1996, it was discovered at a U.S. military base in Iwakuni City, Yamaguchi Prefecture, and then its breeding on the base has become a problem. Since then, they have also been found outside the base. To date, this species has been confirmed in two Prefectures, Shiga and Yamaguchi. It has also been known to appear near trash cans in homes and under indoor tables and chairs.

5.4.4 Latrodectus tredecimguttatus *(Rossi, 1790) (Mediterranean Black Widow Spider)*

Latrodectus tredecimguttatus is native to areas from Europe to Central Asia. Unlike the other spiders described above, this species is more common in grasslands and other open areas than in forests and shrubby areas. The body size of females is

9–18 mm, and males are 4–7 mm long. Its body has a variety of patterns, including red, orange, and white stripes or spots. Usually 13–17 red spots in irregular rows. *Latrodectus tredecimguttatus* has never been confirmed in Japan to date. However, they have α-latorotoxin of almost the same size as other widow spiders and anti-*L. tredecimguttatus* and -RBS α-latorotoxin monoclonal antibodies were effective against other widow spiders. Therefore, if *L. tredecimguttatus* were to invade Japan, it is necessary to be careful about its bites, and the RBS antivenom may be effective in dealing with this problem.

5.4.5 Other Important Widow Spiders

Latrodectus elegans (black widow spider) has been recorded from mainland China, Taiwan, Myanmar, Vietnam, Bhutan, and Nepal. It is thought to be a native species or an early introduction to Japan. This species was considered to be the RBS until 1996, but populations in the Yaeyama Islands and Taiwan were temporarily designated as "*Latrodectus* sp." (Nishikawa and Kanazawa 1996). Later, the scientific name was returned to *L. elegance* (Ono 1996). This species is known from Hateruma, Ishigaki, and Iriomote Islands, Okinawa Prefecture, and at a U.S. military base in Iwakuni City, Yamaguchi Prefecture. Since it is considered a native species, the damage has not been studied in detail by 1995. Still, there are reports of at least three bite cases on Ishigaki Island in 1953, Iriomote Island in 1955, and Hateruma Island in 1995.

Latrodectus hesperus (western black widow spider) is distributed in Israel and western North America, from Mexico to Canada. In Japan, it was found in Gunma Prefecture in 1996, but there have been no reports of its discovery since then. At present, it is unknown whether it has been established in Japan.

5.5 Wasp and Bee

Tetsuhiko Sasaki and Masakazu Nishimura

The order Hymenoptera consists of various insect groups, including wasps, bees, and ants. Wasps and bees are represented using the same Chinese characters, which are also used in Japan (Table 5.3). In countries where wasps and bees are not verbally distinguished, less awareness may exist of their differences. They are similar in that they quickly fly through the air and possibly sting people, but they differ in their morphology and behaviors (Smallheer 2013). Wasps have elongated body shape and smooth bodies without hair. The bodies of bees are rounded and covered with downy hairs that are useful for collecting pollen from the flowers. Bees are vegetarians that collect nectar and pollen from flowers (Fig. 5.16). Nectar is a carbohydrate primarily used for energy production, and pollen is a protein source

Table 5.3 Chinese characters that represent ant, wasp, and bee

English	Japanese	Chinese
Ant	蟻	蚁
Wasp and bee	蜂	蜂
Hornet	雀蜂	胡蜂
Paper wasp	脚長蜂	长脚蜂
Honey bee	蜜蜂	蜜蜂
Bumble bee	丸花蜂	熊蜂[a]
Carpenter bee	熊蜂[a]	木蜂

Note that a common character that means wasp and bee are used for hornet, paper wasp, honey bee, bumble bee and carpenter bee

[a]The same word has different meanings in Japanese and Chinese

Fig. 5.16 The examples of wasps and bees, *Vespa mandarinia* (**a**), *Polistes rothneyi* (**b**), *Apis mellifera* (**c**), and *Bombus ardens* (**d**)

primarily used to grow larvae. Although wasps rely on plants for carbohydrates and are attracted to sweet foods, they are primarily carnivorous and hunt for other insects as protein resources.

Most wasps and bees have solitary life cycles, whereas some groups of Vespidae and Apidae have evolved sociality. Solitary hymenopteran insects rarely sting people. Species that potentially sting people are limited to the social Hymenoptera.

Workers of social species attack enemies using their stings to protect their colonies when the nests are threatened. The females of the social Hymenoptera undergo caste differentiation into queens and infertile workers. Worker stings are evolutionarily derived from ovipositors, which are a part of the female reproductive organ. Males do not have stings; therefore, they never attack humans, regardless of whether they belong to solitary or social species.

The stings of social wasps and bees cause a considerable number of accidents. According to a demographic survey by the Ministry of Health, Labor and Welfare of Japan (MHLW), approximately 10–20 people die from hymenopteran stings each year, which are much greater than those of victims of other animals, such as venomous snakes and bears. Remarkably, in 1984, the number of victims of hymenopteran insects reached 74. When sting accidents occur, identifying the insect species attacking the victim is not always possible. Although the number of deaths caused by each species is unknown, almost all fatal accidents are probably caused by hornets. In this section, hornets are referred to as wasps that belong to the genus *Vespa* in the family Vespidae. Bee stings are rarely fatal to humans. However, honey bees are bred for industrial purposes, and people who work with them, such as beekeepers, often experience stings that can cause allergic reactions. This section focuses on hornets as the most important stinging pests and honey bees in the genus *Apis* from Apidae as an important livestock that is potentially harmful when they sting.

5.5.1 Symptoms Caused by Stings

The sting of a wasp or bee inevitably causes pain owing to physical damage caused by the sting itself and the chemical effect of the venom, of which the chemical effect has a greater impact. A person stung by a honey bee for the first time probably feels stronger pain than that imagined from the tiny body of the honey bee due to the pain-causing components in the venom. The degree of pain from a sting greatly varies depending on the insect species. A sting from a giant hornet causes shocking pain to the victim. One victim said, "*It felt as if a red-hot thick nail was driven into my skin,*" and another said, "*The pain felt like I had broken a bone.*"

A hymenopteran sting usually causes a local inflammatory response that may last only a few minutes or several hours (Burzyńska and Piasecka-Kwiatkowska 2021). Local reactions may be magnified in those allergic to the venom. Swelling accompanied by pain, itching, and fever can spread over an area of 10 cm, and it may take several days for the symptoms to subside. Additionally, stings can induce anaphylaxis, which is the most severe end of the allergic reaction spectrum and may lead to the death of the victim (Cardona et al. 2020). A total of 1161 anaphylactic deaths were reported in Japan between 2001 and 2020. Drug-induced anaphylaxis was the most common, with 452 cases, and Hymenoptera stings were the second most common, with 371 cases. Forty-nine fatalities were attributed to food-induced anaphylaxis (JSA 2022). In addition to anaphylaxis mediated by specific immunoglobulin M, when a person is simultaneously stung by many insects, the toxicity of the venom itself can kill the victim without an immune response. Since 2005, at least nine people have died in Japan, receiving dozens or >100 stings of mass attacks by hornets.

5.5.2 *Life History of Social Wasp*

The family Vespidae is a large group of wasps, comprising approximately 5000 species and classified into six subfamilies. All wasps belonging to the subfamilies Euparagiinae and Masarinae are solitary. The subfamilies Eumeninae and Stenogastrinae include both solitary and social wasps. All species in the subfamilies Polistinae and Vespinae are social (Ross and Matthews 1991).

Social wasps distributed in temperate regions have annual lifecycles. Queens emerge in the fall and have a lifespan of approximately 1 year. Several days after emergence, the future queens leave their natal nests, mate with males from other colonies in the field, and store sperms in their spermathecae. After mating, the future queens enter winter diapause, while males and workers do not survive the winter. The colony annually disappears because all workers die before winter.

The queens emerge from the winter diapause in spring and start forming small colonies (Fig. 5.17a). In many species in the temperate regions, the queen builds a

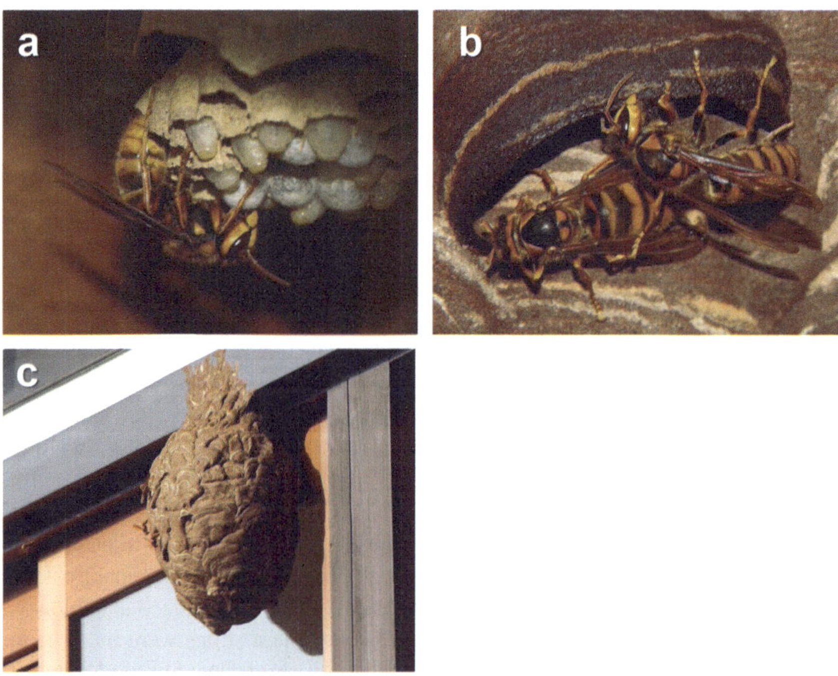

Fig. 5.17 The Japanese yellow hornet, *Vespa simillima*. A small colony founded by the queen in an enclosed space (**a**), workers expanding the next (**b**), a large nest built in an open space after relocation from the initial enclosed location (**c**)

nest alone, lays eggs, and raises the first worker brood during the early stages of colony founding. In some tropical species, multiple queens cooperate to form a colony (Rau 1931; Spradberry 1986). During colony formation, the queen rears exclusively female workers. The mechanism of sex determination in hymenopteran insects is haplodiploidy, wherein females and males develop from fertilized and unfertilized eggs, respectively. The ratio of females to males within a colony can be controlled by determining whether the queen uses the sperm stored in her spermatheca when laying eggs or not. Using this mechanism, social wasps only rear female workers during the nonbreeding season, whereas males only emerge during the breeding season.

After the first worker brood emerges, an increase in the workforce accelerates colony development (Fig. 5.17b, c). During fall, the colony reaches climax and begins to produce males and future queens. The colony-founding queen dies during this season. The number of workers decreases after the queen's death, and the colony comes to an end before the winter arrives.

The number of wasp-stinging incidents tends to increase as the number of workers in the colony increases. The risk increases through the fall when the colonies are at their peak, and workers exhibit rapid and intense defensive responses to disturbances.

5.5.3 Important Social Wasp Species in Asia

All social wasps display defensive reactions to threats to their nests. They have developed sophisticated weapons and defensive behaviors; thus, no harmless species exist. However, some species, such as giant hornets and yellow hornets, are particularly dangerous to humans because of their habits.

Hornets, specifically species of the genus *Vespa*, are primarily found in eastern Asia. *Vespa mandarinia* is one of the largest hornets (approximately 4 cm in length) and is known as the "giant hornet," with a distribution ranging from Malaysia, China, and Japan to southeastern Russia. This species has also recently been inadvertently introduced into North America (Archer 2012; Wilson et al. 2020). They inhabit rich natural environments, such as wooded forests or bamboo forests in rural areas. Their nests are usually built in enclosed spaces such as small underground spaces or tree cavities. *Vespa mandarinia* forms large colonies, with up to several hundred workers in a single nest. Individual workers possess notably high fighting ability owing to their developed mandible, long length, robust sting (Fig. 5.18), and a large volume of venom in their poison sacks. Therefore, their defensive behavior in nest protection is more intense and potentially lethal than other hornet species. People engaged in forestry or agriculture as well as hikers may encounter attacks from this species.

Another giant hornet, *V. soror*, which is closely related to *V. mandarinia* and often called the "southern giant hornet" exists in eastern Asia. Their habitat range is narrower than that of *V. mandarinia*, with its distribution spanning from Vietnam to

Fig. 5.18 The weapons of the giant hornet (*Vespa mandainira*), *mandible* (**a**), and sting (**b**)

southern China (Archer 2012). The southern giant hornet is as dangerous as *V. mandarinia*. Both giant hornet species are notorious pests of beekeeping because hornet workers attack beehives in groups, causing serious damage to honey bees (Matsuura and Sakagami 1973; Ono et al. 2003; Lee 2009; Mattila et al. 2021).

The most common hornet found in Japan is the yellow hornet, *V. simillima*, which is a small hornet with a body length of approximately 2–3 cm. This species is distributed in southeastern Russia, eastern China, Korea, and Japan (Archer 2012). Colonies of *V. simillima* can grow quite large, sometimes containing more than a thousand workers. The yellow hornet is well-adapted to urban environments, and its nesting sites include open spaces, such as tree branches and building eaves, as well as enclosed spaces, such as tree cavities, above ceilings, and under floors. After hibernation, a queen hornet often initiates her colony in an enclosed space in spring, and the colony often relocates as a group to places such as the eaves of buildings when the number of workers reaches approximately 100. The territory of the yellow hornet often overlaps with human settlements because of its nesting habits (Fig. 5.17c), thus increasing the likelihood of stinging incidents by this species.

The yellow-legged hornet (also referred to as the Asian hornet), *V. velutina*, is a species closely related to *V. simillima*. It was originally distributed in Asia, from Afghanistan to Indonesia (Carpenter and Kojima 1997). This hornet forms huge colonies, with nests reaching up to 70 cm in diameter and housing thousands of workers. They possess highly adaptable nesting abilities, and their nests can be found in various environments, ranging from rural to urban areas. Recently, *V. velutina* was accidentally introduced into South Korea, France, and Spain owing to human activities (Kim et al. 2006; Haxaire et al. 2006; Castro and Pagola-Carte 2010). In Japan, *V. velutina* was first recorded on Tsushima Island, located between the Korean Peninsula and Japan, in 2013 and designated as an invasive alien species by MOE, Japan, in 2015 (MOE 2015). Later, the nests of this hornet were found in the Kyushu region, located in the southwestern part of mainland Japan (Minoshima et al. 2015).

Paper wasps, which belong to the subfamily Polistinae, are social wasps commonly found in rural and urban areas of Japan. Although they are generally considered less dangerous than hornets, it is worth noting that the venom composition of paper wasps is similar to that of hornets and can potentially trigger allergic reactions due to immune cross-reactivity, as mentioned later.

5.5.4 *Honey Bee*

Honey bees belong to the genus *Apis*. Nine species of this genus exist worldwide (Raffiudin and Crozier 2007). The European honey bee, *Apis mellifera*, inhabits the African continent and the western part of the Eurasian continent, whereas the eastern honey bee (also known as the Asian honey bee), *A. cerana*, inhabits eastern Eurasia, including India, Southeast Asia, China, Korea, and Japan (Fig. 5.19). The distributions of the other seven species are limited to South and Southeast Asia. Currently, beekeeping is actively practiced in North and South America, as well as Australia. However, honey bees did not originally inhabit these regions. They were introduced from Europe during the Age of Discovery. European honey bees can produce more honey than Eastern honey bees but are also bred in Asian countries for commercial beekeeping. In Japan, the western honey bee was introduced in 1877. Asian honey bees are not suitable for commercial beekeeping because they have lower honey-producing abilities and often abandon their hives when they perceive a deterioration in their living environment. However, in Japan, some people raise Eastern honey bees as a hobby because they are well adapted to the Japanese environment, resistant to diseases, and require minimal care.

5.5.5 *Bees as Pollinators*

In addition to producing honey, honey bees are used to pollinate agricultural crops. Honey bees are well-known essential pollinators for almond production in California (Glenny et al. 2017). Beginning fall of 2006, a considerable disappearance of honey

a b

Fig. 5.19 A queen and workers of the European honeybee, *Apis mellifera* (**a**), and *A. cerana* workers defending their hive against the yellow hornet *Vespa simillima* (**b**)

bees was observed in various parts of the United States for unknown reasons, which was widely reported as a serious problem (Leska et al. 2021). Three-quarters of the world's most productive crops rely on animal pollination, although the degree of dependence considerably varies among crops (Klein et al. 2007). According to the Japan Beekeeping Association, approximately 240,000 honey bee colonies are raised for honey production. In addition, approximately 100,000 colonies are utilized for pollination. Rentals of honey bees for pollination are primarily in high demand for greenhouse strawberry cultivation in Japan (Matsuka and Sakai 1989). Honey bees are gentle, nonaggressive insects, and their stings rarely cause death in humans; however, they are commonly used commercially. Particularly, beekeepers often experience repeated stings, leading to several anaphylaxis cases.

In addition to honey bees, bumble bees are used to pollinate crops. The buff-tailed bumble bee, *Bombus terrestris*, was introduced as a greenhouse pollinator from Europe but was later classified as an invasive alien species in Japan (Matsumura et al. 2004). As an alternative to *B. terrestris*, *B. ignitus*, native to Northeast Asia, is now commercially available in Japan, China, and Korea (Namin et al. 2023).

5.5.6 Life History of Honey Bee

Honey bees always live in colonies. Unlike the founding queens of social wasps or bumble bees, queen bees never live alone (Fig. 5.19a). The sociality of honey bees is more developed than that of wasps and bumble bees in that the queen does not forage or raise larvae, and the roles of the queen and workers are completely separated. A new queen, when produced, is allowed to take over the original nest as the old queen leaves the nest with approximately half of the workers and migrates to the new nesting location. This phenomenon, called swarming, is the colony-level reproduction of honey bees. The honey bee breeding season in temperate regions spans spring to early summer. New queens and males are raised during this period. After emergence, the queen goes on a mating flight, mates in the air with over ten males that come from other nests, and returns to the nest where she was raised. The queen stores the sperm she receives during the mating flight in her spermatheca and gradually utilizes it to continue laying eggs for 2–3 years during her lifespan. Workers typically have a lifespan of 1–2 months. Male bees, with a lifespan of approximately 1 month, are raised during the breeding season and disappear from the nest after autumn. The life history of honey bees is climate-dependent and varies from region to region.

5.5.7 Defensive Behavior and Honey Bee Venom

Honey bees store honey and raise numerous larvae in their hives, making them an attractive food source for other animals. Hornets, particularly giant hornets, are formidable honey bee predators. Natural enemies of honey bees also include large mammals such as bears. Therefore, honey bees have developed strong defense mechanisms to protect their colonies. A notable feature of a honey bee's defensive behavior is that each worker bee can use its sting against an enemy only once in its life. The honey bee sting has a barbed structure at its tip. When a worker bee stings a target, the sting falls out of the bee's abdomen and remains on the victim's skin along with the venom sac, which automatically pulsates and injects all the stored venom inside the victim's body (Fig. 5.20). Workers who lose stings and parts of their internal organs die following the defensive behavior. Considering that thousands of worker bees exist in a colony, even if a small number of worker bees die, the strong defense exhibited by each bee is effective in defending the entire colony.

5.5.8 *Venom Components of Wasp and Honey Bee*

The venom of wasps and honey bees is a complex mixture of low-molecular-weight bioactive substances such as bioactive amines, short peptides with different functions, and proteins such as enzymes and allergens. Two biogenic amines, histamine and serotonin, are commonly present in the venom of hornets belonging to the genus *Vespa* and in paper wasps of the genus *Polistes* (Nakajima et al. 1983). Histamine stimulates inflammation and other immune responses, whereas serotonin acts as an irritant and contributes to pain at the injection site (Weyda and Kodrik 2021). In addition to histamine and serotonin, hornet venom contains large amounts of acetylcholine, which can increase the pain of a sting by stimulating pain receptors (Habermann 1971; Moreno and Giralt 2015).

Several peptides found in wasp venom are composed of 12–50 amino acids, and their total content accounts for up to 70% of the venom's dry weight. (Herrera et al. 2020; dos Santos-Pinto et al. 2018). Based on their biological activities, peptides with varying functions can be categorized into several families, such as

Fig. 5.20 A moment on the stinging of European honey bee (*Apis mellifera*). The sting is falling out from the body (**a**), and the sting is left on the opponent (**b**)

mastoparans, chemotactic peptides, and kinins (Herrera et al. 2020). Mastoparans, the predominant peptide component in hornet venom, play a role in mast cell degranulation, subsequently leading to histamine release (Silva et al. 2017). Chemotactic peptides, the second most important peptide family, promote macrophage and polymorphonuclear leukocyte recruitment to sites of sting injuries (Souza et al. 2005). Kinins, the third most important peptide family, cause pain and vasodilation (Herrera et al. 2020).

Although peptides cause local reactions, wasp venom proteins can induce local tissue damage and systemic allergic responses. The main antigens in wasp venom are hyaluronidase, phospholipase A1, and antigen 5 (Luo et al. 2022). Hyaluronidase catalyzes the hydrolysis of hyaluronic acid, a major component of the extracellular matrix. Disrupting the intercellular connections prevents other toxic components from penetrating the tissues. Phospholipase A1 (PLA1) catalyzes phospholipid hydrolysis by excising sn-1 acyl chains (fatty acids at the 1-position of the glycerol moiety), destabilizing the cell membranes. PLA1 and hyaluronidase synergistically enhance the penetration of toxic components into the tissues. Almost all wasp species belonging to the family Vespidae contain a protein called antigen 5 in their venom (Bazon et al. 2018). Although the biological effects of this protein are unknown, it is frequently the most potent allergen in patients stung by wasps (Luo et al. 2022). The protein components found in wasp venom, particularly PLA1 and antigen 5, exhibit cross-reactivity as antigens between *Vespa* and *Polistes* wasps (Lu et al. 1993). Consequently, individuals stung by a paper wasp may develop an allergic reaction to hornet stings and vice versa.

Honey bee venom is immunochemically distinct from wasp venom (Smallheer 2013). More than 100 peptides and proteins have been identified in bee venom (Van Vaerenbergh et al. 2014). Biogenic amines in the venom include histamine, serotonin, dopamine, and noradrenaline, with histamine being the major component (Abdela and Jilo 2016). A characteristic of honey bee venom is that it contains large amounts of a melittin peptide, accounting for approximately 40–60% of its dry weight. Melittin, a highly basic polypeptide consisting of 26 amino acids, can be inserted into the phospholipid bilayer of the cell membrane. It causes cell membrane disruption and functions as a pain-producing substance (Chen et al. 2016). Phospholipase A2 (PLA2) accounts for approximately 12% of the dry weight of the venom (Burzyńska and Piasecka-Kwiatkowska 2021). This enzyme releases sn-2 acyl chains, including arachidonic acid, from the membrane phospholipids (Lambeau and Gelb 2008). Arachidonic acid is the major precursor of inflammatory lipid mediators such as prostaglandins and leukotrienes (Lambeau and Gelb 2008). Most arachidonic acids in cells are attached to the 2-position of the glycerol moiety in phospholipids, and only a few molecules remain free. Disturbance of the arachidonic acid pathway by PLA2 causes fever, pain, and inflammatory reactions. The two major allergens in honey bee venom are PLA2 and hyaluronidase, whereas melittin is a minor allergen (Burzyńska and Piasecka-Kwiatkowska 2021).

5.5.9 *Desensitization Immunotherapy*

People who have previously experienced hypersensitivity reactions or anaphylaxis to hymenopteran insect stings should avoid further stings. However, avoiding bee stings by beekeepers and farmers who use bees to pollinate crops is unrealistic. Forestry workers are frequently exposed to the risk of stinging by hornets, paper wasps, and other insects. Desensitization immunotherapy with antigens is an effective treatment to reduce the risk for such patients. It is effective in >80% of adult patients, and its effects last for >10 years (JSA 2022). During therapy, the venom extract is repeatedly injected into patients as an antigen, gradually increasing dosages (Burzyńska and Piasecka-Kwiatkowska 2021). The treatment starts with an extremely small antigen amount, which is gradually increased to the equivalent of an actual insect sting over approximately 2 weeks. Subsequently, continuing booster antigen injections every 1 or 2 months for 3 to 5 years is necessary. Although this therapy is time-consuming and expensive, eliminating or reducing allergic reactions greatly improves the patient's quality of life.

5.5.10 *Conclusion*

Insects are the most successful animal group on Earth in terms of species diversification. Carnivorous wasps, especially hornets, occupy the top of the insect food chain hierarchy and play an important ecological role. Vegetarian bees are crucial pollinators of flowering plants. Wasps and bees are essential members of terrestrial ecosystems. Although they may be perceived as potentially dangerous insects due to their venomous stings, only social species sting humans to protect their nests. They are not aggressive but defensive. It would be important to recognize these facts when considering the relationship between stinging hymenopteran insects and humans.

5.6 Chironomid Midge

Kimio Hirabayashi

5.6.1 *Chironomidae*

The nonbiting chironomid midges (Family: Chironomidae, Order: Diptera) include a number of species whose larvae develop in almost all types of landwaters and some soils and have been known to play an important role in the transmission of nutrients along the food web in nature. It has been estimated that the number of species worldwide may be as high as 15,000. This high species diversity has been attributed to the antiquity of the family, relatively low vagility leading to isolation, and evolutionary plasticity (Armitage et al. 1995).

5.6.2 Life Cycle

The life cycle of Chironomidae is usually very simple, i.e., the eggs laid by adult females hatch in water or in soil, the larvae feed on algae, detritus and microorganisms and become pupae, and then leave the water or soil as adults. Males and females copulate, usually in the air, often in swarms, and the females deposit eggs onto water or soil suited to the development of the larvae (first to fourth instar larvae). Adults do not consume food except for honey or water; they use the nutrients accumulated during the larval stage to produce eggs. Figure 5.21 shows the life cycle of Chironomidae.

5.6.3 Information on Chironomids in Asia

Sasa and Kikuchi (1995) reported that information on chironomids in Asia, especially Japan and East Asia, has been very scarce until recently. The oldest records of the chironomid species from this region were made by Kieffer on 12 species from Taiwan in 1912 (Sasa and Kikuchi 1995). After that are records from the Philippines in 1922. From 1933 to 1965, Tokunaga studied and described the chironomid midges of the East Asian and Micronesian regions, when a total of 160 species had already been recorded from the present territory of Japan (Sasa and Kikuchi 1995). More recent contributions to identifying the chironomid fauna of this region have also been made by a number of other chironomid researchers (from Japan, Korea,

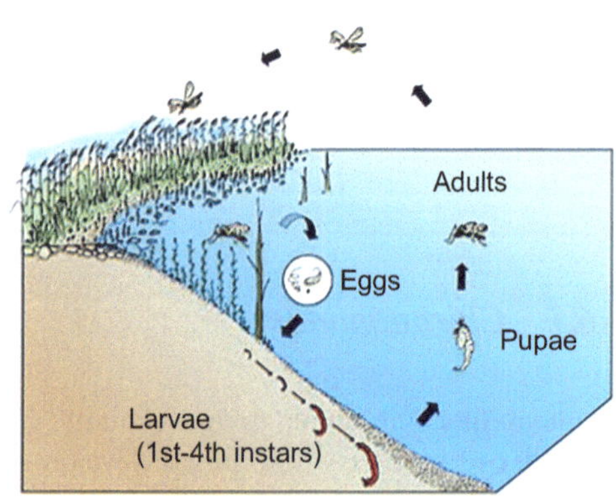

Fig. 5.21 The life cycle of Chironomidae

China, Russia, India, and elsewhere). In 1995, Sasa and Kikuchi published a checklist of Chironomidae (Diptera) of Japan (Sasa and Kikuchi 1995).

5.6.4 Chironomids and Human Life

Chironomidae are known to have various impacts on human life because of the enormous mass and large numbers of chironomid midges breeding in various types of natural environments. Sasa and Kikuchi (1995) noted four points in the relationship between Chironomidae and human life: (1) Chironomid midges as environmental nuisances, (2) Chironomid midges as causative agents of allergic diseases, (3) The role played by chironomid midges in the importance of land water quality, and (4) Chironomid midges as indicators of land water quality. In this section, we focus on (1) and (2).

5.6.4.1 Nuisance and Economic Problems.

Cranston (1995a) noted in a review that massive swarms of chironomid midges emerging from a variety of natural and artificial aquatic habitats distributed worldwide have increasingly posed nuisance and economic impacts and, in some cases, medical problems to waterfront residents, workers, visitors and business and industrial establishments. Ali (1995) reported that the annual economic loss attributable to chironomid problems is estimated to amount to millions of US dollars globally. The suspected trend of global warming may also have some effect on the increase in chironomid midge populations. In warmer waters, the productivity of phytoplankton and benthic algae, which serve as food for the larvae of many chironomid species, increases due to increased utilization of solar energy in the process of photosynthesis, thus creating favorable conditions for chironomid development.

According to Ali (1995), the chironomid species most commonly encountered in heavy midge-producing urban and suburban habitats belong to the genera *Procladius* and *Tanypus* (Tanypodinae), *Cricotopus* and *Propsirocerus* (Orthocladiinae), *Chironomus*, *Baeotendipes*, *Dicrotendipes*, *Glyptotendipes*, *Goeldichironomus*, *Kiefferulus*, *Apedilum*, *Polypedilum*, *Cladotanytarsus* and *Tanytarsus* (Chironomidae). There are many reports from the United States, Western Europe, and northern Africa, but in this section, we focus on the Asian countries of Japan, Singapore, and Saudi Arabia. In Japan, *Chironomus yoshimatsui*, *C. plumosus,* and *Propsirocerus akamusi* are the principal pests. The former is a serious problem for people near sewage ditches in many cities, especially along the streams receiving water from sewage disposal plants. At the same time, the latter two are problems for people near the shores of many eutrophicated lakes and ponds during their breeding season in November and December (Sasa and Kikuchi 1995) (Fig. 5.22). In the reservoirs of Singapore, *Polypedilum nubifer* and *Tanytarsus oscillans* are the predominant nuisance midges. The latter is restricted to Asia, from India to southern

Fig. 5.22 Swarming of *Chironomus plumosus* (above) and resting *Propsiropserus akamusi* midges on a white sheet (below) in the Lake Suwa Area, Japan

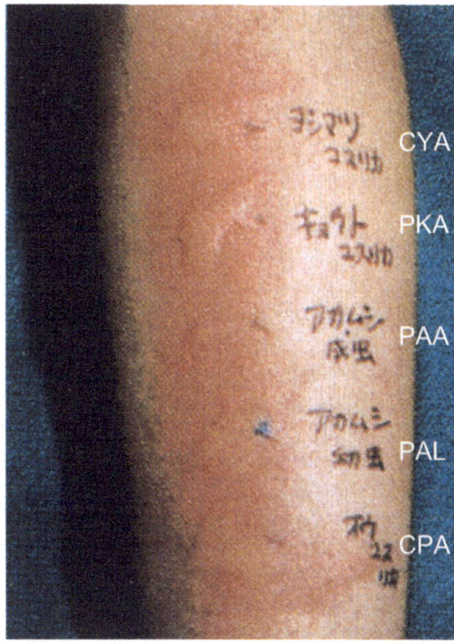

Fig. 5.23 Intracutaneous test: wheal and flare response of the patient. Intradermal injection with extracts of *CYA Chironomus yoshimatsui* adult, *PKA Polypedilum kyotoense* adult, *TPA Propsiropserus akamusi* adult, *TPL Propsiropserus akamusi* larvae, *CPA Chironomus plumosus* adult (Murakami 1989)

Japan (Cranston et al. 2013). In Saudi Arabia, species of *Baeotendipes* pose pest problems. It is likely that several other species of midge emerge at nuisance levels in many countries, but these are presently not reported in the literature (Ali 1995).

5.6.4.2 Medical Impacts

Adults of various chironomid species have been shown to act as allergens for asthmatic patients when their bodies become dusty and inhaled. Such disease has been reported from Sudan (Lewis 1956), where a single species, *Cladotanytarsus* (as *Tanytarsus*) *lewisi*, is implicated. Recently, this disease has become widespread in Japan and India. In Japan, attacks in at least one-third of asthma sufferers are caused by inhalation of dust containing chironomid midges. Adult midges, particularly of *Polypedilum kyotoense*, occur in enormous numbers. Knowing of the medical problems induced in Sudan, Sasa (1985) was stimulated to initiate a study of the possible allergenicity of Japanese nuisance species (Fig. 5.23) (Murakami 1989). The species involved were the Chironominae, *P. kyotoense*, *Tanytarsus oyamai*, *Chironomus plumosus* and *C. salinarius* and an Orthocladiinae, *Propsiropserus akamusi* (Fig. 5.23). In India, two common chironomid species, *C. circumdatus* and *Polypedilum nubifer*, are known for their sensitizing potential as allergens in atopic patients and controls. In one study, skin prick test results revealed that 189 of 198 patients (95.4%) demonstrated sensitization to both chironomid species. Higher levels of total immunoglobulin E (IgE) were observed in atopic subjects than in the control group (Nandi et al. 2014). In Germany, workers producing pet fish food containing freeze-dried blood worms of chironomid larvae develop asthma at a significantly higher rate. For details on occupational allergies and Chironomidae (especially chironomid larvae), see the review of Cranston (1995b).

References

Abdela N, Jilo K (2016) Bee venom and its therapeutic values: a review. Adv Life Sci Technol 44:18–22

Ali A (1995) Nuisance, economic impact, and possibilities for control. In: Armitage PD, Cranston PS, Pinder LCV (eds) The Chironomidae: biology and ecology of non-biting midges. Chapman & Hall, London, pp 339–364

Angulo E, Hoffmann BD, Ballesteros-Mejia L, Taheri A, Balzani P, Bang A, Renault D, Cordonnier M, Bellard C, Diagne C, Ahmed DA, Watari Y, Courchamp F (2022) Economic costs of invasive alien ants worldwide. Biol Invasions 24:2041–2060

Archer ME (2012) Vespine wasps of the world: behaviour, ecology & taxonomy of the Vespinae, 1st edn. Siri Scientific Press, Manchester, pp 251–287

Armitage PD, Cranston PS, Pinder LCV (eds) (1995) The Chironomidae: biology and ecology of non-biting midges. Chapman & Hall, London, 572 pp

Bazon ML, Silveira LH, Simioni PU, Brochetto-Braga MR (2018) Current advances in immunological studies on the Vespidae venom antigen 5: therapeutic and prophylaxis to hypersensitivity responses. Toxins (Basel) 10(8):305

Bertelsmeier C, Ollier S, Liebhold A, Keller L (2017) Recent human history governs global ant invasion dynamics. Nat Ecol Evol 1(7):0184

Brasch J, Schwarz T (2006) 26-year-old male with urticarial papules. J Dtsch Dermatol Ges 4(12):1077–1079

Burzyńska M, Piasecka-Kwiatkowska D (2021) A review of honeybee venom allergens and allergenicity. Int J Mol Sci 22(16):8371

Cardona V, Ansotegui IJ, Ebisawa M, El-Gamal Y, Rivas MF, Fineman S, Geller M, Gonzalez-Estrada A, Greenberger PA, Borges MS, Senna G, Sheikh A, Tanno LK, Thong BY, Turner PJ, Worm M (2020) World allergy organization anaphylaxis guidance 2020. World Allergy Organ J 13(10):100472

Carpenter JM, Kojima JI (1997) Checklist of the species in the subfamily Vespinae (Insecta: Hymenoptera: Vespidae). Nat Hist Bull Ibaraki Univ 1:51–92

Castro L, Pagola-Carte S (2010) *Vespa velutina* Lepeletier, 1836 (Hymenoptera: Vespidae), recolectada en la Península Ibérica. Heteropterus Rev Entomol 10(2):193–196

Centers for Disease Control and Prevention (CDC) (2020) Parasites. Bed Bugs FAQs. https://www.cdc.gov/parasites/bedbugs/faqs.html. Accessed 25 Dec 2023

Chen J, Guan SM, Sun W, Fu H (2016) Melittin, the major pain-producing substance of bee venom. Neurosci Bull 32(3):265–272

Cranston PS (1995a) Introduction. In: Armitage PD, Cranston PS, Pinder LCV (eds) The Chironomidae: biology and ecology of nonbiting midges. Chapman & Hall, London, pp 1–7

Cranston PS (1995b) Medical significance. In: Armitage PD, Cranston PS, Pinder LCV (eds) The Chironomidae: biology and ecology of nonbiting midges. Chapman & Hall, London, pp 365–384

Cranston PS, Ang YC, Heyzer A, Lim RBH, Wong WH, Woodford JM, Meier R (2013) The nuisance midges (Diptera: Chironomidae) of Singapore's Pandan and Bedok Reservoirs. Raffles Bull Zool 61(2):779–793

Delaunay P, Blanc V, Del GP, Levy-Bencheton A, Chosidow O, Marty P, Brouqui P (2011) Bedbugs and infectious diseases. CID 52(2):200–210

dos Santos-Pinto JRA, Perez-Riverol A, Lasa AM, Palma MS (2018) Diversity of peptidic and proteinaceous toxins from social Hymenoptera venoms. Toxicon 148:172–196

Ewan PW (1998) ABC of allergies: venom allergy. BMJ 316:1365–1368
Fletcher CL, Ardern-Jones MR, Hay RJ (2002) Widespread bullous eruption due to multiple bed bug bites. Clin Exp Dermatol 27(1):74–75
Forester L (1995) The behavioural ecology of *Latrodectus hasselti* (Thorell), the Australian redback spider (Araneae Theridiidae): a review. Rec West Aust Mus 52(Suppl):13–24
Foster L (1984) The Australian redback spider (*Letrodectus hasseltii*): its introduction and potential for establishment and distribution in New Zealand. In: Laird M (ed) Commerce and the spread of harmful pests and disease vectors. Praegar Press, New York, pp 273–289
Glenny W, Cavigli I, Daughenbaugh KF, Radford R, Kegley SK, Flenniken ML (2017) Honey bee (*Apis mellifera*) colony health and pathogen composition in migratory beekeeping operations involved in California almond pollination. PLoS One 12(8):e0182814
Goddard J, deShazo RD (2008) Rapid rise in bed bug populations: the need to include them in the differential diagnosis of mysterious skin rashes. South Med J 101:854–855
Goddard J, deShazo R (2009) Bed bugs (*Cimex lectularius*) and clinical consequences of their bites. JAMA 301(13):1358–1366
Habermann E (1971) Venomous animals and their venoms: venomous invertebrates. Academic Press, New York/London, pp 61–93
Hall C (2019) Bee vs wasp sting venom: truth and chemical myths. Chemistry Hall. https://chemistryhall.com/bee-wasp-sting-venom/. Accessed 25 Dec 2023
Hashimoto Y, Yoshimura M, Huang RN (2019) Wasabi versus red imported fire ants: preliminary test of repellency of microencapsulated allyl isothiocyanate against *Solenopsis invicta* (Hymenoptera: Formicidae) using bait traps in Taiwan. Appl Entomol Zool 54(2):193–196
Hashimoto Y, Sakamoto H, Asai H, Yasoshima M, Lin HM, Goka K (2020) The effect of fumigation with microencapsulated allyl isothiocyanate in a gas-barrier bag against *Solenopsis invicta* (Hymenoptera: Formicidae). Appl Entomol Zool 55(3):345–350
Hashimoto Y, Sakamoto H, Asai H, Yasoshima M, Lin HM, Goka K (2022) Effectiveness of allyl isothiocyanate microencapsulated in polyethylene as a repellent against *Solenopsis invicta* (Hymenoptera: Formicidae) infestation of corrugated cardboard boxes in field experiment. Appl Entomol Zool 57(3):257–262
Haxaire J, Tamisier JP, Bouguet JP (2006) *Vespa velutina* Lepeletier, 1836, une redoutable nouveauté pour la faune de France (Hym., Vespidae). Bull Soc Entomol Fr 111(2):194
Herrera C, Leza M, Martínez-López E (2020) Diversity of compounds in *Vespa* spp. venom and the epidemiology of its sting: a global appraisal. Arch Toxicol 94(11):3609–3627
Hifumi T, Fujimi S, Yamagishi T, Arai S, Sawabe K, Yamamoto A, Ato M, Shibayama K, Ginnaga A, Kiriu N, Kato H, Koido Y, Inoue J, Kishikawa M, Abe Y, Kawakita K, Hagiike M, Kuroda Y (2014) Clinical characteristics of redback spider bites. J Intensive Care 2(1):62
Hifumi T, Taki H, Yamamoto A, Ato M, Kiodo Y, Kuroda Y (2015) Update of antivenom supply for redback spider bites in Japan. J Intensive Care 3(1):7
Hiraoka T, Kobayashi M, Sadahiro S, Agui N (1996) Protein components and toxicity of venom gland-extract in the redback widow spiders, *Latrodectus hassettii*, collected in Osaka-City, Japan. Med Entomol Zool 47(3):273–280
Hsu PW, Lee CC, Hsu FC, Tseng SP, Shih CH, Tay JW, Hsiao YC, Yang CCS, Lin CC (2022) Invasion of the little fire ant *Wasmannia auropunctata* (Roger, 1863) (Hymenoptera: Formicidae) in Taiwan. BioInvasions Rec 11(4):864–875
Ibrahim O, Syed UM, Tomecki KJ (2017) Bedbugs: helping your patient through an infestation. Cleve Clin J Med 84(3):207–211
Infectious Agents Surveillance Report (IASR) (1997) Reference: symptoms and response to the redback spider bite. http://idsc.nih.go.jp/iasr/18/211/dj2112a.html. Accessed 1 Sept 2023. (in Japanese)
Inoue MN, Saito-Morooka F, Suzuki K, Nomura T, Hayasaka D, Kishimoto T, Sugimaru K, Sugiyama T, Goka K (2015) Ecological impacts on native ant and ground-dwelling animal communities through Argentine ant (*Linepithema humile*) (Hymenoptera: Formicidae) management in Japan. Appl Entomol Zool 50(3):331–339

Isbister GK (2006) Spider bite: a current approach to management. Aust Prescr 29:156–158

Isbister GK, Fan HW (2011) Spider bite. Lancet 378(9808):2039–2047

Isbister GK, Gray MR (2003) Latrodectism: a prospective cohort study of bites by formally identified redback spiders. Med J Aust 179(2):88–91

Japanese Society of Allergology (JSA) (2022) Anaphylaxis guidelines 2022. https://www.jsaweb.jp/uploads/files/Web_AnaGL_2023_0301.pdf. Accessed 1 Dec 2023

Jerome G, Richard S (2009) Bed bugs (*Cimex lectularrius*) and clinical consequences of their bites. JAMA 301(13):1358–1366

Kanazawa I, Nishikawa Y, Tomizawa O (1996) Introduction of the redback spiders to Japan. Nat Study 42:64–67. (in Japanese)

Keller L, Passera L, Suzzoni JP (1989) Queen execution in the Argentine ant *Iridomyrmex humilis*. Physiol Entomol 14(2):157–163

Khan HR, Rahman MM (2012) Morphology and biology of the bedbug, *Cimex hemipterus* (Hemiptera: Cimicidae) in the laboratory. Dhaka Univ J Biol Sci 21(2):125–130

Kim JK, Choi M, Moon TY (2006) Occurrence of *Vespa velutina* Lepeletier from Korea, and a revised key for Korean *Vespa* species (Hymenoptera: Vespidae). Entomol Res 36(2):112–115

Klein AM, Vaissière BE, Cane JH, Steffan-Dewenter I, Cunningham SA, Kremen C (2007) Importance of pollinators in changing landscapes for world crops. Proc Biol Sci 274(1608):303–313

Kolb A, Needham GR, Neyman KM, High WA (2009) Bedbugs. Dermatol Ther 22(4):347–352

Lambeau G, Gelb MH (2008) Biochemistry and physiology of mammalian secreted phospholipases A2. Annu Rev Biochem 77:495–520

Lee JXQ (2009) A note on *Vespa soror* (Hymenoptera: Vespidae) in Hong Kong. Hong Kong Entomol Bull 1(1):18–22

Leska A, Nowak A, Nowak I, Górczyńska A (2021) Effects of insecticides and microbiological contaminants on *Apis mellifera* health. Molecules 26(16):5080

Lewis DJ (1956) Chironomidae as a pest in the Northern Sudan. Acta Trop 13(2):142–158

Liebold K, Schliemann-Willers S, Wollina U (2003) Disseminated bullous eruption with systemic reaction caused by *Cimex lectularius*. J Eur Acad Dermatol Venereol 17(4):461–463

Lowe S, Browne M, Boudjelas S, De Poorter M (2000) 100 of the world's worst invasive alien species: a selection from the global invasive species database, vol 12. Invasive Species Specialist Group, Auckland. https://portals.iucn.org/library/sites/library/files/documents/2000-126.pdf. Accessed 1 Apr 2023

Lu G, Villalba M, Coscia MR, Hoffman DR, King TP (1993) Sequence analysis and antigenic cross-reactivity of a venom allergen, antigen 5, from hornets, wasps, and yellow jackets. J Immunol (Baltimore, MD, 1950) 150(7):2823–2830

Luo L, Kamau PM, Lai R (2022) Bioactive peptides and proteins from wasp venoms. Biomol Ther 12(4):527

Masetti M, Bruschi F (2007) Bedbug infestations recorded in Central Italy. Parasitol Int 56(1):81–83

Matsuka M, Sakai T (1989) Bee pollination in Japan with special reference to strawberry production in greenhouses. Bee World 70(2):55–61

Matsumura C, Yokoyama J, Washitani I (2004) Invasion status and potential ecological impacts of an invasive alien bumblebee, *Bombus terrestris* L. (Hymenoptera: Apidae) naturalized in Southern Hokkaido, Japan. Global Environ Res 8(1):51–66

Matsumura T, Mashiko R, Sato T, Itokawa K, Maekawa Y, Ogawa K, Isawa H, Yamamoto A, Mori S, Horita A, Ginnaga A, Miyatsu Y, Takahashi M, Taki H, Hifumi T, Sawabe K, Ato M (2018) Venom and antivenom of the redback spider (*Latrodectus hasseltii*) in Japan. Part I. Venom extraction, preparation, and laboratory testing. Jpn J Infect Dis 71(2):116–121

Matsuura M, Sakagami SF (1973) A bionomic sketch of the giant hornet, *Vespa mandarinia*, a serious pest for Japanese apiculture. J Fac Sci, Hokkaido Univ Ser VI Zool 19(1):125–162

Mattila HR, Kernen HG, Otis GW, Nguyen LT, Pham HD, Knight OM, Phan NT (2021) Giant hornet (*Vespa soror*) attacks trigger frenetic antipredator signalling in honeybee (*Apis cerana*) colonies. R Soc Open Sci 8(11):211215

McCrone JD (1964) Comparative lethality of several *Latrodectus* venoms. Toxicon 2(3):201–303

Miller DM, Polanco A (2023) Bed bug biology and behavior. https://www.vdacs.virginia.gov/pdf/bb-biology1.pdf. Accessed 25 Dec 2023

Ministry of the Environment (MOE), Japan (2015) Countermeasures against alien species in Japan. https://www.env.go.jp/nature/intro/. Accessed 1 Dec 2023

Ministry of the Environment (MOE), Japan (2022a) Basic information on RIFA. https://www.env.go.jp/nature/intro/2outline/attention/02_general/index_en.html. Accessed 25 Dec 2023. (in Japanese)

Ministry of the Environment (MOE), Japan (2022b) Cabinet decision on cabinet order designating fire ants as designated invasive alien species requiring urgent action and determining the date of enforcement of the revised invasive alien species act. https://www.env.go.jp/press/press_00829.html. Accessed 1 Apr 2023. (in Japanese)

Ministry of the Environment (MOE), Japan (2022c) Completion of fumigation and insecticidal treatment of containers in which the red imported fire ants were confirmed (follow-up report on the confirmation of the red imported fire ants dated November 2, 2022). https://www.env.go.jp/press/press_00930.html. Accessed 1 Apr 2023. (in Japanese)

Ministry of the Environment (MOE), Japan (2023) List of confirmed cases of the red imported fire ants in 2023. https://www.env.go.jp/nature/dobutsu/2023hiari.html. Accessed 1 Apr 2023. (in Japanese)

Minoshima YN, Yamane S, Ueno T (2015) An invasive alien hornet, *Vespa velutina nigrithorax* du Buysson (Hymenoptera, Vespidae), found in Kitakyushu, Kyushu Island: a first record of the species from mainland Japan. Jpn J Syst Entomol 21(2):259–261

Minton SA, Ori M (1977) Venom diseases (Arachnids). Acta Arachnol 27:299–309. (in Japanese)

Moreno M, Giralt E (2015) Three valuable peptides from bee and wasp venoms for therapeutic and biotechnological use: melittin, apamin and mastoparan. Toxins (Basel) 7(4):1126–1150

Mori S, Horita A, Ginnaga A, Miyatsu Y, Sawabe K, Matsumura T, Ato M, Yamamoto A, Shibayama K, Arai S, Yamagishi T, Takahashi M, Taki H, Hifumi T (2017) Venom and antivenom of the redback spider (*Latrodectus hasseltii*) in Japan. Part II. Experimental production of equine antivenom against the redback spider. Jpn J Infect Dis 70(6):635–641

Mumcuoglu KY (2008) A case of imported bedbug (*Cimex lectularius*) infestation in Israel. Isr Med Assoc J 10(5):388–389

Murakami G (1989) Chironomidae as the new inhalational allergen. Izumi 36:12–13. (in Japanese)

Nakajima T, Yasuhara T, Yoshida N, Takemoto Y, Shinonaga S, Kano R, Yoshida H (1983) The pattern analysis of biologically active amines in some Hymenopteran venoms by high performance liquid chromatography. Jpn J Sanit Zool 34(2):61–71

Nakajima N, Sakamoto Y, Goka K (2019) Rapid detection of the red fire ant *Solenopsis invicta* (Hymenoptera: Formicidae) by loop-mediated isothermal amplification. Appl Entomol Zool 54(3):319–322

Namin SM, Huang J, An J, Jung C (2023) Genetic variation and phylogenetic relationships of commercial populations of *Bombus ignitus* (Hymenoptera, Apidae) with wild populations in Eastern Asia. J Hymenopt Res 96:495–506

Nandi S, Aditya G, Chowdhury J, Das A, Saha GK (2014) Chironomid midges as allergens: evidence from two species from West Bengal, Kolkata, India. Indian J Med Res 139(6):921–926

National Institute for Environmental Studies (NIES) (2020) Latrodectus hasselti. Invasive species of Japan. https://www.nies.go.jp/biodiversity/invasive/DB/detail/70020e.html. Accessed 16 Feb 2024

National Institute for Environmental Studies (NIES) (2023) Lepisiota frauenfeldi. In: Invasive species of Japan. https://www.nies.go.jp/biodiversity/invasive/DB/detail/60580e.html. Accessed 1 Apr 2023

Natsuhara Y (1996) Ecology and clinical toxicology of the widow spiders, *Latrodectus mactans hasselti* and *L geometricus*. Seikatsu Eisei 40(1):13–21. (in Japanese)
Nimorakiotakis B, Winkel KD (2004) Spider bite – the redback spider and its relatives. Aust Fam Physician 33(3):153–157
Nishikawa Y, Kanazawa I (1996) Discovery of *Latrodectus* spiders from Japan and countermeasure to their venoms. Jpn J Environ Entomol Zool 7(4):214–223. (in Japanese)
Ono H (1995) Records of *Latrodectus geometricus* (Araneae: Theridiidae) from Japan. Acta Arachnol 44(2):167–170
Ono H (1996) Scientific and Japanese names of one species of widow spider from the Yaeyama Islands and Taiwan. Acta Arachnol 45(1):108. (in Japanese)
Ono H, Ogata K (2018) Spiders of Japan: their natural history and diversity. Tokai University Press, Kanagawa, 730 pp. (in Japanese)
Ono M, Terabe H, Hori H, Sasaki M (2003) Components of giant hornet alarm pheromone. Nature 424(6949):637–638
Ori M, Shinkai E, Ikeda H (1996) Introduction of widow spiders into Japan. Med Entomol Zool 47(2):111–119. (in Japanese with English summary)
Parola P, Izri A (2020) Bedbugs. N Engl J Med 382(23):2230–2237
Pederson JS, Krieger MJB, Vogel V, Giraud T, Keller L (2006) Native supercolonies of unrelated individuals in the Argentine ant. Evolution 60(4):782–791
Pitts JP, Camacho GP, Gotzek D, Mchugh JV, Ross KG (2018) Revision of the fire ants of the *Solenopsis saevissima* species-group (Hymenoptera: Formicidae). Proc Entomol Soc Wash 120(2):308–411
Raffiudin R, Crozier RH (2007) Phylogenetic analysis of honey bee behavioral evolution. Mol Phylogenet Evol 43(2):543–552
Rau L (1931) The nesting habits of *Polistes rubiginosis*, with special reference to pleometrosis in this and other species of *Polistes* wasps. Psyche 38(4):129–144
Ross KG, Matthews RW (1991) The social biology of wasps, 1st edn. Cornell University Press, Ithaca, pp 7–32
Sakamoto H, Goka K (2021) Acute toxicity of typical ant control agents to the red imported fire ant, *Solenopsis invicta* (Hymenoptera: Formicidae). Appl Entomol Zool 56(2):217–224
Sakamoto H, Goka K (2023) Efficiency of ant-control agents in colony-level oral toxicity tests using *Tetramorium tsushimae* (Hymenoptera: Formicidae) for post-establishment control of the red imported fire ant, *Solenopsis invicta* (Hymenoptera: Formicidae). Appl Entomol Zool 58(1):25–33
Sakamoto Y, Kumagai NH, Goka K (2017) Declaration of local chemical eradication of the Argentine ant: Bayesian estimation with a multinomial-mixture model. Sci Rep 7(1):3389
Sakamoto Y, Hayashi TI, Inoue MN, Ohnishi H, Kishimoto T, Goka K (2019) Effects of fipronil on non-target ants and other invertebrates in a program for eradication of the Argentine ant, *Linepithema humile*. Sociobiology 66(2):227–238
Sansom JE, Reynolds NJ, Peachey RD (1992) Delayed reaction to bed bug bites. Arch Dermatol 128(2):272–273
Sasa M (1985) Chironomid midges as the caused of bronchial asthma. Kankyo Eisei 32(1):8–14. (in Japanese)
Sasa M, Kikuchi M (1995) Chironomidae (Diptera) of Japan. University of Tokyo Press, Tokyo, 333 pp
Sasaki T, Amagai M, Ishikado Y, Goka K (2019) Insecticidal activities of the 'One-Push Eradication System' against Argentine ants concealed in modified import-containers. Pestology 34(1):15–18. (in Japanese with English summery)
Scarupa MD, Economides A (2006) A bedbug bites masquerading as urticaria. J Allergy Clin Immunol 117(6):1508–1509
Shimizu H, Kanazawa I, Nishikawa Y (2014) A report on distributions of widow spiders, the genus *Latrodectus* (Arachnida), and dispersal of redback spider, *L. hasselti*, in Japan. Bull Osaka Mus Nat Hist 68:41–51. (in Japanese with English summary)

Silva JC, Neto LM, Neves RC, Gonçalves JC, Trentini MM, Mucury-Filho R, Smidt KS, Fensterseifer IC, Silva ON, Lima LD et al (2017) Evaluation of the antimicrobial activity of the mastoparan Polybia-MPII isolated from venom of the social wasp *Pseudopolybia vespiceps testacea* (Vespidae, Hymenoptera). Int J Antimicrob Agents 49(2):167–175

Smallheer BA (2013) Bee and wasp stings: reactions and anaphylaxis. Crit Care Nurs Clin North Am 25(2):151–164

Souza BM, Mendes MA, Santos LD, Marques MR, César LMM, Almeida RNA, Pagnocca FC, Konno K, Palma MS (2005) Structural and functional characterization of two novel peptide toxins isolated from the venom of the social wasp *Polybia paulista*. Peptides 26(11):2157–2164

Spradberry JP (1986) Polygyny in the Vespinae with special reference to the hornet *Vespa affinis picea* du Buysson (Hymenoptera Vespidae) in New Guinea. Ital J Zool 20(1):101–118

Stucki A, Ludwig R (2008) Images in clinical medicine: bed bugs. N Engl J Med 359(10):1047

Tatsuki S (2014) The Argentine ant: the strongest invasive animal in history. University of Tokyo Press, Tokyo, 331 pp

Ter Poorten MC, Prose NS (2005) The return of the common bed bug. Pediatr Dermatol 22(3):183–187

Tokyo Metropolitan Government Bureau of Public Health (2014) The redback spider and the white widow spider. https://translation2.j-server.com/LUCHOKENIR/ns/tl.cgi/https://www.hokeniryo.metro.tokyo.lg.jp/kankyo/eisei/yomimono/nezukon/seaka_haiiro.html?SLANG=ja&TLANG=en&XMODE=0&XCHARSET=utf-8&XJSID=0. Accessed 16 Feb 2024

Tschinkel WR (2006) The fire ants. Belkanap Press of Harvard University Press, Cambridge, MA, 752 pp

Tu AT (1996) Redback spider. Chemistry 51(2):101–104. (in Japanese)

Ueda S, Murakami K (2022) A record of *Lepisiota frauenfeldi* (Mayr, 1955) (Hymenoptera, Formicidae) in Osaka Prefecture, Japan. Jpn J Entomol New Ser 25(1):33–36. (in Japanese)

Usinger RL (1966) Monograph of Cimicidae (Hemiptera-Heteroptera). Entomological Society of America. The Thomas Say Foundation, vol VII. College Park, 585 pp

Van Vaerenbergh M, Debyser G, Devreese B, de Graaf DC (2014) Exploring the hidden honeybee (*Apis mellifera*) venom proteome by integrating a combinatorial peptide ligand library approach with FTMS. J Proteome 99:169–178

Vink CJ, Derraik JGB, Phillips CB, Sirvid PJ (2010) The invasive Australian redback spider, *Latrodectus hasseltii* Thorell 1870 (Araneae: Theridiidae): current and potential distributions, and likely impacts. Biol Invasions 13(4):1003–1019

Wang Z, Moshman L, Kraus EC, Wilson BE, Acharya N, Diaz R (2016) A review of the tawny crazy ant, *Nylanderia fulva*, an emergent ant invader in the southern United States: is biological control a feasible management option? Insects 7(4):77

Wetterer JK, Wetterer AL (2006) A disjunct Argentine ant metacolony in Macaronesia and Southwestern Europe. Biol Invasions 8(5):1123–1129

Weyda F, Kodrik D (2021) New functionally ultrastructural details of the honey bee stinger tip: serrated edge and pitted surface. J Apic Res 60(5):875–878

Whyte R, Anderson G (2017) A field guide to spiders of Australia. CSIRO Publishing, Clayton South, 464 pp

Wilson TM, Takahashi J, Spichiger SE, Kim I, Van Westendorp P (2020) First reports of *Vespa mandarinia* (Hymenoptera: Vespidae) in North America represent two separate maternal lineages in Washington state, United States, and British Columbia, Canada. Ann Entomol Soc Am 113(6):468–472

World Spider Catalog (WSC) (2023) Currently valid spider genera and species (2023-09-21). NMBE. https://wsc.nmbe.ch/statistics/. Accessed 1 Sept 2023

Yoshida H, Yoshida M, Iwagami Y, Taki I, Sonoi T, Uchino K, Tanaka T (2003) The population dynamics of *Latrodectus hasseltii* (Araneae: Theridiidae) after control in Osaka Prefecture. Med Entomol Zool 54(4):361–366

Zorrilla-Vaca A, Silva-Medina MM, Escandon-Vargas K (2015) Bedbugs, *Cimex* spp.: their current world resurgence and healthcare impact. Asian Pacific J Trop Dis 5(5):342–352

Chapter 6
Clinical Findings of Arthropod Bite-Related Cutaneous Disorders Without Infections

Masaru Natsuaki

Abstract In this chapter, cutaneous disorders caused by biting arthropods (e.g., centipedes, spiders), stinging arthropods (e.g., bees, ants), and blood-sucking arthropods (e.g., mosquitoes, black flies, fleas, bed bugs, ticks, and mites), contact dermatitis caused by venomous caterpillars and blister beetles, and cutaneous symptoms caused by arthropods infesting human skin (e.g., pediculosis, scabies) are described. In most instances, irritant or allergic inflammatory reactions due to exposure to toxic or salivary gland substances from arthropods can be clinically problematic. In an immediate reaction, an itchy sensation, wheals, or erythema develops approximately 15 min after exposure to the antigen. In severe cases, blood pressure can decrease, potentially leading to anaphylactic shock. In a delayed-type reaction, erythema and papules with itching develop approximately 24–72 h after exposure to the antigen. The diagnosis, treatment, and countermeasures of skin diseases caused by these noxious arthropods are also outlined.

Keywords Allergic reactions · Biting arthropods · Blister beetles · Blood-sucking arthropods · Lice · Salivary gland substances · Scabies · Stinging arthropods · Venomous caterpillars · Venoms

6.1 Introduction

Arthropods cause cutaneous disorders by stinging, biting, other skin contact, sucking blood, and skin infestation. Excluding skin infestations, cutaneous disorders can arise because of (1) physical damage due to stinging and biting, (2) inflammatory reactions due to chemical irritation by toxic substances that are injected into or come into contact with the skin, and (3) allergic inflammatory reactions due to

---

M. Natsuaki (✉)
School of Medicine, Department of Dermatology, Hyogo Medical University, Nishinomiya, Japan
e-mail: natsuaki@hyo-med.ac.jp

K. Sawabe et al. (eds.), *Medical Entomology in Asia*, Entomology Monographs,
https://doi.org/10.1007/978-981-97-4406-0_6

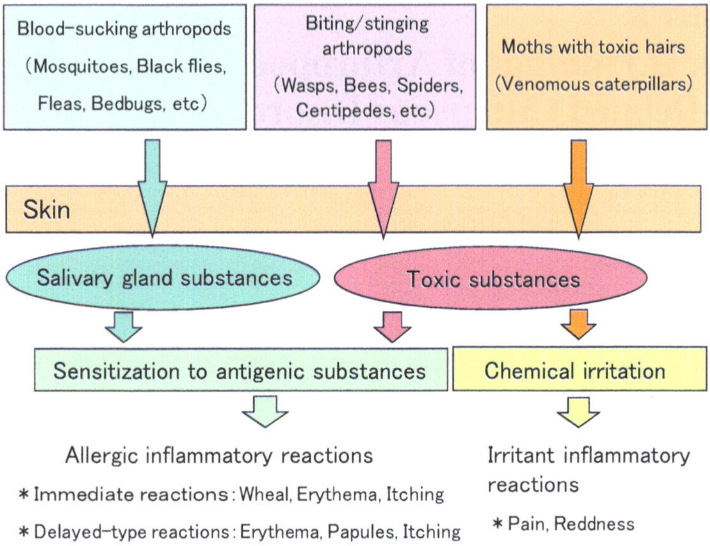

Fig. 6.1 Mechanism of inflammation caused by arthropods

exposure to toxic or salivary gland substances injected during stinging, biting, blood-sucking, or skin contact. In those patients, irritant and allergic inflammatory reactions can be clinically problematic (Fig. 6.1).

For example, bee venom and centipede venom contain pain-producing substances and chemical mediators that cause pain and redness. Enzymes that contribute to tissue damage also serve as antigens, and allergic reactions are triggered if sensitization occurs. Salivary gland substances injected by blood-sucking insects, such as mosquitoes and black flies, rarely irritate the skin but often present as dermatitis, which manifests as an allergic reaction. Arthropods that infest the skin, such as lice, induce inflammatory reactions in the skin by way of allergic reactions to salivary gland substances, body components, and excretory substances.

In addition to causing cutaneous symptoms by the above mechanisms, some arthropods can cause other symptoms because their venom contains neurotoxins. Venomous spiders such as *Latrodectus hasseltii* and some ticks (although none of those found in Japan) possess neurotoxins that can induce sweating, palpitations, increased blood pressure, nerve palsy, and other symptoms. Thus, the pathogenesis of the symptoms depends on the responsible arthropod, and often symptoms are caused by multiple mechanisms.

6.2 Symptoms of Allergic Reactions

Allergic reactions do not occur unless there is sensitization to a substance that serves as an antigen, and an individual's symptoms depend on their sensitization stage or condition. The production of immunoglobulin E antibodies to an antigenic substance causes an immediate reaction, whereas the production of sensitized T lymphocytes causes a delayed-type reaction.

In an immediate reaction, an itchy sensation, wheals, or erythema develops approximately 15 min after exposure to the antigen. In mild cases, these symptoms resolve in 1 to 2 h. However, systemic symptoms can develop and may include respiratory symptoms such as dyspnea and wheezing, gastrointestinal symptoms such as abdominal pain and diarrhea, and cardiovascular symptoms such as palpitations and cyanosis. In particularly severe cases, blood pressure can decrease, potentially leading to anaphylactic shock.

In a delayed-type reaction, erythema and papules develop approximately 24–72 h after exposure to the antigen and are sometimes accompanied by edema. Itching is often severe, but most symptoms resolve within a few days to 2 weeks.

Symptoms vary depending on the causative arthropod and the frequency of stinging, biting, or blood-sucking. Usually, patients experience delayed-type reactions first, which are gradually replaced by immediate reactions upon repeated stimulation by antigens; however, both types of reactions may occur in parallel in the same individual.

6.3 Cutaneous Symptoms Caused by Biting Arthropods

6.3.1 Centipede Bites

Centipedes inject venom into the skin by biting with their maxillipeds (fangs). The most common species of centipede in Japan, *Scolopendra subspinipes mutilans* (Fig. 6.2), is nocturnal and lives in and around gardens and fields and under stones in mountain forests. This type of centipede often enters rooms and bites humans on summer nights, causing intense pain as well as erythema and swelling (Fig. 6.3). In

Fig. 6.2 *Scolopendra subspinipes mutilans*

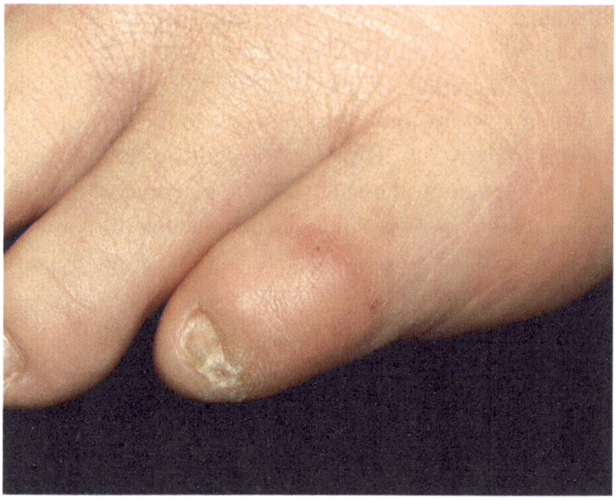

Fig. 6.3 Clinical appearance of centipede bite

Fig. 6.4 *Cheiracanthium japonicum*

rare cases, anaphylactic shock may occur. There is immunological cross-reactivity between bee venom and centipede venom.

6.3.2 Spider Bites (Arachnidism)

6.3.2.1 *Cheiracanthium japonicum* Bites

The spider *Cheiracanthium japonicum* (Fig. 6.4) builds its nest in the grasslands of mountainous areas by curling leaves of pampas grass and other plants, and humans can be bitten on their fingers if they accidentally touch the nest. In some cases, males of *C. japonicum* may wander from their nest, enter a house, and bite humans.

Fig. 6.5 *Latrodectus hasseltii* redback spider)

Cheiracanthium japonicum bites cause intense pain, but inflammatory symptoms are rarely observed.

6.3.2.2 *Latrodectus hasseltii* Bites

Latrodectus hasseltii (Fig. 6.5) is an exotic venomous spider native to Australia that in recent years has spread to almost all areas of Japan (Matsumura et al. 2018). It is often found in gutters and between cement blocks in new residential areas, parks, and parking lots of factories. *Latrodectus hasseltii* venom contains a potent neurotoxin called α-latrotoxin, and humans bitten by this spider experience severe pain, palpitations, sweating, and other symptoms (Natsuaki 2015). Although slight erythema may appear at the site of the bite (Fig. 6.6), the bite mark is almost invisible to the naked eye because the venom fang is small. Pain may also occur at sites other than the bite site; for instance, a patient bitten on the foot by *L. hasseltii* may have abdominal pain. In severe cases, anti-venom therapy is needed. *L. geometricus*, which belongs to the same genus as *L. hasseltii*, has also been found in various parts of Japan, and its bites result in similar symptoms.

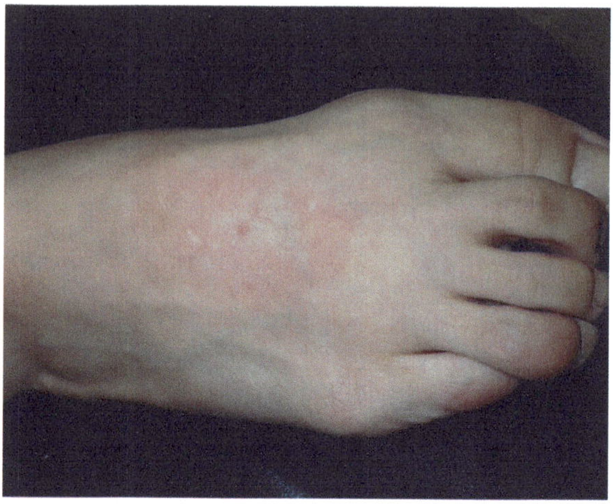

Fig. 6.6 Clinical appearance of redback spider bites

Fig. 6.7 Representative harmful bees in Japan (Upper specimens: *Vespa analis*, *V. mandarinia*, *V. simillima*, Lower specimens: *Vespula flaviceps*, *Polistes jokahamae*, *Apis cerana*)

6.4 Cutaneous Symptoms Caused by Stinging Arthropods

6.4.1 Wasp and Bee Stings

The types of wasps and bees primarily responsible for stings in Japan are hornets, paper wasps, and honeybees (Fig. 6.7). Hornets and paper wasps build their nests from summer to fall, making them more aggressive during this period. Accordingly,

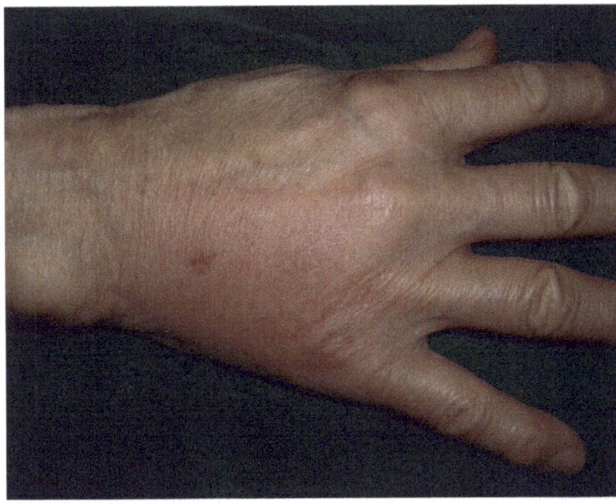

Fig. 6.8 Clinical appearance of wasp stings

humans are often stung by them while hiking, gathering mushrooms, or working in forests or gardens. The wasp *Cephalonomia gallicola*, which lives inside houses, occasionally stings humans, but the number of such cases has declined sharply in recent years thanks to improvements in indoor environments.

In humans stung by wasps and bees, the venom component emanating from the stinger causes pain and redness (Fig. 6.8). These symptoms usually abate within a few hours to a day; however, in humans who have previously been stung, a sting may trigger an allergic reaction because of sensitization to the venom component. In cases in which an immediate reaction takes place, symptoms such as generalized wheals, dyspnea, abdominal pain, and feeling unwell may occur immediately after being bitten, and death can result from anaphylactic shock. In delayed-type reactions, erythema and swelling occur 1 to 2 days after being stung.

6.4.2 Ant Stings

The ant *Brachyponera chinensis* (Fig. 6.9) is widely distributed in Japan and is known to sting people. It sometimes nests in yards and parks and may also enter houses. *Brachyponera chinensis* has a venomous stinger on its tail end, and humans stung by this ant feel mild pain. Depending on a person's predisposition, the sting site may become red and swollen because of an allergic reaction. In rare cases, anaphylactic symptoms may occur.

Solenopsis invicta (Fig. 6.10) is an exotic species native to South America, but since 2017, it has been frequently found in coastal container yards in Japan after being imported with cargo. At present, domestic cases of *S. invicta* stings are few. Humans stung by *S. invicta* experience pain with a burning sensation and erythema due to the alkaloid venom solenopsin, and pustules form the following day (Fig. 6.11). These cutaneous symptoms resolve within a week or two. However,

Fig. 6.9 *Brachyponera chinensis*

Fig. 6.10 *Solenopsis invicta* (fire ant)

Fig. 6.11 Clinical appearance of fire ant stings

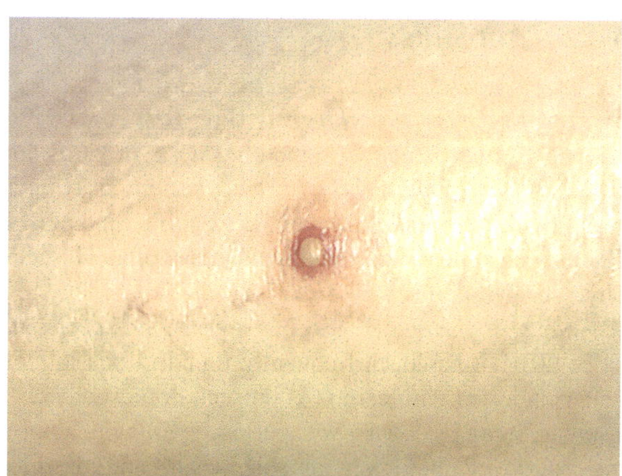

Fig. 6.12 *Agriosphodrus dohrni*

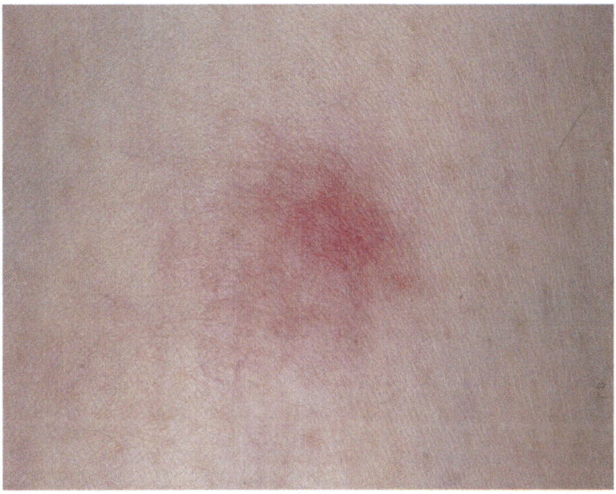

Fig. 6.13 Clinical appearance of triatomine bug stings

sensitization to protein venom may lead to anaphylactic symptoms. Care should be exercised because there is immunological cross-reactivity between *S. invicta* venom and hornet venom (Steigelman and Freeman 2013).

In addition to the above, some other ants, including *Formica yessensis*, bite with their mouthparts, but only a few cases of biting by such ants have occurred in Japan.

6.4.3 Triatomine Bug Stings

Agriosphodrus dohrni (Fig. 6.12), a triatomine bug that stings with its proboscis, is widely distributed in western Japan, and is found around houses and parks. *Agriosphodrus dohrni* inhabits the trunks of cherry trees and zelkova trees, and humans can be stung if they inadvertently grab the bug. Humans stung by *A. dohrni* immediately feel pain and develop erythema and wheals, but these symptoms resolve in 1 to 2 h. In some cases, an allergic reaction follows, which results in itchy erythema (Fig. 6.13) that resolves in 1 to 2 weeks.

Fig. 6.14 *Liocheles australasiae*

Fig. 6.15 *Isometrus maculatus*

6.4.4 Scorpion Stings

In Japan, the scorpions *Liocheles australasiae* (Fig. 6.14) and *Isometrus maculatus* (Fig. 6.15) are found in the Yaeyama Islands, southeast area of Okinawa Prefecture, but they do not often come into contact with humans, and sting cases are rare. Humans stung by the venomous stinger of a scorpion's tail may experience mild pain and erythema, but these symptoms are not severe because of the low toxicity of the venom. It should be noted that on rare occasions, highly venomous species of scorpions are brought into the country with imported cargo.

6.5 Cutaneous Symptoms Caused by Blood-Sucking Arthropods

6.5.1 Diptera Bites

Cutaneous symptoms caused by Diptera bites stem essentially from an immediate or delayed allergic reaction to salivary gland components injected into the skin during blood sucking. The manifestation of cutaneous symptoms depends on the individual's predisposition and the frequency of bites.

6.5.1.1 Mosquito Bites

Mosquitoes live in a range of environments, including indoors and outdoors, and they suck blood from animals and humans. The most common blood-sucking mosquitoes in Japan are *Aedes albopictus* (Fig. 6.16) and *Culex pipiens* (Fig. 6.17). The skin reactions to mosquito bites can be divided into five stages: (1) no reaction (in people with no sensitization to salivary gland substances); (2) delayed-type reaction (Fig. 6.18); (3) immediate reaction (Fig. 6.19) plus delayed-type reaction; (4) immediate reaction; and (5) no response (in people with desensitization to salivary gland substances). As they grow older, humans are believed to transition from stage one to five after being repeatedly bitten by mosquitoes.

Hypersensitivity to mosquito bites refers to a condition in which a mosquito bite causes fever, as well as blisters and blood blisters at the site of blood sucking, followed by ulceration and ensuing scarring (Fig. 6.20). This is a unique condition attributable to chronic active Epstein–Barr virus infection. During the course of the disease, malignant lymphoma and natural killer cell leukemia may occur, leading to a poor prognosis.

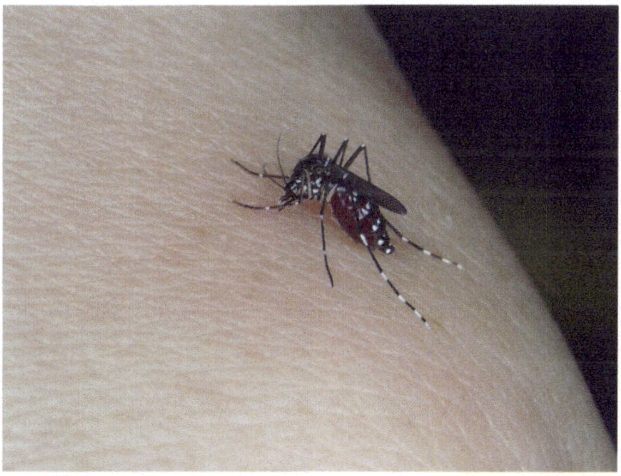

Fig. 6.16 *Aedes albopictus*

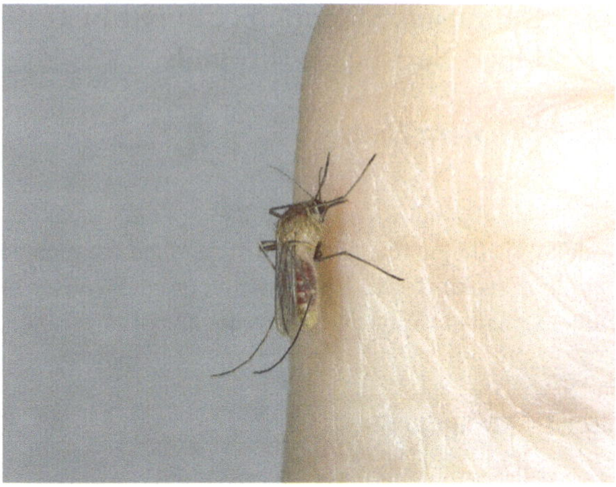

Fig. 6.17 *Culex pipiens*

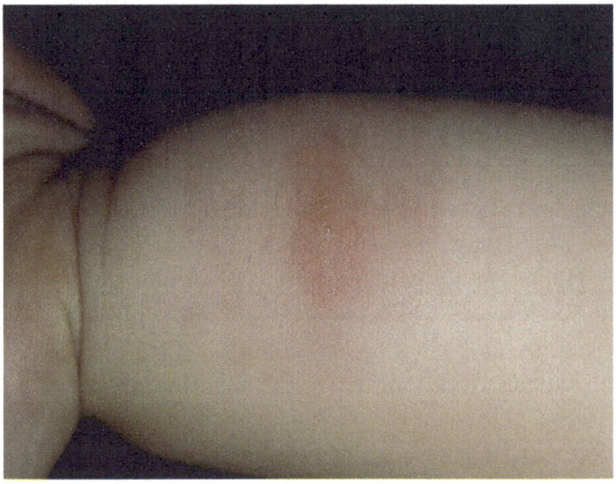

Fig. 6.18 Clinical appearance of delayed-type reaction in mosquito bites

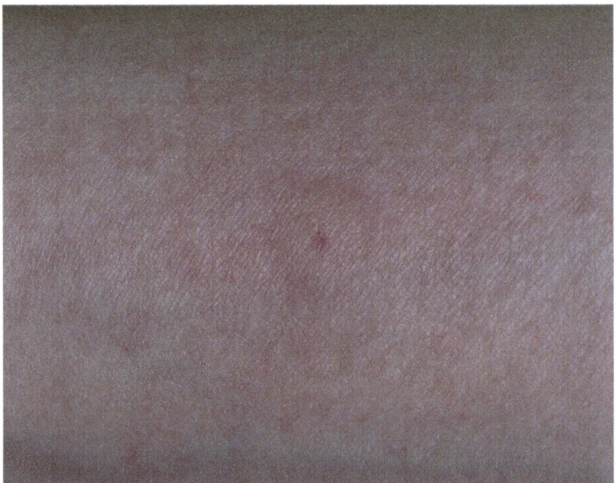

Fig. 6.19 Clinical appearance of immediate reaction in mosquito bites

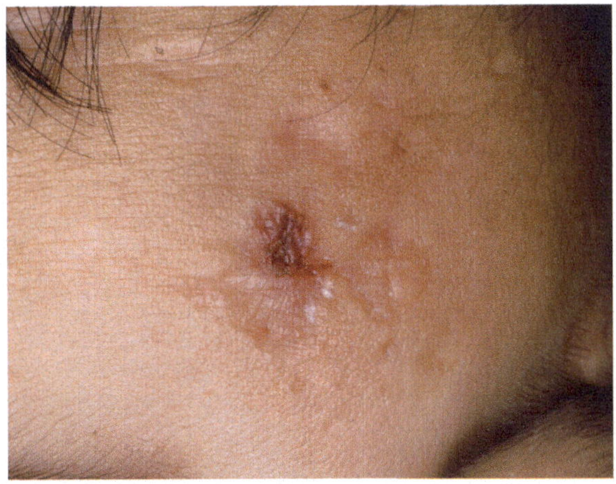

Fig. 6.20 Clinical appearance of hypersensitivity to mosquito bites

Fig. 6.21 *Prosimulium jazonicum*

6.5.1.2 Black Fly Bites

Black flies live along mountain streams and on plateaus and feed mainly in the morning and evening by sucking blood from the exposed skin of human necks and limbs. *Simulium* spp. (*Simulium japonicum, S. arakawae, S. oitanum,* etc.) and *Prosimulium jazonicum* (Fig. 6.21) are responsible for the majority of stings by black flies in Japan. When a black fly sucks blood, humans seldom experience subjective symptoms, and only small bleeding spots are observed (Fig. 6.22). Nevertheless, the next day, redness and swelling appeared, along with intense itching. These symptoms usually resolve within a week or two, but in some cases, severe itching persists and can lead to chronic prurigo (Fig. 6.23).

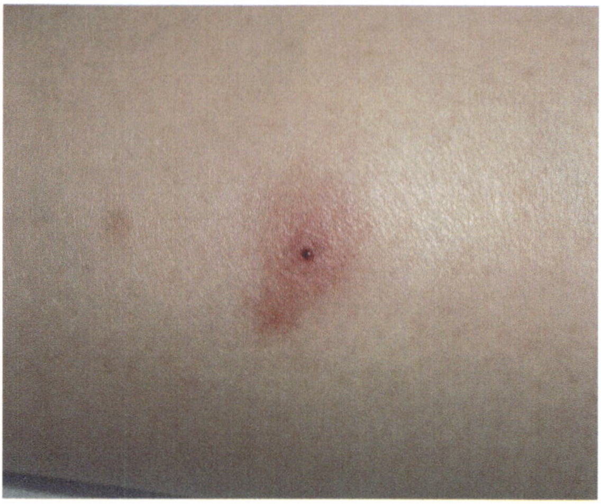

Fig. 6.22 Clinical appearance of blackfly bites

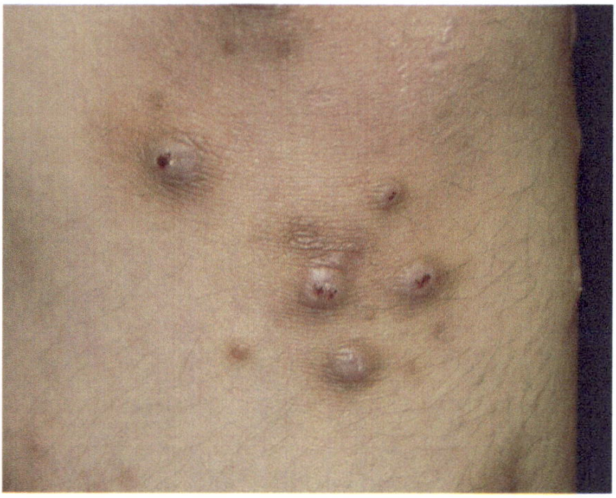

Fig. 6.23 Clinical appearance of chronic prurigo induced by blackfly bites

6.5.1.3 Biting Midge Bites

Among the various types of biting midges, those living in mountain forests (*Culicoides sinanoensis, C. nunomemoguri*) or on the seashore (*C. circumscriptus, Leptoconopus nipponensis*) can harm humans by sucking blood. When biting midges suck blood (Fig. 6.24), humans often feel a mild tingling sensation. Biting midges suck blood mainly from the exposed skin of the neck and limbs, but they can also suck blood by burrowing into the hair of the head or through clothing. Erythematous papules with itching developed at the site of blood sucking (Fig. 6.25). In many cases, biting midges suck blood in swarms; hence, skin rashes at multiple sites are commonly seen in affected individuals.

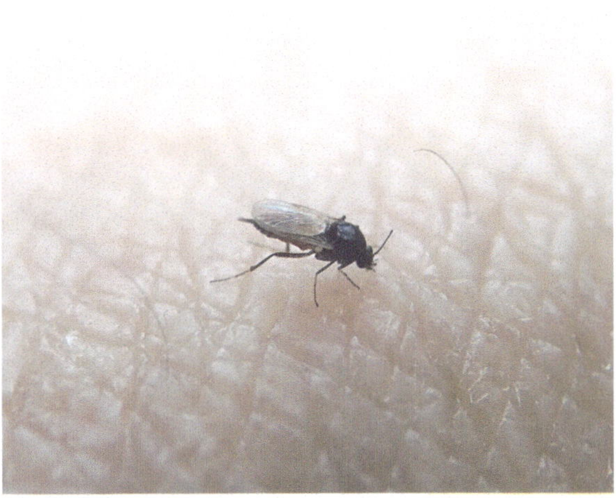

Fig. 6.24 Blood-sucking biting midge (*Leptoconopus nipponensis*)

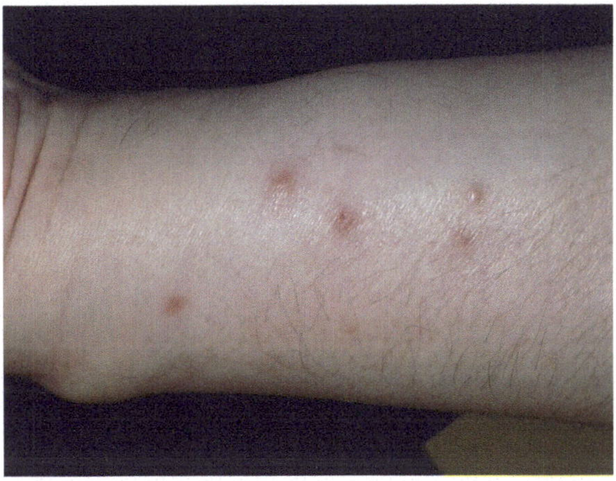

Fig. 6.25 Clinical appearance of biting midge bites

6.5.1.4 Horsefly Bites

Horseflies are found mainly along mountain streams, on plateaus, and in pastures, and they bite people engaged in outdoor recreation or working in mountainous areas. Among other species, *Tabanus iyoensis* (Fig. 6.26) has affected a large number of people in Japan. Horsefly bites cause severe pain because the flies injure the skin when sucking blood. After a day or two, affected individuals have itchy erythema and swelling (Fig. 6.27).

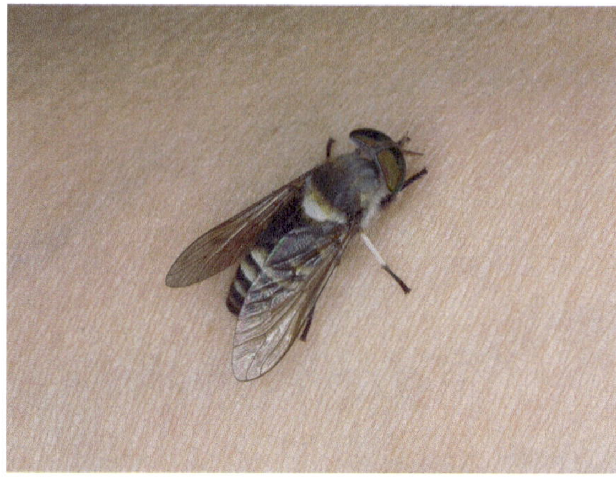

Fig. 6.26 Blood-sucking *Tabanus iyoensis*

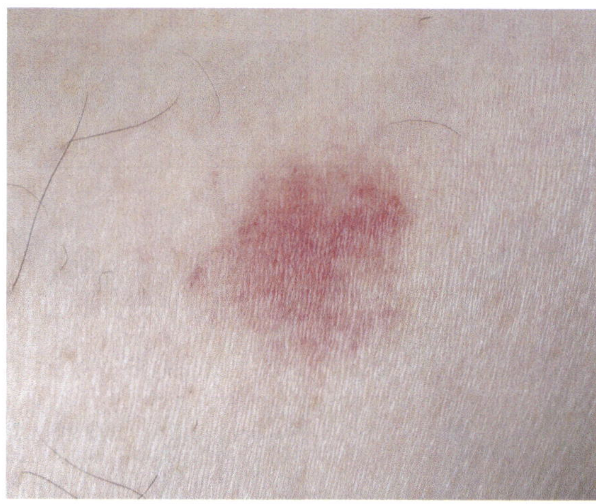

Fig. 6.27 Clinical appearance of horse fly bites

6.5.2 Flea Bites

In Japan, flea bites are commonly caused by blood-sucking *Ctenocephalides felis* (Fig. 6.28), a parasite of stray cats, which live mainly around people's houses. *Ctenocephalides felis* is primarily parasitic on cats, but this type of flea also sucks blood from dogs and humans. *Ctenocephalides felis* also lives on soil in gardens and parks. When humans pass by, the flea jumps onto them and sucks blood from their foot or lower leg. If cats are kept inside a house, *C. felis* on infested animals may bite humans indoors. Flea bites cause a skin rash characterized by solitary, intensely itchy erythematous papules distributed around the lower legs, and blisters form in many cases (Fig. 6.29).

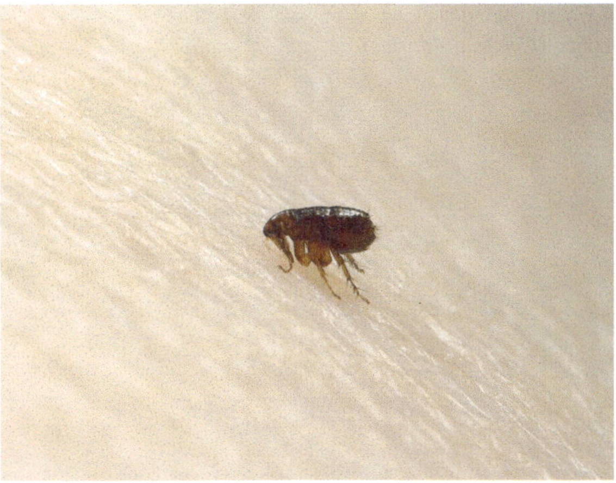

Fig. 6.28 Blood-sucking *Ctenocephalides felis*

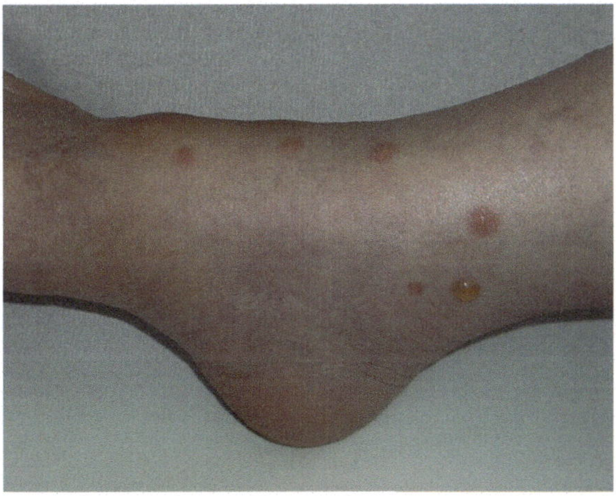

Fig. 6.29 Clinical appearance of flea bites

*6.5.3 Bedbug (*Cimex lectularius*) Bites*

Bedbugs (Fig. 6.30) are blood-sucking insects that live in houses. In recent years, they have been found primarily in lodging facilities and therefore often affect travelers. Increasingly, bedbugs that infest bags and clothing during travel are transported back to travelers' homes, where they breed. Because humans do not develop skin rashes when they are first bitten by bedbugs, they are often unaware that they have been bitten while traveling.

Bedbugs suck blood from exposed skin while humans sleep at night, and skin rashes can therefore be seen on the neck, hands, forearms, lower legs, feet, and other exposed skin areas. Clinically, itchy erythematous papules or erythema is observed at the site of blood sucking. Bedbug bites are characterized by multiple

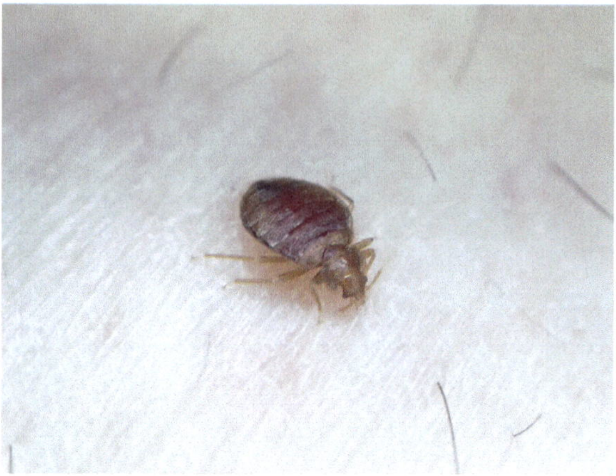

Fig. 6.30 Blood-sucking bedbug (*Cimex lectularius*)

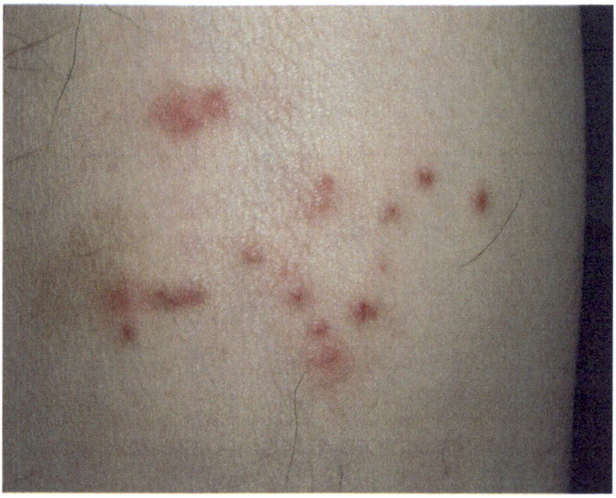

Fig. 6.31 Clinical appearance of bedbug bites

erythematous papules over a small area because the bugs insert and remove their mouthparts one or more times during blood-sucking (Fig. 6.31).

6.5.4 Tick/Mite Bites

6.5.4.1 Tropical Rat Mite *(Ornithonyssus bacoti)* Bites

Tropical rat mites (Fig. 6.32), which are parasites of house mice, invade the interior of houses from mouse nests and suck blood from humans. Skin rashes caused by tropical rat mite bites consist of solitarily distributed red papules that are mainly seen on areas covered by clothing, such as the armpits, lower abdomen, and thighs (Fig. 6.33). Because tropical rat mites suck blood primarily while humans sleep, it

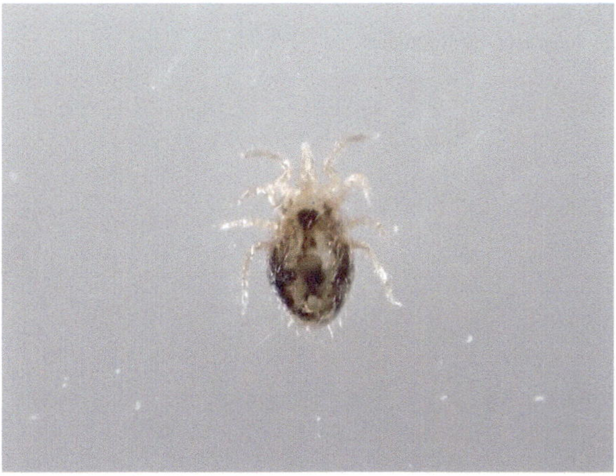

Fig. 6.32 Tropical rat mite (*Ornithonyssus bacoti*)

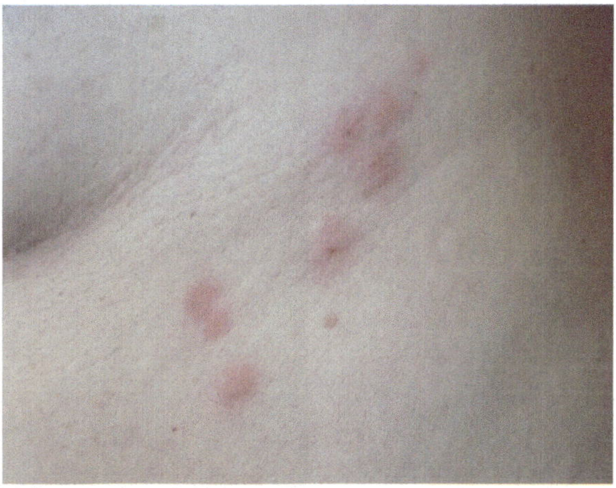

Fig. 6.33 Clinical appearance of tropical rat mite bites

is often impossible to determine what is responsible for the rash. Similar symptoms can occur when mites that parasitize wild birds enter a room from the birds' nest and suck blood from humans.

Cheyletiella and *Pyemotes tritici* can also occasionally bite humans and cause dermatitis, but they are not blood-sucking mites.

6.5.4.2 Tick Bites

Ticks are found mainly in mountain forests, surrounding grasslands, and riverbeds. They parasitize and suck blood from wild animals such as deer and wild boars and often bite humans engaged in outdoor recreation, including hiking and camping, or in agricultural and forestry work in such environments. Japan is home to approximately 40 species of ticks, 10–20 of which are responsible for tick bites in humans.

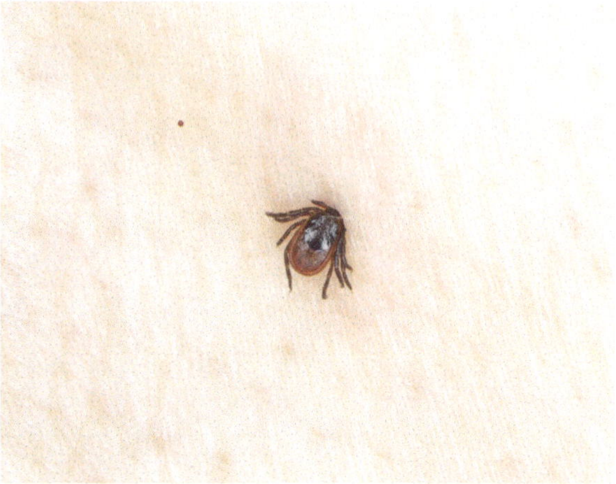

Fig. 6.34 *Ixodes persulcatus* (Adult female)

Fig. 6.35 *Amblyomma testudinarium* (Adult female)

Bites attributable to *Ixodes persulcatus* (Fig. 6.34) and *I. ovatus* are common in northern Japan, whereas bites attributable to *Amblyomma testudinarium* (Fig. 6.35) and *Haemaphysalis longicornis* are common in western Japan (Natsuaki 2021).

When ticks are sucking blood, humans have few subjective symptoms and often do not notice them, particularly when the larvae and nymphs of ticks are responsible because they are so small. After the tick starts to suck blood, its abdomen gradually enlarges over the course of several days, making it easier to find (Fig. 6.36). Often, no skin rash occurs at the site of tick attachment, but occasionally erythema and blisters develop. Erythema larger than 5 cm in diameter associated with the bite of *Amblyomma testudinarium* is called tick-associated rash illness (TARI) and is thought to result from an allergic reaction to salivary gland substances. TARI (Fig. 6.37) should be distinguished from erythema migrans, which is associated with Lyme disease.

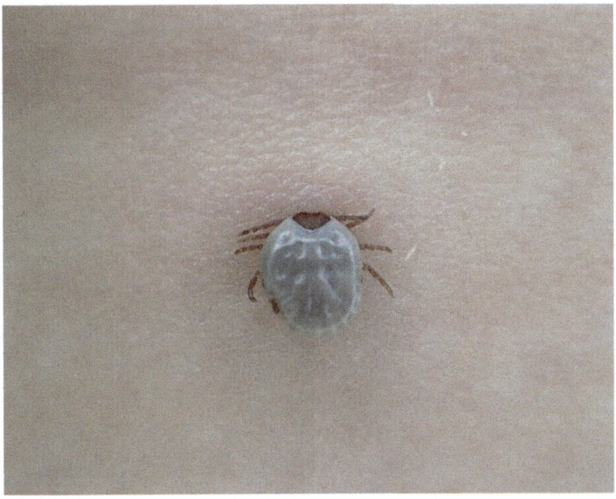

Fig. 6.36 Tick bites (blood-sucking nymph of *Amblyomma testudinarium*)

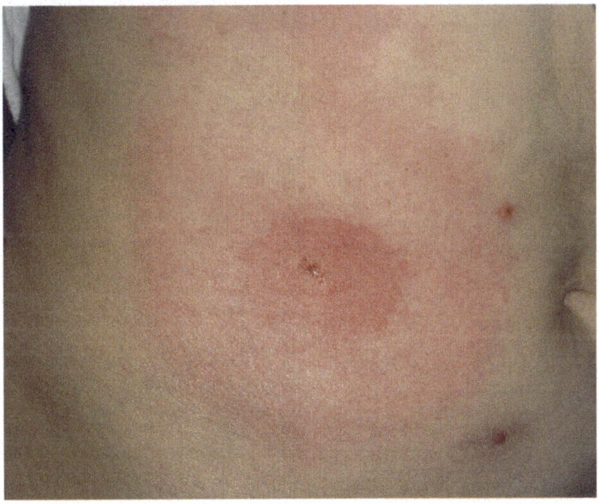

Fig. 6.37 Tick-associated rash illness

In addition to Lyme disease, ticks can transmit other infectious diseases, such as severe fever with thrombocytopenia syndrome and Japanese spotted fever, and due attention should be given to these infections.

6.5.4.3 Trombiculid Mite, Chigger Bites

Trombiculid mites (Fig. 6.38) are found in mountainous areas and riverbeds. Trombiculid mite larvae, or chiggers, can adhere to human skin and cause dermatitis as an allergic reaction (Natsuaki and Takada 2013). Clinically, isolated erythematous papules are observed, but it can be difficult to distinguish trombiculid mite bites from other insect bites because trombiculid larvae are as small as 0.3 mm and

Fig. 6.38 Trombiculid mites (*Leptotrombidium scutellare*)

Fig. 6.39 Larvae of *Euproctis pseudoconspersa*

cannot be visually identified. Trombiculid mites carrying *Rickettsiae* mediate scrub typhus.

6.6 Caterpillar Dermatitis

Caterpillar dermatitis is a form of dermatitis triggered by contact with the larvae of moths with venomous hairs, such as the larvae of *Arna pseudoconspersa* (*Lymantriidae*) (Fig. 6.39) and *Dendrolimus spectabilis* (*Lasiocampidae*) (Fig. 6.40). These larvae have fine venomous hairs that readily fall out and pierce the skin when human skin or clothing comes into contact with them. Dermatitis due to these venomous hairs may result from an immediate or delayed-type allergic

Fig. 6.40 Larva of *Dendrolimus spectabilis*

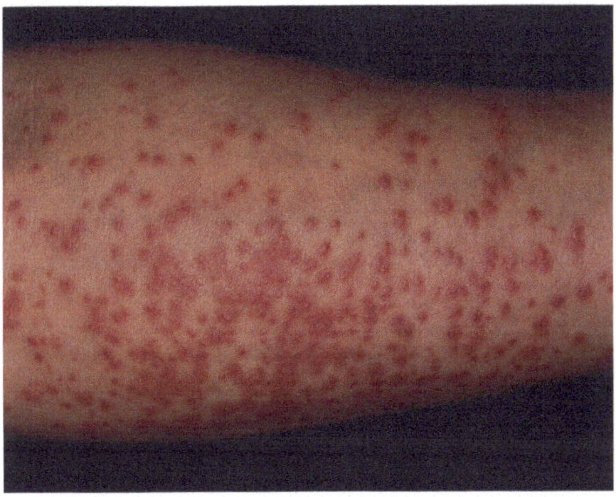

Fig. 6.41 Clinical appearance of caterpillar dermatitis

reaction to the venomous component (Natsuaki 2002). Caterpillar dermatitis is clinically characterized by groups of erythematous papules on the neck and upper limbs, accompanied by intense itching (Fig. 6.41).

The larvae of another type of caterpillar, *Parasa lepida* (*Limacodidae*) (Fig. 6.42), have venomous spines through which venom is injected when contact with human skin occurs. Chemical irritation from venom produces immediate pain and wheals, which are followed by dermatitis due to an allergic reaction to the venom.

Fig. 6.42 Larva of *Parasa lepida*

Fig. 6.43 *Paederus fuscipes*

6.7 Cutaneous Symptoms Caused by Venomous Beetles

The body fluids of the beetle *Paederus fuscipes* (Fig. 6.43) contain pederin, which causes linear dermatitis (Fig. 6.44) with pustules when it comes into contact with the skin. Because of a substance called cantharidin in their body fluids, *Xanthochroa waterhousei* (Fig. 6.45), *Meloe franciscanus*, and *Epicauta gorhami* (blister beetles) can cause bullous dermatitis (Fig. 6.46) when they come into contact with the skin.

Carabidae (*Geadephaga*) and *Damaster blaptoides* squirt methacrylate-containing fluids from their bellies. *Pheropsophus jessoensis*, a species of ground beetle, also injects hot gas from its belly, eliciting pain and inflammation when it comes into contact with the skin.

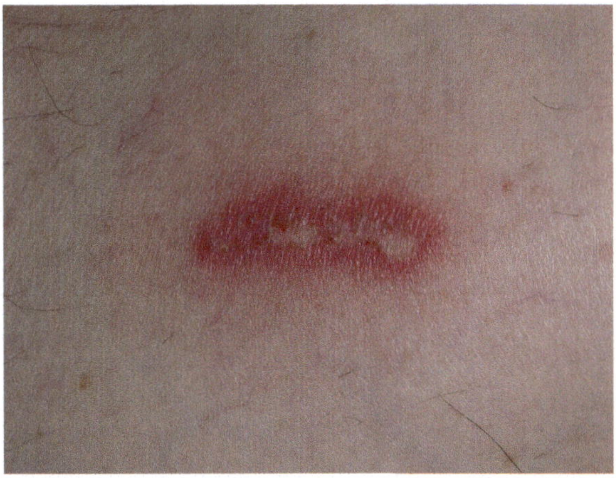

Fig. 6.44 Clinical appearance of linear dermatitis

Fig. 6.45 *Xanthochroa waterhousei*

6.8 Cutaneous Disorders Resulting from Contact with Other Arthropods

6.8.1 Stink Bugs

Stink bugs are considered by the general public to be a nuisance because of their foul odor. Dermatitis may occur when stink bug secretions come into contact with human skin. Some types of stink bugs (*Halyomorpha halys*, *Menida scotti*, etc.) enter homes to overwinter, and stepping on them leaves a brownish pigmentation on the soles of feet.

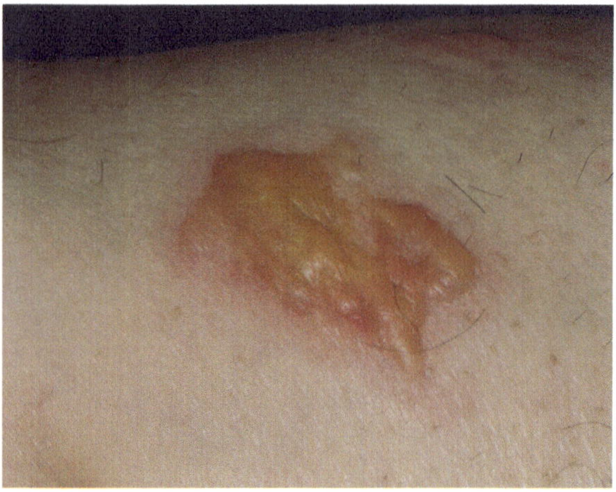

Fig. 6.46 Clinical appearance of bullous dermatitis

6.8.2 Whip Scorpions

Whip scorpions eject a liquid containing acetic acid that causes mild dermatitis. In Japan, the Amami Islands are home to the whip scorpion *Typopeltis stimpsonii*, and the Yaeyama Islands are home to *Typopeltis crucifer*. However, humans are rarely affected by these whip scorpions.

6.8.3 Millipedes

Although it is rare for humans to be attacked by millipedes, some millipedes spray venom that causes inflammation when it comes into contact with the skin.

6.9 Cutaneous Symptoms Caused by Arthropods Infesting Human Skin

6.9.1 Pediculosis capitis

Head lice, *Pediculus humanus capitis* (Fig. 6.47), is a parasite of the hair that causes itching and inflammation of the scalp. In the case of infestation, the eggs are attached to the hair, mainly on the back of the head. Head lice is transferred primarily through hair contact between humans and can be prevalent among schoolchildren and infants. Therefore, information on preventing and treating the infestation should be provided to schools, kindergartens, and childcare centers.

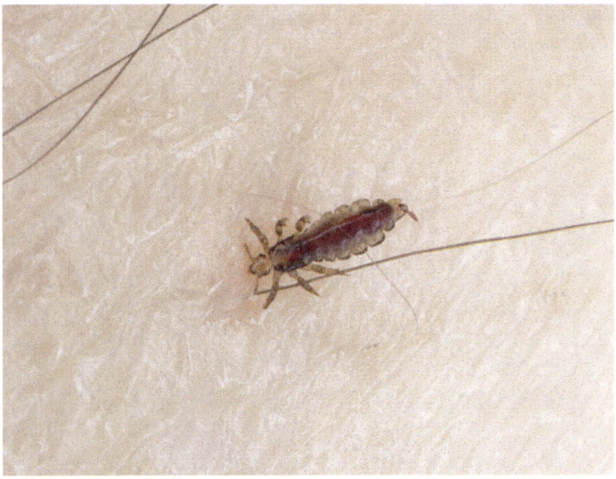

Fig. 6.47 Head lice (*Pediculus humanus capitis*)

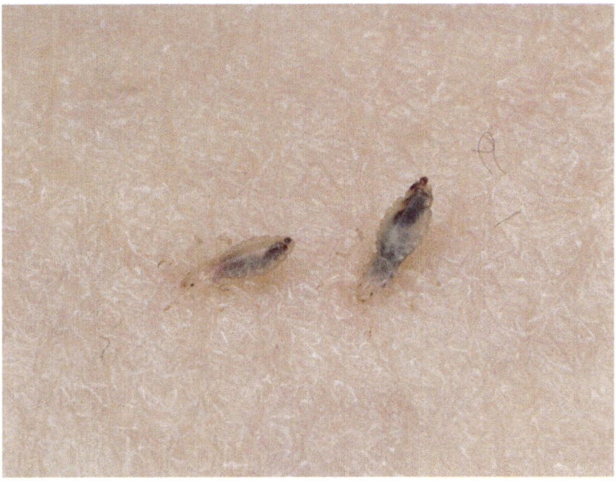

Fig. 6.48 Body lice (*Pediculus humanus corporis*)

6.9.2 Pediculosis corporis

Body lice, *Pediculus humanus corporis* (Fig. 6.48), infests clothing and sucks blood from human skin. It can be prevalent among homeless people. Pruritic papules are seen on the skin of affected individuals.

6.9.3 Pediculosis pubis

Pubic lice, *Pthirus pubis* (Fig. 6.49), is a pubic hair parasite that is transmitted via sexual contact. The chief symptom is itching of the pubic area. Pubic lice and their eggs are found at the base of the pubic hair of affected individuals.

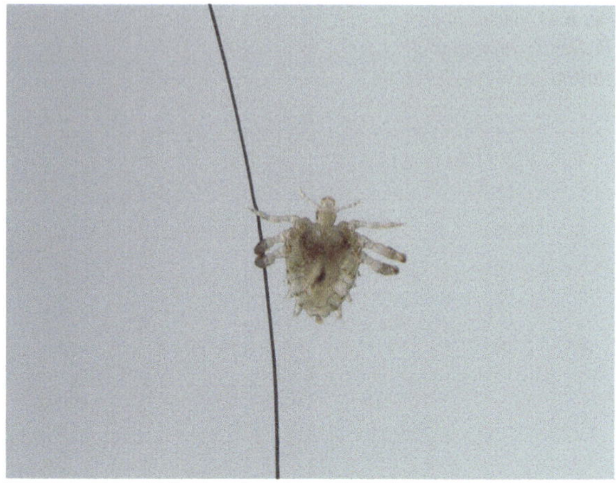

Fig. 6.49 Pubic lice (*Pthirus pubis*)

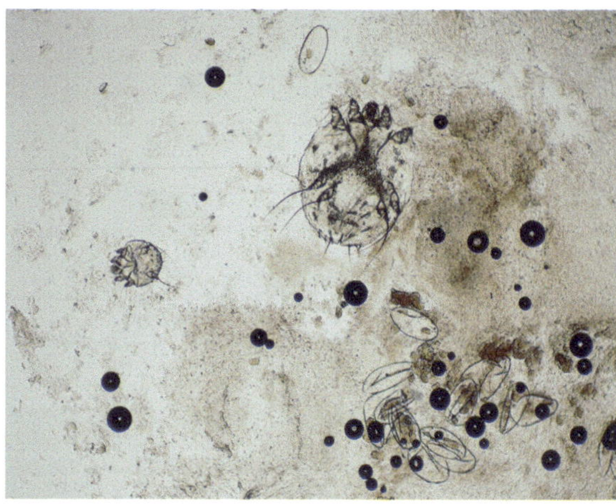

Fig. 6.50 *Sarcoptes scabiei*

6.9.4 Scabies

Scabies is caused by a parasitic infestation of the skin with *Sarcoptes scabiei* (Fig. 6.50) and has long been focused on as a sexually transmitted disease. However, in recent years, scabies have often been found in facilities and homes where infections are transmitted via care for older adults. Clinically, extremely itchy erythematous papules are seen in the axillae, lower abdomen, pubic area, and interdigital areas. Scabies are also characterized by the presence of burrows on the palms, wrists, and interdigital areas (Fig. 6.51).

Crusted scabies, a severe form of the disease, can occur in older adults and immunocompromised patients and cause systemic redness and hyperkeratosis (Fig. 6.52). Crusted scabies are much more contagious than common scabies.

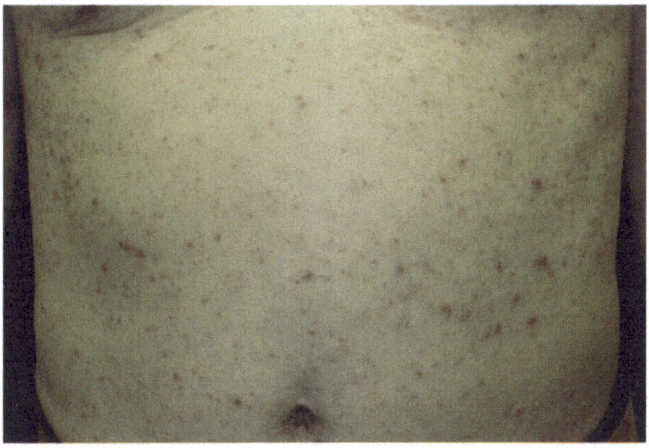

Fig. 6.51 Mite burrows of scabies

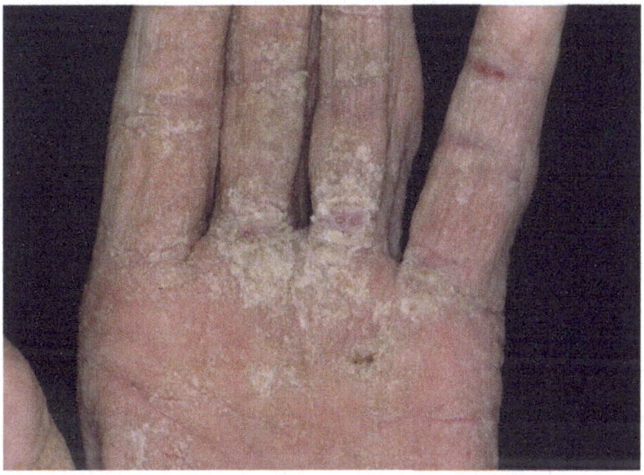

Fig. 6.52 Clinical appearance of crusted scabies

6.9.5 Demodicosis

Demodex folliculorum (Fig. 6.53) is a type of mite that inhabits the hair follicles of the human face. Infestation with a large number of *D. folliculorum* gives rise to demodicosis. Clinically, erythematous papules and pustules, as well as demodectic acne, frequently occur on the face of affected individuals.

6.9.6 Cutaneous Symptoms Caused by House Dust Mites

House dust mites do not directly damage human skin. However, they can cause allergic rhinitis and allergic conjunctivitis. Cases of anaphylactic shock have been reported after ingesting *Dermatophagoides* spp. breeding in foods (such as flour, especially okonomiyaki flour in Japan).

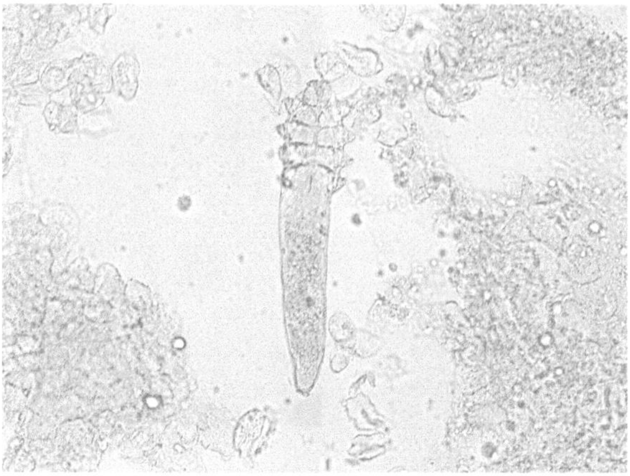

Fig. 6.53 *Demodex folliculorum*

6.9.7 Cutaneous Myiasis

Some types of flies, such as *Dermatobia hominis*, infest the skin in countries outside of Japan, and cases of cutaneous myiasis have occurred in Japanese people who have traveled abroad and developed the disease after returning home. There have been rare cases of myiasis where flies lay eggs in lesions such as skin ulcers, resulting in infestation with maggots.

6.10 Diagnosis and Treatment of Arthropod-Related Cutaneous Disorders and Countermeasures

6.10.1 Diagnosis

Cutaneous disorders caused by arthropods are relatively easy to diagnose if the responsible arthropod is identified. Otherwise, clinicians must draw conclusions about the possible causative arthropod by taking a detailed medical history, determining where the patient has been and how the symptoms have developed over time, and considering preexisting conditions, family history, and clinical characteristics, such as the type and distribution of skin rash. To enable accurate diagnoses, clinicians should be familiar with the distribution, habitat, and biology of various arthropods that can cause skin lesions. To confirm the diagnosis, the responsible arthropod needs to be captured and identified.

6.10.2 Treatment

In general, most arthropod-induced skin rashes are associated with allergic reactions. Itchy eruptions caused by insect bites are treated with corticosteroid ointments. Pain resulting from an irritant inflammatory reaction to venom components is first treated with topical cooling.

If anaphylactic symptoms are observed after a bee sting, adrenaline is injected intramuscularly. Individuals at risk of developing anaphylactic symptoms from bee stings are prescribed self-injected adrenaline (EpiPen) for prophylaxis.

In individuals bitten by *Latrodectus hasseltii*, the neurotoxin causes severe pain and a range of other symptoms, including sweating and elevated blood pressure; if mild, the symptoms can be treated with sedatives and opioid analgesics. Antivenom serum is necessary in severe cases, but it is not easily obtainable in Japan.

Pediculosis responds well to powders and shampoos containing phenothrin. A lice comb is also used to physically remove lice and their eggs. In recent years, *Pediculus humanus capitis* that is resistant to phenothrin (a pyrethroid) has been increasing in some parts of the country, so a dimethicone lotion must be used for such pyrethroid-resistant strains. Ivermectin lotions are not used in Japan. Patient education is important in treating pediculosis and preventing its spread.

Scabies are treated either orally with ivermectin or topically with phenothrin lotion, and in severe cases, both may be used in combination. No permethrin cream is used in Japan.

6.10.3 Countermeasures

Appropriate measures should be taken against each arthropod that can cause cutaneous disorders. Such measures include, for example, staying away from wasp nests, avoiding skin exposure during outdoor activities, and not touching plants that are known to host venomous caterpillars. Appropriate use of repellents is important for protection against blood-sucking arthropods, including mosquitoes, black flies, and ticks.

References

Matsumura T, Mashiko R, Sato T, Itokawa K, Maekawa Y, Ogawa K, Isawa H, Yamamoto A, Mori S, Horita A, Ginnaga A, Miyatsu Y, Takahashi M, Taki H, Hifumi T, Sawabe K, Ato M (2018) Venom and antivenom of the redback spider (*Latrodectus hasseltii*) in Japan. Part I. Venom extraction, preparation, and laboratory testing. Jpn J Infect Dis 71(2):116–121

Natsuaki M (2002) Immediate and delayed-type reactions in caterpillar dermatitis. J Dermatol 29(8):471–476

Natsuaki M (2015) Redback spider (*Latrodectus hasseltii*) and the spider bite. J Kochi Med Assoc 20:27–31. (in Japanese)

Natsuaki M (2021) Tick bites in Japan. J Dermatol 48(4):423–430

Natsuaki M, Takada N (2013) Clinical and histopathological study of dermatitis caused by experimental bites of the larval trombiculid mite, *Leptotrombidium scutellare*. Med Entomol Zool 64(1):17–19. (in Japanese)

Steigelman DA, Freeman TM (2013) Imported fire ant allergy: case presentation and review of incidence, prevalence, diagnosis, and current treatment. Ann Allergy Asthma Immunol 111(4):242–245

Chapter 7
New Technologies for Analyzing and Manipulating the Genomes of Medically Important Arthropods

Kentaro Itokawa

Abstract Over the past two decades, biological science has witnessed significant technological breakthroughs both in the analysis and manipulation of the genomes of animals and plants. The advent of next-generation sequencers has facilitated the parallel sequencing of a massive number of DNA fragments in a single instrument. This resulted in a substantial reduction in the cost of DNA sequencing by orders of magnitude and democratized genome analysis to even ordinal laboratories. Additionally, the development of programmable artificial nucleases, exemplified by the CRISPR/Cas9 system, has realized heritable and precise gene manipulation across a broad spectrum of organisms. These groundbreaking innovations have drastically transformed approaches in various biological research fields, particularly for non-model organisms. Medical entomology, like many other branches of biology, has also received substantial benefits from these technological advancements. In this chapter, we will review the technical details of these cutting-edge technologies and explore their applications in the study and control of medically significant arthropods.

Keywords Gene drive · Gene knock-in · Gene knockout · Genome · Genome editing · High-throughput DNA sequencing

7.1 Introduction

The completion of the genome assembly project for the *Anopheles gambiae* PEST strain was announced in 2002 (Holt et al. 2002). Subsequently, additional draft genomes of two mosquito vectors, the *Aedes aegypti* L. (Nene et al. 2007) and *Culex quinquefasciatus* Say (Arensburger et al. 2010), were also completed. These major achievements marked the beginning of the genomics era for medically important

K. Itokawa (✉)
Department of Medical Entomology, National Institute of Infectious Diseases, Tokyo, Japan
e-mail: itokawa@niid.go.jp

K. Sawabe et al. (eds.), *Medical Entomology in Asia*, Entomology Monographs,
https://doi.org/10.1007/978-981-97-4406-0_7

arthropod vectors. Back then, genome sequencing projects used shotgun sequencing strategies, which included bacterial cloning of DNA fragments and Sanger sequencing (Sanger et al. 1977). Meanwhile, a radically different approach to DNA sequencing, now known as next-generation sequencer (NGS), gradually emerged in the early 2000s. NGS enables faster, cheaper, and more comprehensive DNA sequencing compared to traditional Sanger sequencing. Assembling an insect genome no longer requires expensive and labor-intensive projects and has become an achievable goal for small research groups.

Aside from the paradigm shift in DNA sequencing, there was another breakthrough in the field of genome manipulation techniques from the early 2010s, which was highlighted by the Nobel Prize in Chemistry in 2020 "for the development of a method for genome editing." Unlike transposons used in conventional transgenic experiments, artificial programmable nucleases can be designed to target almost any genes in the genome in a specific manner. The method, so-called genome editing, uses such nucleases to disrupt or rewrite sequences in the genome in vivo. Genome editing is especially useful for gene knockout and knock-in experiments to study gene function, particularly in non-model organisms. In addition, applications for genetical control of insect population are actively tested in recent years.

7.2 Advances in Genome Sequencing of Medically Important Arthropods

Although Sanger sequencing still remains a routine DNA sequencing method in many laboratories, the massively parallel sequencing technology, NGS, has almost completely replaced the electrophoresis-based sequencing for mid to large sequencing projects. NGS basically does not rely on bacterial cloning of DNA molecules, and it conducts sequencing of more than millions of fragments parallelly within a single small reactive chamber which is usually called "flow cell." NGS has reduced the sequencing cost per base as much as order of from 4 to 5 since 2004 (Fig. 7.1). The first commercially available NGS machine, the GS20 by 454 Life Sciences,

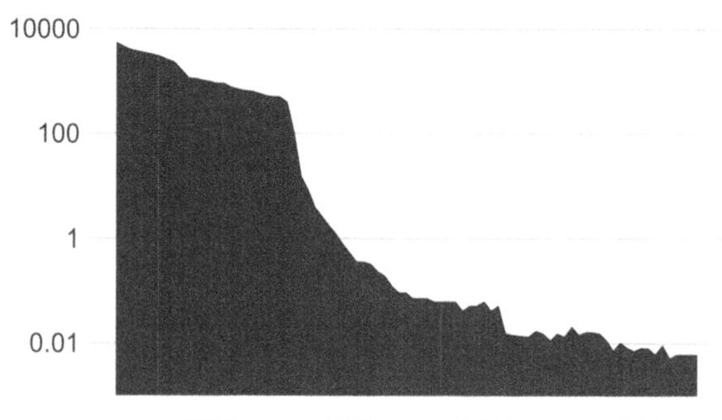

Fig. 7.1 Cost (in $USD) per raw mega-base of DNA sequencing in each year. (Data was from Wetterstrand, K.A. "DNA Sequencing Costs: Data from the NHGRI Genome Sequencing Program (GSP)". Accessed 1 July 2023. http://www.genome.gov/sequencingcostsdata)

appeared in 2004. GS20 and other models of 454 Life Sciences sequencers utilized emulsion PCR and pyrosequencing as their core technologies (Margulies et al. 2005). In emulsion PCR, individual DNA molecules are captured and amplified on the surface of micron-sized beads in vitro. This process corresponds to the cloning of DNA fragments using *Escherichia coli* in the classical shotgun approach. Pyrosequencing detects incorporated nucleotides during DNA synthesis by release of the pyrophosphate, which is measured as a luminescent reaction. The reaction occurs in parallel in each well of microfabricated picolitre-sized reaction vessels. Thus, sequences for a massive number of DNA fragments are obtained in a single run. Although Roche Diagnostics, which acquired 454 Life Sciences, discontinued their sequencers in 2013, a similar approach based on emulsion PCR is still used for the Ion Torrent Systems sequencers, which use hydrogen ions instead of chemoluminescence as a signal of nucleotide incorporation. Currently, however, series of sequencers manufactured by Illumina captured most of the market. Illumina's technology includes bridge amplification that clonally amplifies DNA fragments directly on a solid surface in a flow cell. The clonal population of DNA fragments (clusters) is subject to a sequencing-by-synthesis (SBS) reaction using reversible dye-terminators attached with a fluorescent dye corresponding to each nucleotide. The dye-labeled nucleotide is incorporated at each cycle of the strand extension by DNA polymerase and the fluorescence of each cluster is detected with an optical sensor.

The NGS technologies that appeared in the early the years are often considered as the second-generation sequencing technologies. The second-generation sequencers usually involve clonal amplification of DNA fragments. Massive number of those clones are sequenced in parallel, and the sequence accuracy is relatively high (>99%). The length of DNA fragments to be sequenced reaches 600 bp or shorter. Because of this last feature, they are also referred to as "short read sequencers." The second-generation sequencers are especially suitable for assembling a small genome containing little portion of repetitive elements such as those of viruses or some species of bacteria. Even for organisms with more complex genomes, they are often used for genotyping single-nucleotide polymorphisms if a reference genome assembly is already available or for transcriptome analysis, in which the number of cDNA reads is counted as a proxy for the expression levels of each transcript. When assembling eukaryotic genomes de novo, however, it is usually impossible to obtain assemblies with high contiguity from short reads alone, because they cannot resolve long repeats, such as transposons, and tend to produce highly fragmented assemblies.

"Third-generation" sequencing technologies were developed to sequence longer DNA fragments. The most notable characteristic is that they sequence single DNA molecules directly without clonal amplification. By mid-2023, two third-generation sequencing platforms are commercially available. Pacific Biosciences (PacBio) utilizes a technology known as single molecule real-time (SMRT) sequencing. In SMRT sequencing, template DNA is replicated by an immobilized DNA polymerase in each zero-mode waveguide (ZMW) nanowell array. The incorporation of fluorescent-labeled nucleotides during the synthesis of complementary strands emits different signals according to bases, and these signals are recorded in real time as a movie. In contrast, another commercialized third-generation technology by

Oxford Nanopore Technologies (ONT) uses nanoscale protein pores that are formed on an artificially lipid layer. Each DNA molecule thread through the pore with a support of an energy consuming "motor protein" at a speed of >400 nt/s. The DNA occupying the pore cavity disrupts the electric current in a sequence-dependent manner which is recorded by the sensor array residing at the bottom of the pore. The raw current signal is finally decoded into DNA sequence by a deep learning algorithm, this process is called "base calling." The ONT technology also enables the direct sequencing of native RNA. Third-generation sequencers have markedly extended the limit of read length to ~100 kb for PacBio and even exceeded mega base limit for ONT. Initially, there were significant drawbacks associated with these single molecule DNA sequencing technologies compared with the second-generation sequencers, particularly higher cost per base pair and lower accuracy. However, the continued development on both chemical and algorithmic sides has drastically improved the performance of both technologies. For example, PacBio's Circular Consensus Sequencing (CCS) approach uses closed circular DNA templates, around which polymerases keep traveling and replicating DNA strand. As a result, multiple reads are obtained from the same DNA molecule and a consensus sequence of these raw single path reads, which is called HiFi reads, nominally provides ~99.9% accuracy. The accuracy of ONT sequencers also continues to improve with updates to the pore and base calling model. The most up-to-date chemistry R10.4.1 (as of July 2023) and base calling model provides high modal accuracy (~99.0%) as a single molecular read. In addition, the "duplex" approach, in which both plus and minus strands of the DNA are subsequently sequenced, provides further improved accuracy (~99.9%).

During the de novo assembly process, overlaps between reads are computationally identified and merged to create longer contiguous sequences as "contigs." Contigs represent the best reconstruction of continuous stretches of the original genome that can be assembled from the available reads. Although contigs are usually much longer than raw reads, they are much shorter than the chromosomes particularly when complex genomes of animals or plants are assembled. This is because some repetitive regions in complex genomes are too long to resolve even with current long-read technologies. High-throughput chromosome conformation capture (Hi-C) is a sequencing-based approach used to analyze the three-dimensional organization of chromatin within the cell nucleus. In the Hi-C experiment, DNA is cross-linked to capture physical interactions between distant genomic regions. The cross-linked DNA is digested, and the ends of the physically proximal fragments are ligated together. These chimeric DNA fragments are sequenced by NGS, resulting in a dataset representing the interactions between proximal chromosomal regions. Dudchenko et al. (2017) demonstrated that Hi-C can be used to order and orient existing contigs/scaffolds into the three chromosomes of *Ae. aegypti* and *Cx. quinquefasciatus*. As of 2023, Hi-C has become the most feasible method to finalize genome assembly at the chromosome level for many organisms.

Advancements in sequencing technology over the past two decades have resulted in a substantial proliferation in the number of available insect genome assemblies. As of July 2023, there are 235 genome assemblies of 95 arthropod species of medical importance in NCBI GenBank database (Table 7.1). Of these, there were 139

Table 7.1 Available genome assemblies of medically important arthropods

Group	Species	Number of Assembly	Chromosome	Scaffold	Contig	Representative_genome
Mosquitoes	*Armigeres subalbatus*	1	0	1	0	GCA_024139115.1
Mosquitoes	*Wyeomyia smithii*	1	1	0	0	GCF_029784165.1
Mosquitoes	*Malaya genurostris*	1	1	0	0	GCA_030247185.1
Mosquitoes	*Sabethes cyaneus*	3	2	0	1	GCF_943734655.1
Mosquitoes	*Culex pipiens*	3	2	0	1	GCF_016801865.2
Mosquitoes	*Topomyia yanbarensis*	1	1	0	0	GCA_030247195.1
Mosquitoes	*Culex quinquefasciatus*	3	1	2	0	GCF_015732765.1
Mosquitoes	*Aedes koreicus*	1	0	0	1	GCA_024533555.1
Mosquitoes	*Aedes aegypti*	11	4	5	2	GCF_002204515.2
Mosquitoes	*Uranotaenia lowii*	1	1	0	0	GCF_029784155.1
Mosquitoes	*Aedes albopictus*	9	0	5	4	GCF_006496715.2
Mosquitoes	*Anopheles funestus-like* sensu *Spillings* et al....	1	1	0	0	GCA_016170035.1
Mosquitoes	*Anopheles parensis*	1	1	0	0	GCA_016254315.1
Mosquitoes	*Anopheles punctulatus*	1	0	0	1	GCA_000956255.1
Mosquitoes	*Anopheles bwambae*	1	0	1	0	GCA_007018985.1
Mosquitoes	*Anopheles merus*	4	2	2	0	GCF_017562075.2
Mosquitoes	*Anopheles longipalpis*	1	1	0	0	GCA_016170015.1
Mosquitoes	*Anopheles stephensi*	11	8	3	0	GCF_013141755.1
Mosquitoes	*Anopheles fontenillei*	1	0	1	0	GCA_008817895.1
Mosquitoes	*Anopheles maculipalpis*	2	1	0	1	GCF_943734695.1
Mosquitoes	*Anopheles farauti No. 4*	1	0	0	1	GCA_000956215.1
Mosquitoes	*Anopheles rivulorum*	1	1	0	0	GCA_018908115.1

(continued)

Table 7.1 (continued)

Group	Species	Number of Assembly	Chromosome	Scaffold	Contig	Representative_genome
Mosquitoes	*Anopheles cracens*	1	0	1	0	GCA_002091845.1
Mosquitoes	*Anopheles marshallii*	2	1	0	1	GCF_943734725.1
Mosquitoes	*Anopheles* sp. *NFL-2015*	1	1	0	0	NA
Mosquitoes	*Anopheles funestus*	4	2	1	1	GCF_943734845.2
Mosquitoes	*Anopheles vaneedeni*	1	1	0	0	GCA_016170025.1
Mosquitoes	*Anopheles darlingi*	5	1	1	3	GCF_943734745.1
Mosquitoes	*Anopheles koliensis*	1	0	0	1	GCA_000956275.1
Mosquitoes	*Anopheles farauti*	3	0	2	1	GCA_000473445.2
Mosquitoes	*Anopheles epiroticus*	1	0	1	0	GCA_000349105.1
Mosquitoes	*Anopheles sinensis*	4	0	4	0	GCA_000441895.2
Mosquitoes	*Anopheles quadriannulatus*	1	0	1	0	GCA_000349065.1
Mosquitoes	*Anopheles melas*	2	0	2	0	GCA_000473525.2
Mosquitoes	*Anopheles bellator*	2	2	0	0	GCA_943735745.2
Mosquitoes	*Anopheles maculatus*	2	0	2	0	GCA_002091835.1
Mosquitoes	*Anopheles christyi*	1	0	1	0	GCA_000349165.1
Mosquitoes	*Anopheles albimanus*	5	3	1	1	GCF_013758885.1
Mosquitoes	*Anopheles cruzii*	1	1	0	0	GCF_943734635.1
Mosquitoes	*Anopheles culicifacies*	1	0	1	0	GCA_000473375.1
Mosquitoes	*Anopheles ziemanni*	3	2	0	1	GCA_943734765.2
Mosquitoes	*Anopheles coluzzii*	12	9	1	2	GCF_943734685.1
Mosquitoes	*Anopheles arabiensis*	2	1	1	0	GCF_016920715.1
Mosquitoes	*Anopheles atroparvus*	4	1	1	2	GCA_914969975.1
Mosquitoes	*Anopheles minimus*	1	0	1	0	GCA_000349025.1

Mosquitoes	*Anopheles moucheti*	2	1	0	1	GCF_943734755.1
Mosquitoes	*Anopheles coustani*	2	1	0	1	GCA_943734705.1
Mosquitoes	*Anopheles dirus*	1	0	1	0	GCA_000349145.1
Mosquitoes	*Anopheles aquasalis*	3	1	1	1	GCF_943734665.1
Mosquitoes	*Anopheles nili*	3	1	1	1	GCF_943737925.1
Mosquitoes	*Anopheles gambiae*	8	4	3	1	GCF_000005575.2
Ticks	*Rhipicephalus sanguineus*	2	2	0	0	GCF_013339695.2
Ticks	*Haemaphysalis longicornis*	5	2	1	2	GCA_013339765.2
Ticks	*Hyalomma marginatum*	7	0	0	7	GCA_910592705.1
Ticks	*Hyalomma asiaticum*	2	2	0	0	GCA_013339685.2
Ticks	*Ixodes scapularis*	6	0	2	4	GCF_016920785.2
Ticks	*Amblyomma americanum*	1	0	1	0	GCA_030143305.1
Ticks	*Amblyomma maculatum*	1	0	1	0	GCA_023969395.1
Ticks	*Rhipicephalus annulatus*	1	0	0	1	GCA_013436015.1
Ticks	*Dermacentor andersoni*	3	0	2	1	GCF_023375885.1
Ticks	*Dermacentor silvarum*	2	2	0	0	GCF_013339745.2
Ticks	*Rhipicephalus microplus*	5	1	4	0	GCF_013339725.1
Ticks	*Ixodes persulcatus*	2	0	2	0	GCA_013358835.2
Ticks	*Ixodes ricinus*	2	0	2	0	GCA_000973045.2
Tsetse flies	*Glossina palpalis*	1	0	1	0	GCA_000818775.1
Tsetse flies	*Glossina austeni*	1	0	1	0	GCA_000688735.1
Tsetse flies	*Glossina morsitans*	2	0	1	1	GCA_001077435.1
Tsetse flies	*Glossina pallidipes*	1	0	1	0	GCA_000688715.1
Tsetse flies	*Glossina brevipalpis*	1	0	1	0	GCA_000671755.1

(continued)

Table 7.1 (continued)

Group	Species	Number of Assembly	Chromosome	Scaffold	Contig	Representative_genome
Tsetse flies	*Glossina fuscipes*	3	0	3	0	GCF_014805625.2
Sand flies	*Lutzomyia longipalpis*	2	1	1	0	GCF_024334085.1
Sand flies	*Phlebotomus papatasi*	3	2	1	0	GCF_024763615.1
Sand flies	*Phlebotomus perniciosus*	2	0	1	1	GCA_918844115.2
Sand flies	*Phlebotomus argentipes*	1	0	1	0	GCA_947086385.1
Biting midges	*Culicoides sonorensis*	4	0	2	2	GCA_900002565.1
Biting midges	*Culicoides tainanus*	1	0	0	1	NA
Biting midges	*Forcipomyia taiwana*	1	0	0	1	GCA_949837095.1
Body louse	*Pediculus humanus*	2	0	1	1	GCF_000006295.1
Other flies	*Musca domestica*	3	0	2	1	GCF_000371365.1
Other flies	*Stomoxys calcitrans*	1	0	1	0	GCF_001015335.1
Other flies	*Philornis downsi*	1	0	1	0	GCA_019455685.1
Other flies	*Polietes domitor*	2	1	1	0	GCA_947397865.1
Other flies	*Muscidae* sp.	1	0	1	0	GCA_029304375.1
Other flies	*Haematobia irritans*	1	0	1	0	GCA_003123925.1
Flea	*Ctenocephalides felis*	1	0	1	0	GCF_003426905.1
Bed Bugs	*Cimex lectularius*	4	0	4	0	GCF_000648675.2
Bed Bugs	*Cimex hemipterus*	1	0	0	1	NA
Kissing Bugs	*Triatoma infestans*	1	0	1	0	GCA_011037195.1
Kissing Bugs	*Rhodnius prolixus*	3	0	2	1	GCA_000181055.3
Chiggers	*Leptotrombidium deliense*	2	0	2	0	GCA_003675905.2
Chiggers	*Leptotrombidium pallidum*	1	0	0	1	NA
Cockroaches	*Periplaneta americana*	3	0	3	0	GCA_025594305.2
Cockroaches	*Blattella germanica*	3	0	3	0	NA
Cockroaches	*Diploptera punctata*	1	0	0	1	GCA_030220185.1
Scabies mites	*Sarcoptes scabiei*	3	1	2	0	GCA_020844145.1

assemblies for 51 species of mosquitoes (Culicidae), 39 assemblies for 13 species of ticks (Ixodida), 9 assemblies for 6 species of Tsetse flies (Glossinidae), 8 assemblies for 4 species of sand flies (Phlebotominae), 6 assemblies for 3 species of biting midges (Ceratopogonidae), 2 assemblies for louse (Pediculidae), 1 assembly for flea (Pulicidae), 5 assemblies for 2 species of bedbugs (Cimicidae), 5 assemblies for 2 species of kissing bugs (Reduviidae), 3 genomes for 2 species of chiggers (Trombiculidae), 7 assemblies for 3 species of cockroaches (Blattidae and Ectobiidae), and 1 assembly for scabies mite (Sarcoptidae). VectorBase (Giraldo-Calderón et al. 2015; Greene et al. 2007; Giraldo-Calderón et al. 2022) is another database hosting genome assemblies of arthropod vectors. VectorBase provides an interactive web interface that enables users to readily browse and access genome, gene model, transcriptome, genotype data, and more without cumbersome command-line operations. In 2019, VectorBase is merged with EuPathDB (Aurrecoechea et al. 2017) to form VEuPathDB.org which provides comprehensive support for both invertebrate vectors and eukaryotic pathogens (protists and fungi) (Amos et al. 2022).

The exponential increase in publicly available genome assemblies has provided researchers with enhanced methodologies for addressing pivotal biological inquiries within medically significant arthropods. These notable accomplishments include, for example, the identification of sex-determining genes in mosquitoes (Liu et al. 2020; Hall et al. 2015; Krzywinska et al. 2016; Criscione et al. 2016), which has been a long-standing enigma that has eluded biology for over half a century (Gilchrist and Haldane 1947). A large-scale whole genome sequencing project of *Anopheles gambiae* (Ag1000G) revealed its detailed genetic diversity and the ongoing selection on voltage-gated sodium channel gene, which encodes a receptor for pyrethroid insecticides (Clarkson et al. 2021).

7.3 Advances in Genetic Engineering for Medically Important Arthropods

In addition to the advancement of technologies to analyze genomes, there was another leap-frogging innovation in technologies used to manipulate the genomes of organisms. Classically, methods of making heritable changes in the genomes of medically important arthropods were limited to random mutagenesis by radiation, chemical mutagens, or transposons. Although transposons, such as *piggy-bac*, were widely used to integrate heterologous genes into insect genomes, they lacked control over the position of insertion. The genome editing technologies exploit artificial endonucleases (Fig. 7.2) with recognition sequences long enough to target only one desired site within the genome, and importantly, these recognition sequences are programmable. These artificial nucleases can induce DNA double-strand breaks (DSBs) at the targeted site in the genome. DSBs are subject to several distinct repair processes. Non-homologous end-joining (NHEJ) repair DSBs by re-ligating the

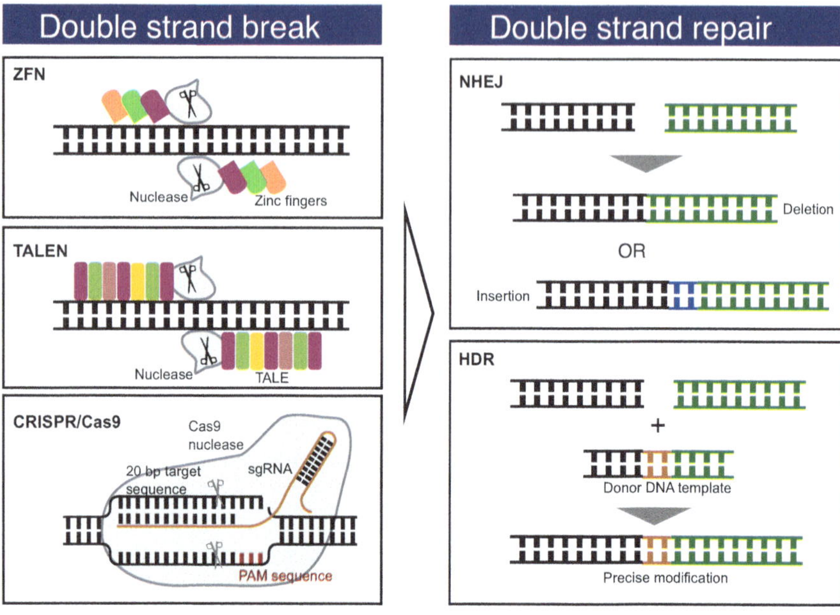

Fig. 7.2 Genome editing platforms and mechanisms for DSB repair with endogenous DNA. (This figure is modified from Li et al. (2020). Used under CC-BY 4.0 International license. Accessed 1 July 2023. http://creativecommons.org/licenses/by/4.0/)

broken ends without using a homology template. NHEJ is a relatively error-prone process, which often leaves a small insertion or deletion at the cut site. If these mutations occur in the protein coding sequence, the gene could be functionally disrupted (knockout). In contrast, homology-directed repair (HDR) utilizes a homologous template (e.g., homologous chromosome or sister chromatid) to repair DSBs. The HDR pathway may be exploited for more accurate manipulations of a genome, such as integrating DNA construct into a target site or replacing a specific nucleotide to another one (knock-in). For knock-in experiments, a donor DNA template, in formats such as plasmids or single-strand oligos, having sequence arms homologous to the flanking sequence of the targeted site is co-injected with a nuclease programmed to cut the same target site. Sequences in the donor template will be incorporated into the designated target site of the genome as a result of the HDR process.

Zinc finger nuclease (ZFN) is an artificially constructed protein containing a zinc finger domain with a DNA binding function and a nuclease domain from the *Fok* I endonuclease (Kim et al. 1996). The zinc finger domain of ZFN can be tailored to bind to a desired sequence of DNA and cuts that sequence in the genome. Although ZFN was an early pioneer in the field of genome editing, the specificity of the zinc finger domains after DNA binding was highly context-dependent and required intensive selection. For this reason, ZFN was superseded by other artificial nucleases that were simpler and cheaper to develop.

Transcription activator-like effector nuclease (TALEN) is another artificial programmable nuclease (Miller et al. 2011). Like ZFN, TALEN consists of a DNA binding domain (TAL effector domain) fused with *Fok* I nuclease domain (Li et al. 2011; Christian et al. 2010). TAL effector domain was originally discovered in TAL effector proteins in plant pathogenic bacteria. TAL effectors bind to specific promoter sequences in plant genes to regulate their expression. The sequence specificity of TAL effectors is attributed to the TAL effector domain (Boch et al. 2009; Moscou and Bogdanove 2009), which has a modular structure with tandem repeats of nearly identical amino acid sequence unit (monomer) with length of 33–35 residues. Each monomer contains two consecutive amino acid residues known as repeat-variable residues or RVDs in the middle. RVDs determine the specificity of the monomer to either base of DNA (A, T, G or C), and the order of the monomers in a TAL effector domain corresponds with the sequence of the target DNA. Thus, compared with ZFN, the sequence specificity of TALEN is more predictable and easier to program. Genome editing was further simplified by the clustered regularly interspaced short palindromic repeats (CRISPR) and CRISPR-associated protein 9 (Cas9) system (Jinek et al. 2012). CRISPR was originally reported in *E. coli* (Ishino et al. 1987) and is now considered a defensive system of bacteria and archaea against parasitic viruses or plasmids (Wiedenheft et al. 2012). Cas9 is an RNA-guided DNA nuclease associated with CRISPR in *Streptococcus pyogenes*. Naturally, Cas9 forms a complex with two RNA molecules called the CRISPR RNA (crRNA) and the trans-activating crRNA (tracrRNA). Cas9 causes a DSB if the DNA sequence is complementary to its own crRNA. In an engineered CRISPR/Cas9 system, crRNA and tracrRNA are often fused to form a short guide RNA (sgRNA) (Jinek et al. 2012). As of 2023, the CRISPR/Cas9 system is the most popular genome editing tool because of its low cost and flexibility.

Genome editing has made genetic manipulation accessible for non-model organisms, including medically important insects. One useful application of genome editing is reverse genetics studies, in which researchers disrupt a gene of interest, and observe the phenotypic consequences to understand its biological function. For example, CRISPR/Cas9 and TALEN were used to show a causative relationship between cytochrome P450 (CYP9M10) in *Cx. quinquefasciatus* and pyrethroid insecticides. In *Cx. quinquefasciatus*, overexpression of cytochrome *CYP9M10* was suspected to be responsible for strong resistance against pyrethroid insecticides (Komagata et al. 2010; Itokawa et al. 2010, 2011). When the *CYP9M10* gene was knocked out in a resistance strain, the resistance level decreased approximately 100-fold (Itokawa et al. 2016). CRISPR/Cas9 was also used to establish the role of mosquito genes in sex determination (Hall et al. 2015; Liu et al. 2020) or susceptibility to *Plasmodium* parasite (Dong et al. 2018; Yamamoto et al. 2018). For evaluating the phenotypic consequence of some mutations, such as gain-of-function mutations, knock-in rather than gene knockout experiments are often required. The gene knock-in technique using oligonucleotides as repair donor templates can be used to create designated point mutations in genome; however, the ability to obtained desired genome edited insect is much lower compared with that for gene knockout. Samantsidis et al. (2019, 2020) used a transgenic *Drosophila melanogaster* strain endogenously expressing the Cas9 protein under the control of a germline-specific promoter to knock-in point mutations on voltage-gated sodium channel (*para*) gene of the fly. Using this approach, the consequence of some point mutations seen in orthologous genes of pyrethroid-resistance *A. aegypti* and other insect pests on insecticide susceptibility phenotype were evaluated.

To cause a heritable mutation in the genome, DSB and subsequent repair processes must occur in germline cells, which eventually form gametes (sperms or eggs). For insect species, microinjection of synthesized Cas9 protein/mRNA and sgRNA during early embryonic development is often employed. However, this approach often suffers extremely low efficiency of knockout or knock-in depending on the targeted gene, probably due to the somatic mosaicism in G0 individual. Controlling Cas9 gene expression using germline-specific promoters can enhance the production of heritable mutations (Ren et al. 2013). Therefore, transgenic mosquitoes expressing Cas9 under a germline promoter are developed (Li et al. 2017; Feng et al. 2021). Microinjection into insect embryos is also technically complicated or even infeasible for some species. Shirai et al. (2022) recently showed that injecting sgRNA and the Cas9 protein complex into the haemocoel of the adult German cockroach *Blattella germanica* during vitellogenesis resulted in heritable mutations in the progeny.

In addition to being as useful tools for biologists, artificial endonucleases are being harnessed to control or manipulate wild populations of medically important vectors. Homing endonuclease genes (HEGs) are naturally occurring selfish genetic factors which autonomously spread populations of host organisms (Burt and Trivers 2008). The rapid spread of HEGs is called "gene drive." This phenomenon is the result "super Mendelian inheritance," in which a HEG in a heterozygote genotype is inherited to offspring more frequently than expected from Mendel's low (i.e., 50%). A HEG encodes a endonuclease whose recognition sequence is long and occuring only once in a whole genome of the host. HEGs are inserted into their own recognition site of the host genome. In heterozygous cells, the endonuclease encoded by HEG cuts the intact recognition site in the homologous chromosome. Once this DNA break is repaired by HDR, HEG is incorporated into the targeted chromosome, which results in conversion of the genotype into homozygous state (homing). Burt (2003) proposed using HEG to introduce and spread a desirable gene, such as one that prevents transmission of pathogens, into a population of pest animals. This strategy may also be used to reduce or even eliminate the population if spreading HEG has a deleterious effect to fecundity of a target species. For population reduction, the target gene of HEG should be recessively lethal or cause sterility when its function is disrupted. In addition, if this effect is specific only to females, the drive of HEGs can occur more rapidly because male can spread HEGs. For example, HEG targeting X-chromosome-specific ribosomal RNA genes in *Anopheles* mosquitoes during the spermatogenesis in male meiosis destroys X-chromosome in the gametes. As a result, the sex ratio of the progeny is extremely biased to male (Galizi et al. 2014). Importantly, researchers can now engineer the CRISPR/Cas9 system, instead of naturally occurring nucleases, into a selfish genetic factor behaving like a HEG (Galizi et al. 2016). Hammond et al. (2016) identified three protein-coding genes in *An. gambiae* that exhibit desirable recessive female-specific lethality and are potential targets for synthetic HEGs. Unfortunately, HDR is not a very accurate process; it often creates mutated alleles instead of perfect homing. These mutated alleles can no longer be targeted by homing nucleases because they have their lost recognition sequences. Such alleles resistance against homing can even preexist naturally as standing genetic variation. If there is an allele that is resistant to homing while retaining the original gene function in the population, it could enjoy positive selection as the frequency of the HEG allele increases (Burt 2003). This ultimately results in the elimination of an HEG allele from the population and failure of population manipulation. Kyrou et al. (2018) established an *An. gambiae* strain that expresses synthetic homing gene targeting the boundary of intron 4 and exon 5 of the *doublesex* gene (*dbsx*), which plays an important role in determining sexual dimorphism. Because exon 5 of *dbsx* is only used in female-specific transcript variant, the disruption of this sequence causes infertility only in female. In a caged population experiment, a rapid drive of the construct and reduction of total egg production were observed, which finally resulted in the complete collapse of the population. The region targeted in *dbsx* is highly conserved at the nucleotide level among mosquitoes, probably because of some functional constraint. Because of this constraint, it is considered that the system is more robust against emergence and spread of resistant alleles. In addition, maternal deposition of Cas9 protein (and/or Cas9-gRNA complex) to embryos seems to cause fitness reduction of offspring probably due to somatic mutations. Therefore, the optimization of the Cas9 promoter to strictly control expression in germline may be important (Hammond et al. 2021). Furthermore, although population manipulation/elimination by gene drive is an active field in public health, the uncertainties of this new technology also raised large concerns from both academia and the public. Thus, techniques that can revert the spread of gene-drive in population when any adverse effects are recognized are desired. Taxiarchi et al. (2021) created transgenic *An. gambiae* expressing a phage-derived anti-CRISPR protein (AcrIIA4). Germline expression of *AcrIIA4* inhibited the homing activity of the driving constructs by restoring their inheritance to near Mendelian rates. In a caged population experiment, the single release of the *AcrIIA4* transgenic mosquitos suppressed the spread of a preexisting homing allele and prevented population collapse.

References

Amos B, Aurrecoechea C, Barba M, Barreto A, Basenko EY, Bażant W, Belnap R et al (2022) VEuPathDB: the eukaryotic pathogen, vector and host bioinformatics resource center. Nucleic Acids Res 50(D1):D898–D911

Arensburger P, Megy K, Waterhouse RM, Abrudan J, Amedeo P, Antelo B, Bartholomay L et al (2010) Sequencing of *Culex quinquefasciatus* establishes a platform for mosquito comparative genomics. Science 330(6000):86–88

Aurrecoechea C, Barreto A, Basenko EY, Brestelli J, Brunk BP, Cade S, Crouch K et al (2017) EuPathDB: the eukaryotic pathogen genomics database resource. Nucleic Acids Res 45(D1):D581–D591

Boch J, Scholze H, Schornack S, Landgraf A, Hahn S, Kay S, Lahaye T, Nickstadt A, Bonas U (2009) Breaking the code of DNA binding specificity of TAL-Type III effectors. Science 326(5959):1509–1512

Burt A (2003) Site-specific selfish genes as tools for the control and genetic engineering of natural populations. Proc Biol Sci 270(1518):921–928

Burt A, Trivers R (2008) Genes in conflict: the biology of selfish genetic elements. Belknap Press of Harvard University Press, Cambridge, MA, 602 pp

Christian M, Cermak T, Doyle EL, Schmidt C, Zhang F, Hummel A, Bogdanove AJ, Voytas DF (2010) Targeting DNA double-strand breaks with TAL effector nucleases. Genetics 186(2):757–761

Clarkson CS, Miles A, Harding NJ, O'Reilly AO, Weetman D, Kwiatkowski D, Donnelly MJ, *Anopheles gambiae* 1000 Genomes Consortium (2021) The genetic architecture of target-site resistance to pyrethroid insecticides in the African malaria vectors *Anopheles gambiae* and *Anopheles coluzzii*. Mol Ecol 30(21):5303–5317

Criscione F, Qi Y, Tu Z (2016) GUY1 confers complete female lethality and is a strong candidate for a male-determining factor in *Anopheles stephensi*. elife 5:e19281

Dong Y, Simões ML, Marois E, Dimopoulos G (2018) CRISPR/Cas9-mediated gene knockout of *Anopheles gambiae* FREP1 suppresses malaria parasite infection. PLoS Pathog 14(3):e1006898

Dudchenko O, Batra SS, Omer AD, Nyquist SK, Hoeger M, Durand NC, Shamim MS et al (2017) De novo assembly of the *Aedes aegypti* genome using Hi-C yields chromosome-length scaffolds. Science 356(6333):92–95

Feng X, López Del Amo V, Mameli E, Lee M, Bishop AL, Perrimon N, Gantz VM (2021) Optimized CRISPR tools and site-directed transgenesis towards gene drive development in *Culex quinquefasciatus* mosquitoes. Nat Commun 12(1):2960

Galizi R, Doyle LA, Menichelli M, Bernardini F, Deredec A, Burt A, Stoddard BL, Windbichler N, Crisanti A (2014) A synthetic sex ratio distortion system for the control of the human malaria mosquito. Nat Commun 5:3977

Galizi R, Hammond A, Kyrou K, Taxiarchi C, Bernardini F, O'Loughlin SM, Papathanos PA, Nolan T, Windbichler N, Crisanti A (2016) A CRISPR-Cas9 sex-ratio distortion system for genetic control. Sci Rep 6:31139

Gilchrist BM, Haldane JBS (1947) Sex llinkage and sex determination in a mosquito, *Culex molestus*. Hereditas 33:175–190

Giraldo-Calderón GI, Emrich SJ, MacCallum RM, Maslen G, Dialynas E, Topalis P, Ho N, Gesing S, VectorBase: Consortium, Madey G, Collins FH, Lawson D (2015) VectorBase: an updated bioinformatics resource for invertebrate vectors and other organisms related with human diseases. Nucleic Acids Res 43(DI):D707–D713

Giraldo-Calderón GI, Harb OS, Kelly SA, Rund SS, Roos DS, McDowell MA (2022) VectorBase.org updates: bioinformatic resources for invertebrate vectors of human pathogens and related organisms. Curr Opin Insect Sci 50:100860

Greene JM, Collins F, Lefkowitz EJ, Roos D, Scheuermann RH, Sobral B, Stevens R, White O, Di Francesco V (2007) National Institute of Allergy and Infectious Diseases bioinformatics resource centers: new assets for pathogen informatics. Infect Immun 75(7):3212–3219

Hall AB, Basu S, Jiang X, Qi Y, Timoshevskiy VA, Biedler JK, Sharakhova MV, Elahi R, Anderson MA, Chen XG, Sharakhov IV, Adelman ZN, Tu Z (2015) Sex determination. A male-determining factor in the mosquito *Aedes aegypti*. Science 348(6240):1268–1270

Hammond A, Galizi R, Kyrou K, Simoni A, Siniscalchi C, Katsanos D, Gribble M et al (2016) A CRISPR-Cas9 gene drive system targeting female reproduction in the malaria mosquito vector *Anopheles gambiae*. Nat Biotechnol 34(1):78–83

Hammond A, Karlsson X, Morianou I, Kyrou K, Beaghton A, Gribble M, Kranjc N, Galizi R, Burt A, Crisanti A, Nolan T (2021) Regulating the expression of gene drives is key to increasing their invasive potential and the mitigation of resistance. PLoS Genet 17(1):e1009321

Holt RA, Subramanian GM, Halpern A, Sutton GG, Charlab R, Nusskern DR, Wincker P et al (2002) The genome sequence of the malaria mosquito *Anopheles gambiae*. Science 298(5591):129–149

Ishino Y, Shinagawa H, Makino K, Amemura M, Nakata A (1987) Nucleotide sequence of the Iap gene, responsible for alkaline phosphatase isozyme conversion in *Escherichia coli*, and identification of the gene product. J Bacteriol 169(12):5429–5433

Itokawa K, Komagata O, Kasai S, Okamura Y, Masada M, Tomita T (2010) Genomic structures of Cyp9m10 in pyrethroid resistant and susceptible strains of *Culex quinquefasciatus*. Insect Biochem Mol Biol 40(9):631–640

Itokawa K, Komagata O, Kasai S, Masada M, Tomita T (2011) Cis-acting mutation and duplication: history of molecular evolution in a P450 haplotype responsible for insecticide resistance in *Culex quinquefasciatus*. Insect Biochem Mol Biol 41(7):503–512

Itokawa K, Komagata O, Kasai S, Ogawa K, Tomita T (2016) Testing the causality between CYP9M10 and pyrethroid resistance using the TALEN and CRISPR/Cas9 technologies. Sci Rep 6:24652

Jinek M, Chylinski K, Fonfara I, Hauer M, Doudna JA, Charpentier E (2012) A programmable dual-RNA-guided DNA endonuclease in adaptive bacterial immunity. Science 337(6096):816–821

Kim YG, Cha J, Chandrasegaran S (1996) Hybrid restriction enzymes: zinc finger fusions to fok I cleavage domain. Proc Natl Acad Sci USA 93(3):1156–1160

Komagata O, Kasai S, Tomita T (2010) Overexpression of cytochrome P450 genes in pyrethroid-resistant *Culex quinquefasciatus*. Insect Biochem Mol Biol 40(2):146–152

Krzywinska E, Dennison NJ, Lycett GJ, Krzywinski J (2016) A maleness gene in the malaria mosquito *Anopheles gambiae*. Science 353(6294):67–69

Kyrou K, Hammond AM, Galizi R, Kranjc N, Burt A, Beaghton AK, Nolan T, Crisanti A (2018) A CRISPR–Cas9 gene drive targeting doublesex causes complete population suppression in caged *Anopheles gambiae* mosquitoes. Nat Biotechnol 36(11):1062–1066

Li T, Huang S, Jiang WZ, Wright D, Spalding MH, Weeks DP, Yang B (2011) TAL nucleases (TALNs): hybrid proteins composed of TAL effectors and fokI DNA-cleavage domain. Nucleic Acids Res 39(1):359–372

Li M, Bui M, Yang T, Bowman CS, White BJ, Akbari OS (2017) Germline Cas9 expression yields highly efficient genome engineering in a major worldwide disease vector, *Aedes aegypti*. Proc Natl Acad Sci USA 14(49):E10540–E10549

Li H, Yang Y, Hong W, Huang M, Wu M, Zhao X (2020) Applications of genome editing technology in the targeted therapy of human diseases: mechanisms, advances and prospects. Signal Transduct Target Ther 5(1):1

Liu P, Jin B, Li X, Zhao Y, Gu J, Biedler JK, Tu ZJ, Chen XG (2020) Nix is a male-determining factor in the Asian tiger mosquito *Aedes albopictus*. Insect Biochem Mol Biol 118:103311

Margulies M, Egholm M, Altman WE, Attiya S, Bader JS, Bemben LA, Berka J et al (2005) Genome sequencing in microfabricated high-density picolitre reactors. Nature 437(7057):376–380

Miller JC, Tan S, Qiao G, Barlow KA, Wang J, Xia DF, Meng X, Paschon DE, Leung E, Hinkley SJ, Dulay GP, Hua KL, Ankoudinova I, Cost GJ, Urnov FD, Zhang HS, Holmes MC, Zhang L, Gregory PD, Rebar EJ (2011) A TALE nuclease architecture for efficient genome editing. Nat Biotechnol 29(2):143–148

Moscou MJ, Bogdanove AJ (2009) A simple cipher governs DNA recognition by TAL effectors. Science 326(5959):1501

Nene V, Wortman JR, Lawson D, Haas B, Kodira C, Tu ZJ, Loftus B et al (2007) Genome sequence of *Aedes aegypti*, a major arbovirus vector. Science 316(5832):1718–1723

Ren X, Sun J, Housden BE, Hu Y, Roesel C, Lin S, Liu LP et al (2013) Optimized gene editing technology for *Drosophila melanogaster* using germ line-specific Cas9. Proc Natl Acad Sci USA 110(47):19012–19,017

Samantsidis GR, O'Reilly AO, Douris V, Vontas J (2019) Functional validation of target-site resistance mutations against sodium channel blocker insecticides (SCBIs) via molecular modeling and genome engineering in Drosophila. Insect Biochem Mol Biol 104:73–81

Samantsidis GR, Panteleri R, Denecke S, Kounadi S, Christou I, Nauen R, Douris V, Vontas J (2020) What I cannot create, I do not understand': functionally validated synergism of metabolic and target site insecticide resistance. Proc Biol Sci 287(1927):20200838

Sanger F, Nicklen S, Coulson AR (1977) DNA sequencing with chain-terminating inhibitors. Proc Natl Acad Sci USA 74(12):5463–5467

Shirai Y, Piulachs MD, Belles X, Daimon T (2022) DIPA-CRISPR is a simple and accessible method for insect gene editing. Cell Rep Methods 2(5):100215

Taxiarchi C, Beaghton A, Don NI, Kyrou K, Gribble M, Shittu D, Collins SP, Beisel CL, Galizi R, Crisanti A (2021) A genetically encoded anti-CRISPR protein constrains gene drive spread and prevents population suppression. Nat Commun 12(1):3977

Wiedenheft B, Sternberg SH, Doudna JA (2012) RNA-guided genetic silencing systems in bacteria and archaea. Nature 482(7385):331–338

Yamamoto DS, Sumitani M, Hatakeyama M, Matsuoka H (2018) Malaria infectivity of xanthurenic acid-deficient anopheline mosquitoes produced by TALEN-mediated targeted mutagenesis. Transgenic Res 27(1):51–60

Chapter 8
Pesticide Resistance

Shinji Kasai and Tomoyuki Hashimoto

Abstract In the Indochina Peninsula, dengue vector mosquitoes that exhibit very strong pyrethroid resistance have recently been found. Crossing-over events have resulted in multiple mutations on the target site of insecticide, leading to this strong resistance. *Aedes aegypti* L. carrying such genes is distributed in Vietnam and Cambodia, and there are concerns about the impact of their future movement and spread on dengue control. On the other hand, in Japan, one recent trend of insecticide is room-temperature volatile ingredients that can be treated in spaces. Such new insecticide types possibly work as repellents and therefore they are expected to avoid rapid development of insecticide resistance. Trends in other insecticides that control cockroaches, fire ants, and Argentine ants are also discussed.

Keywords *Aedes aegypti* · *Aedes albopictus* · Argentine ant · Cockroach · Dengue · kdr · Fire ant · Knockdown resistance · L982W · Pyrethroid · VGSC · Voltage-gated sodium channel · Volatile insecticide

8.1 Introduction

A super insecticide-resistant dengue vector mosquito has emerged in the Indochina peninsula, in recent years. *Aedes aegypti* (L.) is distributed in the tropics and subtropics, while *Aedes albopictus* (Skuse) is widely distributed from tropical to temperate regions. Both are common in Asia and they transmit various arboviruses, including dengue, chikungunya, and Zika viruses. The larvae may be found in used

S. Kasai (✉)
Department of Infectious Diseases, National Institute of Infectious Diseases, Tokyo, Japan
e-mail: kasacin@niid.go.jp

T. Hashimoto
Environmental Biology & Living Environmental Department, Japan Environmental Sanitation Center, Kawasaki, Kanagawa, Japan
e-mail: tomoyuki_hashimoto@jesc.or.jp

K. Sawabe et al. (eds.), *Medical Entomology in Asia*, Entomology Monographs,
https://doi.org/10.1007/978-981-97-4406-0_8

tires, abandoned bathtubs, toilets, and other artificial small water containers made of plastic or polystyrene foam. *Ae. aegypti* is also found in vases or other water-filled containers inside houses, which adults tend to occupy. In contrast to *Ae. albopictus*, which sucks blood from various mammals and birds, *Ae. aegypti* prefers human bloodsucking (anthropophilic). Therefore, *Ae. aegypti* has a higher chance of insecticide exposure and development of insecticide resistance.

As with other mosquito-borne infectious diseases, vector control is necessary for controlling *Aedes*-borne infections. Insecticide use is the primary method of vector control. There are many classes of insecticides, which require not only acute and high toxicity to insects but also a high level of safety to humans and animals. Pyrethroid insecticides meet these requirements and play a major role in the control of various vector arthropods. The continuous use of pyrethroids, however, has resulted in the development of insecticide resistance in mosquitoes, making their control more difficult. Because of its high affinity toward humans, *Ae. aegypti* has been exposed to insecticides inside houses and has developed resistance.

This chapter presents the latest research on insecticide resistance and new methods for controlling fire ants and cockroaches in Japan.

8.1.1 Pyrethroid Susceptibility Test in Adult Aedes Mosquitoes

Assessing insecticide susceptibility levels is important to confirm their efficacy and to select appropriate agents. There are three main methods for evaluating the insecticide susceptibility of adult mosquitoes: WHO bottle bioassays (WHO 2022a), WHO tube tests (WHO 2022b), and topical application (Kasai et al. 2014). For dengue vector mosquitoes, in which control by spatial spraying is the primary treatment method, evaluation by topical application is preferable. For malaria-transmitting mosquitoes (i.e., *Anopheles*) and sand flies, insecticides are applied by residual spraying on a wall surface; therefore, the filter paper contact method may be more realistic for evaluating susceptibility. We conducted a test using the insecticide-susceptible strain HKM of *Ae. albopictus* and determined the LD_{99} value for permethrin, which is one of the most widely used pyrethroids. For field populations of *Ae. aegypti*, we administered LD_{99} and $10 \times LD_{99}$ permethrin doses and observed mortality 24 h after treatment. The results indicated that 100% of the susceptible SMK strain died, whereas 17 populations collected in Vietnam and Indonesia exhibited resistance, with mortality less than 20% at LD_{99} (Fig. 8.1). Ten percent or more of the tested insects survived when these 17 populations were treated with a $10 \times LD_{99}$, suggesting that a much higher dose of insecticide is required to completely control these resistant populations (Kasai et al. 2022).

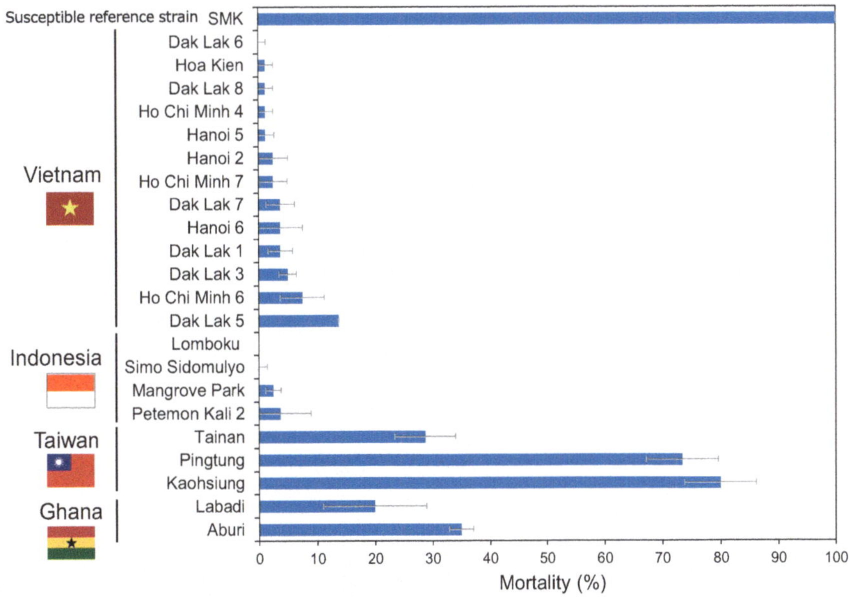

Fig. 8.1 Permethrin susceptibility of 22 field-collected populations of *Ae. aegypti* collected from Vietnam, Indonesia, Taiwan, and Ghana. Permethrin (5.9 ng) was topically applied to each adult mosquito and mortality was assessed 24 h after the treatment. Error bars indicate SEs for the mortalities of each population (four replications)

8.1.2 Pyrethroid Resistance Resulting from Mutations in the Target Protein

The target of the pyrethroids is the voltage-gated sodium channel (VGSC). Insects that develop resistance to pyrethroids often harbor mutations in VGSCs. This is also called knockdown resistance or simply kdr and is the major mechanism of pyrethroid resistance (Scott 2019). In *Ae. aegypti*, mutations such as V1016G, F1534C, I1011M, and V410L, have been reported to confer resistance (Smith et al. 2016) and shown to cause reduced VGSC susceptibility in electrophysiological studies. Therefore, we genotyped one Vietnamese resistant population (Hoa Kien) including surviving and dead individuals after a 10 × LD_{99} of permethrin bioassay, for segment 6 of domain II and segment 6 of domain III in VGSC. Of the 37 individuals who died, 33 were heterozygous for F1534C and L982W. In contrast, 40 of the 41 surviving individuals were homozygous for L982W (Fig. 8.2). L982W is a mutation discovered in 2003 in the Vietnamese *Ae. aegypti*, but it has not received significant attention as a kdr mutation. The L982 allele is located close to a typical kdr allele V1016G in domain II; therefore, it can be detected by genotyping procedures for these mutant alleles.

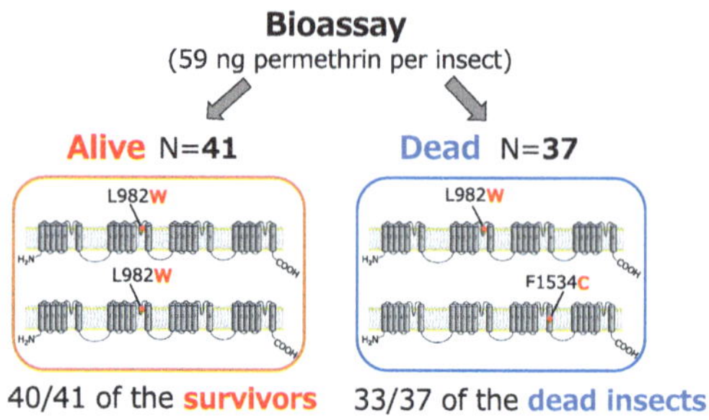

Fig. 8.2 Genotyping of segment 6 of domain II and segment 6 of domain III of *Ae. aegypti* VGSC in Hoa Kien population from Vietnam. Permethrin (59 ng) was topically applied to adult mosquitoes, and survived and dead mosquitoes were genotyped individually

8.1.3 Establishing Resistant Strains Harboring Homozygous VGSC Mutations

A simple, routine genotyping method raised the possibility that L982W may be responsible for a higher level of resistance compared with F1534C. We established strains homozygous for mutant VGSCs from *Ae. aegypti* collected from Vietnam and Singapore. Each pupa was isolated in an Eppendorf tube and allowed to hatch. DNA was isolated from one hind leg of each hatched individual and genotyped for VGSC amino acids S66, L199, A434, F213, L982, S989, T1385, and V1703, to create populations homozygous for VGSC mutations. Eventually, 1594 individuals were individually genotyped in 12 separate steps, resulting in 10 resistant strains. In our previous studies, we established a new method to analyze all VGSC coding regions (>6000 bps) in the genome by combining a target capture method with next-generation sequencing (Itokawa et al. 2019). By analyzing the full VGSC coding regions for at least 8 individuals for each strain, we confirmed that each of the 10 strains contained a single VGSC haplotype (Kasai et al. 2022).

8.1.4 Pyrethroid Resistance Levels of 10 Resistant Strains

The 10 established resistant strains were tested for permethrin susceptibility and compared for resistance levels. A range of permethrin doses were administered and the LD_{50} values were calculated based on the dose-mortality response. Resistance ratios (RRs) were calculated by dividing the LD_{50} of each strain by the LD_{50} of the susceptible SMK strain. The RRs of 3 strains harboring F1534C exhibited fivefold to eightfold resistance, the RRs of 3 strains harboring V1016G were 36–102-fold, and the RRs of 2 strains harboring L982W were 52- and 75-fold (Table 8.1). Notably, the RR of the FTWC strain, which contains both L982W and F1534C, was 1039-fold. Furthermore, the PGC strain harboring both V1016G and F1534C also

Table 8.1 Ten resistant strains established and their levels of permethrin resistance

Grouping	Strain name	Mutation(s)	Resistance ratio	Resistance ratio (+PBO[a])
With F1534C	F1534C	F1534C	8.0	3.6
With F1534C	FC66	S66F + F1534C	5.0	5.5
With F1534C	FC213	L213F + F1534C	6.1	6.1
With V1016G	GY	V1016G + D1763Y	36	27
With V1016G	PGG	S989P + V1016G + V1703G	50	18
With V1016G	PG	S989P + V1016G	102	48
With L982W	FTW	L199F + A434T + L982W	52	33
With L982W	FWI	L199F + L982W + T1385I	75	47
Double kdr mutations	PGC	S989P + V1016G + F1534C	165	177
Double kdr mutations	FTWC	L199F + A434T + L982W + F1534C	1039	300

Adapted from Kasai et al. (2022)
[a]*PBO* piperonyl butoxide, an inhibitor of cytochrome P450 monooxygenases

exhibited a high resistance ratio of 165-fold (Table 8.1). These RRs did not change significantly following pretreatment with piperonyl butoxide (PBO), an inhibitor of the detoxifying enzyme cytochrome P450 monooxygenases (Table 8.1). In the FTWC strain, the RR became 300-fold from 1039-fold under PBO pretreatment conditions, indicating that detoxification activity by P450 monooxygenases contributes to the resistance (Kasai et al. 2022).

8.1.5 Concomitant Mutations V1016G and F1534C

V1016G was reported in 2003 from *Ae. aegypti* populations collected from Thailand and Indonesia (Brengues et al. 2003), whereas F1534C was reported in 2010 from an *Ae. aegypti* population collected from Thailand (Yanola et al. 2010). Populations with each amino acid substitution alone have been identified; however, individuals with both have not been identified for many years. We artificially established VGSCs harboring both mutations, expressed them in *Xenopus* oocytes, and electrophysiologically examined their sensitivity to pyrethroids. The relative decrease in sensitivity to both permethrin and deltamethrin was V1016G + F1534C > V1016G > F1534C (Hirata et al. 2014). If mosquitoes carrying both V1016G and F1534C type VGSC were to emerge in the field, their control with pyrethroids would be extremely difficult. About the same time we published these data, *Ae. aegypti* with a VGSC exhibiting V1016G and F1534C was identified in Myanmar (Kawada et al. 2014). Subsequently, similar cases have been reported in Singapore (Pang et al. 2015), Malaysia (Ishak et al. 2015), Indonesia (Wuliandari et al. 2015), and Saudi Arabia (Al Nazawi et al. 2017). We established a strain homozygous for V1016G + F1534C

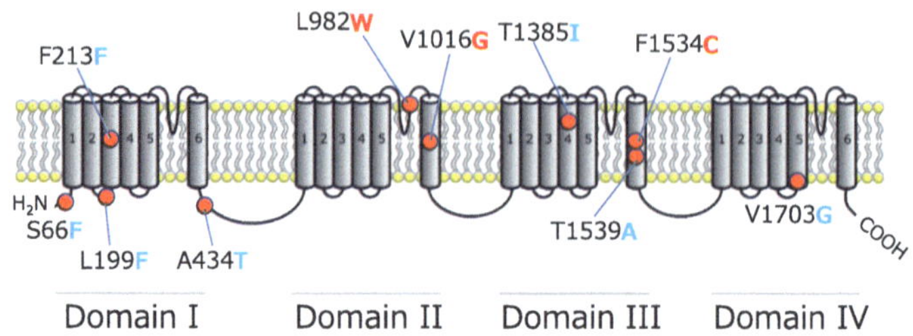

Fig. 8.3 Seven newly found synonymous mutations of *Ae. aegypti* VGSC found by the analysis of full-length coding region. L982W, V1016G, and F1534C are mutations found in previous papers and confer pyrethroid resistance

and showed for the first time that resistance levels are markedly higher compared with populations harboring either mutation alone.

8.1.6 Molecular Model Analysis

In the process of establishing 10 resistant strains, seven amino acid mutations (S66F, L199F, F213F, A434T, T1385I, T1539A, and V1703G) were identified for the first time in the VGSC of *Ae. aegypti* (Fig. 8.3). A molecular model analysis was performed to determine whether these mutations confer insecticide resistance. Twelve mutations were newly found in *Ae. aegypti* from Vietnam and Singapore were located away from the permethrin-binding region. In contrast, L982W, V1016G, and F1534C were located in the permethrin-binding region, suggesting that these amino acid mutations are strongly associated with decreased susceptibility to pyrethroids. Similarly, mutations L199F and A434T found in the FTWC strain were unlikely to be involved in resistance. Because L199F is always accompanied by L982W, this does not rule out the possibility that this mutation is necessary for VGSC to maintain its function as a channel. Furthermore, a docking simulation analysis suggested that L982W alone inhibits permethrin binding, but the double mutations, L982W and F1534C, would further inhibit its binding to VGSC. In 2023, electrophysiological studies of VGSCs expressed in *Xenopus* oocytes provided direct evidence that L982W is involved in pyrethroid resistance (Wu et al. 2023); however, definitive results on the synergistic effect of L982W and F1534C were not provided. Further studies are needed on the synergistic effects of multiple amino acid substitutions, including the involvement of L199F.

8.1.7 Super-Insecticide-Resistant Genes Are the Result of Recombination Events

Analysis of the full-length coding region of the VGSC gene revealed several synonymous mutations as well as non-synonymous mutations that cause amino acid substitutions. The double-mutant *Ae. aegypti* VGSC did not result from the

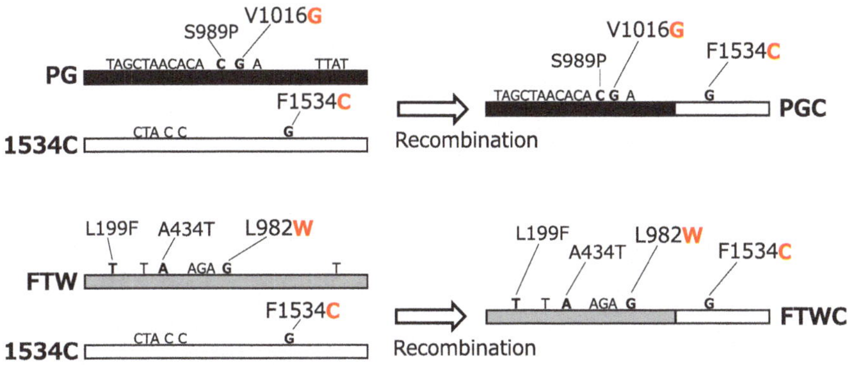

Fig. 8.4 Super-pyrethorid-resistant alleles were generated by recombination (crossing over) events. Small characters indicate nucleotide mutations found in each haplotype, indicating traces of evolution. Bold letters indicate non-synonymous mutations, causing amino acid substitutions

accumulation of point mutations, but from recombination between two haplotypes (Fig. 8.4). In nature, recombination occurs during meiosis in germ cells, and amino acid mutations involved in resistance can accumulate efficiently and rapidly.

8.1.8 Frequency of kdr Genes in Ae. aegypti *in the Indochina Peninsula*

The *kdr* gene frequencies of *Ae. aegypti* collected from northern (Hanoi), highland (Dak Lak), and southern (Ho Chi Minh City) Vietnam were analyzed. The frequency of V1016G, which has been recognized as a major kdr mutation in Asia, was low at 10.8% in Hanoi and 0.3% in Dak Lak. It was not identified in 89 individuals in Ho Chi Minh City. In contrast, L982W, which was considered to confer similar level of resistance as V1016G based on insecticidal bioassays, accounted for more than 79% of the mosquitoes in these regions, indicating that many mosquitoes in Vietnam already harbor L982W (Fig. 8.5). Furthermore, a certain frequency of individuals with the L982W + F1534C double mutation was observed, including >28.4% in Hanoi. An analysis of 51 individuals captured in Phnom Penh, Cambodia, in 2019 revealed that L982W + F1534C accounted for 71.6% and V1016G + F1534C accounted for 18.6% of the population. This indicates that genes conferring super-resistance account for more than 90% of the total in the capital of Cambodia (Fig. 8.5).

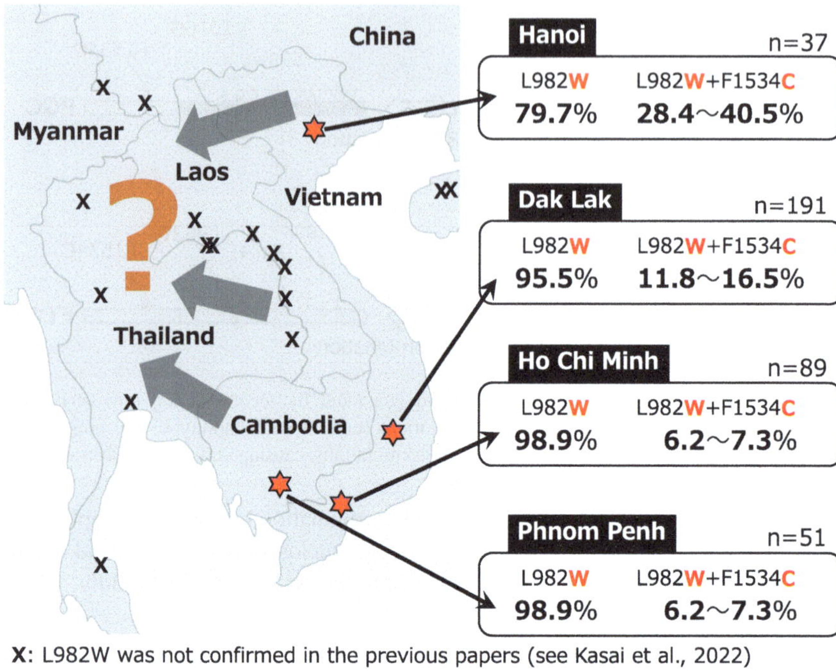

Fig. 8.5 Frequency of L982W and L982W + F1534C alleles of VGSC in *Ae. aegypti* collected from Vietnam and Cambodia. X denotes the location where L982W was not confirmed in *Ae. aegypti* although it was genotyped

8.1.9 *Threat of Multiple Mutations in VGSC of* Ae. aegypti *in the Indochina Peninsula*

To date, several studies have focused on the L982 allele of *Ae. aegypti* collected in the Indochina Peninsula; however, L982W has never been confirmed outside of Vietnam or Cambodia (Fan et al. 2020; Marcombe et al. 2022; Shimono et al. 2021). It will be of interest to determine the actual status of L982W in neighboring Laos, Thailand, Malaysia, Myanmar, and China (Fig. 8.5). Although VGSCs of the V1016G + F1534C type have already been identified in Myanmar, Singapore, Malaysia, Indonesia, Saudi Arabia, and Thailand, the distribution of L982W + F1534C, which causes stronger resistance compared with each single mutation, is of more interest. If these multiple VGSC mutant-carrying *Ae. aegypti* become widely distributed in Southeast Asia and increase in frequency, they will pose a major threat to the control of dengue fever and other arbovirus infections.

8.1.10 Reconsideration of the Previous Genotyping Results

Kawada et al. captured *Ae. aegypti* throughout Vietnam from 2006 to 2008 and examined the correlation between pyrethroid susceptibility and the *kdr* genes (Kawada et al. 2009a, b). F1534C mutation was found at some frequency, but there was no strong correlation with pyrethroid susceptibility. In this study, L982W was not recognized as a kdr gene. Since the involvement of L982W in pyrethroid resistance was reported, the previous data were reexamined and a strong correlation between L982W and pyrethroid susceptibility was observed (Kawada et al. 2023). The allele frequencies of L982W in *Ae. aegypti* collected between 2006 and 2008 were 0, 91.1, and 66.7% in the north, highlands, and south Vietnam, respectively, which were lower compared with those of *Ae. aegypti* collected in Vietnam in 2016 (Hanoi 79.7%, Dak Lak 95.5%, Ho Chi Minh 98.9%). It is presumed that mosquitoes were selected by pyrethroid exposure during this period.

8.1.11 Pyrethroid Resistance and kdr Mutations in Ae. albopictus

As previously mentioned, *Ae. albopictus* is relatively less susceptible to pyrethroid insecticides compared with *Ae. aegypti* because its ecology is different and it is not frequently exposed to insecticides. Nevertheless, some populations have developed resistance. F1534C, which confers relatively low levels of pyrethroid resistance, was first identified in the Singapore population of *Ae. albopictus* (Kasai et al. 2011), which was reported from several other countries. Eight years later, V1016G, which confers a higher level of resistance, was identified in *Ae. albopictus* samples collected from Italy, Vietnam, and China (Kasai et al. 2019; Zhou et al. 2019). In China, *Ae. albopictus* with multiple kdr mutations (F1534C/S/L and I1532T) has been identified at a relatively high frequency (Wei et al. 2021). In 2022, VGSC carrying V1016G were found throughout Europe, including Bulgaria, France, Georgia, Malta, Romania, Spain, Switzerland, and Turkey (Pichler et al. 2022). Therefore, the possibility of detecting L982W or multiple kdr mutations in *Ae. albopictus* in the future cannot be ruled out. In addition, with the development of vast worldwide transportation networks, there is an increasing risk that dried viable eggs of resistant populations will be carried across borders. In recent years, the distribution of *Ae. albopictus* has expanded in various parts of Europe. The European population of *Ae. albopictus* is believed to have originated in Asia, where it has shown varying levels of susceptibility to insecticides, particularly pyrethroids. In Turkey, for example, the western part of the country experienced an expansion of *Ae. albopictus*, but some populations have been shown to exhibit reduced susceptibility to pyrethroids (Kasai et al., unpublished data). It will be important to determine whether the resistant population in Turkey, which is often classified as West Asian, was introduced from Europe, such as Romania and Bulgaria, or from Asian regions. The development of logistics for invasion surveillance and vector mosquito control methods are needed. It is hoped that future population genetic studies will clarify this issue.

8.1.12 Knockdown Resistance Genes and Fitness Cost

L982W was identified at a high frequency in *Ae. aegypti* collected from Vietnam and Cambodia. It is located on the linker between the fifth and sixth segments of the second domain of VGSC. This region is important for ion selectivity and permeability (Silva and Scott 2020). In fact, the 31 amino acids surrounding L982W are completely conserved among more than 80 insects. We crossed the FTWC strain carrying the L982W + F1534C mutations with a susceptible SMK strain and repeated the self-crossing for up to 15 generations. In this process, the allele frequency of L982W + F1534C was examined, and its allele frequency decreased over generations, from 50% in F1 to 14% in the 15th generation (Uemura et al. 2023). The resistance gene could not be maintained in an environment without pyrethroid selection pressure, thus a fitness cost existed. In the future, it will be important to determine the factors responsible for the fitness cost.

8.1.13 Conclusions

In recent years, attempts have been made worldwide to use *Wolbachia* bacteria and genetically modified mosquitoes to reduce the density of mosquitoes or reduce arbovirus susceptibility to change the vector competence of mosquitoes. However, these techniques are still in their infancy and there are still several issues to address in terms of safety, effect on the nature, and cost. Therefore, we may have to continue to control vector mosquitoes with insecticides for a while longer. Pyrethroid insecticides are the most commonly used agents because of their safety and high efficacy against wild-type mosquitoes. Future studies are needed to analyze the frequency of these resistance genes in detail in Southeast Asia, particularly in the Indochina Peninsula. It is also important to analyze the physiology and ecology of mosquitoes carrying resistant genes, including their viability and virus transmission ability.

8.2 Evolution of Pesticides and Recent Case of Their Usage

8.2.1 Change of Pest Control and the Position of Pyrethroids in Household Pesticides in Japan

In Japan, pesticides for medically important pests are regulated by the Pharmaceutical and Medical Device (PMD) Act under the jurisdiction of the Ministry of Health, Labor and Welfare. The medically important pests in the PMD Act include house flies, mosquitoes, cockroaches, fleas, lice, bed bugs, tropical rat mites, house dust mites, and ticks. Until around 1960, the principal purpose of pesticides was vector control, such as mosquitoes and flies, from the standpoint of public health (Seki 2009). This significance has diminished with the spread of sewage systems and other environmental improvements. Furthermore, sanitation in the living environment has kept city dwellers away from vectors and harmful pests, and the types and importance of pests have changed drastically. Accordingly, the classification and formulation of pesticides have become more complex. In such a situation, formulations containing pyrethroid ingredients have remained the mainstay in household pesticides for a long time, accounting for more than 90% of the market share (Katsuda 2006).

8.2.2 Advancement of Room-Temperature Volatile Pyrethroids

Pyrethroids exhibit high lethal and knockdown activity at low doses, while their toxicity to humans and animals is extremely low. New pyrethroid ingredients have been continuously developed in response to the demand for household pesticides. One recent trend is room-temperature volatile ingredients that can be treated in spaces and are also intended to repel flying insects. Existing empenthrin has been known as a volatile ingredient at room temperature. Pyrethroid ingredients with relatively high volatility and higher performance have been developed since the 2000s (Table 8.2).

Transfluthrin developed by Bayer AG was approved in Japan in 2001 as an active ingredient (Shinjo et al. 2021). Compared to empenthrin, transfluthrin exhibits highly lethal and immediate effects against common mosquitoes and houseflies. It has been formulated as a household pesticide for flies, mosquitoes, and cockroaches, as a resin transpiration formulation, aerosol, mosquito coil, liquid, and fan formulation. Metofluthrin was developed by Sumitomo Chemical with the goal of searching for a component that surpasses d-allethrin and prallethrin and was approved in 2005 (Ujihara et al. 2004). It has been shown to be highly effective against *Culex pipiens*, *Aedes albopictus*, and *A. aegypti*, and liquid, fan, aerosols, and transpiration formulations have been marketed to date. In addition, profluthrin is a compound discovered in the process of the development of metofruthrin (Ujihara et al. 2010). There

Table 8.2 Highly volatile pyrethroids launched in Japan

Common name	Launched year in Japan	Vapor Pressure (mPa)	References
Empenthrin	1993	14 (23.6 °C)	14th pesticide manual
Transfluthrin	2001	0.4 (20 °C)	14th pesticide manual
Metofluthrin	2005	1.96 (25 °C)	14th pesticide manual
Profluthrin[a]	2004	10.3 (25 °C)	Ujihara et al. (2010)

Adapted from Ujihara et al. (2010) and Tomlin (2006)

[a]Profluthrin is not approved as a hygienic pesticide in Japanese law, PMD Act

are no products of profluthrin for mosquitoes or cockroaches under the PMD Act, but there are hanging transpiration agents for clothing and nuisance insects. These room-temperature volatile pyrethroids are expected to be effective as spatial insect repellents, and new products are launched every year (Fig. 8.6) (Insecticide Research Group Meeting 2022).

Nowadays, the resistance of pyrethroids to house flies, mosquitoes, cockroaches, bed bugs, etc., is a global concern. Reduced sensitivity due to sodium channel mutations is considered the most crucial factor. Yet another known factor is detoxification metabolism by cytochrome P450 enzymes. Pyrethroid components with a phenoxybenzyl ring, such as permethrin and deltamethrin, have been shown to be decomposed from its ring by P450 enzymes. In contrast, transfluthrin, which does not have a phenoxybenzyl ring in its structure, has been suggested to be less susceptible to detoxification metabolism by the cytochrome P450 (Horstmann and Sonneck 2016).

Such highly volatile pyrethroids are expected to have the suppression effect of the blood-sucking of indoor mosquitoes. Although its high volatility at room temperature makes it difficult to apply to mosquito nets used in areas where malaria is endemic, long-term inhibition of blood-sucking is expected due to the continuous release of components (Kawada et al. 2020). Long-term ingredient release can lead to resistance development due to constant exposure to mosquitoes. It is suggested that the inhibition of the blood-sucking behavior of metofluthrin in pyrethroid-resistant *Aedes aegypti* is due to lethality and knockdown effects rather than spatial avoidance behavior (Ritchie and Devine 2013). However, there is a point of view

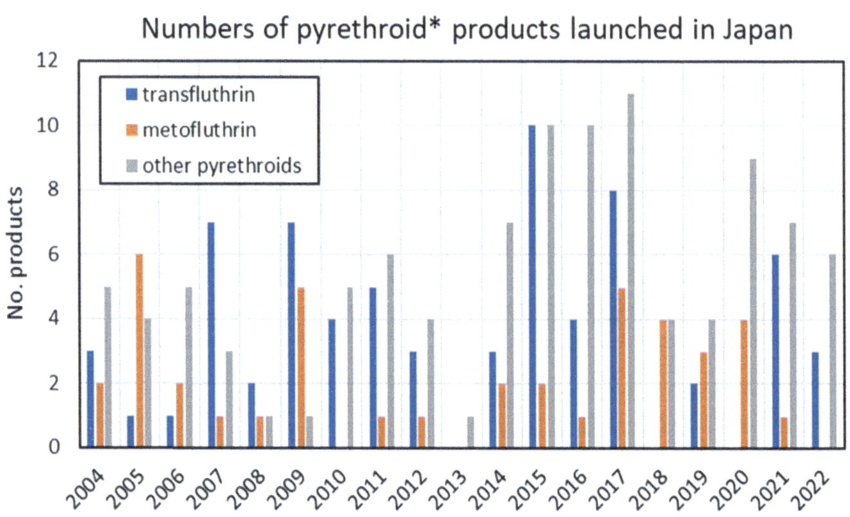

Fig. 8.6 Numbers of PMD-Act approved pyrethroid products launched in Japan. Products containing pyrethrins and etofenprox were counted on "other pyrethroids." Products containing transfluthrin or metofluthrin with other pyrethroid were counted as transfluthrin or metofluthrin, respectively. Products for only household pesticides were counted. (Adapted from Ujihara et al. 2010)

that spatial repellent could allow mosquitoes the opportunity to escape, only inducing mortality if insects stay in the treated space (Britch et al. 2020). Along with elucidating the mechanism of metabolic resistance in the highly volatile pyrethroids, research on practical usage is being developed.

In Japan, where mosquito-borne diseases such as dengue and West Nile fever are not prevalent, local governments should implement measures against mosquito larvae. Personal measures should focus on removing its breeding source in the yard and controlling the adult mosquitoes. Pyrethroids, which can be volatilized into the air, are the main insecticides used for this purpose, and demand for these products will probably continue in the future.

8.2.3 Expectations and Challenges of Pesticides on Cockroach Control

As mentioned above, only the PMD Act-approved chemicals can be used to control cockroaches in Japan. The residual and spatial spray of organophosphorus and pyrethroid formulations were the main method for controlling cockroaches till about 1990 in Japan. Meanwhile, in the US, toxic baits containing hydramethylnon were switched from conventional formulations for cockroach control (Abe 1995). Boric acid dumplings containing 15–40% boric acid have also been used in Japan as a toxic bait for a long time, but they were not widely applied for German cockroach control because of their low lethal effect (Ogawa et al. 2006). The newest formulation at that time, hydramethylnon bait, was highly palatable to German cockroaches. The toxic baits were gradually introduced into the household formulations due to easy handling and less indoor pollution. Moreover, a combination of syringe-type applicators and bait guns greatly contributed to the widespread as the Pest Control Operator (PCO) use formulations (hygienic formulations). Subsequently, other ingredients such as fipronil, dinotefulan, and indoxacarb have been added to the lineup, and the toxic baits have become accepted by Japanese PCO. Today, the PCO can choose from various baits, but it has become clear that different baits have different feeding properties and lethal efficacy against field-collected cockroaches (Fig. 8.7) (Shinjo et al. 2006). Although the toxic bait treatment is superior to the conventional formulations in terms of the efficiency of chemical treatment operations, as it has become more widely used to control cockroaches, new problems have arisen. It has been reported that cockroaches develop behavioral resistance to avoid ingesting D-glucose in the baits, called glucose aversion (Wada-Katsumata et al. 2014). In addition, physiological resistance has also been reported (Wang et al. 2004). For these reasons, new formulations other than baits have been desired.

Ingredients in the hygienic and household pesticides, except the baits, were organophosphorus, pyrethroids, and carbamates. Many of these were approved before the 1990s. It is recommended that pesticides with different modes of action be used in rotation to suppress the development of resistance in theory. However, the

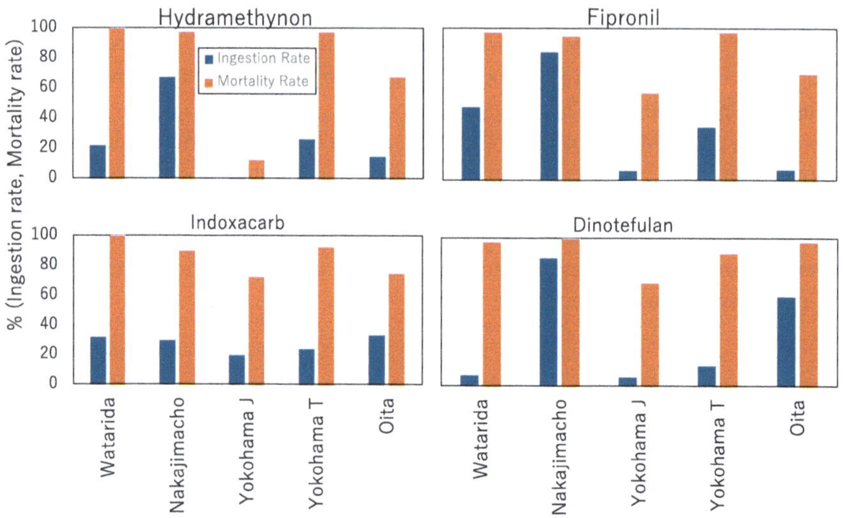

Fig. 8.7 Difference of susceptibility of five colonies of German cockroach on four bait formulations. Toxic bait and nontoxic bait were placed together and allowed to be consumed freely by the cockroaches (20 females and 20 males), and the subsequent lethal situation was monitored. Ingestion rate = (Reduction weight of toxic bait/Reduction weight of nontoxic + toxic bait) × 100, Mortality Rate was observed 10 days after treatment. (Adapted from Shinjo et al. 2006)

PCO had only a few options available, and they were not able to change the pesticides under annual contracts. As a result, the same pesticides have been treated repeatedly, and the resistance has become a common problem in many buildings and restaurants.

Under such circumstances, a new ingredient, broflanilide with a new mode of action, was released in 2021. Broflanilide inhibits the transmission of chloride ions to nerve cells by acting on GABA receptors in nerve cells, and its mechanism of action was categorized in group 30 of the International Resistance Action Committee's (IRAC) classification (Katsuta et al. 2019). This active ingredient indicated high lethal activity in the laboratory tests against organophosphate- and pyrethroid-resistant strains (Table 8.3). And 100% control was achieved in field tests at restaurants where control had been difficult in the past (Hashimoto and Kazuma 2023). It was notable that contact with the residue surface for only 10 min propagated the lethal effect from exposed cockroaches to unexposed cockroaches in the laboratory tests (Fig. 8.8). The flowable formulation of this ingredient is applied against cockroaches, bed bugs, fleas, and house dust mites in Japan. PCO has already begun using it to control the German cockroach. Because of its formulation, it requires a hand sprayer and room preparation for the treatment and more time for spraying than bait formulations. It is interesting how far Japanese PCO will introduce it instead of the bait, as it is expected to be highly effective.

Table 8.3 Mortality and knockdown status of four colonies of German cockroach when in contact with residue of three formulations for 6 h. Values were average of three replications with ten females

Formulation	AI(mg)/m^2	Colony	% KD	% Mortality	
			360 min	d1	d2
Broflanilide FL	50	Watarida	0	80.0	100
		Kamogawa	0	86.7	100
		Hamamatsucho	0	90.0	100
		Daishi	0	83.3	100
Propetamphos EC	150	Watarida	16.7	73.3	80.0
		Kamogawa	6.7	63.3	63.3
		Hamamatsucho	0	0	0
		Daishi	0	0	3.3
Permethrin EC	250	Watarida	13.3	26.7	40.0
		Kamogawa	0	0	0
		Hamamatsucho	3.3	3.3	3.3
		Daishi	0	0	0

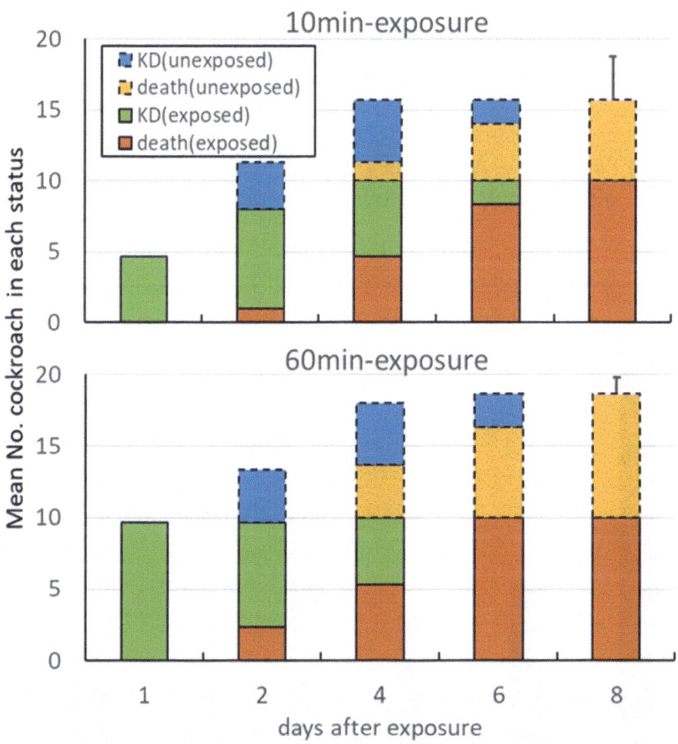

Fig. 8.8 Lethality and knockdown status after mixing German cockroaches (ten females; organophosphorus- and pyrethroid-resistant) exposed to broflanilide-treated surfaces (50 mg/m^2) for 10 or 60 min with unexposed cockroaches (ten females). Values were average of three replications

8.2.4 The Front Line of Invasive Alien Ants in Japan

Fire ants, including *Solenopsis invicta* and *S. geminata* have been found in Japan since 2017. The fire ants are poisonous and aggressive to humans, making them harmful pests in the countries where they were established. These South American native ants are currently expanding their distribution in Asian countries. In Japan, 92 cases have been confirmed by the end of 2022 (Fig. 8.9). The ants were found in the container yards of harbor areas between April and October in most cases. They were also found in 43 marine containers. Of these containers, 37 containers departed from China, and six containers were from other countries respectively (Fig. 8.10) (MOE 2024a). Fortunately, the fire ants found in Japan were exterminated on each occasion and have not yet established themselves.

On the other hand, the Argentine ant, *Linepithema humile*, is not a harmful pest, but a nuisance. The ants were first discovered in Japan in 1993. While they were successfully exterminated in some areas, they became established in the Setouchi region (Tatsuki 2008). The Ministry of Environment has created a list of designated invasive alien species that includes these ants and has taken control actions when detected (MOE 2024b).

A primary countermeasure is to keep ants out of the marine containers in the country of departure. To prevent the ants from entering the containers, applying pesticide or repellent to the containers is effective. However, it is not easy to treat the chemicals in the containers due to the national regulations on using chemicals depending on the country. Approaches to countermeasure differs from country to country. It is also difficult to ask other countries to take countermeasures by

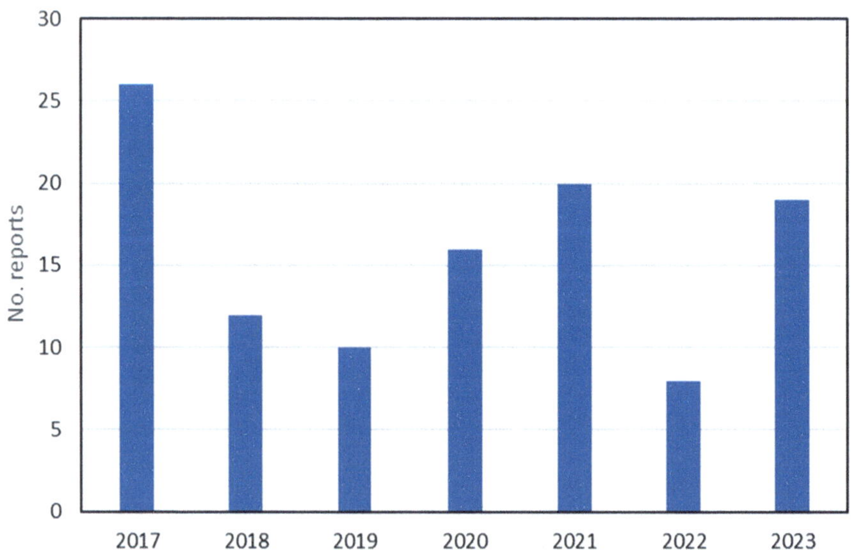

Fig. 8.9 Confirmation reports of the fire ants in Japan. (Adapted from MOE 2024a)

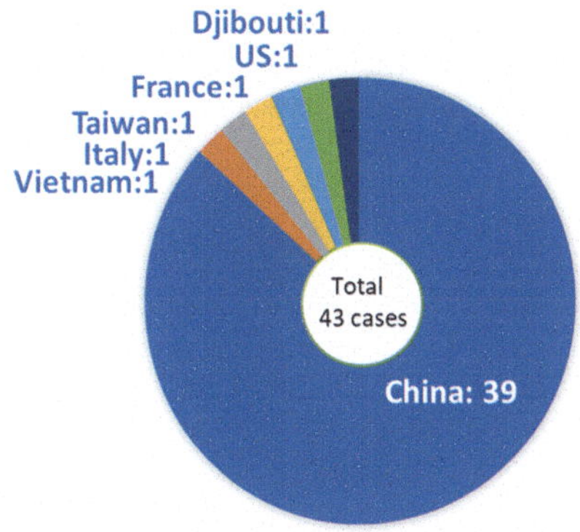

Fig. 8.10 Countries of departure of marine containers in which the fire ants were found in Japan (2017–2023). (Adapted from MOE 2024a)

themselves. As a result, what Japanese importers can do is border measures, including inspection, detection, and elimination of invasive alien ants.

Baits and liquid formulations are mainly used as the elimination agents against the fire ants. The baits are for foraging ants and the colony, and the liquid pesticides are applied to direct spray or residual spray. The liquid formulations containing pyrethroid ingredients work quickly, but if the ant population is high, run the risk of spreading them around.

Regarding the baits, since fire ant mounds are still rare in Japan, the broadcast application is adopted in the fire ant control. The broadcast baiting allows the foraging ants to access the baits easily. Unlike cockroaches, worker ants find the bait, carry it back to their colony, and feed it to the larvae, other workers, and queens. Thus, the operator can quickly and completely control large areas.

As for household insecticides, two Japanese bait formulations containing hydramethylnon and fipronil as active ingredients have been shown to exhibit a high extermination effect against fire ants (Yasudai et al. 2022). Recently, the susceptibility of *S. invicta* to pyrethroids and new-type agents was investigated, and fipronil showed the lowest LD_{50} (Table 8.4) (Sakamoto and Goka 2021). Cytochrome P450 detoxification of fipronil has been shown in the fire ants (Xiong et al. 2019), but apparent pesticide resistance in field fire ant colonies is not yet known. Social ants are less likely to develop insecticide resistance due to their lower genetic diversity. Even if some workers survive after pesticide treatment, they usually cannot oviposit. Killing the queen in the nest is difficult, but pesticide resistance is unlikely to occur. Therefore, it is imperative to feed the ants well. Similar to the cockroach countermeasures, preference for baits would be an important factor.

8.2.5 Conclusions

Pesticides for medically important pests in Japan have changed in accordance with changes in the pest fauna. Vectors have become less important in the past, but recently there has been concern about the prevalence of emerging and reemerging

Table 8.4 Susceptibility of *Solenopsis invicta* to pesticide active ingredients by topical application within 24 h

Chemicals	LD50 (ng/ant) at 24 h
Transfluthrin	9.6
Prallethrin	10.9
Phenothrin	10.9
Permethrin	13.6
Metofluthrin	14.5
Pyrethrin	137.0
Fipronil	1.1
Thiamethoxam	6.4
Indoxacarb	109.6
Imidacloprid	84.3
Hydramethylnon	>100

Adapted from Sakamoto and Goka (2021)

infectious diseases. Some species of pests are not expected to be controlled by conventional pesticides. In addition, the invasive alien species has created a need for new responses. The concept of control strategies that do not rely on pesticides is spreading among Japanese PCOs, but pesticides still remain an important means of control measures. Considering the time and expense involved in developing new insecticides, they need to be used in such a way that they do not develop resistance to existing insecticides.

References

Abe Y (1995) Point treatment technology for household and epidemic prevention insecticides. Jpn J Environ Entomol Zool 7:32–36. (in Japanese)

Al Nazawi AM, Aqili J, Alzahrani M, McCall PJ, Weetman D (2017) Combined target site (*kdr*) mutations play a primary role in highly pyrethroid resistant phenotypes of *Aedes aegypti* from Saudi Arabia. Parasit Vectors 10(1):161

Brengues C, Hawkes NJ, Chandre F, McCarroll L, Duchon S, Guillet P, Manguin S, Morgan JC, Hemingway J (2003) Pyrethroid and DDT cross-resistance in *Aedes aegypti* is correlated with novel mutations in the voltage-gated sodium channel gene. Med Vet Entomol 17(1):87–94

Britch SC, Linthicum KJ, Kline DL, Aldridge RL, Golden FV, Wittie J, Henke J, Hung K, Gutierrez A, Snelling M, Lora C (2020) Transfluthrin spatial repellent on US military materials reduces *Culex tarsalis* incursion in a desert environment. J Am Mosq Control Assoc 36(1):37–42

Fan Y, O'Grady P, Yoshimizu M, Ponlawat A, Kaufman PE, Scott JG (2020) Evidence for both sequential mutations and recombination in the evolution of kdr alleles in *Aedes aegypti*. PLoS Negl Trop Dis 14(4):e0008154

Hashimoto T, Kazuma T (2023) Insecticidal efficacy of Broflanilide against German cockroach, *Blattella germanica* (Dictyoptera: Blattellidae): field and laboratory evaluations. Med Entomol Zool 74(4):133–138

Hirata K, Komagata O, Itokawa K, Yamamoto A, Tomita T, Kasai S (2014) A single crossing-over event in voltage-sensitive Na^+ channel genes may cause critical failure of dengue mosquito control by insecticides. PLoS Negl Trop Dis 8(8):e3085

Horstmann S, Sonneck R (2016) Contact bioassays with phenoxybenzyl and tetrafluthrin pyrethroids against target-site and metabolic resistant mosquitoes. PLoS One 11(3):e0149738

Insecticide Research Group Meeting (2022) List of newly launched insecticides. https://server51.joeswebhosting.net/~js4308/insecticide/publications (in Japanese). Accessed 30 Dec 2022

Ishak IH, Jaal Z, Ranson H, Wondji CS (2015) Contrasting patterns of insecticide resistance and knockdown resistance (*kdr*) in the dengue vectors *Aedes aegypti* and *Aedes albopictus* from Malaysia. Parasit Vectors 8:181

Itokawa K, Sekizuka T, Maekawa Y, Yatsu K, Komagata O, Sugiura M, Sasaki T, Tomita T, Kuroda M, Sawabe K, Kasai S (2019) High-throughput genotyping of a full voltage-gated sodium channel gene via genomic DNA using target capture sequencing and analytical pipeline MoNaS to discover novel insecticide resistance mutations. PLoS Negl Trop Dis 13(11):e0007818

Kasai S, Ng LC, Lam-Phua SG, Tang CS, Itokawa K, Komagata O, Kobayashi M, Tomita T (2011) First detection of a putative knockdown resistance gene in major mosquito vector, *Aedes albopictus*. Jpn J Infect Dis 64(3):217–221

Kasai S, Komagata O, Itokawa K, Shono T, Ng LC, Kobayashi M, Tomita T (2014) Mechanisms of pyrethroid resistance in the dengue mosquito vector, *Aedes aegypti*: target site insensitivity, penetration, and metabolism. PLoS Negl Trop Dis 8(6):e2948

Kasai S, Caputo B, Tsunoda T, Cuong TC, Maekawa Y, Lam-Phua SG, Pichler V, Itokawa K, Murota K, Komagata O, Yoshida C, Chung HH, Bellini R, Tsuda Y, Teng HJ, Filho JLL, Alves LC, Ng LC, Minakawa N, Yen NT, Phong TV, Sawabe K, Tomita T (2019) First detection of a *Vssc* allele V1016G conferring a high level of insecticide resistance in *Aedes albopictus* collected from Europe (Italy) and Asia (Vietnam), 2016: a new emerging threat to controlling arboviral diseases. Euro Surveill 24(5):1700847

Kasai S, Itokawa K, Uemura N, Takaoka A, Furutani S, Maekawa Y, Kobayashi D, Imanishi-Kobayashi N, Amoa-Bosompem M, Murota K, Higa Y, Kawada H, Minakawa N, Cuong TC, Yen NT, Phong TV, Keo S, Kang K, Miura K, Ng LC, Teng HJ, Dadzie S, Subekti S, Mulyatno KC, Sawabe K, Tomita T, Komagata O (2022) Discovery of super-insecticide-resistant dengue mosquitoes in Asia: threats of concomitant knockdown resistance mutations. Sci Adv 8(51):eabq7345

Katsuda S (2006) Overview of household pesticides 1. In: General remarks of household pesticides III. Household Pesticide Industry Association of Japan, pp 1–9. (in Japanese)

Katsuta H, Nomura M, Wakita T, Daido H, Kobayashi Y, Kawahara A, Banba S (2019) Discovery of broflanilide, a novel insecticide. J Pestic Sci 44(2):120–128

Kawada H, Higa Y, Komagata O, Kasai S, Tomita T, Thi Yen N, Loan LL, Sanchez RA, Takagi M (2009a) Widespread distribution of a newly found point mutation in voltage-gated sodium channel in pyrethroid-resistant *Aedes aegypti* populations in Vietnam. PLoS Negl Trop Dis 3(10):e527

Kawada H, Higa Y, Nguyen YT, Tran SH, Nguyen HT, Takagi M (2009b) Nationwide investigation of the pyrethroid susceptibility of mosquito larvae collected from used tires in Vietnam. PLoS Negl Trop Dis 3(3):e391

Kawada H, Oo SZ, Thaung S, Kawashima E, Maung YN, Thu HM, Thant KZ, Minakawa N (2014) Co-occurrence of point mutations in the voltage-gated sodium channel of pyrethroid-resistant *Aedes aegypti* populations in Myanmar. PLoS Negl Trop Dis 8(7):e3032

Kawada H, Nakazawa S, Shimabukuro K, Ohashi K, Kambewa EA, Pemba DF (2020) Effect of metofluthrin-impregnated spatial repellent devices combined with new long-lasting insecticidal nets (Olyset® Plus) on pyrethroid-resistant malaria vectors and malaria prevalence: field trial in south-eastern Malawi. Jpn J Infect Dis 73(2):124–131

Kawada H, Higa Y, Kasai S (2023) Reconsideration of importance of the point mutation L982W in the voltage-sensitive sodium channel of the pyrethroid resistant *Aedes aegypti* (L.) (Diptera: Culicidae) in Vietnam. PLoS One 18(5):e0285883

Marcombe S, Shimell K, Savage R, Howlett E, Luangamath P, Nilaxay S, Vungkyly V, Baby A, King M, Clarke J, Jeffries C, Jojo J, Lacey E, Bhatty F, Mabika D, Dela Cruz A, Fisher C, Mbadu M, Despiniadis I, Brey PT, Thammavong P, Jones AK (2022) Detection of pyrethroid resistance mutations and intron variants in the voltage-gated sodium channel of *Aedes* (*Stegomyia*) *aegypti* and *Aedes* (*Stegomyia*) *albopictus* mosquitoes from Lao People's Democratic Republic. Med Vet Entomol 36(4):424–434

Ministry of the Environment (MOE) (2024a) Basic concept of fire ant control ver.4.0. https://www.env.go.jp/nature/intro/2outline/attention/file/20230403_kangaekata_ver4.pdf. Accessed 17 Jan 2024

Ministry of the Environment (MOE) (2024b) Information on red imported fire ant (RIFA), a designated invasive alien species. https://www.env.go.jp/nature/intro/2outline/attention/hiari.html. Accessed 17 Jan 2024

Ogawa K, Murakami Y, Iguchi T, Kurihara Y (2006) Repellent and attractant insecticides. In: General remarks of household pesticides III. Household Pesticide Industry Association of Japan, pp 45–50. (in Japanese)

Pang SC, Chiang LP, Tan CH, Vythilingam I, Lam-Phua SG, Ng LC (2015) Low efficacy of delthamethrin-treated net against Singapore *Aedes aegypti* is associated with *kdr*-type resistance. Trop Biomed 32(1):140–150

Pichler V, Caputo B, Valadas V, Micocci M, Horvath C, Virgillito C, Akiner M, Balatsos G, Bender C, Besnard G, Bravo-Barriga D, Bueno-Mari R, Collantes F, Delacour-Estrella S, Dikolli E, Falcuta E, Flacio E, Garcia-Perez AL, Kalan K, Kavran M, L'Ambert G, Lia RP, Marabuto E, Medialdea R, Melero-Alcibar R, Michaelakis A, Mihalca A, Mikov O, Miranda MA, Muller P, Otranto D, Pajovic I, Petric D, Rebelo MT, Robert V, Rogozi E, Tello A, Zitko T, Schaffner F, Pinto J, Della Torre A (2022) Geographic distribution of the V1016G knockdown resistance mutation in *Aedes albopictus*: a warning bell for Europe. Parasit Vectors 15(1):280

Ritchie SA, Devine GJ (2013) Confusion, knock-down and kill of *Aedes aegypti* using metofluthrin in domestic settings: a powerful tool to prevent dengue transmission? Parasit Vectors 6:262

Sakamoto H, Goka K (2021) Acute toxicity of typical ant control agents to the red imported fire ant, *Solenopsis invicta* (hymenoptera: Formicidae). Appl Entomol Zool 56:217–224

Scott JG (2019) Life and death at the voltage-sensitive sodium channel: evolution in response to insecticide use. Annu Rev Entomol 64:243–257

Seki N (2009) Life improvement program, "better life without mosquitoes and flies": lessons learned from the postwar development in Japan with particular focus on community participation and international assistance. J Int Health 24(1):1–11. (in Japanese)

Shimono T, Kanda S, Lamaningao P, Murakami Y, Darcy AW, Mishima N, Inthavongsack S, Soprasert O, Xaypangna T, Nishiyama T (2021) Phenotypic and haplotypic profiles of insecticide resistance in populations of *Aedes aegypti* larvae (Diptera: Culicidae) from central Lao PDR. Trop Med Health 49(1):32

Shinjo G, Koizumi T, Sakuma R, Kawase M, Tanikawa T, Mizuno S (2006) Control effect of gel bait products against the German cockroach. Health, Labor and Welfare Science Research. [Chief Researcher; TANAKA, I.], pp 80–103. (in Japanese)

Shinjo G, Ito Y, Mizutani K (2021) Introduction to pesticides. In: Pest control in living environment, 6th edn. Japan Environmental Sanitation Center, pp 243–325. (in Japanese)

Silva JJ, Scott JG (2020) Conservation of the voltage-sensitive sodium channel protein within the Insecta. Insect Mol Biol 29(1):9–18

Smith LB, Kasai S, Scott JG (2016) Pyrethroid resistance in *Aedes aegypti* and *Aedes albopictus*: important mosquito vectors of human diseases. Pestic Biochem Physiol 133:1–12

Tatsuki S (2008) An invasive alien ant, "the Argentine ant", its distribution, ecology and control. Jpn J Environ Entomol Zool 19(1):39–45

Tomlin CDS (2006) Pesticide Manual 14th ed. BCPC, 1349 pp

Uemura N, Furutani S, Tomita T, Itokawa K, Komagata O, Kasai S (2023) Concomitant knockdown resistance allele, L982W + F1534C, in *Aedes aegypti* has the potential to impose fitness costs without selection pressure. Pestic Biochem Physiol 193:105422

Ujihara K, Mori T, Iwasaki T, Sugano M, Shono Y, Matsuo N (2004) Metofluthrin: a potent new synthetic pyrethroid with high vapor activity against mosquitoes. Biosci Biotechnol Biochem 68(1):170–174

Ujihara K, Sugano M, Nakada K, Iwakura K, Nishihara K, Katoh H (2010) Discovery and development of profluthrin (fairytale), a new active ingredient of moth proofer. Sumitomo Chemical Co., Ltd. https://www.sumitomo-chem.co.jp/english/rd/report/files/docs/02_fairytale_e.pdf. Accessed 7 Jan 2024

Wada-Katsumata A, Silverman J, Schal C (2014) Sugar aversion: a newly-acquired adaptive change in gustatory receptor neurons in the German cockroach. Comp Physiol Biochem 31(4):220–230. (in Japanese)

Wang C, Scharf ME, Benett GW (2004) Behavioral and physiological resistance of the German cockroach to gel baits (Blattodea: Blattellidae). J Econ Entomol 97(6):2067–2072

Wei Y, Zheng X, He S, Xin X, Zhang J, Hu K, Zhou G, Zhong D (2021) Insecticide susceptibility status and knockdown resistance (*kdr*) mutation in *Aedes albopictus* in China. Parasit Vectors 14(1):609

World Health Organization (WHO) (2022a) Standard operating procedure for testing insecticide susceptibility of adult mosquitoes in WHO bottle bioassays. World Health Organization. https://www.who.int/publications/i/item/9789240043770. Accessed 6 Jan 2024

World Health Organization (WHO) (2022b) Standard operating procedure for testing insecticide susceptibility of adult mosquitoes in WHO tube tests. World Health Organization. https://www.who.int/publications/i/item/9789240043831. Accessed 6 Jan 2024

Wu H, Qian J, Xu Z, Yan R, Zhu G, Wu S, Chen M (2023) Leucine to tryptophane substitution in the pore helix IIP1 confer sodium channel resistance to pyrethroids and DDT. Pestic Biochem Physiol 194:105490

Wuliandari JR, Lee SF, White VL, Tantowijoyo W, Hoffmann AA, Endersby-Harshman NM (2015) Association between three mutations, F1565C, V1023G and S996P, in the voltage-sensitive sodium channel gene and knockdown resistance in *Aedes aegypti* from Yogyakarta. Indonesia Insects 6(3):658–685

Xiong T, Qiu XH, Ling SQ, Liu JL, Zeng XN (2019) Interaction of fipronil and the red imported fire ant (Solenopsis invicta): toxicity differences and detoxification responses. J Insect Physiol 115:20–26

Yanola J, Somboon P, Walton C, Nachaiwieng W, Prapanthadara L-A (2010) A novel F1552/C1552 point mutation in the *Aedes aegypti* voltage-gated sodium channel gene associated with permethrin resistance. Pestic Biochem Physiol 96(3):127–131

Yasudai R, Matsubara A, Hsu PW, Lee CC, Lin CC, Yang CCS (2022) Laboratory and field evaluations of two bait formulations against the invasive fire ant, *Solenopsis invicta* (hymenoptera: Formicidae). J Econ Entomol 115(2):624–630

Zhou X, Yang C, Liu N, Li M, Tong Y, Zeng X, Qiu X (2019) Knockdown resistance (*kdr*) mutations within seventeen field populations of *Aedes albopictus* from Beijing China: first report of a novel V1016G mutation and evolutionary origins of *kdr* haplotypes. Parasit Vectors 12(1):180

Chapter 9
Climate Changes and Adaptation

Osamu Komagata and Mayuko Yonejima

Abstract The chapter presents a comprehensive review of the impacts of climate change on vector-borne diseases, with a particular focus on how variations in temperature and precipitation affect the behaviors and distributions of arthropod vectors such as mosquitoes and ticks. The impact of climatic conditions on the population dynamics of vectors was reviewed, with a specific emphasis on the expansion of *Aedes albopictus* in Japan and its wider consequences for public health. The analysis indicated that if temperatures increase and precipitation patterns change, the distribution of disease-carrying organisms can expand, hence potentially elevating the risk of disease transmission to people. This scenario highlighted the necessity for a more comprehensive comprehension of the adaptive mechanisms of vectors in response to fluctuating environmental conditions.

Furthermore, the significance of holistic approaches that encompass environmental management, efficient public policy, and educational initiatives in addressing vector-borne diseases was emphasized. In order to address the complex issues presented by climate change and the spread of vector-borne diseases, it is essential that many scientific and policy-making fields collaborate. An integrative approach is considered essential for the development of successful methods to mitigate and adapt to changing environmental conditions in order to safeguard public health.

Keywords *Aedes albopictus* · Climate change · Distribution expansion · Global warming · Northern region · Risk map · Tohoku region

O. Komagata (✉)
Department of Infectious Diseases, National Institute of Infectious Diseases, Tokyo, Japan
e-mail: komagata@niid.go.jp

M. Yonejima
Faculty of Humanities and Social Sciences, Kumamoto University, Kumamoto, Japan
e-mail: yonejima@kumamoto-u.ac.jp

K. Sawabe et al. (eds.), *Medical Entomology in Asia*, Entomology Monographs,
https://doi.org/10.1007/978-981-97-4406-0_9

9.1 Introduction

Climate change has significant implications for the transmission and spread of vector-borne diseases because it impacts the biogeography and ecology of insect vectors (Filho et al. 2019). Pathogens and parasites in human populations cause illnesses known as vector-borne diseases. Arthropods such as mosquitoes, ticks, and fleas transmit vector-borne diseases. These diseases disproportionately affect the poorest populations in tropical and subtropical regions. As climate change advances, it is imperative to employ countermeasures against arthropods to prevent the spread of vector-borne diseases (WHO 2020).

The phenomenon of climate change is a growing concern on a global scale and has significant implications for a range of ecosystems and human welfare. The objective of this chapter is to investigate the association between climate change and arthropod vectors of infectious diseases, such as mosquitoes and ticks. Understanding the dynamics of disease transmission and devising effective public health interventions necessitate a comprehensive understanding of the implications of climate change on these vectors.

Climate change refers to the long-term alterations in temperature, precipitation patterns, and other climatic variables that arise from human activities, particularly the release of greenhouse gases into the atmosphere. Changes within ecosystems have the potential to significantly impact that may have noteworthy consequences, influencing the distribution and behavior of various organisms, including vectors.

Mosquitoes and ticks are poikilotherms; therefore, it is assumed that they are particularly sensitive to climatic variations. They are transmitting diseases such as malaria, dengue fever, Zika virus, Lyme disease, and others. Therefore, the sensitivity of these vectors to climate change has implications for the transmission of these diseases.

Assuming the current trends in global warming continue, several outcomes can be anticipated: Climate change has a complex effect on arthropods that transmit diseases. The development and survival rates of vectors are directly impacted by increasing temperatures, which in turn affect their population dynamics and geographic distribution. Because of higher temperatures, the life cycles of arthropod vectors progress faster, which means that more diseases are being spread. Furthermore, modified precipitation patterns have an impact on the reproduction locations and accessibility of water, which subsequently affect the populations of vectors.

Investigating the correlation between climate change and arthropod vectors that transmit infectious diseases is crucial for preserving public health. The transmission of infectious diseases through vectors imposes a considerable burden on global societies, impacting a huge number of people annually. By comprehending the influence of climate change on the geographical range and conduct of disease-carrying organisms, we can predict alterations in the spread of illnesses and execute prompt and specific measures.

As climate change progresses, the importance of this knowledge increases significantly. This is due to the possibility of arthropod vectors spreading to new areas and introducing diseases to previously unaffected regions. Through the examination of the effects of climate change on vectors, it is possible to determine regions that are susceptible to disease outbreaks. This knowledge can be utilized to efficiently allocate resources for the purpose of preventing or reducing the impact of these outbreaks. Moreover, comprehending the adaptive mechanisms of vectors in response to varying environments aids in formulating tactics to disrupt disease transmission cycles.

In summary, the investigation of climate change and its impact on arthropod carriers of contagious illnesses is a crucial area of study that has significant implications for the human welfare of the general population. Through a thorough understanding of the intricate relationships among climate, vectors, and diseases, it is possible to devise preemptive strategies, enhance monitoring systems, and execute specific interventions to mitigate the effects of these diseases on populations worldwide. The investigation of the topic and collaboration across different fields are crucial to finding new and effective solutions for the challenges caused by climate change and vector-borne diseases.

9.2 Temperature and Vector Population

The association between temperature and vector population dynamics is an essential component of comprehending the impact of climate change on the distribution, population size, and behavior of insect vectors. The impact of temperature on vector species is crucial as it affects their growth, survival, and reproductive abilities, which in turn determine their population size and distribution patterns.

The population size and geographic distribution of arthropod vectors can be significantly impacted by rising temperatures brought on by climate change. Higher temperatures can hasten the maturation rates of immature stages (such as larvae and pupae), leading to a quicker population increase and more adults emerging. Larger vector populations may then develop in regions where they were previously constrained by temperature issues. Warmer weather can also increase the habitat that vectors can live in, enabling them to expand their geographic range to previously unsuitable higher latitudes or elevations. As a result, regions that were previously free of some vector-borne diseases may now be susceptible to transmission as a result of shifting temperature patterns.

All of the stages of the life cycle of a vector, like egg formation, larval growth, pupation, and adult survival, are significantly influenced by temperature. Higher temperatures can hasten the maturation of eggs and immature stages, shortening the period of time needed for vectors to finish their life cycle. More generations per year may result from this shortened life cycle, which would increase the size of the population as a whole. Contrarily, high death rates or a disruption in the timing of the life cycle of the vector due to the availability of suitable hosts or breeding locations might have a negative impact on vector populations during extreme temperature events like heatwaves.

The ability of insect vectors to reproduce and survive is influenced by temperature. In general, warmer temperatures boost vector reproductive rates, increasing egg production and mating success rates. The survival rates of vector species are also influenced by temperature, with certain species more tolerant of a given range of temperatures than others. The competitive advantage among vector species can therefore change in response to temperature changes, possibly favoring those that are better suited to warmer climates. This may modify the composition of the vector species and the dynamics of the transmission of vector-borne diseases.

In many cases, it is assumed that temperature affects the dynamics of vector population transmission. For instance, increasing temperatures have broadened the range of anopheline mosquitoes, the main carriers of the disease, in areas where malaria is endemic. The risk of malaria transmission has grown as a result of this expansion in formerly untouched areas. Similar to how the spread of malaria has been linked to rising temperatures (Pinault and Hunter 2011; Tantely et al. 2012; Novikov and Vaulin 2014), the primary carriers of dengue fever, *Aedes* mosquitoes, are more likely to reproduce and survive in warmer climates. These demonstrate the connection between temperature and the distribution, abundance, and capacity for disease transmission of insect vectors.

An increase in temperature doesn't necessarily indicate a simple change. According to Hashizume et al. (2012), a significant malaria resurgence in the East African highlands in the 1990s was brought on by the Indian Ocean Dipole (IOD). The Indian Ocean Dipole (IOD) is an ocean–atmosphere coupling that impacts the region's climate variability. Positive IOD events are characterized by warmer sea surface temperatures in the western Indian Ocean and cooler temperatures in the east, whereas negative IOD events are characterized by cooler sea surface temperatures in the west and warmer temperatures in the east. The evidence supporting the role of the IOD as a driver of malaria in Africa comes from various studies and observations. One piece of evidence is the association between positive IOD events and increased rainfall in the East African highlands, which creates more breeding sites for mosquitoes and leads to higher malaria transmission. Additional studies have indicated that the patterns of association between the IOD and rainfall in East Africa are stronger and more regular. Furthermore, atmospheric circulation models have suggested that Indian Ocean sea surface temperatures (SSTs), which are influenced by the IOD, have a greater impact on East African rainfall than Pacific SSTs. This suggests that the IOD plays a significant role in influencing the climatic conditions that are conducive to malaria transmission. It is important to note that while the IOD is a significant driver of malaria dynamics in the East African highlands, it is not the only influence. Other climatic factors, such as El-Nino Southern Oscillation (ENSO), as well as non-climatic factors like control measures and human mobility have also been implicated in malaria incidence. Overall, the evidence suggests that the IOD, through its influence on rainfall patterns, is an important factor in driving malaria outbreaks in Africa, particularly in the highland regions.

Examples of the geographical spread of *Aedes albopictus* (Skuse) in Japan are discussed in the paragraphs that follow.

9.3 Spreading of *Ae. albopictus* to the North in the Tohoku (Northern) Region of Japan

Aedes albopictus is a species of mosquito that is frequently observed in Japan. It adapted to survive in artificial habitats. *Ae. albopictus* did not inhabit the northern region due to the cold temperatures during winter. Prior to 1968, the species was discovered in Sendai City in Miyagi, which was the northernmost location (Kamimura 1968). Although *Ae. albopictus* had been found as far north as Sendai, Nawrocki and Hawley (1987) quoted the Kamimura report and said, "Although *Ae. albopictus* has been found as far north as Sendai, we know of no data on its overwintering capabilities there." There was still not enough information about how well it could hibernate in Sendai in 1987. After that, Kurihara conducted a study to examine how Ae. *albopictus* was distributed in various cities in northern Japan between 1993 and 1996. (Kurihara et al. 1997). They collected larvae and adult mosquitoes of the *Ae. albopictus* species in urban areas of the Tohoku region during early autumn seasons from 1993 to 1996. Ae. *albopictus* hadn't been found in previous surveys conducted in cities including Ishinomaki, Shiogama, Furukawa, Sakata, Yonezawa, and Honjo, but it was found in the survey. They advocated that *Ae. albopictus* was currently not widely distributed in the inland Tohoku region. However, there were indications that it may gradually spread in the future due to human transportation.

Kurihara et al.'s research was continued by his colleague Kobayashi. In the northeast region of Japan, Kobayashi et al. (2002) surveyed a field to collect information on the distribution of *Ae. albopictus*. Collecting mainly mosquito larvae from various locations and identifying them by species was part of the survey. The survey data was then used to generate distribution maps of *Ae. albopictus* throughout the region. At this time, Nihei, Kobayashi's colleague, innovated geographic information systems (GIS) for the analysis of Ae. albopictus geographic distribution. They analyzed mesh climate data, provided by the National Institute of Agro-Environmental Sciences, on GIS to analyze the temperature conditions under which *Ae. albopictus* could overwinter (the subsequent mesh climate data in Japan is still available as of 2023 as the Agro-Meteorological Grid Square Data provided from National Agriculture and Food Research Organization). Kobayashi et al. (2002) finally found a simple explanation for the distribution of this mosquito species. It is dependent on an annual average temperature of at least 11 °C, which acts as a developmental threshold. Subsequently, it was documented that the same temperature conditions are also applicable in Europe (ECDC 2009). They and their colleagues continued to survey. In 2015, near the northern end of Honshu (the main island),

Aomori City was included in the spread of *Ae. albopictus* (Maekawa et al. 2020) (Fig. 9.1)

The increase in temperature caused by climate change is creating suitable conditions for the growth and distribution of *Ae. albopictus*, which is causing its migration towards the north of Japan. Because of climate change, it is expected that *Ae. albopictus* will expand its geographical distribution to previously unoccupied areas. The success of such mosquitoes in overwintering is probably dependent on the springtime temperature (Komagata et al. 2017). Cool spring temperatures prevent mosquitoes from breaking dormancy and becoming active. *Aedes albopictus* hibernates in eggs during the winter. The duration of the potential dormancy is uncertain. That period has a limit.

Kobayashi et al. (2002) described that the mosquito's movement towards the north is facilitated by precipitation patterns, particularly when the annual rainfall

Fig. 9.1 Northern distribution of *Aedes albopictus* in Tohoku district, Honshu Island, Japan. If mosquitoes were collected at the same location for more than 2 years, they were regarded as distributed. Each circle represents the distribution of *Ae. albopictus*, and its color indicates when it was initially identified: white before the year 2000, gray between the years 2001 and 2010, and black after 2011. The estimated northern boundaries of the distribution in 2010 (Upper) and 2000 (Lower) are represented by the black lines. The gray area is located outside of the Tohoku district. As of 2022, no *Ae. albopictus* has been found in Hakodate or any other city on Hokkaido Island. (Created based on data from Kurihara et al. (1997), Kobayashi et al. (2002), Sato et al. (2012), and Maekawa et al. (2020))

exceeds 1000 mm. This suggests that precipitation could promote the development of favorable environments for the growth and distribution of *Ae. albopictus*. Further investigation is necessary to obtain a complete comprehension of the correlation between precipitation and the dispersal of this disease carrier. The significance of understanding the impact of climate change on the spread of *Ae. albopictus* and other disease-carrying organisms are highlighted by them. This understanding is crucial for devising efficient measures to control disease outbreaks and safeguard public health. The study conducted emphasizes the influence of human conduct on the transmission of *Ae. albopictus*. The transportation of used tires has been identified as a contributing factor to the proliferation of mosquito larvae. Mogi and Tuno (2014) analyzed additional data, including distribution data from Kobayashi et al. (2002). Since the beginning of the twentieth century, the overwintering season in northern Japan has become increasingly suitable for *Ae. albopictus*, whereas the reproductive season has become less so. Retrospective analyses revealed that these criteria were met in the majority of northern Honshu's coastal lowlands prior to the accelerated temperature increase after the mid-1980s and the first records of this species after 1990; at some sites, the temperature criteria were met during or before the turn of the twentieth century. After 1990, expansion of the thermally appropriate range was only supported for inland regions and the northernmost Pacific coast. The expansion rate was estimated to be 26 km per decade. They also evaluate the effects of rain and precipitation. Rainfall occurs throughout the year in Japan, with annual totals rarely falling below 1000 mm and typically being significantly higher. *Aedes albopictus* thrives in regions with annual precipitation greater than 800 millimeters. In northern Honshu, neither an increase nor a decrease in summer precipitation has been observed. Northern Japan receives abundant snowfall, which may be advantageous for embryos that overwinter (Figs. 9.2, 9.3, and 9.4).

Hirabayashi et al. (2020) observed the elevational distribution of *Ae. albopictus* in the inland mountain region of the Nagano and Yamanashi prefectures, Japan. *Ae. albopictus* is found at higher elevations, and the research has produced substantial findings suggesting that temperature has an impact on where it can be located. Additionally, the study identified potential habitats for *Ae. albopictus* at high elevations. Adult *Ae. albopictus* were not captured at locations with elevations exceeding 800 m. Furthermore, larvae were not observed to be present at altitudes exceeding 728 m. The locations where *Ae. albopictus* was observed in a hilly inland area and exhibited an average yearly air temperature between 9.4 °C and 15.28 °C, while the mean air temperature during the months of July to September ranged from 19.9 °C to 25.08 °C. It is the first report of *Ae. albopictus* in a mountainous inland region in Japan. Future climate change-induced temperature increases may cause vectors to become more active at higher altitudes. Additional long-term field studies are desirable, and their findings may contribute to an improved understanding of the relationship between altitude and mosquito fauna (Fig. 9.5).

Human activity, particularly global transportation such as international old tire (it stores water, which is often the suitable egg-laying container for *Ae. albopictus*.) trade, is primarily responsible for the spread of *Ae. albopictus* populations in new geographic locations. Known for its adaptability and stubbornness, this

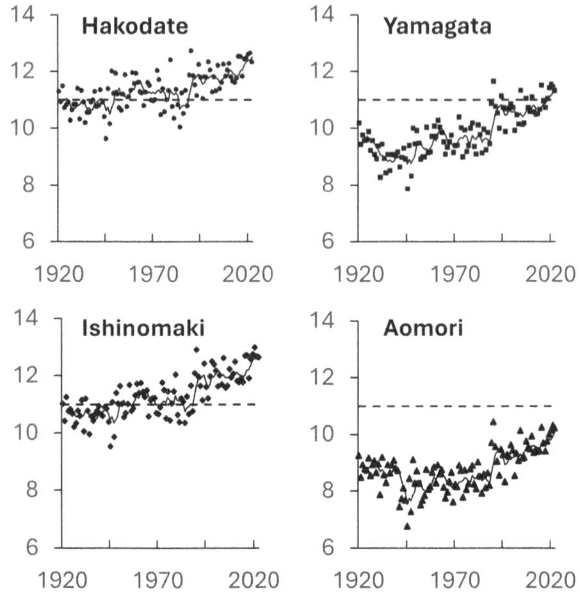

Fig. 9.2 Annual average temperature (°C) in four cities over the past 100 years. The data were obtained from the Japan Meteorological Agency (JMA, https://www.data.jma.go.jp/gmd/risk/obsdl/index.php). Each symbol represents the annual mean temperature, and the solid line represents the 5-year moving average. The dashed lines are the estimated habitable temperature (11 °C, Kobayashi et al. (2002)) for *Ae. albopictus*. Hakodate is situated in proximity to the southernmost point of Hokkaido Island. Presently, there is no evidence to suggest the presence of *Ae. albopictus* in this region. Aomori is situated in the northernmost area where the mosquito population has been confirmed. Yamagata and Ishinomaki are situated south of the Hakodate and Aomori. They are used by the JMA to observe long-term temperature trends. *Aedes albopictus* already inhabits the regions. Although annual average temperatures have increased over the long term, annual temperatures have fluctuated substantially

cosmopolitan species' function as a vector for a number of infectious diseases, such as dengue fever and the Zika virus, has made it a serious public health problem. Creating successful public health initiatives requires an understanding of the complex interactions between human behavior—such as international trade and travel—and the spread of disease vectors. These approaches should take into account long-term ecological and environmental concerns, especially those impacted by climate change, in addition to immediate mosquito control efforts. It has been demonstrated that changes in habitats and weather patterns brought about by climate change influence mosquito populations' dispersion and abundance by fostering an environment that is more conducive to their proliferation. Furthermore, efforts to control *Ae. albopictus* are made more difficult by the emergence of new habitats

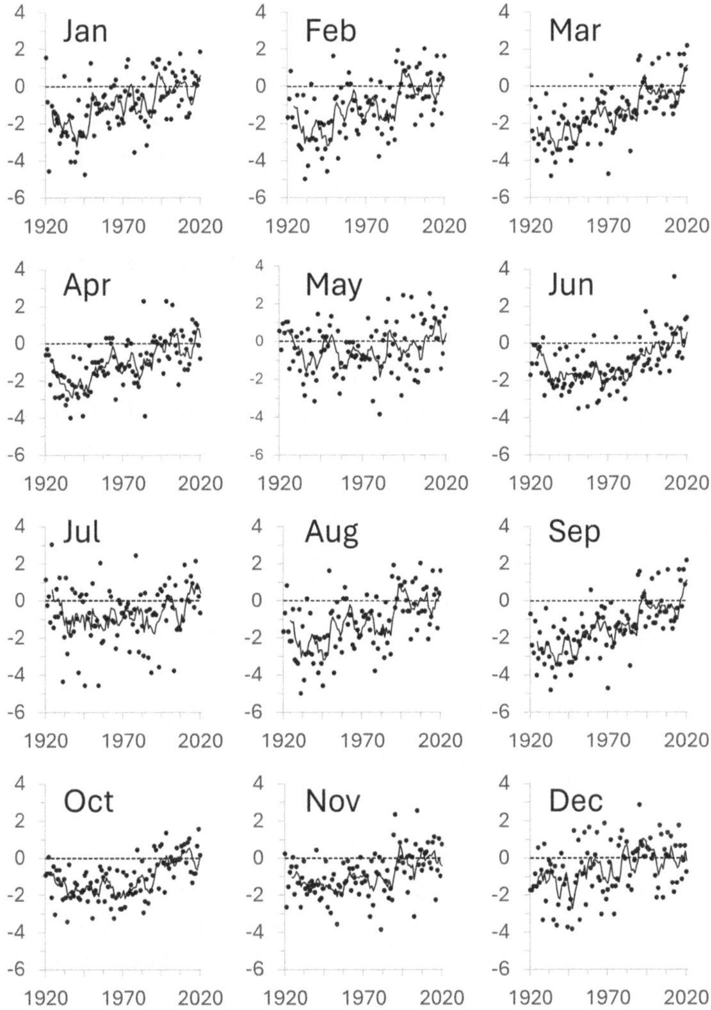

Fig. 9.3 The dots represent monthly average temperature (°C) deviations in Aomori city, while the lines represent the 5-year moving averages. The deviation is the difference between the monthly mean temperature and the standard mean temperature for a given month. The reference mean temperature for each month has been calculated by averaging the data from 1991 to 2020. (Original data were obtained from the Japan Meteorological Agency (JMA))

through urbanization and changes in land use. Public health solutions must be diverse, combining not just conventional techniques like habitat reduction and the use of insecticides but also cutting-edge strategies like public awareness campaigns, community education, and the encouragement of personal protective measures. Furthermore, further study is required to comprehend how mosquito populations are changing in response to alterations in their surroundings and human activity.

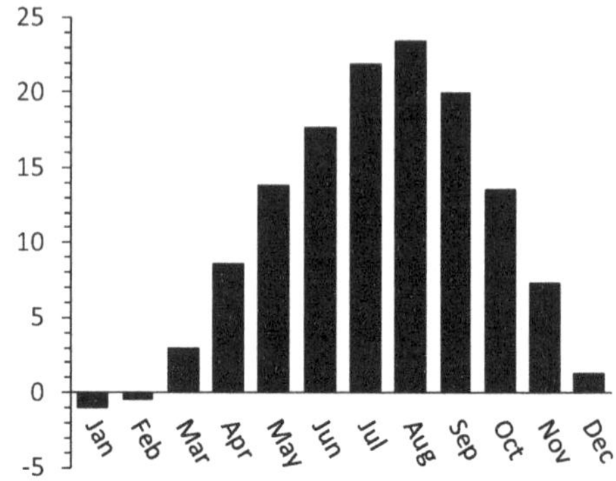

Fig. 9.4 The reference mean temperature for each month at Aomori city had been calculated by averaging the data from 1991 to 2020. The formula for calculation that the Japan Meteorological Agency (JMA) uses. (Original data were obtained from JMA)

Targeted and long-lasting strategies can be created by combining socioeconomic and ecological research. The goal of these tactics ought to be to lessen the negative effects of *Ae. albopictus* on public health, especially in areas where the diseases it transmits are most prevalent. To effectively address this global health concern, collaboration at the local, national, and international levels is vital. In addition to aiding in the current *Ae. albopictus* spread control, such all-encompassing strategies will assist in getting ready for any future mosquito-borne disease epidemics.

Nihei et al. (2014) discussed the competing ecological niches between *Ae. albopictus* and *Ae. japonicus japonicas* (Theobald) in areas of the *Ae. albopictus* invasion. Its presence was first reported in Yamagata city, northeast Honshu, Japan in 2000, where previously only *Ae. j. japonicus* was found. Within 2 years, *Ae. albopictus* became dominant, comprising over 80% of the larval colonies in the area. This species now infests new residential areas of Yamagata, while the original mosquito is more common near the mountains with higher vegetation indices. Contrary to expectations, *Ae. j. japonicus* has not been completely displaced by *Ae. albopictus*; both species coexist, often in the same water bodies. The competitive dynamics between these two species, especially in a warming environment, warrant close monitoring.

9.4 Risk Mapping of Dengue Fever Outbreaks in Japan by Retrospective Research

It was previously mentioned that dengue fever infections could arise in Japan due to the ecology of vector mosquitoes. Although there had been cases of foreign visitors contracting dengue fever soon after leaving Japan, it was unclear whether the patients had contracted the virus there (Kobayashi et al. 2014). In 2014, Tokyo experienced its first endemic dengue virus outbreak in 70 years, originating in

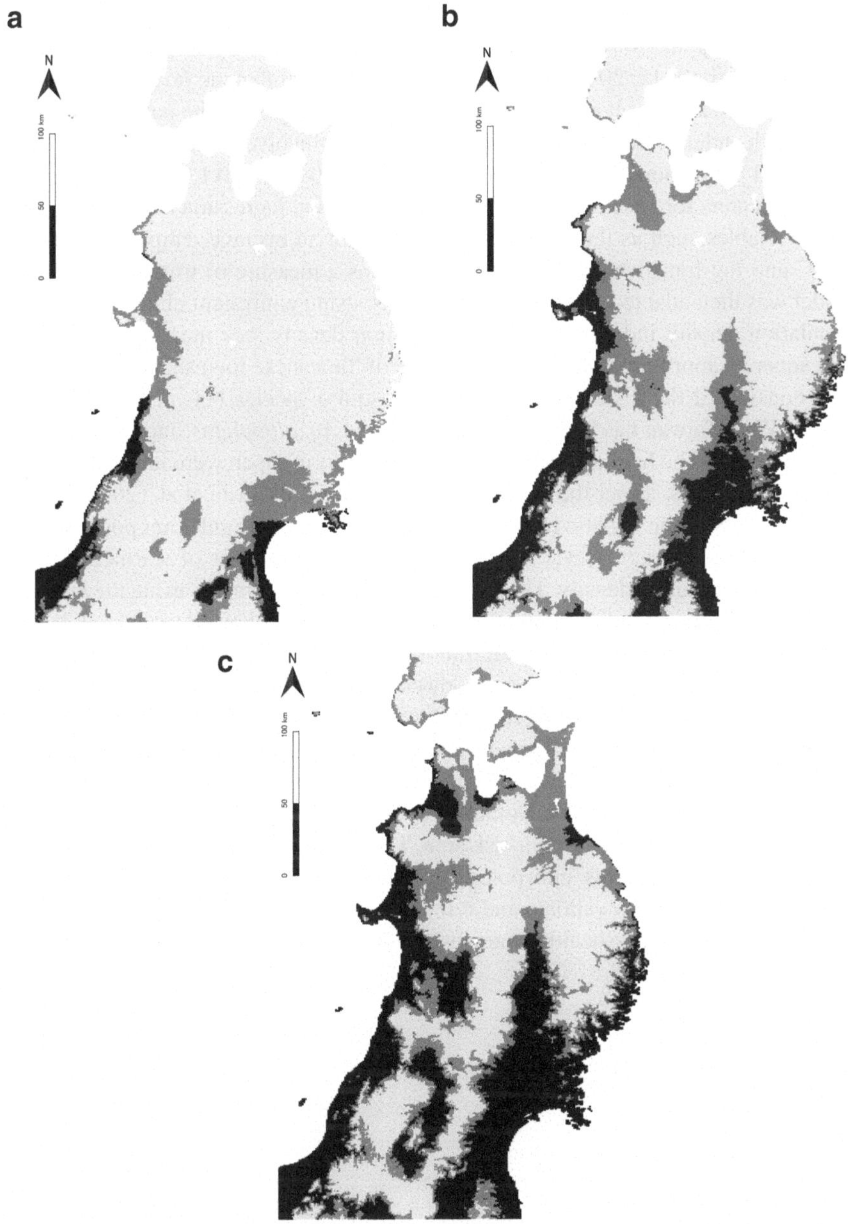

Fig. 9.5 The three maps (**a**, **b**, and **c**) of the Tohoku region in Japan depict the annual mean temperature at a resolution of 1 km. Within the graph's color scheme, a black-and-white gradient represents temperature variations. The deeper black color represents a higher temperature. (**a**) represents the year 2015, (**b**) represents the year 2014, and (**c**) represents the year 1980. *Aedes albopictus*, a species of mosquito, was confirmed to overwinter in Aomori City for the first time in 2015. The black area in the legend represents regions with an annual average temperature of 11 °C

Japan. The irregular outbreaks had sparked significant interest in comprehending and elucidating the susceptibility of society to vector-borne illnesses in Japan. Yonejima et al. (2015, 2018) tried to create risk maps for dengue fever outbreaks in Japan by utilizing recent outbreak data from Taiwan (Fig. 9.6), the nation closest to Japan with endemic dengue fever. The methodology involved using Taiwanese data on annual indigenous dengue incidence from 2010 and 2011, collected by the Taiwan Centers for Disease Control. The binary logistic regression models incorporated variables such as the number of months with an average temperature above 18 °C and log-transformed population density as a measure of urbanization. This model was then adapted to Japanese conditions, using equivalent climatic data and population density indices. In Japan, population density was measured in 10-km grid squares, approximating the average size of Taiwanese townships. In addition, they considered the distinctions in vector mosquito species between Taiwan and Japan. While Taiwan has both *Ae. aegypti* (L.) and *Ae. albopictus*, Japan only hosts *Ae. albopictus*. The habitat regions for *Ae. albopictus* in Japan were identified using climate conditions, as per the method established by Kobayashi et al. (2002).

The results of the logistic regression analysis revealed a significant positive correlation between dengue fever incidence and both the number of warmer months and higher population density. The risk map was created for Japan, using the adapted model and climatic conditions suitable for *Ae. albopictus*, showed varied risk levels across different regions. Hokkaido, Tohoku, and mountainous areas of the main islands were identified as having the lowest risk, primarily due to their climatic conditions. In contrast, the Ryukyu Islands and central metropolitan areas exhibited the highest risk. The high risk in metropolitan areas was attributed to urban climates conducive to mosquito habitats and increased chances of virus entry or infection due to high population density. Notably, Yoyogi Park, the site of the 2014 outbreak, was marked as a moderately high-risk area. They concluded that while the risk map provides valuable insights into potential dengue fever hotspots in Japan, further research is necessary to validate and refine the mapping accuracy. This research is pivotal in guiding public health strategies and interventions to prevent future dengue fever outbreaks in Japan.

Fig. 9.5 (continued) or higher. This temperature range provides the necessary conditions for the overwintering of *Ae. albopictus*. The island's center region is mountainous, has low temperatures because of its high elevation, and is home to a small population. *Aedes albopictus* was only found in a few cities such as Morioka on level ground in the mountains; no other distribution in the mountains was recorded. (The Ministry of Land, Infrastructure, Transport, and Tourism provided the map data used in this study, while the JMA)

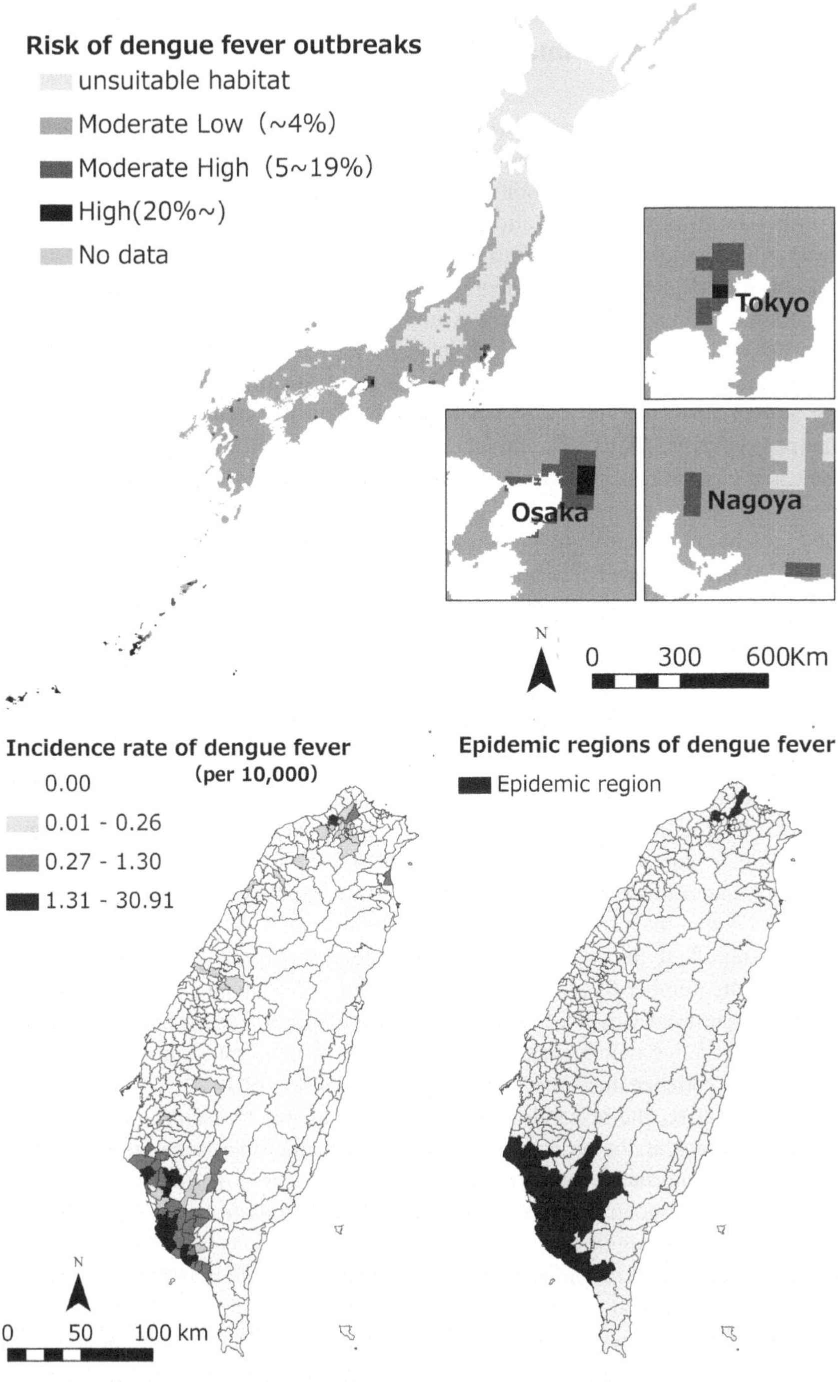

Fig. 9.6 Risk map for dengue fever outbreaks in Japan. (This risk map was developed based on the methodology of Yonejima et al. (2015), utilizing data from recent dengue fever outbreaks in Taiwan, the closest nation to Japan with endemic dengue fever)

9.5 Local Government-Assessed Arthropod-Borne Disease Risks Due to Climate Change

The impacts of climate change will vary depending on local climate, topography, and socioeconomic conditions. In addition, since adaptation to the effects of climate change will lead to the protection of local livelihoods and regional development, it is important to promote adaptation efforts that are suited to the region. Based on these considerations, from 2017 to 2019, the Ministry of the Environment in Japan implemented regional projects about climate change adaptation (NIES 2018). In this context, Kanagawa Prefecture was studied primarily with respect to risk assessment of vector-borne mosquitoes. Kanagawa Prefecture is a large urban area in Japan situated to the west of Tokyo. The city boasts a substantial international harbor and an abundance of tourist landmarks. Several risk maps were proposed for that project (NIES 2019). The project focused on the fact that rising temperatures can cause the outbreak of both adult mosquitoes to occur earlier and longer, thereby increasing the period during which people encounter them (the period during which they suck blood) and potentially increasing the risk of infectious disease. Therefore, the following two types of maps were developed. One is a "habitat duration map." A map that evaluates how much the habitat duration of both adult mosquitoes lengthens as the temperature rises. The other is a "risk control priority map." A map that evaluates the areas that need to be prioritized for control of infectious disease vector mosquitoes based on the "habitat potential" of both adult mosquitoes (habitat duration and green areas where they live) and the "population potential" that could encounter them.

According to the MRI-CGCM3 (Meteorological Research Institute-Coupled General Circulation Model 3) model, the adult *Ae. albopictus* in Kanagawa Prefecture is expected to have a duration of 4.5–5 months in most areas today. However, under the Representative Concentration Pathways (RCP) 2.6 scenario, it is predicted that the number of areas with a duration of 5–5.5 months will increase from the middle to the end of the twenty-first century. Similarly, under the RCP8.5 scenario, the number of areas with a duration of 5–5.5 months will also increase during the period. By the mid-twenty-first century, some areas are predicted to have a duration of 5.5–6 months, and by the end of the twenty-first century, some places may even experience durations exceeding 6 months (Fig. 9.7).

To summarize, the research performed in Kanagawa Prefecture highlights the urgent requirement for proactive and location-specific approaches to reduce the effects of climate change on diseases transmitted by vectors. The creation of habitat duration and risk control priority maps, using projections from the MRI-CGCM3 model, etc., offers significant tools for identifying locations with heightened risk caused by prolonged mosquito activity periods. These maps not only illustrate the possibility of extended mosquito seasons under several climatic scenarios; but also underscore the significance of focused control strategies in regions with abundant mosquito breeding grounds and population density. Given the ongoing impact of

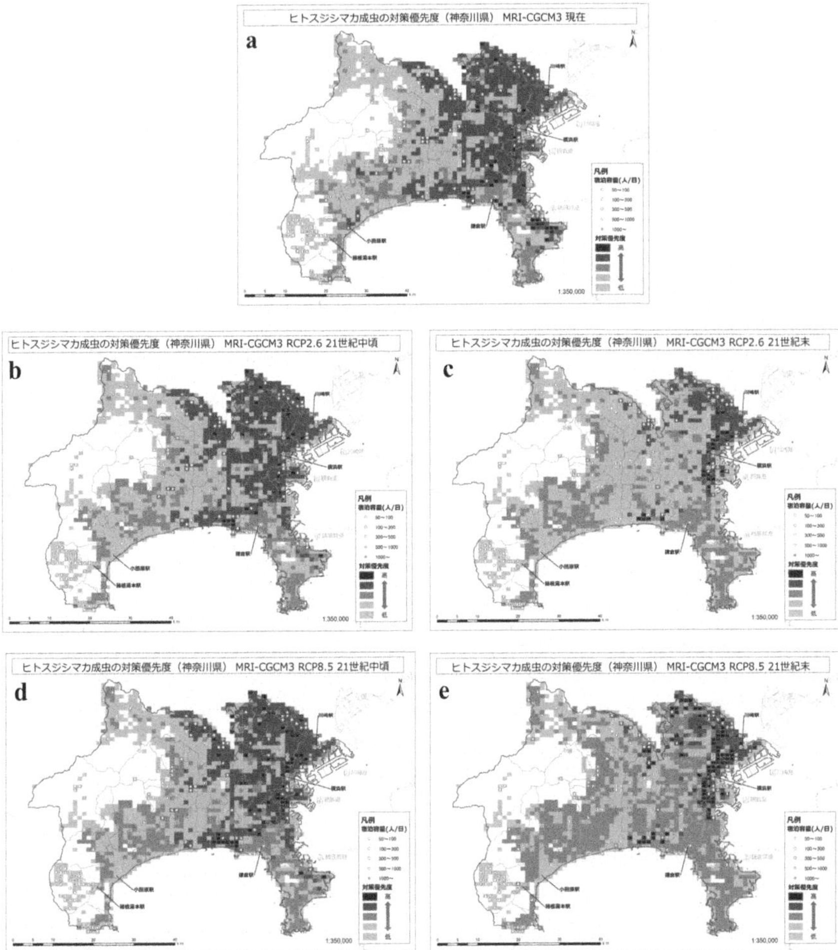

Fig. 9.7 *Aedes albopictus* risk map created by Kanagawa Local Government. A darker color indicates a higher risk for mosquitoes, high risk indicates higher requirements for vector control measures. Only five of the risk maps developed for the project are shown. The future climate was forecasted by the climate forecast model as MRI-CGCM3 (Meteorological Research Institute-Coupled General Circulation Model 3) with Representative Concentration Pathways (RCP). (**a**) present (2019), (**b**) mid-twenty-first century (+RCP2.6), (**c**) end-twenty-first century (+RCP2.6), (**d**) mid-twenty-first century (+ RCP8.5), (**e**) end-twenty-first century (+RCP8.5). (Other maps in the full report (in Japanese) of the project from A-PLAT website (https://adaptation-platform.nies.go.jp/conso/report/2-4.html))

climate change on environmental conditions, it is crucial to adopt region-specific strategies to effectively manage and mitigate the threat of vector-borne diseases. This is necessary to protect public health and ensure regional stability.

9.6 Comprehensive Strategies for Controlling Vector-Borne Diseases in Climate Change

Several primary approaches to controlling arthropods that transmit vector-borne infections are as follows:

9.6.1 *Enhanced Vector Control*

To prevent vector-borne infectious diseases in the context of climate change, enhanced effective strategies include integrated vector management, intersectoral collaboration, vector surveillance, environmental monitoring, and taking into account the impact of environmental factors on disease transmission.

9.6.2 *Public Policy*

Strengthening public policy design by integrating vector surveillance and control with environmental management is crucial. This involves coordinating efforts among various sectors, mobilizing resources, and ensuring strong leadership in the health sector.

9.6.3 *Environmental Management*

Implementing sustainable land-use practices, enhancing water resource management, and placing a high priority on biodiversity preservation are essential steps to lowering the risk of vector-borne diseases brought on by deforestation, urbanization, agricultural expansion, water scarcity, and loss of biodiversity.

9.6.4 *Personal Protection*

Since there are few vaccines available to protect against vector-borne diseases, avoiding insect bites is the best method to prevent them. This can be accomplished by using insect repellents, donning protective clothing, and taking other preventative measures against insect bites.

9.6.5 Education and Awareness

Public health requires raising awareness about vector-borne diseases and their prevention methods. This includes educating communities on the significance of private environmental management to prevent arthropod reproduction.

9.6.6 Research and Innovation

Developing new pesticides, formulations, and novel tools for vector control can help advance the prevention and control of vector-borne diseases. Additionally, understanding the impact of climate change on vector-borne diseases can help inform public health research and policy. Successful development of vaccines and novel medications for vector-borne infectious diseases might reduce the importance of vector control.

9.7 Conclusion

The thorough analysis of the effects of climate change on vector-borne illnesses, with specific emphasis on the proliferation of *Ae. albopictus* in Japan; highlights the urgent requirement for adaptive and proactive approaches in public health. The research findings presented in this chapter provide compelling evidence of a strong association between escalating temperatures, modified precipitation patterns, and the proliferation of vector habitats, resulting in heightened susceptibility to diseases such as dengue fever. The development of risk maps and the evaluation of local government interventions, as observed in Kanagawa Prefecture, offer valuable scientific insights into the regions that are most susceptible to vector-borne diseases. The results of this study underscore the significance of incorporating environmental management, public policy, personal protection, and education as integral components of a holistic strategy for vector control. Furthermore, continuous scientific investigation and advancement in vector control and disease prevention are imperative to proactively address the ever-changing obstacles presented by climate change. As vectors undergo evolutionary adjustments in response to shifting climatic conditions and extend their geographical range, it becomes crucial to enhance international and regional partnerships in order to efficiently alleviate the public health hazards linked to vector-borne illnesses.

References

European Centre for Disease Prevention and Control (ECDC) (2009) ECDC technical report: development of *Aedes albopictus* risk maps. pp 4.

Filho WL, Scheday S, Boenecke J, Gogoi A, Maharaj A, Korovou S (2019) Climate change, health and mosquito-borne diseases: trends and implications to the Pacific Region. Int J Environ Res Public Health 16(24):5114

Hashizume M, Chaves LF, Minakawa N (2012) Indian Ocean dipole drives malaria resurgence in East African highlands. Sci Rep 2:269

Hirabayashi K, Nihei N, Kobayashi M, Tsuda Y, Sawabe K (2020) Elevational distribution of the Asian tiger mosquito, *Aedes albopictus*, in the inland mountain area of Nagano and Yamanashi prefectures, Japan. J Am Mosq Control Assoc 36(1):1–10

Kamimura K (1968) The distribution and habit of medically important mosquitoes of Japan. Jpn J Sanit Zool 19(1):15–34

Kobayashi M, Nihei N, Kurihara T (2002) Analysis of northern distribution of *Aedes albopictus* (Diptera: Culicidae) in Japan by geographical information system. J Med Entomol 39(1):4–11

Kobayashi M, Komagata O, Yonejima M, Maekawa Y, Hirabayashi K, Hayashi T, Nihei N, Yoshida M, Tsuda Y, Sawabe K (2014) Retrospective search for dengue vector mosquito *Aedes albopictus* in areas visited by a German traveler who contracted dengue in Japan. Int J Infect Dis 26:135–137

Komagata O, Higa Y, Muto A, Hirabayashi K, Yoshida M, Sato T, Nihei N, Sawabe K, Kobayashi M (2017) Predicting the start of the *Aedes albopictus* (Diptera: Culicidae) female adult biting season using the spring temperature in Japan. J Med Entomol 54(6):1519–1524

Kurihara T, Kobayashi M, Kosone T (1997) The northward expansion of *Aedes albopictus* distribution in Japan. Med Entomol Zool 48(1):73–77

Maekawa Y, Higa Y, Sawabe K, Kasai S (2020) Expansion of *Aedes albopictus* distribution. Infectious Agents Surveillance Report (IASR) 41(6):4–5

Mogi M, Tuno N (2014) Impact of climate change on the distribution of *Aedes albopictus* (Diptera: Culicidae) in northern Japan: retrospective analyses. J Med Entomol 51(3):572–579

National Institute for Environmental Studies (NIES), Japan (2018) Regional adaptation consortium project. https://adaptation-platform.nies.go.jp/conso/overview/index.html. Accessed 28 Dec 2023

National Institute for Environmental Studies (NIES), Japan (2019) Report2–4 assessing the risk of arthropod-borne infectious diseases due to climate change. https://adaptation-platform.nies.go.jp/conso/report/2-4.html. Accessed 28 Dec 2023

Nawrocki SJ, Hawley WA (1987) Estimation of the northern limits of distribution of *Aedes albopictus* in North America. J Am Mosq Control Assoc 3(2):314–317

Nihei N, Komagata O, Mochizuki K, Kobayashi M (2014) Geospatial analysis of invasion of the Asian tiger mosquito *Aedes albopictus*: competition with *Aedes japonicus japonicus* in its northern limit area in Japan. Geospat Health 8(2):417–427

Novikov YM, Vaulin OV (2014) Expansion of *Anopheles maculipennis* s.s. (Diptera: Culicidae) to northeastern Europe and northwestern Asia: causes and consequences. Parasit Vectors 7:389

Pinault LL, Hunter FF (2011) New highland distribution records of multiple *Anopheles* species in the Ecuadorian *Andes*. Malar J 10:236

Sato T, Matsumoto F, Abe T, Nihei N, Kobayashi M (2012) Reacent distribution of *Aedes albopictus* in Iwate Prefecture, northern Japan and analysis of climatic conditions for northern limit of geographic distribution by geographic information systems. Med Entomol Zool 63(3):195–204

Tantely ML, Rakotoniaina J-C, Tata E, Andrianaivolambo L, Fontenille D, Elissa N (2012) Modification of *Anopheles gambiae* distribution at high altitudes in Madagascar. J Vector Ecol 37(2):402–406

World Health Organization (WHO) (2020) Fact sheet: vector-borne diseases. https://www.who.int/news-room/fact-sheets/detail/vector-borne-diseases. Accessed 28 Dec 2023

Yonejima M, Nakaya T, Chan TC, Nihei N, Kobayashi M (2015, Risk mapping of dengue fever outbreaks in Japan, 2015) Paper presented at the KAGIS fall conference 2015 and international symposium on GIS, Busan, Republic of Korea, 5 November, 2015

Yonejima M, Nakaya T, Yasumoto S, Chan TC (2018) Estimating the distribution of potential risks of dengue fever outbreaks in Japan. In: Annual Meeting of the Association of Japanese Geographers. Spring 2018. https://doi.org/10.14866/ajg.2018s.0_000217

Chapter 10
Special Topics from Asian Countries

Kyoko Sawabe, Tenzin Wangdi, Pradya Somboon, Vongphayloth Khamsing, Moritoshi Iwagami, Siew Hwa Tan, Khatanbaatar Igori, Basu Dev Pandey, Kouichi Morita, Jiamei Sun, Astri Nur Faizah, Yusuf Ozbel, Tran Vu Phong, Vu Sinh Nam, Hwa-Jen Teng, Han-Hsuan Chung, Pai-Shan Chiang, and Shiu-Ling Chen

Abstract This chapter introduces topics of medical entomology and vector-borne diseases (VBDs) in 12 Asian countries. In particular, malaria, dengue fever, and Japanese encephalitis, which are mosquito-borne diseases, and a sense of urgency about the spread of serious damage is common among Asian countries. Several countries also announced information from scrub typhus (tsutsugamushi disease) from Bhutan, Indonesia, Japan, Laos, Nepal, and Taiwan, SFTS from China, Taiwan, and Japan, and leishmaniasis from Bhutan, Indonesia, Nepal, and Thailand. There

K. Sawabe (✉) · A. N. Faizah
Department of Medical Entomology, National Institute of Infectious Diseases, Tokyo, Japan
e-mail: sawabe@niid.go.jp; astrinf@niid.go.jp

T. Wangdi
Vector Borne and Climate Sensitive Diseases, Ministry of Health, Thimphu, Bhutan

P. Somboon
Department of Parasitology, Faculty of Medicine, Chiang Mai University, Chiang Mai, Thailand
e-mail: pradya.somboon@cmu.ac.th

V. Khamsing
Medical Entomology-Vector Borne Disease Laboratory, Institut Pasteur du Laos, Vientiane, Lao PDR
e-mail: k.vongphayloth@pasteur.la

M. Iwagami
Department of Tropical Medicine and Malaria, Research Institute National Center for Global Health and Medicine, Tokyo, Japan

Parasitology Laboratory, Institut Pasteur du Laos, Vientiane, Lao PDR
e-mail: miwagami@ri.ncgm.go.jp

S. H. Tan
Genetics and Molecular Biology, Institute of Biological Sciences, Faculty of Science, University of Malaya, Kuala Lumpur, Malaysia

K. Igori
School of Veterinary Medicine, Mongolian University of Life Sciences, Ulaanbaatar, Mongolia
e-mail: khatanbaatar@muls.edu.mn

K. Sawabe et al. (eds.), *Medical Entomology in Asia*, Entomology Monographs,
https://doi.org/10.1007/978-981-97-4406-0_10

are also concerns about increased Zika virus infection in Southeast Asian countries (Indonesia, Laos, Malaysia, Thailand, and Vietnam). Flea-borne diseases (plague) are also important VBDs in Taiwan and Vietnam. From Japan, the occurrence of redback spider bites and outbreaks of highly pathogenic avian influenza (HPAI) are introduced. Indonesia introduced the situation of livestock losses due to VBDs, and Thailand introduced the problem of insecticide resistance. In Turkey, located in West Asia, *Hyalomma* ticks that do not inhabit East and Southeast Asia are distributed, and Crimean-Congo hemorrhagic fever (CCHF), which is caused by the bite of this tick, has been introduced as an important VBD. The development of forensic entomology in Malaysia, and the use of various biological control methods in Vietnam are also well described.

It should be noted that the spread of VBD and other pests is not limited to the countries listed here, but may already be a problem throughout Asia. This information is valuable in understanding the current status of VBDs in Asia and will be expected to contribute to future efforts to address VBDs.

Keywords Arbovirus · DNA-based identification · Emerging zoonosis · Forensic entomology · Global warming · Insecticide resistance · Mosquito · Neglected tropical diseases (NTDs) · Tick · Vector-borne diseases (VBDs) · Vector control

10.1 Introduction

Although there are different ways to classify countries, there are about 50 countries and regions in Asia. The Asian countries are long from east to west, from west Asia to east Asia, and also extend widely from north to south, from the tropics to the cold region. Therefore, there are great differences in climate and vegetation, and the

B. D. Pandey · K. Morita
DEJIMA Infectious Disease Research Alliance (DIDA) and Vaccine Research Development Center (VRDC), Institute of Tropical Medicine, Nagasaki University, Nagasaki, Japan
e-mail: basudevpandey@nagasaki-u.ac.jp; moritak@nagasaki-u.ac.jp

J. Sun
Weifang People's Hospital, Weifang, Shandong, China

Y. Ozbel
Department of Parasitology, Ege University, Izmir, Turkey
e-mail: yusuf.ozbel@ege.edu.tr

T. V. Phong · V. S. Nam
Medical Entomology and Zoology Department, National Institute of Hygiene and Epidemiology, Hanoi, Vietnam

H.-J. Teng · H.-H. Chung · P.-S. Chiang · S.-L. Chen
Center for Diagnostics and Vaccine Development, Centers for Disease Control, Taipei, Taiwan
e-mail: hhc@cdc.gov.tw; pschiang@cdc.gov.tw; SL@cdc.gov.tw

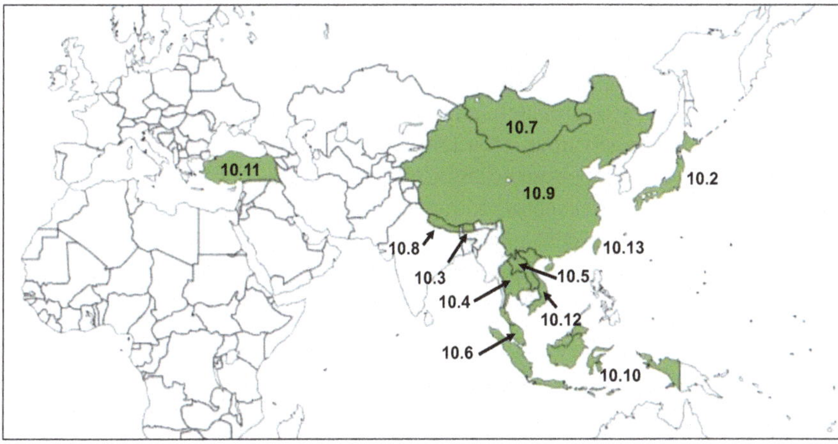

Fig. 10.1 Asian countries and regions introduced in this Chap. 10. The numbers on the map are chapter numbers

distribution of arthropods and the damage they cause also vary greatly. In other words, we have to consider that the damage caused by medically important arthropods is not just a matter for the country concerned, but is strongly influenced by the conditions in neighboring countries. Especially for imported vector-borne diseases (VBDs), it is necessary to quickly assess the epidemic situation in neighboring countries and exchange information in real time.

In this chapter, researchers from each country present the latest information on the 12 Asian countries shown in Fig. 10.1. It will provide a clear understanding of the differences and similarities in the medical entomological situation in Asian countries. Unfortunately, it was not possible to cover all Asian regions in this chapter. In this chapter, information has been presented mainly for East and Southeast Asian countries, but we hope that you will notice differences in medical entomology, for example, between East and West Asia, and between continents and islands.

We expect that information about countries that could not be introduced here will be available through the authors of each country that wrote in this chapter.

10.2 Japan

Kyoko Sawabe

The Communicable Disease Prevention Law enacted in 1897 was enforced as the New Law on Infectious Diseases (NLID) revised on 1 April 1999 (MHW 1999). Approximately half of the 44 diseases classified as Infectious Disease (ID) Type IV in NLID are vector-borne diseases (VBDs). Mosquito-borne diseases are the most common among these VBDs, followed by tick-borne diseases. In recent years, most cases of VBDs in Japan have been imported, except Japanese encephalitis. However, as exemplified by the 2014 autochthonous dengue outbreak, the threat of foreign infectious diseases in Japan continues to increase. Japan is long from north to south and has a rich diversity of climates and environments. It is one of the few countries in the world with a wide variety of arthropods. Although there have been frequent invasions of medically important arthropods from overseas in recent years, we must remember that the potential vector arthropods for these VBDs are already present in Japan. Here, we review recent trends in mosquito-borne and tick-borne diseases, other arthropod-borne diseases, and various types of damage involving arthropods.

10.2.1 Mosquito-Borne Diseases

There are more than 3700 species of mosquitoes in the world and approximately 110 species in Japan. Mosquitoes are the most threatening animals to humans, causing more than 700,000 deaths annually. They are zoonotic vectors that transmit pathogens to animals other than humans. Although many mosquito-borne diseases exist worldwide, including Japanese encephalitis (JE) and dengue fever (DF), this section outlines five mosquito-borne diseases classified as ID-Type IV diseases under the NLID in Japan. The number of reported cases is shown in Fig. 10.2.

Most of the mosquito-borne diseases reported in Japan in recent years, except JE, have been imported cases. After 2020, the global pandemic of the novel coronavirus infection led to severe restrictions on travel to and from overseas, and imported cases have uniformly plummeted. Cutting off the flow of people and goods to and

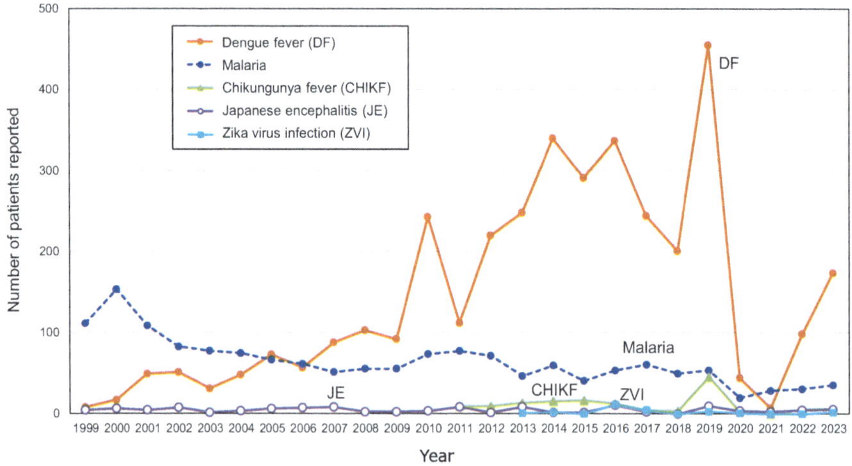

Fig. 10.2 Changes in the annual number of reported cases of major mosquito-borne diseases classified as Infectious Disease (ID) Type IV under the New Law on Infectious Diseases (NLID). The number of patients was taken from the Infectious Disease Weekly Report (IDWR) of the National Institute of Infectious Diseases (NIID)

from overseas will prevent the entry and spread of mosquito-borne infectious diseases in Japan; however, it will be difficult to continue controlling pathogens when the government targets 60 million foreign visitors to Japan by 2030. As mentioned in the Introduction, we must remember that the disease and its potential vectors are already present in Japan.

10.2.1.1 Malaria

Japan used to be plagued by "Indigenous malaria" and "War malaria" The former is endemic to the vivax malaria in Hokkaido and the five prefectures surrounding Lake Biwa between the Meiji and early Showa periods. The latter was an outbreak of falciparum malaria in the Yaeyama Islands, Okinawa Prefecture, before and after World War II. Reported malaria cases in Japan reached 28,000 annually in 1945 and 1946, with over 7000 cases of vivax malaria reported up to the end of the 1950s. Surprisingly, at least seven cases of falciparum malaria were reported between 1947 and 1951 in Rubeshibe, Hokkaido, in northern Japan. Malaria cases have decreased drastically since then, and endemic malaria was first eliminated in the 1960s. Since then, all malaria-reported cases have been imported cases, and fewer than 70 imported cases have occurred annually over the past 10 years. In 2022, 31 cases were reported, all imported (IDWR 2022).

Anopheline mosquitoes are the most important malarial vectors, with approximately 460 species worldwide, 12 of which are distributed in Japan. The larvae generally inhabit paddy fields, lakes, and rivers. Among the species recorded in Japan, *An. sinensis* Wiedemann is the most widespread anopheline species in East Asia and Japan, except in Hokkaido. This species is considered a major vector of vivax malaria. *Anopheles lesteri* Baisas & Hu is thought to have been involved in the falciparum malaria epidemic in Hokkaido in the 1940s. The current distribution is limited to Hokkaido and Okinawa prefectures. *Anopheles yaeyamaensis* Somboon & Harbach in Somboon et al. (formerly *An. minimus* species E), which inhabits the Sakishima Islands, Okinawa, is a vector mosquito for falciparum malaria. *Anopheles belenrae* Rueda, first described in South Korea in 2005, has also been found in the Kushiro area of Hokkaido, and *An. japonicus* Yamada inhabits the Tokara Islands. Because these two species are vectors of vivax malaria overseas, if these Japanese species are capable of transmitting *Plasmodium vivax*, then there are five species of malaria vectors in Japan (Somboon et al. 2001, 2010; Sawabe et al. 2003, 2021; Imanishi et al. 2018).

10.2.1.2 Japanese Encephalitis

Japanese encephalitis virus (JEV) was first isolated from humans in 1934, and its genotypes are divided into I to V. JE is distributed throughout Asia. Still, in recent years, JE cases have been confirmed in Papua New Guinea, Australia, and Angora in 2016, raising concerns about its spread to regions other than Asia. In the early

1990s, a genotype shift occurred from III to I throughout Asia, including Japan. In 1924, Japan recorded an epidemic involving more than 6000 people, with a mortality rate of more than 60%. Since 1992, the number of patients has decreased rapidly because of changes in breeding style and has been hovering at approximately 10 per year since 1992.

There were more than 30 vector species, including experimentally confirmed vector species and dominant species in the region. Among these three important species in Asia are *Culex tritaeniorhynchus* Giles, *Cx. pseudovishnui* Colless, and *Cx. vishnui* Theobald, which belongs to the *Culex vishnui* subgroup. Larvae grow in large water areas such as paddy fields and ponds, and adults prefer to feed on the blood of large animals. *Culex tritaeniorhynchus* overwinters as adults in Japan but migrates from the continent every summer. This finding was prompted by research findings that Japanese populations of *Cx. tritaeniorhynchus* have different nucleotide sequences of the mitochondrial cytochrome c oxidase subunit I (*COI*) gene than those of overseas populations (Arai et al. 2022). Recently, infection experiments have revealed individuals with high susceptibility to the virus in *Ae. japonicus* (Theobald) (Faizah et al. 2020). This invasive mosquito may play an important role as a vector mosquito for JE in overseas areas where JE vectors such as the *Culex vishnui* subgroup are not present.

10.2.1.3 Dengue Fever and Severe Dengue

Approximately half of the world's population is now at risk of dengue, and more than 100 countries in tropical and subtropical regions of the world are now endemic. An estimated 100–400 million infections, with more than 25,000 classified as severe dengue occur each year (WHO 2023a). In Japan, more than 200,000 cases have occurred in the 3 years since 1942, but there have been only imported cases to date, with a few exceptions. In 2013, a suspected case of domestic infection among German travelers was reported, and in 2014, 162 cases of autochthonous dengue infection were reported, mainly in parks in Tokyo. Additionally, 2 autochthonous dengue cases were reported in 2019. The number of patients in Japan is increasing, with more than 300 imported cases reported annually. Worldwide, the primary vector mosquito is *Aedes aegypti* (L.); however, in Japan, *Ae. albopictus* (Skuse) is an important vector species.

Aedes albopictus is widely distributed in the tropical and temperate regions of the world and inhabits many regions of Japan, except Hokkaido. Their eggs are resistant to dryness, and when they were first confirmed in Houston, USA, in 1984, it was reported that the eggs were attached to the inside of old tires and exported from Japan. Larval growth requires an annual average temperature of 11 °C or higher (Kobayashi et al. 2002), but owing to the effects of global warming, the distribution area is moving northward yearly. Establishments were confirmed in Morioka City in 2010 (Sato et al. 2012) and in Aomori City in 2016 (Maekawa et al. 2020), and the distribution area within Aomori Prefecture is expanding further. The current concern is the invasion and establishment of this species in Hokkaido.

Aedes aegypti is the most important dengue vector mosquito worldwide. This species is distributed throughout tropical regions worldwide, especially in urban areas. In Japan, it was collected from Chichi Jima Island in the Ogasawara Islands in 1916, Okinawa Prefecture in 1931, and Kumamoto Prefecture from 1944 to 1947, during a dengue epidemic. It became extinct in Japan after Ishigaki Island in the 1970s (Tanaka et al. 1979). Still, since 2012, eggs and larvae have been found almost yearly at multiple international airports (Sukehiro et al. 2013). Partly because of the effects of global warming, there is a high possibility that this species will become established in Japan in the future. If *Ae. aegypti*, which has a serious problem of insecticide resistance worldwide, invades and settles in Japan, it will be difficult to eradicate.

10.2.1.4 Chikungunya Fever

In 2006, a returnee from Sri Lanka became the first imported case of infection among Japanese individuals; since then, approximately 10 imported cases have been reported every year. The main vector of the chikungunya virus is *Ae. aegypti* and *Ae. albopictus*. These two species are also vectors of dengue virus, as mentioned above. Surprisingly, during the 2005–2006 outbreak on the French Reunion Island that infected 270,000 people (including approximately 250 deaths), a mutant strain, E1-A226V, emerged in the body of *Ae. albopictus* which multiplied more than 100 times faster than the original strain in *Ae. aegypti* (Schuffenecker et al. 2006; Tsetsarkin et al. 2007). Outbreaks in Italy in 2007, Indian Ocean Island countries in 2005–2006, and the Mediterranean coast of France in 2011 were caused by *Ae. albopictus*. This suggests that the invasion of the mutant strain, E1-A226V, threatens Asian countries, especially Japan, where only *Ae. albopictus* is found.

10.2.1.5 Zika Virus Infection

In 2007, 31 cases were reported in Yap, the Federated States of Micronesia. Still, only a dozen cases were reported until 2013, when approximately 10,000 cases occurred in French Polynesia, and the first cases imported to Japan were from this region. The most important vector of the Zika virus is *Ae. aegypti*, but *Ae. albopictus* and several *Culex* species have also been reported to be susceptible to Zika virus (Huang et al. 2016; Fernandes et al. 2020).

10.2.1.6 West Nile Fever

West Nile fever began with a sudden outbreak of cases in New York City in 1999, and the virus spread throughout the United States in just 4 years. Because they feed on the blood from both avians and mammals, they are likely bridge vectors for the

West Nile virus (WNV) and other avian-derived pathogens in humans, making their control extremely difficult.

In the United States, the most important vectors of WNV are several *Culex* species (e.g., *Cx. pipiens* L., *Cx. quinquefasciatus* Say, and *Cx. tarsalis* Coquillet) (Anderson et al. 2004; Kilpatrick et al. 2006; Godsey et al. 2012). In Japan, *Cx. p. pallens* Coquillett, which is a subspecies of *Cx. p. pipiens*, is a common species found throughout Japan, particularly in residential areas. Adults *Cx. p. pallens* begin their activity at dusk before sunset and are attracted by animal odors and carbon dioxide to enter indoor areas to suck blood during the night. The larvae occur in rainwater troughs and sewage. In addition, *Cx. p.* form *molestus* Forskal, which is the biological variant of *Cx. p. pipiens*, is also widely distributed in Japan. Both *Cx. p. pallens* and *Cx. p.* form *molestus* are flexible in their host-feeding habits, broadly preferring avian and mammalian hosts (Sawabe et al. 2010). Therefore, it was inferred that these two *Cx. pipiens* species are the most important potential vectors of the WNV in Japan.

10.2.2 Tick-Borne Diseases

In Japan, 49 species of ticks belonging to five genera are widespread in various environments. Many tick-borne diseases and their vectors have been endemic in Japan for a long time. Their movement is closely related to the migratory range of their host wild animals. Because large mammals are important blood-sucking sources for ticks. It has been suggested that the population growth and migration of large mammals, such as Japanese deer and wild boars, may influence the prevalence of tick-borne diseases. Thus, ticks can be vectors of zoonotic diseases involving livestock, wildlife, and humans. The number of cases of tick-borne disease reported in Japan since 1999 is shown in Fig. 10.3. Although tsutsugamushi is not a tick, it is often compared to ticks because of its morphological similarity. In this section, tsutsugamushi and tsutsugamushi-borne diseases, scrub typhus (ST), are also outlined.

As mentioned above, the distribution of ticks is closely related to the distribution of host wildlife and the plants they feed on. Future warming trends will alter the vegetation in the tick habitat and affect the behavior of the tick host animal. Furthermore, the spread of tick-borne diseases is feared. For example, wild boars were first identified in Toyama Prefecture in 1988, and a bite case of *Amblyomma testudinarium* Koch was reported in 2010; the first case of severe fever with thrombocytopenia syndrome (SFTS) was reported in 2022. Additionally, in Chiba Prefecture, wild boar and deer populations have increased rapidly since 1980. Japanese spotted fever (JSF) was reported in 1987; in 2017, the first case of SFTS was reported in the Kanto area. In recent years, exotic tick-borne pathogens such as Jingmenviruses have been discovered in Japan (Kobayashi et al. 2021a). In addition, the first cases of Yezo virus and Oz virus infection have emerged (Kodama et al. 2021; NIID 2023a), and close monitoring pressure must be applied to accompany tick-borne diseases in Japan.

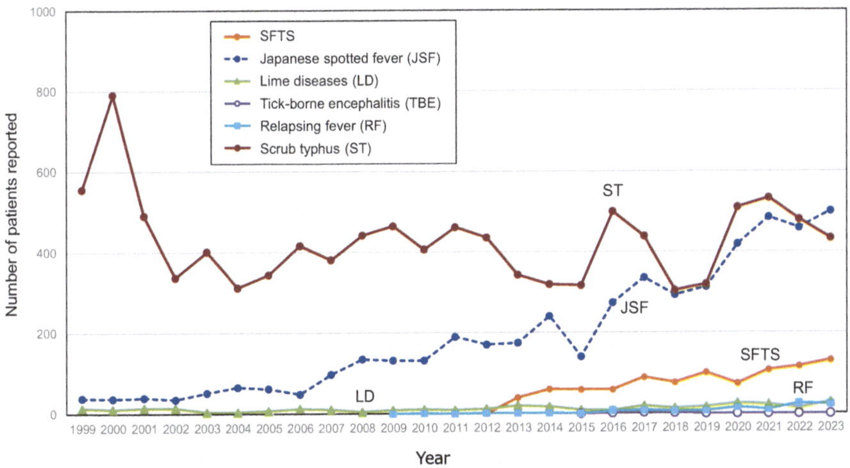

Fig. 10.3 Changes in the annual number of reported cases of major tick-borne diseases include tsutsugamushi-born disease, classified as Infectious Disease (ID) Type IV under the New Law on Infectious Diseases (NLID). The number of patients was taken from the Infectious Disease Weekly Report (IDWR) of the National Institute of Infectious Diseases (NIID)

10.2.2.1 Severe Fever with Thrombocytopenia Syndrome

Severe fever with thrombocytopenia syndrome (SFTS) is a tick-borne disease caused by a novel bunyavirus in the genus phlebovirus, SFTS virus (SFTSV). In the new classification of the International Committee on Taxonomy of Viruses (ICVT) in 2018, names of the family, genus, and virus of SFTSV were changed to *Huaiyangshan banyangvirus*, which belongs to the family Phenuiviridae, the genus *Banyangvirus*. However, hereafter we referred to it as "SFTSV." SFTSV was first discovered in China in 2009 (Yu et al. 2011), followed by Japan, South Korea, Vietnam, Taiwan, the United States, and many other countries. In Japan, the SFTS virus was isolated from a woman living in Yamaguchi Prefecture with no travel history who died of an unknown cause in the fall of 2012 and was reported as the first case in Japan in January 2013 (Takahashi et al. 2014). Since then, patients with SFTS have emerged mainly in western Japan, and from 2005 to July 2023, 900 patients (including 101 deaths) have been reported in 29 prefectures in western Japan and other regions (NIID 2023b). In contrast, wild animals positive for viral antibodies have been confirmed in areas of eastern Japan where no patients have been infected, and viral genes have been detected in several tick species, raising concerns about the spread of the epidemic in the future.

The SFTS virus gene was detected in two tick species, *Haemaphysalis longicornis* Neumann and *Rhipicephalus microplus* (Canestrini), in China and was isolated from Korean *H. longicornis*. However, the vector species of the SFTS in Japan remain unknown. SFTS is a tick-borne zoonosis; however, there have been reports of contact transmission from companion animals (e.g., dogs and cats) to humans (Nam et al. 2020; Tsuru et al. 2021) in Japan. Human-to-human transmission of SFTS has also occurred in other countries (Chen et al. 2013).

10.2.2.2 Japanese Spotted Fever

Japanese spotted fever (JSF) was first reported in 1984 as rickettsiosis caused by *Rickettsia japonica* and is more frequently observed from early summer to autumn in Tokushima Prefecture (Mahara et al. 1985). Before 1998, JSF was confined to parts of Kyushu, Shikoku, the Kinki region, Kanagawa, and Chiba prefectures, and 10–20 JSF cases were reported per year in 1994. However, the outbreak area has further expanded, and the number of patients has increased, with 481 patients reported annually in 2022 (IDWR 2022). Rickettsioses caused by the same genus are widely distributed worldwide, including Rocky Mountain spotted fever in North America, Mediterranean spotted fever in Mediterranean coastal areas, and Queensland tick typhus in Australia. The main symptoms of JSF are fever and rash accompanied by headaches, fever, and malaise. Tick bites were often observed. These symptoms are similar to scrub typhus (ST), and a distinction should be made between JSF and ST. JSF-vector ticks are thought to be *H. flava* Neumann, *H. longicornis*, and *Ixodes ovatus* Neumann. Although *R. japonica* is usually transmitted from parent ticks to their offspring (transovarial transmission), some wild animals carry or are naturally infected with *R. japonica*. It is also known that there is a transmission cycle from reservoirs to humans via ticks. Rodents and wild deer are important reservoirs in this cycle.

10.2.2.3 Lyme Disease

Lyme disease (LD) is a tick-borne disease caused by a zoonotic bacterium (spirochete) transmitted by Ixodid ticks in nature, with wild rodents and small birds serving as reservoirs. In Japan, most cases of LD occur after the bite of *I. persulcatus* Schulze. This tick inhabits Hokkaido and Honshu, Shikoku, and Kyushu mountainous areas. In Hokkaido and parts of the Aomori Prefecture, it is also found on plains, except in urban areas. The first case of LD in Japan was reported in 1986, and several hundred cases have been reported to date. Particularly in the Hokkaido and Nagano prefectures, with approximately 10–20 cases annually and 14 in 2022 (IDWR 2022).

10.2.2.4 Tick-Borne Encephalitis

Tick-borne encephalitis (TBE) is caused by the TBE virus (TBEV) belonging to the same genus as flaviviruses, such as dengue virus (DENV) and JE virus (JEV). Still, they are located in a rather distant cluster in the molecular phylogenetic tree. The TBEV group includes Russian spring-summer encephalitis (Far Eastern-type tick-borne encephalitis, RSSE), Central European TBE, Kyasanur Forest disease, louping illness, Omsk hemorrhagic fever, Poissant encephalitis, Alkhumra, and Langat, all of which are characterized by high fatality rates and severe symptoms. The symptoms of encephalitis may begin with headaches, fever, nausea, and vomiting, followed by mental confusion, coma, and convulsions. The fatality rate is estimated to be 30% and paralysis often persists even after recovery.

The RSSE virus (RSSEV) was first isolated from encephalitis patients in Far Eastern Russia in 1937 and was found to be transmitted by *I. ovatus* Neumann, and *I. persulcatus*. In Russia, 4000 to 10,000 cases occur annually. In 1993, the first confirmed TBE case (RSSE-type) was reported in the southern portion of Hokkaido on the northern island of Japan (Takashima et al. 1997). TBE has been isolated in this area from dogs, ticks, *I. ovatus*, and wild rodents, *Apodemus speciosus* (Temminck) (Takeda et al. 1998, 1999). However, antibody-positive wild rodents were also found in Honshu and Hokkaido. The second case of TBE in Japan was reported in July 2016, 23 years later, and the third and fourth cases were reported in 2017 (Takahashi et al. 2020). These results suggest that the TBE is endemic in Japan.

10.2.2.5 Relapsing Fever

Relapsing fever (RF) is caused by the *Borrelia* spirochetes, transmitted by mites of the genus *Ornithodorus* and lice, with small rodents and birds as reservoirs. *Ornithodoros* mites are found in Africa, the Americas, parts of Europe, and the Middle East. RF is characterized by a recurrent fever that repeats several times between the febrile and nonfebrile phases. If no treatment is administered, the mortality rate is estimated to be as high as 30%; however, tetracyclines are effective. Although RF outbreaks have been reported in the Americas, Africa, the Middle East, and parts of Europe, the number of imported cases has recently increased in Japan (25 cases were reported in 2022) (IDWR 2022).

10.2.2.6 Scrub Typhus

Scrub typhus (ST) is a rickettsial infection caused by a rickettsia, *Orientia tsutsugamushi*, and is transmitted by three species of the *Leptotrombidium* tsutsugamushi in Japan: *L. akamushi* Brumpt, *L. scutellare* Nagayo, Miyagawa, Mitamura, Tamiya et Tenjin, and *L. pallidum* Nagayo, Mitamura et Tamiya. The occurrence of ST is closely related to the time of year when tsutsugamushi larvae are active. More than 90% of the patients present with fever, thrush stings, rash, malaise, and headache. Swelling of lymph nodes near the stings or throughout the body may occur. The fatalities are high if treatment is delayed. The domestic Japanese vectors of this disease are *L. akamushi*, *L. scutellare*, and *L. pallidum*, all with a prevalence rate of 0.1–3%. It is known that tsutsugamushi do not take up the rickettsiae through bitting reservoir animals but pass it only by transovarial transmission to the next generation. Domestic cases have increased in recent years; however, they are also widely distributed in neighboring Asian and Southeast Asian regions and should be monitored as an imported infectious disease.

10.2.3 Current Topics in Medical Entomology in Japan

10.2.3.1 Previously Unfocused Relationships Between Arthropods and Pathogens

Several arthropods transmit pathogens, including horseflies, biting midges, sandflies, lice, and eye gnats, many of which infect mammals other than humans and are important vectors of zoonotic diseases that cause extensive damage, especially in livestock. It has been suggested that house flies, *Musca domestica* L., were involved in the enterohemorrhagic *Escherichia coli* O157 (EHEC) outbreak in Saga Prefecture in 1996 (Iwasa et al. 1999; Kobayashi et al. 1999), and that mechanical transmission by blow flies, *Calliphora nigribarbis* Vollenhoven, and *Aldrichina grahami* (Aldrich), caused serious problems during the highly pathogenic avian influenza (HPAI) epidemic that occurred in Kyoto Prefecture in 2005 (Sawabe et al. 2006, 2009). Recent advances in genetic analysis techniques using next-generation sequencing have also revealed previously unknown relationships between arthropods and pathogens; for example, *Trypanosoma* sequences were detected in tick (*H. flava*) and horse fly (*Tabanus rufidens* Bigot) (Kobayashi et al. 2020), and a novel flavivirus, Tabanus rufidens flavivirus (TrFV), was detected in *T. rufidens* (Kobayashi et al. 2021b). We look forward to further research in progress.

10.2.3.2 Global Warming Affects Arthropods in Japan

With globalization, the scale and speed of the movement of people and goods have increased dramatically. This has resulted in an associated increase in the spread and global distribution of arthropods and the infectious disease-causing pathogens that they transmit. Although Japan is a temperate climate zone, global warming increases the likelihood that pest species that have not previously inhabited Japan will invade and establish in Japan. Thus, the number of invasive alien species entering Japan is increasing significantly and there is an urgent need to strengthen control measures.

Global warming trends in recent years have been severe and have greatly affected the ecology of native species of mosquitoes and ticks. For example, it directly affected *Ae. albopictus*, moving its distribution northward Maekawa et al. (2020). The distribution range has also expanded to high latitudes, and the invasion of Hokkaido is a concern. In recent years, *Ae. aegypti* has been frequently introduced overseas, and the warming trend threatens the establishment and expansion of its distribution in Japan. In addition, vector species and potential vectors for JE and *Cx. tritaeniorhynchus* and *Armigeres subalbatus* (Coquillett) were first collected from Hakodate in 2019 (Maekawa et al. 2020). Climate change is also closely related to vegetation and wildlife distribution and activity. Future warming trends will directly affect vegetation and the behavior of wildlife and host animals. These environmental changes expand the habitats of ticks and other vector insects and alter their activities.

10.2.3.3 Diverse Damage Caused by Medically Important Arthropods and Animals

Medically important animals cause a wide variety of damage and transmit infectious diseases. For example, salivary gland components injected during bloodsucking induce dermatitis and stinging and biting injure the skin. Spiders and scorpions also have lethal venoms, and bee and centipede stings can cause allergic symptoms and anaphylactic shock in severe cases. The shape and form of scorpions and spiders, as well as the number of dung beetles and ants, can cause discomfort to people (unpleasant pests); some species, such as chironomids and house dust mites, are allergens that can cause asthma and urticaria. Shellfish, which serve as intermediate hosts for parasites; rats and pigeons, which transmit infectious diseases; and woodlice and cockroaches (economic pests), which devour wood and grain and cause economic losses, are also in the categories of medical entomology in Japan.

Damage caused by invasive arthropods had a spotlight. Argentine ants, *Linepithema humile* (Mayr), were confirmed in Hiroshima Prefecture in 1993. The first case of the redback spider, *Latrodectus hasselti*, Thorell, bite occurred in 1997, and it was necessary to create domestically produced anti-venom. Fire ants, *Solenopsis invicta* Buren, 1972, invaded Osaka Prefecture in 2020, and fire ants, *Wasmannia auropunctata* (Roger), a highly poisonous South American species, were first discovered in Okayama Prefecture in 2023. The invasion of exotic animals has also been serious. Since a raccoon escaped from a zoo in Aichi Prefecture in 1967, the number of raccoons in Japan has been increasing. Populations of the masked palm civet and Reeves' muntjac are also increasing rapidly. It is believed that the increase in the population of these exotic animals has significantly impacted SFTS outbreaks.

The damage caused by these medically important animals is no longer limited to Japan but has become a global threat. In this sense, we need to understand the global situation, especially in Asia, in real time. In contrast, Japan has experienced the eradication of malaria and lymphatic filariasis, drastically reduced the number of JE patients, and successfully controlled houseflies in the metropolis of Tokyo. Our history provides valuable information on the Asian region. Therefore, it is imperative to continue exchanging and sharing information.

10.3 Kingdom of Bhutan

Tenzin Wangdi

Wedged between China to the north and India to the south, Bhutan, with a total surface area of 38,390 km^2, lies on the eastern end of the mighty Himalayas. The country, with a population of 763,249, experiences three distinct climatic conditions: subtropical in the southern foothills and inner Himalayan valleys, temperate in the central mountainous region, and alpine conditions with year-round snow-capped mountains on the main Himalayan summits.

The country is administratively divided into 20 districts and 205 subdistricts. Among the vector-borne diseases, malaria, dengue, leishmaniasis, and scrub typhus are endemic to Bhutan. Currently, many of the vector-borne diseases are confined or endemic to the seven southern districts that experience subtropical climatic conditions. However, in recent times, outbreak of dengue has been reported from a few northern nonendemic districts (Tsheten et al. 2021). In the years to come, it is foreseeable that Bhutan will see a northward expansion of vector-borne diseases, establishing new hotspots and putting vulnerable populations at risk.

10.3.1 Malaria

Malaria has been recorded in Bhutan since 1965 (Fig. 10.4). Since the first survey conducted by the National Malaria Eradication Programme of India in 1962, *Plasmodium falciparum* and *P. vivax* were the only two parasite species detected and have remained largely unchanged until now. *Anopheles minimus* was found to be the dominant vector species during the 1962 survey, and one specimen was also reported to be sporozoite-positive (WHO 2010a).

Based on a survey from 2007 to 2018, 30 Anophelines have been recorded in Bhutan (Namgay et al. 2018). Although no vectors have been implicated for malaria in Bhutan, except for *An. minimus* in 1962, incrimination of vectors in the current setting is extremely difficult owing to a very low parasite prevalence in the country (Wangchuk et al. 2019). However, based on circumstantial evidence, viz., association with cases, preponderance over other species during the transmission season, propensity toward indoor biting and resting observed during malaria case investigations, *An. vagus, An. culicifacies s.l., An. minimus s.l.,* and *An. pseudowillmori* are suspected to be malaria vectors in recent times. While *An. pseudowillmori* has the widest distribution from the southern plains and foothills to the interior northern mountainous region, adult *An. minimus* was only collected from Tenpaling village in Samtse, bordering India.

In recent times, *An. dirus s.l.* has been added to the list of known Anophelines in Bhutan. *An. dirus s.l.*, the primary vector in the forested areas of Southeast Asia (Gould et al. 1966), was found breeding in elephant footprints under the canopy of tall trees in the southern foothills of Sarpang District (Fig. 10.5). *Anopheles barianensis* James was detected from a tree hole in Thimphu District. Two species of the Baileyi Complex (Somboon et al. 2022a), and four species of the Lindesayi Complex (Somboon et al. 2022b) were newly described from Bhutan, bringing the total *Anopheles* species to 38. In addition, *An. sarpangensis* Somboon, Namgay & Harbach was described as a new species of the Barbirostris Complex found in Bhutan (Somboon et al. 2024).

In one survey conducted in Sarpang District (Naszeerah 2015) *An. vagus* and *An. pseudowillmori* were predominantly found to be biting during the first 6 h of the

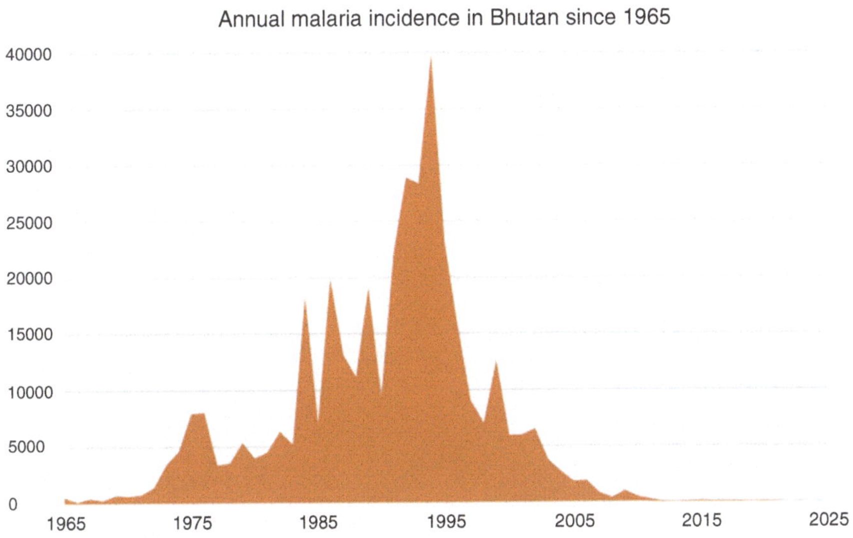

Fig. 10.4 Annual incidence of malaria in Bhutan since 1965

Fig. 10.5 A house where a malaria case was detected in Sarpang District of Bhutan. The majority of malaria cases are reported from socioeconomically disadvantaged households during near elimination

night whereas, *An. culicifacies s.l.* were mostly found biting during the second half of the night. All three species were largely collected from cattle-baited collections with only a few from indoor human landing and resting collections. The high coverage of indoor residual spraying (IRS) and long-lasting insecticidal nets (LLINs) in malaria-risk areas of Bhutan could have been detrimental in changing their behavior (Fig. 10.6).

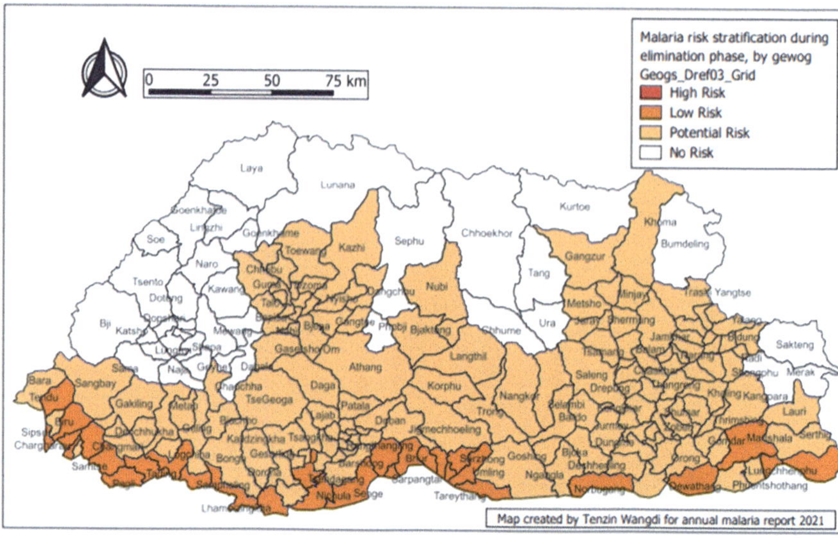

Fig. 10.6 Map of Bhutan showing malaria risk stratification at subdistrict level during elimination

10.3.2 Dengue Fever

Dengue was first recorded in Bhutan with a massive outbreak in the city of Phuntsholing bordering India in 2004. Since then, the city has reported few cases every year with seasonal outbreaks occuring every 2- to 3-year cycle. The city is now considered dengue endemic, with cases reported throughout the year. From 2017, clusters of dengue cases were reported from other bordering towns in the southern parts of the country with the presence of *Ae. aegypti* confirmed. To date, the largest epidemic of dengue occurred in 2019, with case reporting from 19 of the 20 districts, and the virus having spread to northern nonendemic regions, mostly from Phuntsholing (Tsheten et al. 2021) (Fig. 10.7). Except for DENV-4, all the other three serotypes have been reported, with a shift in serotype dominance resulting in outbreaks (Zangmo et al. 2015). Phylogenetic analysis of complete gene envelopes of DENV-2 (Cosmopolitan genotype) from 2004 and DENV-3 (genotype III) from 2005 to 2006 indicated that the virus was introduced from India (Dorji et al. 2009).

Both *Aedes aegypti* and *Ae. albopictus* are prevalent in Bhutan, with the former likely introduced from neighboring countries through various trade and transport means and the latter most likely native to the country. *Aedes aegypti* is restricted to urban areas of the southern plains and is found mostly breeding indoors in water containers where water scarcity is common (Fig. 10.8). *Aedes albopictus*, however, has a widespread distribution throughout the country and is found in both urban and rural settings. During the massive 2019 dengue outbreak, *Ae. aegypti* was found to have spread to a few small towns of northern districts (Mongar and Trashiyangtse). While climate change could have contributed to their survival and reproduction,

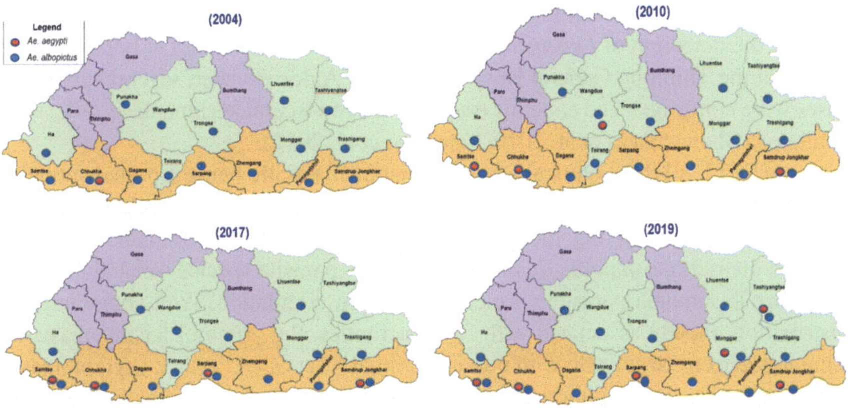

Fig. 10.7 District map of Bhutan showing the distribution of dengue vector and spread of *Ae. aegypti* from 2004 to 2019

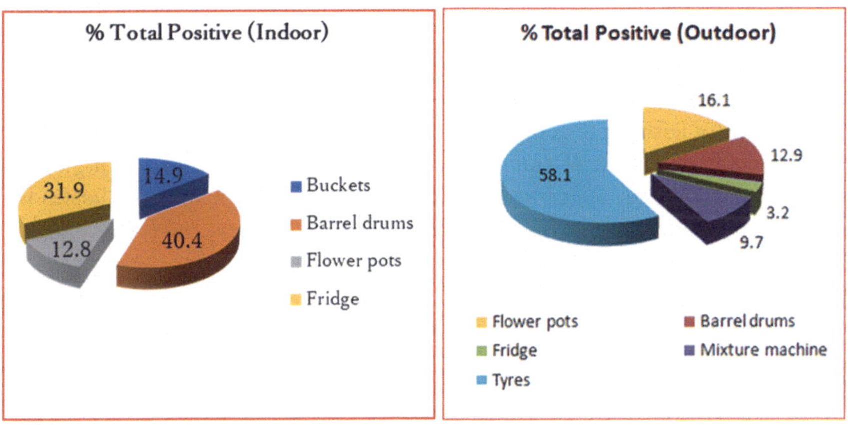

Fig. 10.8 Characteristics of *Ae. aegypti* breeding (indoor and outdoor) in Phuntsholing, Bhutan

unchecked transport of goods and ease of travel could have directly led to the northward expansion of the vector species. However, with massive social campaigns and community mobilization, the vector species was not found in the two districts during subsequent surveys.

10.3.3 Leishmaniasis

The earliest known case of leishmaniasis in Bhutan was detected in a man from Neykorlog village in Mongar district. The case came to light during a sand fly surveillance in 2019, where he reported to the survey team that he had suffered from

Kala-azar disease. On examining his medical record, his diagnosis was confirmed through the detection of *Leishmania donovani* bodies (LD bodies) in bone marrow biopsy at the Thimphu General Hospital in 1994. A decade later, several suspected cases were confirmed, and endemic transmission of visceral leishmaniasis (VL) was validated (Bhattacharya et al. 2010; Yangzom et al. 2012).

As of 2023, all three clinical forms of leishmaniasis are being reported sporadically, and the annual incidence of VL remained below 1 per 10,000 population in the reported districts, meeting the elimination threshold. Mongar remains the most affected district. *Leishmania donovani* is the only parasite species isolated thus far, but other leishmania parasite species could be in circulation. With sporadic case distribution, mostly associated with forest activities, the origin of leishmaniasis in Bhutan seems to be sylvatic with possible animal reservoirs. Further, asymptomatic screening of whole communities using rK39 strip during case investigations yielded not a single positive case.

Based on the survey records of sand flies from 2019 to 2022 in areas where leishmaniasis was reported, six species of *Phlebotomus* and four species of *Sergentomyia* were identified (Tenzin et al. 2024). *Phlebotomus argentipes*, considered the principal vector of VL in the Indian subcontinent, appears to be rare, collected in low numbers only from Kalapang village in the Mongar district. It becomes obvious that some other unknown sand fly species could be playing the role of vector in the transmission of leishmaniasis in Bhutan. The sand flies that have been recorded thus far include *P. chinensis. P. kiangsuensis, P. longiductus, P. major, P. neglectus, S. iyengari, S. arboris, S. barraudi,* and *S. taianensis*. The potential candidates for vector incrimination may include *P. longiductalus, P. chinensis,* and *P. major*, based on evidence from China (Guan et al. 2016; Leng 1997).

10.3.4 Scrub Typhus

Bhutan falls under the infamous *Tsutsugamushi* triangle. However, the chigger-borne disease was first reported in 2009 (Wangchuk et al. 2012). The disease failed to come under the scanner of the clinicians, although many undifferentiated fevers have been reported. Clinicians were least aware of the rickettsial disease prior to the confirmation of several undifferentiated fevers in 2009 (Wangchuk et al. 2012). In an outbreak of scrub typhus in a remote school in central Bhutan in 2014 that resulted in two young deaths, the investigation team noted low awareness among the general public about the disease and delayed health-seeking behavior as the immediate cause of death (Tshokey et al. 2017). What strain of *tsutsugamushi* is in circulation has yet to be explored. The incidence of the disease is highest in the southern subtropical districts, although cases were reported from northern temperate districts as well. Strong seasonality and rural agricultural settings are some of the characteristics of the disease reported in Bhutan (Dorji et al. 2019).

In an attempt to identify the reservoir host in one of the hotspot areas for scrub typhus in Bhutan, one study identified the presence of four species of rodents from four genera, but none were found infected with the bacteria (Phuentshok et al. 2018). There is no report of any studies performed on the vector in Bhutan.

10.4 Kingdom of Thailand

Pradya Somboon

Thailand is a tropical country that has faced public health problems caused by vector-borne diseases, including malaria, dengue and dengue hemorrhagic fever, chikungunya fever, Zika virus infection, Japanese encephalitis, lymphatic filariasis, and leishmaniasis. Malaria and dengue are the most important diseases that occur annually, whereas the others are sporadic, with occasional outbreaks in some years. Refer to Chap. 1 for details of taxonomic information on mosquitoes.

10.4.1 Malaria

Malaria was the leading cause of mortality in Thailand, resulting in over 30,000 deaths a year and over 200,000 morbidity cases annually in the late 1940s. Following the national malaria control program starting in the 1950s, which depended mainly on DDT residual spray, case detection, and chemotherapy, the incidence of malaria cases decreased dramatically. Elimination of malaria transmission was successful in many areas, particularly in the plains, coinciding with the disappearance of most mosquito vector species in the plains. In recent years, a significant reduction in malarial cases has also occurred in other countries of the Greater Mekong subregion (GMS), including Cambodia, China (Yunnan), Laos, Myanmar, and Vietnam, falling by 76% between 2010 and 2018, and malaria mortality fell by 95% over the same period (WHO 2020). In 2012, there were 33,835 reported cases, which decreased to 2951 cases in 2021. In 2022, however, the number of cases increased to 8223 (as of September); approximately 40% of cases were immigrants from neighboring countries (MOPH 2022). *Plasmodium falciparum* was the predominant species over *P. vivax*, approximately 80% in 1965, but *P. vivax* is now predominant (approximately 80%). *Plasmodium malariae* and *P. ovale* are rare in the country. Intense malaria transmission in Thailand is confined to forest and foothill areas along the Thai-Myanmar and Thai-Cambodian borders, usually with two transmission peaks. The first peak appears in the early rainy season (June–July) when the vector populations start to rise, coinciding with increased agricultural activities (e.g., rice cultivation) of people. The second peak occurs in the early cool season (October–November), related to harvesting rice and increased forest activities. A study in northwest Thailand revealed that people engaged in their agricultural and forest activities received approximately 3- and 12-times higher risk of acquiring malaria than those in their residential villages (Somboon et al. 1998).

Of approximately 74 *Anopheles* species occurring in Thailand, *An. dirus, An. baimaii*, and *An. minimus* are regarded as primary vectors of human malarial parasites, while *An. aconitus, An. epiroticus,* some species of *An. maculatus* group (*An. maculatus*, *An. pseudowillmori*, and *An. sawadwongporni*) and *An. wejchoochotei* (the Barbirostris Complex) are regarded as secondary vectors (Rattanarithikul et al. 2006). Most vector species play a role in the transmission of malarial parasites in forest and forest foothill areas, except for *An. epiroticus,* which is the vector along coastal areas and on islands. Although resistance to DDT has rarely been detected in the vectors *An. minimus* and *An. dirus*, they have changed their behavior, avoiding resting on sprayed areas due to the excito-repellency of DDT. Pyrethroids (permethrin and deltamethrin) were introduced to replace DDT usage by the late 1990s (Somboon et al. 1995), and they are currently used for impregnating bednets, long-lasting insecticide bednets, and, to a lesser extent, residual spraying. Community-wide use of pyrethroid-treated bednets resulted in a reduction in malaria morbidity but did not generally show a mass-killing effect against vectors, probably due to their exophagic and zoophilic behavior (Somboon et al. 1995). *Anopheles* vector species in Thailand are generally susceptible to pyrethroids, but resistance to pyrethroids has been reported in some areas (Somboon et al. 2003; Sumarnrote et al. 2017).

Despite problems of multidrug-resistant *P. falciparum* malaria over the decades, there has been a significant reduction in malaria mortality and morbidity in recent years, and the country recently verified 37 provinces as malaria-free. As a result, Thailand committed to eliminating malaria by 2024. However, this goal is challenged by several factors: (1) immigrants from neighboring countries often carry malarial parasites and introduce malaria infections to local people on the Thai side, (2) Thai people who engage in their forest and agriculture activities across the borders often return with malarial infections and introduce the parasites to their communities, (3) foci of transmission occur not only in village settlements but also in forest and remote farm hut areas where transmission is more intense (Somboon et al. 1998), (4) increased problems on resistance of *P. falciparum* and *P. vivax* to multiple malarial drugs (WHO 2022), and (5) there is an increasing trend of zoonotic malaria transmission, but there is no specific strategy for control. Several species of monkeys are infected with simian malaria in Thailand, but how people acquire the infection is not clearly known (Fungfuang et al. 2020).

In recent years, zoonotic infections with *P. knowlesi*, a common simian malarial parasite, were regularly reported, with 142 cases in 2022. A few cases of *P. cynomogli*, another simian malarial parasite, were also reported (Putaporntip et al. 2021, 2022; Saingam et al. 2022). Each of the simian parasites was found alone or mixed with a human malarial parasite. Apart from *P. falciparum* and *P. vivax*, a number of *P. knowlesi*-infected cases were found among travelers returning from Thailand and other countries in Southeast Asia (Froeschl et al. 2018). The vectors of simian malaria in Thailand are not known. Elsewhere, species of the Leucosphyrous Group are predominantly reported to be positive for *P. knowlesi*, i.e., *An. balabacensis, An. dirus, An. cracens, An. hackeri, An. introlatus, An. latens* and *An. leucosphyrus.* Other possible *Anopheles* vector species include *An. barbirostris s.l.* and *An.*

donaldi (The Barbirostris Complex), *An. roperi* and *An. collessi* (the Umbrosus Group), and *An. sundaicus* (van de Straat et al. 2022). *Anopheles kochi* is a suspected vector of *P. knowlesi* in Singapore (Jeslyn et al. 2011). The role of these mosquito vectors in transmitting simian parasites to humans varies depending on their anthropophilic behavior.

10.4.2 Dengue and Dengue Hemorrhagic Fever

Dengue and dengue hemorrhagic fever were first reported in Thailand in 1950 and mainly confined to Bangkok. The first large outbreak of the disease occurred in 1958, with 2158 cases and 300 deaths (Rojanapithayakorn 1998). Since 1965, the disease has been reported from all regions of the country, and over 100,000 cases per year were often reported. Despite over 60 years of dengue control using insecticide-based measures, i.e., thermal fogging spray, ultra-low-volume (ULV) spray, and larvicide (temephos sand granules) and breeding site reduction via community participation campaigns, dengue outbreaks are still an important public health problem. The number of cases varied year by year. In 2019 and 2020, a total of 131,157 and 72,130 dengue cases were reported, respectively. The daytime-biting mosquitoes, *Aedes aegypti* and *Ae. albopictus,* are important vectors. The former is common in urban and semi-urban settings, whereas the latter predominates in rural and forest areas.

Control of the vector *Ae. aegypti* based on insecticide application is not effective due to insecticide resistance. There are two major mechanisms in *Ae. aegypti* populations in Thailand: knockdown resistance (*kdr*) and, to a lesser extent, metabolic enzyme-based resistance (Yanola et al. 2010; Somwang et al. 2011). Knockdown resistance is associated with single or multiple mutations of the voltage-gated sodium channel (VGSC) gene and is an important mechanism for resistance to pyrethroids and DDT (Hemingway and Ranson 2000). Three major mutations of the VGSC gene have been detected in *Ae. aegypti* populations in Thailand, F1534C, V1016G, and S989P, which can be detected by DNA sequencing and PCR-based assays (Yanola et al. 2011; Stenhouse et al. 2013; Saingamsook et al. 2017). The F1534C mutation, a phenylalanine to cysteine substitution at position 1534 in domain IIIS6 of the VGSC gene, is associated with resistance to type I pyrethroid (permethrin). The V1016G mutation, a valine to glycine substitution at position 1016 in IIS6 of the VGSC gene, is associated with both type I and type II (deltamethrin) pyrethroids. The S989P mutation, a serine to proline substitution at position 989 in IIS6, is often found with the V1016G mutation. A combination of F1534C + V1016G + S989P mutations in a heterozygote genotype results in an additive effect of pyrethroid resistance (Plernsub et al. 2016a). However, laboratory studies revealed a fitness cost positively related to the level of resistance (Plernsub et al. 2013; Saingamsook et al. 2019). The pyrethroid-susceptible PMD strain (no *kdr* gene) had the highest fitness compared with the two pyrethroid-resistant strains (PMD-R, homozygous 1534C allele, and UPK-R, homozygous 1016G allele) and the hybrid. These fitness costs include shorter wing length, reduced egg hatchability, shorter female lifespan, and decreased viability of eggs after storage, whereas no effect was observed on the blood feeding rate. The effects on developmental time, egg production, and blood ingestion were variable among *kdr* genotypes. The UPK-R strain, which displayed the highest pyrethroid resistance level, had significantly prolonged larval development times, resulting in increased adult emergence times compared to the other genotypes (PMD, PMD-R, and hybrid). This indicates that the 989P + 1016G kdr alleles may be particularly detrimental. Metabolic enzyme-based resistance in *Ae. aegypti* populations in Thailand is principally associated with cytochrome P450 monooxygenases (P450s), aldehyde hydrogenases, and glutathione-S-transferases (Prapanthadara et al. 2005; Somwang et al. 2011; Lumjuan et al. 2014; Choovattanapakorn et al. 2017). This mechanism can be reduced or inhibited by metabolic inhibitors, such as piperonyl butoxide (PBO), a monooxygenase inhibitor. However, metabolic inhibitors have no effects on *kdr*. The presence of both resistance mechanisms in *Ae. aegypti* populations in Thailand have hampered the efficacy of the current pyrethroid-based control approaches using a thermal fogging spray with deltamethrin plus PBO (Plernsub et al. 2016b) and commercial household insecticide aerosol sprays (Nachaiwieng et al. 2021). *Aedes albopictus* is generally susceptible to most insecticides, but the F1534C *kdr* mutation has been detected in this species in many countries (Zulfa et al. 2022). This finding would cause a problem in controlling this vector species using pyrethroid-based methods.

One strategy to cope with pyrethroid-resistant mosquitoes is to use another insecticide with a different mode of action. Pyriproxyfen (PPF), an insect growth regulator, is normally used at very low concentrations to control mosquitoes' immature stages. At high concentrations, it has a reproductive effect on exposed female mosquitoes. In the laboratory, Suttana et al. (2019) demonstrated that tarsal contact with 333 mg AI PPF/m^2 for 1 min was sufficient to cause over 95% emergence inhibition (EI) in the progeny of exposed *Ae. aegypti* females. Exposure for 5 min completely inhibited fecundity. Since *Ae. aegypti* adult mosquitoes are generally drawn towards darker areas, application of PPF spray on a dark surface, such as artificial resting boxes, is a potential method to control *Ae. aegypti.* In the laboratory, the PPF-treated resting box was very attractive to *Ae. aegypti* and led to 94% EI of exposed females, either before or after blood feeding. The resting box was also attractive for male mosquitoes. Exposed males could transfer sufficient PPF to virgin females via copulation to cause approximately 90% EI. Additionally, PPF-exposed gravid females from the treated resting box were able to disseminate sufficient PPF to small larvae containing cups to reduce adult emergence by 50%. Based on 10 min of exposure, the residual effect of PPF-treated resting boxes (over 80% EI) was observed over a 4-month period. The portable nature of the resting boxes and their low cost per unit imply that they could be deployed in high numbers and in a wide variety of properties, with minimal disruption to residents. Further evaluation under natural field conditions is needed, as there are several factors involved, including the presence of other dark objects that can compete with resting boxes.

It is well known that many insect vector species of medical and veterinary importance are found abundantly in animal shelters. In these areas, they often rest for a period of time on objects around the animals both before and after blood feeding. A trial for controlling *Anopheles* and *Culex* mosquitoes by using PPF (666 mg/m^2)-treated polypropylene sheets and resting boxes placed in areas where animals are held was carried out by Khemrattakool et al. (2019). Exposed female mosquitoes (only 10 min on sprayed sheets and overnight in resting boxes) revealed significant reductions in oviposition rates, egg hatchability, pupation, and adult emergence. PPF has an advantage over neurotoxic insecticides, which can pose hazards to animals. Therefore, PPF treatment on the surface material and resting boxes in areas where animals are held is a potential approach for controlling mosquito vectors of malaria, dengue, Japanese encephalitis, and other insect-borne diseases. Further evaluation in wild populations of vectors is needed.

Another potential method which may cope with the problem of insecticide resistance and help control dengue transmission is the use of *Wolbachia*, an intracellular maternally transmitted endosymbiont, which is found naturally in many organisms, including mosquitoes (O'Neill 2018). Due to the unique traits of *Wolbachia* that cause cytoplasmic incompatibility, some strains are useful to humans to reduce transmission by mosquito vectors of human diseases (O'Neill et al. 2019). Some strains of *Wolbachia* may reduce the longevity and fitness of vectors. The bacteria may also inhibit the development, replication, or dissemination of important mosquito-borne pathogens, including dengue, chikungunya, yellow fever, West Nile, and Zika viruses as well as malarial parasites (Kittayapong et al. 2019). Field trials for the control of *Ae. aegypti* and reducing dengue transmission by releasing infected *Wolbachia* mosquitoes to the wild have been carried out in several countries with promising results (Sadanandane et al. 2022). Kittayapong et al. (2018, 2019) conducted a field trial in Thailand by releasing irradiated *Wolbachia*-infected *Aedes aegypti* males. After 6 months of release, a significant reduction in the mean egg hatch rate and the mean number of females per household was achieved in the treatment areas when compared to the control areas.

10.4.3 Chikungunya Fever

In Thailand, the first CHIKV outbreak was reported in Bangkok in 1958, the same year as the first large dengue outbreak in Thailand, caused by the Asian genotype (Hammon et al. 1960). No outbreaks of CHIKV occurred for over 50 years until a dramatic reemergence in 2008–2009, which affected mainly the southern region of the country, especially in rubber plantation and orchard areas where *Ae. albopictus* is predominant. However, the number of cases was low in 2010 and continuously declined until several years later, when the large outbreak started again in 2018 in the south and spread nationwide in 2019. Approximately 15,000 confirmed cases were reported between 2018 and 2019, with morbidity rates of 5.40 and 19.73 per 100,000 population, respectively (Khongwichit et al. 2021).

10.4.4 Zika Virus Infection

In Asia, the virus was isolated in Malaysia from *Ae. aegypti* mosquitoes in 1966 and later reported in neighboring countries (Duong et al. 2017). The first appearance of Zika in Thailand was reported in 2006, but it was probably introduced years earlier (Ruchusatsawat et al. 2019). Of the 3089 samples from 1717 symptomatic individuals tested between January 2016 and December 2017, 368 were confirmed to have Zika virus infection. Cases of Zika virus infection were reported throughout the year and from 29 of the 76 Thai provinces. Since ZIKV shares the same *Aedes* vector species, mainly *Ae. aegypti* and *Ae. albopictus,* coinfections with dengue, and chikungunya viruses often occur.

10.4.5 Japanese Encephalitis

In 1969, the first large epidemic of JE occurred in Chiang Mai and other areas in northern Thailand (Pandey et al. 1999). Since then, Thailand has been an endemic area for JE, reporting 1500 to 2500 cases annually in the 1970s and 1980s (Sudjaritruk et al. 2022). To reduce the burden of disease, the Thailand Ministry of Public Health (MOPH) introduced stepwise vaccination against JEV in children in the 1990s. Approximately half of the population living in a JEV-endemic area demonstrated seroprotection against JEV (Sudjaritruk et al. 2022). The annual number of JE cases in Thailand has been low in recent years. Transmission of JE is related to the rainy season. Several *Culex* mosquitoes are the vectors of JEV, particularly *Cx. vishnui* subgroup, e.g., *Cx. tritaeniorhynchus*, *Cx. vishnui*, and *Cx. pseudovishnui.* The virus has been isolated from other species, including *Cx. bitaeniorhynchus, Cx. fuscocephala, Cx. gelidus, Cx. infula*, and *Cx. quinquefasciatus* (Rattanarithikul et al. 2005). A study in Chiang Mai Province revealed that the populations of *Culex* larvae in rice fields varied depending on the rice growing activity, with a peak during ploughing and a dramatic reduction after transplanting until harvesting. Uncultivated rice fields appeared to be significant breeding sites throughout the year (Somboon et al. 1989). Recently, Arai et al. (2022) recognized two cryptic species in *Cx. tritaeniorhynchus* based on *COI* sequences. This species in Thailand is genetically similar to the specimens found in the Philippines, mainland China and Taiwan, Vietnam, Okinawa (Japan) and some parts of South Korea and Kyushu (Japan) but distinct from the specimens in Honshu (Japan).

10.4.6 Lymphatic Filariasis

Thailand is endemic for lymphatic filariasis (LF) caused by the filarial worms *Brugia malayi* and *Wuchereria bancrofti.* The first survey for LF was conducted during 1949–1952 in several provinces of southern Thailand (Iyengar 1953) and found a microfilarial positive rate averaging 21.0% with an elephantiasis of approximately 5%. All cases were *B. malayi* infections, except 1 case infected with *W. bancrofti*, which was considered an imported case from China. Elephantiasis of one or both legs was the most common, followed by elephantiasis of one or both arms to a lesser extent. Hydrocele was rare. The disease was recognized as being of public health importance in 1953. Several species of mosquitoes were positive for various stages of *B. malayi* larvae; four species of *Mansonia* mosquitoes, *Ma. annulifera, Ma. uniformis, Ma. indiana*, and *Ma. dives* (as *Ma. longipalpis*) and five species of *Anopheles, An. barbirostris s.l., An. montanus* (as *An. albotaeniatus*), *An. nigerrimus, An. sinensis*, and *An. umbrosus* (Iyengar 1953). Laboratory studies revealed that several species of *Anopheles* were susceptible to *B. malayi, An. peditaeniatus, An. crawfordi, An. nigerrimus, An. argyropus*, and *An. pursati,* whereas *An. nitidus* and *Ae. aegypti* (Thailand strain) were refractory to the parasite (Saeung et al. 2013). Cats and dogs appear to be reservoir hosts of *B. malayi* in Thailand (Chansiri et al. 2002; Wongkamchai et al. 2013; Satjawongvanit et al. 2019). The periodicity of *B. malayi* microfilariae in cats is nocturnally subperiodic with a peak around midnight; however, a diurnally subperiodic form was observed in a human host with a peak at 6.00 (Chansiri et al. 2002).

Wongsathuaythong et al. (1963) reported many cases of bancroftian filariasis found in Narathiwat Province, southern Thailand, not far from the Thai–Malaysian border. Harinasuta et al. (1970) reported that bancroftian filariasis was common in rural and forest villages of Kanchanaburi Province, western Thailand along the Thai–Myanmar border, with a maximum infection rate of 27%. Infective larvae of *W. bancrofti* were detected in the *Aedes niveus* subgroup, which was later identified as a new species, *Ae.* (*Downsiomyia*) *harinasutai* (Knight 1978), but first or second larvae were found in *Ae. harveyi, An. maculatus s.l., An. minimus s.l.*, and *An. vagus*. In addition, the first and second larvae were also detected in *Ae. desmotes* and *Ma. dives* (Gould et al. 1982). In the laboratory, several species supported the development of nocturnally subperiodic *W. bancrofti* microfilariae to the infective stage, *An. maculatus* (73.1%), *An. minimus* (61.0%), *An. dirus* (24.7%), *Aedes* (*Downsiomyia*) sp. (27.3%), *Ae. desmotes* (24.2%), *Cx. quinquefasciatus* (19.2%), *Ma. uniformis* (9.2%), and *Ae. albopictus* (0.01%), but *Ae. aegypti* and *Armigeres subalbatus* were refractory to the parasite (Pothikasikorn et al. 2008). Humans are apparently the only animals naturally infected with *W. bancrofti*. No reservoir host is known. However, experimental infections revealed that leaf monkeys are susceptible to this parasite (Harinasuta et al. 1981). The appearance of microfilaria of *W. bancrofti* in the peripheral blood of humans in Kanchanaburi Province of Thailand is nocturnally subperiodic—more abundant during the nighttime than daytime, with an average peak of microfilaremia during 20:00 h for *W. bancrofti* (Harinasuta et al. 1970). However, a variant form with a peak of microfilaremia at 18:00 h was reported from Tak Province by Khamboonruang et al. (1987), as found in 1 case reported by Harinasuta et al. (1970).

The impact of mass administration of diethylcarbamazine citrate (DEC) treatment was carried out from 1978 to 2001 (Rojanapanus et al. 2019). The National Programme for Elimination of Lymphatic Filariasis was launched in 2001. In late

2017, WHO concluded that Thailand achieved criteria for eliminating lymphatic filariasis as a public health problem (Rojanapanus et al. 2019).

10.4.7 Zoonotic Filariasis

Animal filarial worms that commonly cause aberrant infections in humans in Thailand include *Dirofilaria immitis*, *D. repens*, and *B. pahangi*. Domestic dogs and cats are common reservoir hosts of these parasites, but the prevalence in these animals varies depending on the locations and time of surveys. Human dirofilariasis typically manifests as either subcutaneous nodules, mostly by *D. repens*, or as lung parenchyma disease, mostly by *D. immitis* (CDC 2022). In Chiang Mai Province, the prevalence of *D. immitis* infection in stray dogs was approximately 47% (Choochote et al. 1987) but was approximately 12% in a recent study (Kaikuntod et al. 2021). In southern Thailand, the overall prevalence of *D. immitis* microfilaria infection in dogs and cats was 24.1% (95/394) and 36.4% (32/88), respectively (Kamyingkird et al. 2017). Recent studies in Bangkok revealed that the prevalence rates of *D. immitis* in dogs and cats were 58% and 20%, respectively (Satjawongvanit et al. 2019; Thengchaisri et al. 2022).

Several species of mosquitoes serve as vectors of *D. immitis*. In Kanchanaburi Province, third-stage larvae of *D. immitis* were found in *Ae. albopictus*, *Ae. caecus*, *Ae. vexans*, *An. barbumbrosus*, *An. campestris*, *An. philippinensis*, and *An. vagus* (Harinasuta et al. 1970). In Chiang Mai Province, natural infection with *Dirofilaria* larvae was found in *Ma. uniformis* (Choochote et al. 1986). In addition, mosquito species feeding on *D. immitis*-positive dogs, including *Ma. uniformis, Cx. quinquefasciatus, Ae. albopictus*, and *Ae. aegypti,* produced infective larvae, but *Ar. subalbatus* was refractory to the parasite (Choochote et al. 1987). However, *Ar. subalbatus* was reported as the vector of *D. immitis* in southern Thailand (Intarapuk and Bhumiratana 2021). Experimental infections revealed that *Ae. aegypti, Ae. togoi*, and *Cx. quinquefasciatus* are susceptible to *D. immitis* (Riyong et al. 2000; Tiawsirisup and Nithiuthai 2006).

Brugia pahangi is known only in Asia. The microfilariae of this species are morphologically similar to those of *B. malayi.* Recent studies reported approximately 23% and 8% prevalence in dogs in Bangkok and Chiang Mai Provinces, respectively (Satjawongvanit et al. 2019; Kaikuntod et al. 2021). Infected humans may develop signs and symptoms of lymphatic filariasis but rarely produce microfilaraemia. This animal parasite occasionally causes infection in Thai children, as summarized by Intarapuk and Bhumiratana (2021); infected patients usually develop fever with or without pathology. Mosquito species *Ma. annulata* and *Ma. dives* are known to be natural vectors of *B. pahangi* in Malaysia (Wharton 1962). Laboratory experiments in Thailand showed that *Ma. uniformis* was susceptible to both *B. malayi* and *B. pahangi*, whereas refractory species included *Ma. indiana* and *Ma. annulifera* (Choochote et al. 1986). *Armigeres subalbatus* is the vector of *B. pahangi* in southern Thailand and elsewhere (Intarapuk and Bhumiratana 2021).

10.4.8 Leishmaniasis

The first report of endemic human leishmaniasis in Thailand came in 1918, but its origin was not clear (Leelayoova et al. 2017; Hustedt et al. 2022). Since then, several leishmaniasis cases have been detected in Thailand, but they were considered imported cases, mainly from the Middle East, until 1999, when the first confirmed autochthonous case was reported, but the species of the parasite was unknown (Thisyakorn et al. 1999). To date, over 23 autochthonous cases have been reported and confirmed by parasitological and/or molecular methods. Transmission of leishmaniasis in Thailand is most likely to be a zoonotic cycle, in which animal reservoirs play an important role in disease transmission in nature. The most common species were *Leishmania martiniquensis* and *L. orientalis*. The only exception is 1 case of *L. infantum* reported in 2008, but it was suspected that the infection may have been acquired elsewhere. Some previous cases were reported as *L. siamensis* (a *nomen nudum*), but they were later identified molecularly as *L. martiniquensis* or *L. orientalis* (Jariyapan et al. 2018). *Leishmania martiniquensis* can cause either visceral or disseminated cutaneous leishmaniasis or both, whereas *L. infantum* causes visceral leishmaniasis. *Leishmania orientalis* is a new species described from Thailand, causing cutaneous leishmaniasis in immunocompetent individuals and disseminated cutaneous leishmaniasis accompanying visceral leishmaniasis in immunocompromised individuals (Jariyapan et al. 2018). The vectors of leishmaniasis in Thailand are not completely known. DNA of *L. martiniquensis* (as *L. siamensis*) was detected in *Sergentomyia barraudi*, *Se. khawi*, and *Se. gemmea* in the south, suggesting that sandflies are potential vectors (Kanjanopas et al. 2013; Chusri et al. 2014). Interestingly, *L. martiniquensis* DNA was detected in three samples of 106 biting midge *Culicoides mahasarakhamense* specimens tested (Sunantaraporn et al. 2021). Laboratory experiments revealed that *L. orientalis* was unable to establish infection in the sandfly *Lutzomyia longipalpis,* but it successfully established infection in *Culicoides sonorensis,* suggesting that biting midges are potential vectors of these parasites (Chanmol et al. 2019). DNA of *L. martiniquensis* (as *L. siamensis*) was detected in the blood, liver, and spleen of black rats (*Rattus rattus*) captured around the patients' house, which could serve as a natural reservoir host (Chusri et al. 2014).

10.5 Lao People's Democratic Republic

Vongphayloth Khamsing and Moritoshi Iwagami

Lao People's Democratic Republic (Lao PDR) is situated in Southeast Asia (SE Asia), bordered by China, Myanmar, Cambodia, Vietnam, and Thailand. The country comprises 148 districts, divided into 18 provinces (Fig. 10.9), with a total population of 7.4 million in 2022. Among lower-middle-income countries, Lao PDR is

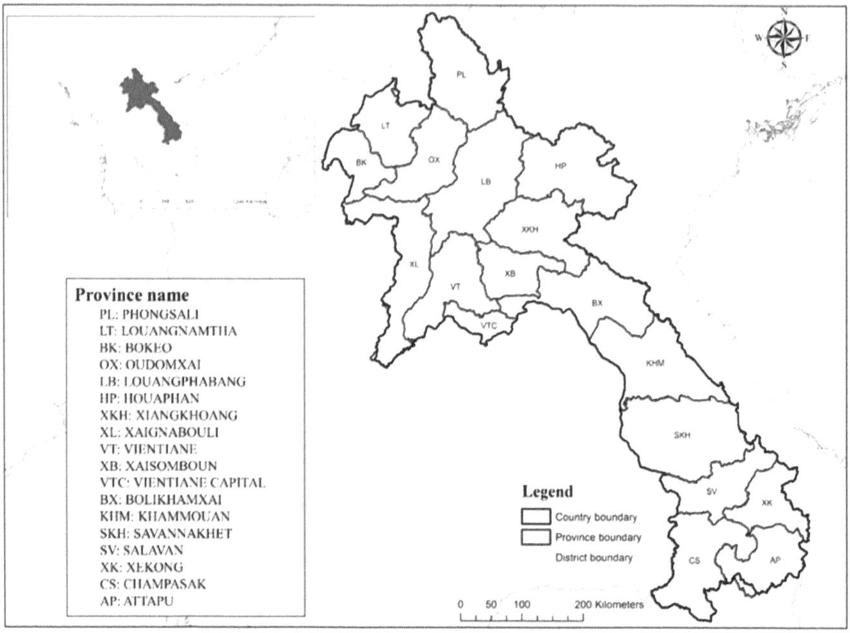

Fig. 10.9 Administrative map of Lao PDR showing provinces and districts and their location in Southeast Asia

one of the fastest-growing economies in Asia (World Bank 2023). Tourism is growing rapidly and plays a vital role in the economy. The forest area is decreasing due to economic development (Matsumoto-Takahashi et al. 2023), and there are concerns about changes in the ecosystem. Lao PDR had been called a land-locked country, but the country has been transformed from a land-locked to a land-linked economic country by connecting a high-speed railway from Kunming, China to Vientiane Capital, Lao PDR. This railway is expected to carry millions of passengers per year by 2030 (World Bank 2020).

10.5.1 Vector-Borne Diseases

Vector-borne diseases (VBDs) are one of the major public health problems of Lao PDR. Like other lower-middle-income countries, VBDs are often neglected in the country because of a lack of tools for diagnosis as well as an appropriate surveillance system. Among VBDs, mosquito-borne diseases are well known to be prevalent in Lao PDR including parasitic diseases: malaria and lymphatic filariasis, arboviral diseases: dengue fever, chikungunya, Japanese encephalitis (JE), and Zika. Other VBDs such as rickettsial diseases transmitted by chigger mites (Tsutsugamushi), fleas, and ticks are also known to occur in Lao PDR, but the clinical epidemiology of these infections is still poorly characterized.

10.5.2 Mosquito-Borne Parasitic Diseases

In the last decade, malaria incidence in Lao PDR markedly decreased from an estimated 46,141 cases in 2012 to 2305 cases in 2022 (WHO 2023b). Recently, the transmission occurred mostly in remote, hilly, and forested areas in the southern part of the country. *Plasmodium vivax* and *P. falciparum* are the two main causative agents of malaria. A total of 42 species of *Anopheles* mosquitoes has been reported in the country (Motoki et al. 2019b), of which four species are recognized as primary vectors of malaria in Lao PDR, namely *An. dirus s.l.* (complex species), *An. minimus s.l.*, *An. maculatus s.l.* Theobald s.l., and *An. jeyporiensis* James (Suwonkerd et al. 2013). There is very limited information on sibling species and ecology such as breeding habitats of malaria vectors in Lao PDR. However, the principal vectors have been known associated with forests (Garros et al. 2006; Obsomer et al. 2007). Deforestation in Lao PDR showed a negative impact on malaria transmission (Matsumoto-Takahashi et al. 2023). Existing data on vector adaptation to land use and climate change are still limited.

Lymphatic filariasis (LF) is caused by a parasite *Wuchereria bancrofti* transmitted by mosquitoes, especially by *Culex quinquefasciatus* Say. Nothing is known about the vector of LF in Lao PDR. Before 2023, LF was endemic in the two southern provinces: Attapu and Sekong (Dickson et al. 2017). However, the World Health Organization (WHO) declared the elimination of LF as a public health problem in Lao PDR in October 2023 (WHO 2023c).

10.5.3 Mosquito-Borne Arboviral Diseases

The first dengue hemorrhagic fever (DHF) cases were reported in Lao PDR in 1979. Since then, dengue fever has been considered endemic in the country with the circulation of the four DENV serotypes (DENV-1, -2, -3, and -4) (Fukunaga et al. 1994). *Aedes aegypti* (L.) and *Ae. albopictus* (Skuse) are known to be the two main dengue vectors. *Aedes aegypti* is considered the primary vector in Lao PDR as it is the human host. *Aedes aegypti* is widely distributed in the country, and its expansion was correlated with urbanization areas (Hiscox et al. 2013). *Aedes albopictus* is known as a native species in SE Asia in which it is widely distributed and species complex. In this region, *Ae. albopictus* coexists with populations of six other species within the *Ae. albopictus* Subgroup (Minard et al. 2017). It is known to be widely distributed in Lao PDR (Motoki et al. 2019a).

In Lao PDR, *Aedes aegypti* was suspected as the main vector responsible for the chikungunya outbreak in the southern part of the country in 2012 (Somlor et al. 2017), and a vector competence test also showed that Laotian *Ae. aegypti* was capable of transmitting the yellow fever virus (Lataillade et al. 2020). *Aedes aegypti* has also been known as the principal vector of Zika virus (ZIKV). In Lao PDR, ZIKV neutralization tests among blood donors revealed the circulation of the virus with low prevalence in the country (Pastorino et al. 2019).

Japanese encephalitis virus (JEV) has been known to be circulated in Lao PDR (Mayxay et al. 2020). However, available data on JEV vectors in the country are still limited. *Culex tritaeniorhynchus* Giles is widely distributed in Lao PDR (Motoki et al. 2019b) and it is known as the principal vector of JEV in SE Asia (Pearce et al. 2018).

10.5.4 Other Vector-Borne Diseases in Lao PDR

Rickettsial diseases caused by *Orientia tsutsugamushi* (scrub typhus), *Rickettsia typhi* (murine typhus), and other members of the spotted fever group (SFG) have been known as causative agents of fever in Lao PDR (Phongmany et al. 2006). However, the clinical epidemiology of these infections was still poorly characterized. Chigger mites, fleas, and ticks are known as vectors of *Orientia tsutsugamushi*, *Rickettsia typhi*, and SFG respectively. Existing data on the taxonomy, ecology, and distribution of these vectors in Lao PDR are very limited, although progress has been made recently in tick studies in the country (Taylor et al. 2016; Vongphayloth et al. 2022).

10.6 Malaysia

Siew Hwa Tan

Malaysia is a Southeast Asian country located in the equatorial region, characterized by a tropical climate with high temperatures and humidity throughout the year. Malaysia is prone to various vector-borne diseases, where insects serve as carriers of pathogens, leading to significant public health challenges. The major diseases of concern are dengue fever, chikungunya fever, Zika virus infection, and malaria, which are transmitted by mosquitoes. Medical entomology in Malaysia is a crucial field of study and research dedicated to understanding the role of insects and other arthropods in the transmission of pathogens to humans and their use in medicine. Additional subfields of medical entomology encompass forensic and veterinary entomology, which employ insects for various purposes.

10.6.1 Aedes Mosquitoes and Arboviral Diseases

In Malaysia, dengue fever, chikungunya, and Zika are recognized as important arboviral diseases, posing considerable public health challenges. These diseases are primarily transmitted to humans through the bite of infected *Aedes* mosquitoes, with *Aedes aegypti* (L.) and *Ae. albopictus* (Skuse) the main vectors responsible for their transmission. The dengue virus (DENV) and Zika virus (ZIKV) fall under the Flaviviridae family, *Flavivirus* genus, while chikungunya virus (CHIKV) is an RNA virus in the *Alphavirus* genus of the Togaviridae family. In Malaysia, *Ae. aegypti* serves as the principal vector for DENV, CHIKV, and ZIKV, while *Ae. albopictus* assumes the role of a secondary vector (Ahmad et al. 2020). Nevertheless, in certain contexts, particularly in rural areas and rubber plantations, *Ae. albopictus* can function as the primary vector for these diseases (Dieng et al. 2010).

Dengue fever is endemic in Malaysia, constituting a major health concern, with documented annual fatalities within the country. The disease's occurrence involves periodic outbreaks that affect various regions, and its prevalence has displayed an increasing trend over the years. However, the number of dengue cases in Malaysia displays variability from year to year, influenced by environmental factors, weather patterns, and public health interventions. Particularly high incidences of dengue have been reported in urban and semi-urban areas (MOH 2010). The primary vector for dengue virus transmission in urban areas is *Ae. aegypti*, while *Ae. albopictus* is more prevalent in rural and forested areas (Ahmad et al. 2018).

The first recorded outbreak of chikungunya in Malaysia occurred in late 1998 (AbuBakar et al. 2007), followed by subsequent surges in cases with cluster outbreaks observed in various years: 2008 (4271 cases), 2009 (5430 cases), 2010 (804 cases), 2017 (270 cases), 2019 (990 cases), and 2020 (2601 cases) (MSIDC 2020). Diverging from dengue fever, chikungunya fever is distinguished by the presence of chronic, severe joint and muscle pain. Persistent arthralgia and joint swelling are prevalent long-term manifestations following infection with CHIKV. This disease has raised growing concern among public health authorities, attributed to the influence of climate change, urbanization, and global travel, collectively contributing to the dissemination of vectors and diseases.

In 2016, Malaysia reported its first locally transmitted cases of ZIKV, triggering heightened surveillance and control measures. ZIKV infection during pregnancy has been associated with the occurrence of congenital abnormalities, resulting in microcephaly in newborns. Moreover, the virus can also be transmitted through sexual contact (Counotte et al. 2018; Mead et al. 2018) and blood transfusions (Musso et al. 2016; Magnus et al. 2018). Despite concerns regarding outbreaks of Zika virus infection, Malaysia has not reported any outbreaks to date.

Malaysia has been actively engaged in research and surveillance activities to understand the distribution, abundance, behavior, and ecology of disease vectors. This information is crucial for developing effective control strategies. To surmount the dissemination of arbovirus diseases, a comprehensive and integrated strategy encompassing persistent vector control, public health campaigns, robust surveillance and monitoring systems, international cooperation, and dedicated research efforts for the development of efficacious vaccines and antiviral treatments is imperative.

10.6.2 Forensic Entomology in a Medical Context

Forensic entomology represents a scientific discipline that employs insect-related information for legal investigations (Hall 2001). This field involves a comprehensive, multidisciplinary exploration encompassing scientific domains such as taxonomy, ecology, biology, toxicology, and physiology. The applications of forensic entomology can be broadly classified into three primary areas: (i) urban entomology, addressing insect infestations (e.g., termites) in human environments such as residential dwellings; (ii) stored product entomology, concerning pests affecting food products; and (iii) medicolegal entomology, which focuses on insects implicated in criminal cases, including murder, suicide, and rape. Medicolegal entomology is also referred to as medico-criminal or forensic medical entomology and involves the application of insect knowledge from crime scenes to aid in crime resolution (Hall 1990).

Forensic medical entomology is commonly referred to as forensic entomology, despite the broader connotation of the term "forensic entomology." This distinction arises due to the paramount importance of forensic medical entomology within the three fields, as it plays a crucial role in legal investigations. Insects recovered from corpses discovered at crime scenes provide valuable information that can be interpreted as evidence and utilized in the investigative process. These insects prove particularly useful in determining the time of death or postmortem interval (PMI), as well as the location and cause of death in medicolegal entomology. Entomological samples, especially those collected after 72 h of death, provide a simpler and more accurate estimation of PMI compared to traditional pathology methodologies (Greenberg and Kunich 2002). Notably, fly larvae retrieved from corpses offer a convenient means for interpreting forensic information by studying the species, larval age, and fly life cycle (Lee 1989).

10.6.3 Advancement of Forensic Entomology in Malaysia

In recent years, forensic entomology has attained increasing significance and recognition in Southeast Asia, with particular emphasis on Malaysia and Thailand. In Malaysia, the Institute for Medical Research (IMR) in Kuala Lumpur plays a prominent and central role as the primary institution responsible for examining forensic entomological samples recovered from human corpses. The majority of forensic entomological specimens in Malaysia are directed to the IMR for species identification, with a primary objective of estimating the PMI in relevant cases (Lee 1996a, b). The pioneering documentation of forensic entomology in Malaysia was conducted by Reid (1953), followed by subsequent investigations by Lee and Cheong (1982) and Lee et al. (1984).

In 1989, Lee compiled a comprehensive report that provided valuable insights into forensically significant entomological specimens recovered from human corpses during the period spanning 1973 to 1988. The report consisted of 101 samples obtained from various hospitals in Malaysia, wherein *Chrysomya megacephala* (Fabricius) and *Chrysomya rufifacies* (Macquart) represented the predominant fly species (76.2%). Subsequently, between 1993 and 1996, as the number of forensic entomological specimens received by the IMR increased, Lee (1996b) reported and analyzed insect species associated with 95 human corpses. In his report, *C.*

megacephala and *C. rufifacies* remained the predominant fly species, constituting 74.7% of the total, while *Sarcophaga* spp., *Lucilia* spp., *Hermetia* spp., and *Ophyra* spp. were identified for the first time. Following these investigations, in 2004, Lee and colleagues performed an extensive review of 448 forensic entomological specimens received by the Unit of Medical Entomology, IMR, from 1972 to 2002. This study revealed the existence of 18 cyclorrhapha fly species retrieved from human corpses, with *C. megacephala* being the most frequently encountered species (215 cases, 47.99%), followed by *C. rufifacies* (132 cases, 19.46%).

Given the remarkable achievements and contributions of IMR in the field of forensic entomology, other hospitals, and universities have actively pursued investigations in this area. Distinguished institutions such as Kuala Lumpur Hospital, Hospital Universiti Kebangsaan Malaysia, Ipoh Hospital, Penang Hospital, and Forensic Entomology Laboratory Universiti Kebangsaan Malaysia have commenced reviewing forensic entomology cases within Malaysia (Omar et al. 1994b; Hamid et al. 2003; Salleh et al. 2007, 2008; Syamsa et al. 2010; Kumara et al. 2012; Abdullah et al. 2017). Consequently, more comprehensive analyses and evaluations, indicating the estimation of the PMI, have been conducted in forensic entomology cases (Salleh et al. 2007, 2008; Rajagopal et al. 2008; Syamsa et al. 2010; Zuha and Baba 2016).

Over the course of reviewing Malaysian forensic entomology cases, prominent associations with human corpses were observed among fly families belonging to Calliphoridae, Muscidae, and Sarcophagiadae, followed by Phoridae, Stratiomyidae, Syrphidae, and Milichiidae. The documented dipteran fly species are listed in Table 10.1. In addition to Diptera, Coleoptera were also collected from human cadavers (Lee 1989, 1996b; Lee et al. 2004; Kumara et al. 2009a).

Table 10.1 List of forensically important dipteran fly species recovered from human corpses in Malaysia

Family	Species	References
Calliphoridae	*Calliphora vicina* Robineau-Desvoidy	Lee (1989, 1996a)
	Calliphora sp. or spp.	Lee (1989, 1996a), Lee et al. (2004)
	Chrysomya bezziana Villeneuve	Lee (1989, 1996a), Lee et al. (2004)
	Chrysomya megacephala (Fabricius)	Reid (1953), Lee (1989, 1996a, b), Lee et al. (1984, 2004), Hamid et al. (2003), Salleh et al. (2007, 2008), Rajagopal et al. (2008), Kumara et al. (2010, 2012), Syamsa et al. (2010, 2015), Kavitha et al. (2013), Abdullah et al. (2017, 2022).
	Chrysomya nigripes Aubertin	Lee (1996a, b), Hamid et al. (2003), Lee et al. (2004), Salleh et al. (2008), Kavitha et al. (2013), Abdullah et al. (2017)
	Chrysomya pinguis (Walker)	Lee et al. (2004), Kavitha et al. (2013)

(continued)

Table 10.1 (continued)

Family	Species	References
	Chrysomya rufifacies (Macquart)	Lee (1989, 1996a, b), Lee et al. (2004), Hamid et al. (2003), Salleh et al. (2007, 2008), Rajagopal et al. (2008), Kumara et al. (2010, 2012), Syamsa et al. (2010, 2015), Kavitha et al. (2013), Abdullah et al. (2017, 2022)
	Chrysomya villeneuvi Patton	Lee (1996a, b), Hamid et al. (2003), Lee et al. (2004), Salleh et al. (2008), Kumara et al. (2012), Kavitha et al. (2013)
	Chrysomya sp. or spp.	Lee (1989, 1996a, b), Lee et al. (2004)
	Hemipyrellia ligurriens (Wiedemann)	Hamid et al. (2003), Lee et al. (2004), Kumara et al. (2012), Kavitha et al. (2013)
	Hemipyrellia tagaliana (Bigot)	Kumara et al. (2012)
	Hemipyrellia sp.	Lee et al. (2004)
	Lucilia cuprina (Wiedemann)	Syamsa et al. (2010), Abdullah et al. (2017)
	Lucilia sp. or spp.	Lee (1989, 1996a, b), Lee et al. (2004)
Muscidae	*Hydrotaea spinigera* Stein	Hamid et al. (2003), Lee et al. (2004), Nazni et al. (2008), Abdullah et al. (2017)
	Hydrotaea sp. or spp. (=*Ophyra*)	Lee (1996b), Lee et al. (2004), Kumara et al. (2012)
	Synthesiomyia nudiseta (van der Wulp)	Omar et al. (1994b), Hamid et al. (2003), Lee et al. (2004), Salleh et al. (2008), Kumara et al. (2010, 2012), Syamsa et al. (2012, 2015), Kavitha et al. (2013), Abdullah et al. (2017)
Sarcophagidae	*Boettcherisca peregrina* (Robineau-Desvoidy)	Zuha and Baba (2016)
	Liopygia ruficornis (Fabricius)	Kumara et al. (2012), Kavitha et al. (2013)
	Parasarcophaga (*Liosarcophaga*) *dux* (Thomson)	Zuha and Baba (2016)
	Seniorwhitea princeps (Wiedemann)	Kumara et al. (2012)
	Sarcophaga sp. or spp.	Lee (1989, 1996a, b), Lee et al. (2004), Salleh et al. (2007, 2008), Kumara et al. (2012)
	Sarcophagidae	Hamid et al. (2003), Kumara et al. (2010), Syamsa et al. (2010, 2015), Abdullah et al. (2017)
Phoridae	*Megaselia curtineura* (Brues)	Thevan et al. (2010), Kumara et al. (2012)
	Megaselia scalaris (Loew)	Kumara et al. (2012), Kavitha et al. (2013), Syamsa et al. (2015), Abdullah et al. (2017)
	Megaselia spiracularis Schmitz	Thevan et al. (2010), Kumara et al. (2012)
	Megaselia spp.	Kumara et al. (2012)

Table 10.1 (continued)

Family	Species	References
	Phoridae	Lee (1989)
Piophilidae	*Piophila casei* (L.)	Nazni et al. (2008), Kumara et al. (2012)
Stratiomyidae	*Hermetia illucens* (L.)	Hamid et al. (2003), Lee et al. (2004), Salleh et al. (2007), Rajagopal et al. (2008), Kavitha et al. (2013)
	Hermetia sp. or spp.	Lee and Cheong (1982), Lee (1989, 1996a, b), Lee et al. (2004)
Syrphidae	*Eristalinus arvorum* (Fabricius)	Heo et al. (2019)
	Eristalis tenax (L.)	Salleh et al. (2007)
	Eristalis spp.	Lee (1994, 1996a), Lee et al. (2004), Abdullah et al. (2017, 2022)
Milichiidae	*Desmometopa* sp.	Kumara et al. (2010, 2012)

10.6.3.1 Calliphoridae and Muscidae

The family Calliphoridae holds significant importance in forensic entomology, particularly in relation to the estimation of the PMI (Smith 1986). In Malaysia, *C. megacephala* and *C. rufifacies* represent the most predominant and noteworthy fly species in forensic entomology. They are among the primary colonizers of the corpse. Alongside Calliphoridae, flies belonging to the families Muscidae and Sarcophagidae also play a crucial role in forensic investigations. The muscid fly, *Hydrotaea* spp., *Synthesiomyia nudiseta* (van der Wulp), and sarcophagid flies are frequently associated with corpses discovered indoors (Omar et al. 1994b; Hamid et al. 2003; Kumara et al. 2012; Syamsa et al. 2015). In addition, Sarcophagidae and *S. nudiseta* are known to infest corpses in high-rise buildings (Syamsa et al. 2012, 2015; Abdullah et al. 2017).

10.6.3.2 Sarcophagidae

In recent years, there has been an increasing recognition of the importance of Sarcophagidae in the discipline of forensic entomology. These flies frequently act as the primary colonizers of a cadaver (Byrd and Castner 2010), highlighting their critical role as important forensic indicators. However, the morphological identification of Sarcophagidae presents challenges due to the high similarity in anatomical characteristics in both larval and adult stages (Smith 1986; Wells et al. 2001; Byrd and Castner 2010). Distinguishing Sarcophagidae at the species level, and even at the genus level, proves to be difficult. Unambiguous species identification of Sarcophagidae can only be achieved through the examination of male genitalia (Rohdendorf 1937; Kano et al. 1967; Nandi 2002; Kurahashi et al. 2021). In contrast, identifying female Sarcophagidae poses notable challenges, as only certain genera can be identified even with the assistance of genitalia. Identification of

Sarcophagidae in their larval stages may not be feasible, particularly at earlier developmental stages (Byrd and Castner 2010).

An alternative approach to achieve species identification involves rearing the immature stages to adulthood. However, this method has limitations, as it allows for the unambiguous identification of only male flies. In Malaysia, the identification of Sarcophagidae in forensic cases is predominantly limited to the genus level due to a lack of expertise and a local taxonomic key (Lee et al. 2004; Salleh et al. 2007; Kumara et al. 2010; Syamsa et al. 2010, 2015; Abdullah et al. 2017). Historically, the study of Sarcophagidae in Malaysia has been relatively insufficient. The first taxonomic key for Sarcophagidae in Peninsular Malaysia was established by Singh et al. (1979), documenting 8 genera and 15 species. After 42 years, Kurahashi et al. (2021) developed a comprehensive taxonomic key exclusively for Sarcophagidae in Sarawak, East Malaysia, featuring detailed morphological character illustrations to facilitate identification (Fig. 10.10). Their research documented a total of 24 genera and 59 species, including 2 new genera, *Omarisca* and *Zulisca*, discovered in mangrove environments, as well as 8 new species. The development of this local taxonomic key for Sarcophagidae denotes a significant advancement, facilitating the identification process of Sarcophagidae species of forensic importance in future investigations.

10.6.3.3 Morphology-Based Identification

In general, the determination of the PMI commences with the identification of the species retrieved from the cadaver. Precise species identification is an essential prerequisite for accurate interpretations during investigative processes. Conventionally, entomological specimens are identified based on their morphological characteristics. However, this procedure can be intricate, particularly when dealing with specimens at early immature stages (e.g., eggs and first instar larvae), where their morphological features are not yet fully developed, or in instances where taxonomic keys are unavailable. The situation can be further complicated when dealing with sarcophagid specimens, as species within this family display a high resemblance to each other. Therefore, facilitating species identification often necessitates rearing immature specimens to adulthood. Nonetheless, this procedure is time-consuming and subject to the suitability of environmental conditions, and not all specimens successfully reach maturity. Additionally, specimens obtained from crime scenes may exhibit damage or lack intact features, resulting in a scarcity of distinguishing characteristics for proper identification.

10.6.3.4 DNA-Based Identification

To overcome the challenges associated with morphological identification, DNA-based techniques have emerged as a promising alternative. The application of DNA sequence data, especially mitochondrial DNA (mtDNA), has witnessed significant

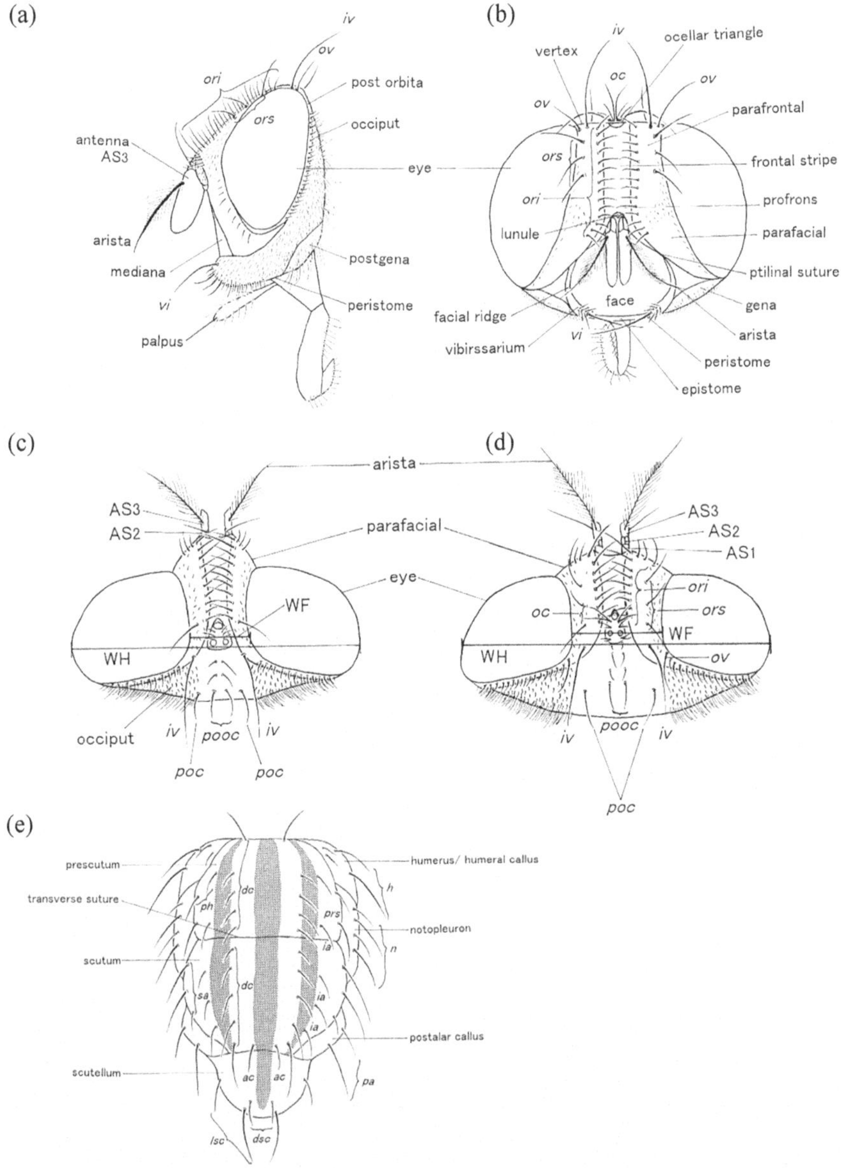

Fig. 10.10 General morphology of Malaysian Sarcophagidae species, *Boettcherisca karnyi* (Hardy). (**a**) head, lateral view, (**b**) head, anterior view, (**c**) male head, dorsal view, (**d**) female head, dorsal view, (**e**) thorax, dorsal view, (**f**) thorax, lateral view, (**g**) female abdomen, dorsal view, (**h**) male abdomen, dorsal view, (**i**) female abdomen, ventral view, (**j**) male abdomen, ventral view, (**k**) fore leg, dorsal view, (**l**) mid-leg, dorsal view, (**m**) hind leg, dorsal view, (**n**) wing, dorsal view, (**o**) male genitalia, lateral view, (**p**) female genitalia, caudal/posterior view, with an inset of spermatheca, one of three illustrated, (**q**) aedeagus, lateral view. Adapter from Kurahashi et al. (2021)

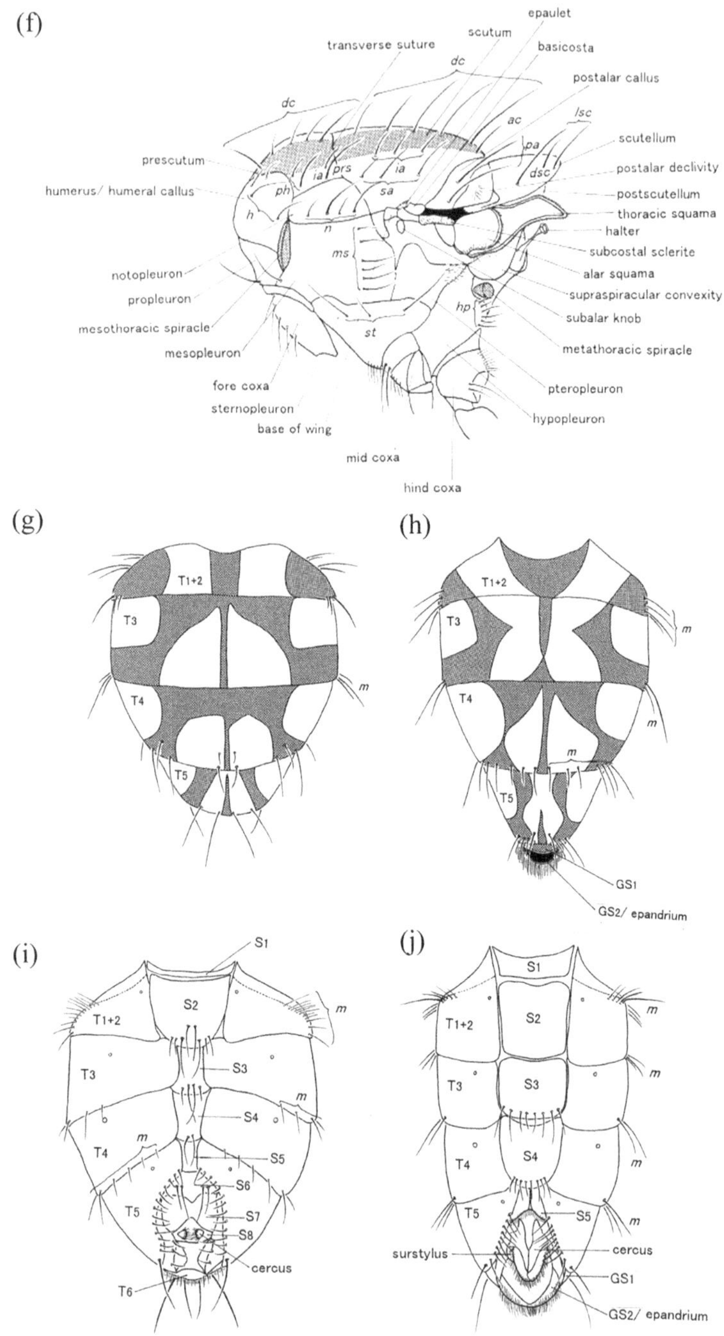

Fig. 10.10 (continued)

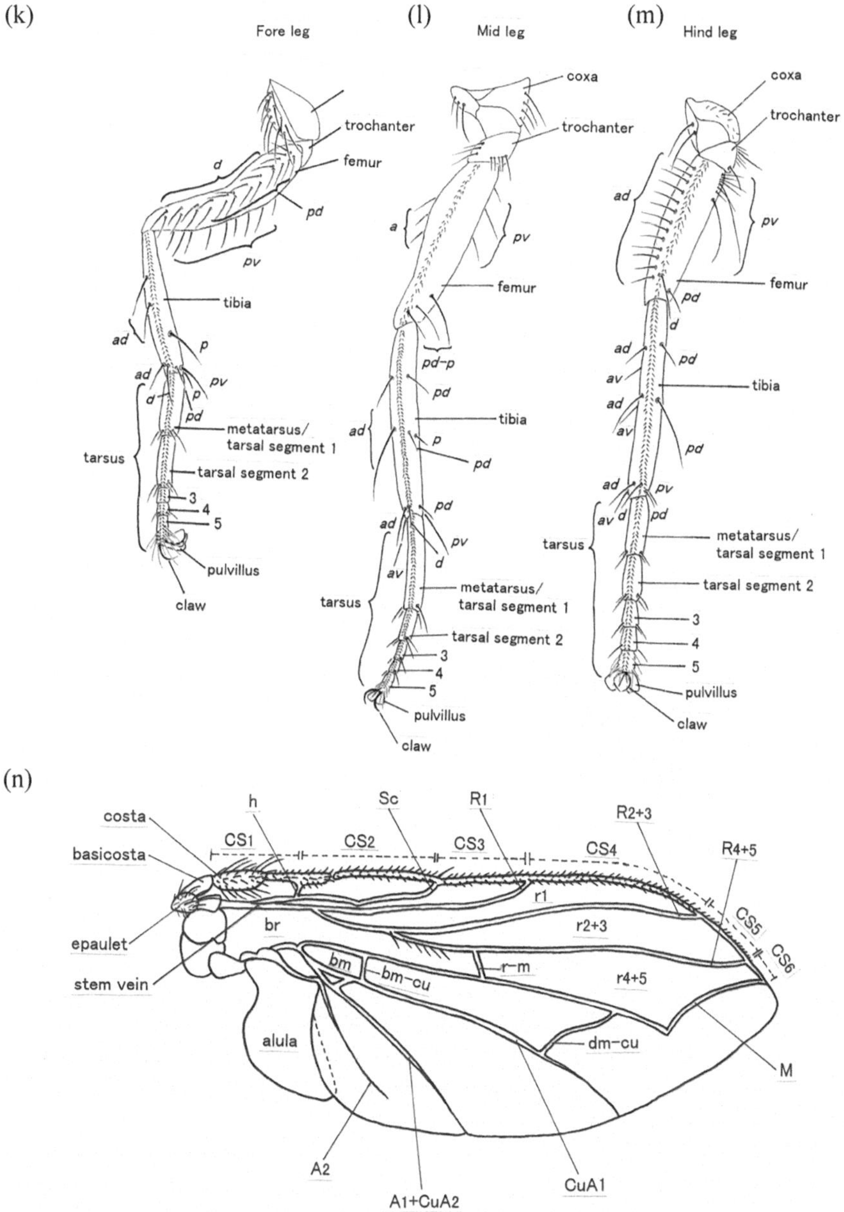

Fig. 10.10 (continued)

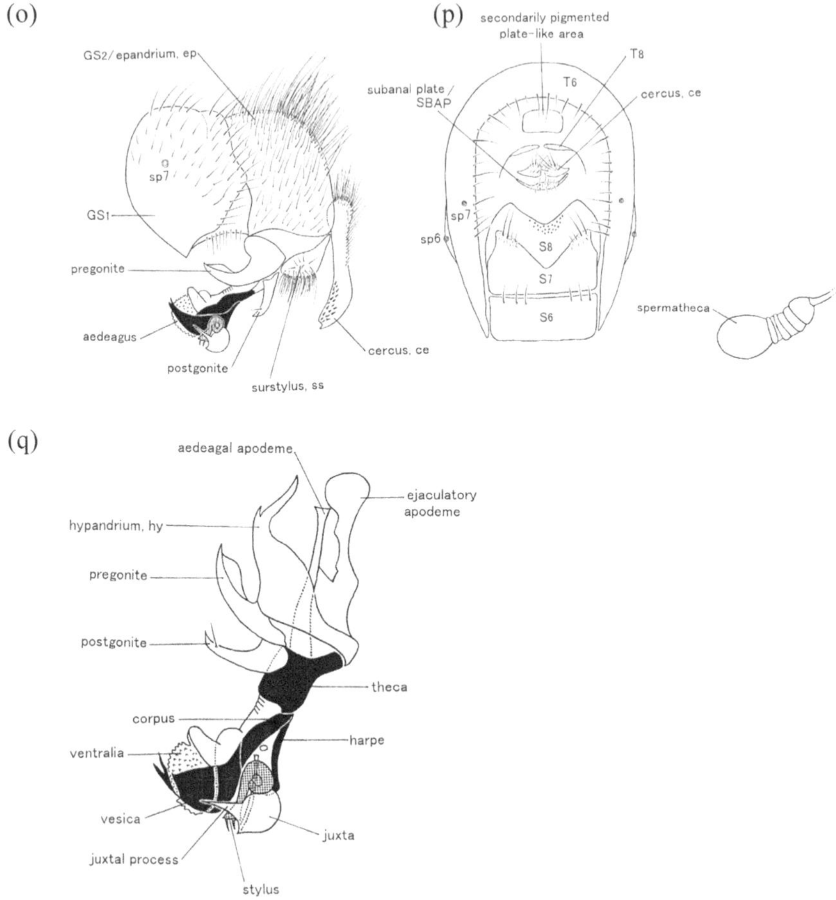

Fig. 10.10 (continued)

advancements in discriminating between species. Notably, the cytochrome oxidase gene subunit I (*COI*) stands out as a renowned DNA marker of mtDNA, extensively employed as a DNA-barcoding system in taxonomy (Hebert and Gregory 2005). Researchers have successfully utilized this marker to distinguish various sarcophagid species from different regions (Table 10.2).

In Malaysia, Tan et al. (2009) were pioneers in the application of DNA-based identification for two predominant forensically significant fly species, *C. megacephala* and *C. rufifacies*. Subsequently, Tan et al. (2010b) reported, for the first time, the feasibility of DNA-based identification for 17 Malaysian Sarcophagidae species. The scope of their study encompassed all sarcophagid species relevant to forensic cases in Malaysia. Beyond *COI*, several other genetic markers, including 12S and 16S ribosomal DNA (rDNA), cytochrome oxidase gene subunit II (*COII*), period gene, and internal transcribed spacer (ITS), have been explored for species

Table 10.2 Identification of Sarcophagidae flies based on *COI* DNA data from different countries

Reference	*COI* (bp)	Country	Number of species
Wells et al. (2001)	783	USA Egypt Polynesia Denmark	9 1 1 1
Saigusa et al. (2005)	304	Japan	2
Tan et al. (2010b)	1534 + (*COII*-692 bp)	Malaysia Indonesia Japan	17 2 1
Meiklejohn et al. (2011)	658	Australia	16
Guo et al. (2011, 2012a, 2014)	278, 272, 700	China	23, 7, 9
Khoso et al. (2015)	1250	Malaysia	5
Zhang et al. (2015)	1534 + (*COII*-692 bp)	China	17
Sharma et al. (2015)	465	India	10
Algalil and Zambare (2017)	658	India	3
Jafari et al. (2019)	1534 + (*COII*-692 bp)	Iran	10
Samerjai et al. (2020)	1534 + (*COII*-692 bp)	Thailand	14

identification within the Sarcophagidae family (Song et al. 2008; Tan et al. 2010b; Guo et al. 2011, 2012b, 2014; Roziah et al. 2015; Zhang et al. 2015). In the study conducted by Roziah et al. (2015), two DNA regions, 12S and 16S rDNA, along with ITS, were employed to distinguish three sarcophagid species in the genus *Boettcherisca* in Malaysia, one of which is the forensically significant species, *B. peregrina* (Robineau-Desvoidy).

DNA-based identification provides a rapid and precise system that could significantly enhance forensic investigations. Another molecular approach used for species identification is PCR-restriction fragment length polymorphism (PCR-RFLP), which allows for species differentiation within 6 h (Ratcliffe et al. 2003). Initially, a DNA fragment containing unique informative sites, such as mtDNA or rDNA, is amplified through PCR. Subsequently, the PCR amplicon is digested using selected restriction enzymes, yielding multiple fragments that serve as markers and are visualized from gel electrophoresis. The resulting gel banding patterns are then employed for species identification. However, this method has certain limitations. It may not be suitable for simultaneously differentiating a large number of species. Additionally, multiple restriction enzymes may be required for species discrimination. Care must be exercised to avoid intraspecific variations at the restriction enzyme digestion site, as this could lead to misidentification. In the study carried out by Tan et al. (2010a), SspI was utilized to differentiate between two closely related Malaysian sarcophagid species, namely, *Iranihindia martellata* (Senior-White) and *Liopygia ruficornis* (Fabricius). Both species are synanthropic, possessing orange-colored antennal segments and palpi. *Liopygia ruficornis* is recognized as an important sarcophagid species extensively recorded in forensic cases in Malaysia, Thailand, and Taiwan.

10.6.4 Species Checklists and Geographical Distribution

In forensic entomology, the larvae collected from a corpse serve as a valuable source of information, offering clues regarding the crime's location and the potential movement of the body. Therefore, obtaining local data on species diversity, checklists, and geographic distribution becomes imperative before applying this information to draw relevant inferences in forensic investigations. Comprehensive checklists of Malaysian Calliphoridae have been prepared (Kurahashi et al. 1997, Kurahashi and Leh 2007, 2009). However, faunistic studies on Malaysian Sarcophagidae have encountered limitations primarily associated with species identification. Earlier researchers have made scarce documentation on Sarcophagidae (Senior-White et al. 1940; Singh et al. 1979). Nonetheless, notable progress was made in 2009 and 2021 when two exhaustive checklists of Malaysian Sarcophagidae from Peninsular Malaysia and Sarawak, East Malaysia, were compiled (Kurahashi and Tan 2009; Kurahashi et al. 2021). These comprehensive studies listed 20 genera and 45 species from Peninsular Malaysia and a total of 24 genera and 59 species from Sarawak. It is worth noting that Malaysia consists of two noncontiguous regions, Peninsular Malaysia (on the Malay Peninsula) and East Malaysia (on Borneo), separated by the South China Sea. The fly species found in these regions exhibit variations influenced by their ecological and geographical distribution. For example, *Leucomyia alba* (Schiner) was exclusively collected from Peninsular Malaysia, while *Parasarcophaga quinqueramosa* (Sugiyama) was found solely in Sarawak (Kurahashi and Tan 2009; Kurahashi et al. 2021). To achieve more precise local interpretations, it is imperative to broaden the scope of geographical distribution studies to include other states and habitats. The comprehensive studies on sarcophagid species checklists and geographic distribution data obtained from these efforts will provide invaluable insights for criminal investigations.

10.6.5 Larval Growth Rate and Species Succession Pattern in Decomposition

Forensic entomology in Malaysia has a well-established historical foundation, with extensive investigations and research conducted on various aspects of insects. These studies include larval morphology and identification (Omar 2002; Ahmad Firdaus et al. 2010; Heo et al. 2011, 2015a, b; Salleh et al. 2014; Maramat et al. 2021), insect succession on animal carcasses (Lee and Marzuki 1993; Omar et al. 1994a; Heo et al. 2008a, b, 2011; Ahmad et al. 2011; Maramat and Rahim 2015; Silahuddin et al. 2015; Abdullah et al. 2016; Zuha et al. 2017), and larval growth development (Chen et al. 2008; Kumara et al. 2009b, 2013; Zuha et al. 2012; Khalid et al. 2015). The studies specifically dedicated to Sarcophagidae are the works of Heo et al. (2015a) and Kumara et al. (2013), who presented valuable findings on the larval morphology of *Boettcherisca* species and the growth development of *Parasarcophaga dux* (Thomson), respectively. Despite the emphasis on Diptera, it is essential to acknowledge and explore the significance of other forensic insects, such as Coleoptera, as they are equally important as flies. Unfortunately, these insects have often been underestimated or overlooked. Greater attention should be directed toward insect diversity in high-rise buildings and aquatic environments. Further research and data collection in these areas is imperative before applying the findings to densely populated residential areas, such as Kuala Lumpur, and criminal cases occurring in aquatic settings.

10.6.6 Forensic Entomotoxicology

Forensic entomotoxicology involves the investigation of toxins or drugs found in insects that feed on cadavers, representing an emerging research domain in Malaysia. Numerous studies have been conducted employing drugs as a basis to assess their impact on the larval development of forensically significant species (Rashid et al. 2008, 2010, 2012, 2013; Mahat et al. 2009, 2012, 2014). The presence of drugs and toxins in cadavers can influence the growth and development of larvae that consume the decomposing remains, potentially affecting the estimation of the PMI (Gosselin et al. 2011). Qualitative detection of drugs and toxins in a cadaver can be accomplished through the analysis of larval samples or npupariums. Insect samples have been demonstrated to be a more sensitive, stable, and robust method compared to putrefactive materials and autopsy tissues (Kintz et al. 1990a, b).

10.6.7 Conclusion

In Malaysia, arbovirus diseases transmitted by mosquito vectors, such as dengue, chikungunya, and Zika, constitute a primary focus of concern in the field of medical entomology. These diseases bear significant mortality rates and pose a substantial threat to human lives. The government and health authorities in Malaysia have implemented various vector control measures to reduce the spread of diseases. These include the use of insecticides, larvicides, *Wolbachia*-based control strategies, and public awareness campaigns. Forensic entomology, on the other hand, exhibits substantial potential, with foundational research serving as a critical basis for establishing reliable and robust applications of this field in legal investigations. There is a pressing need to further enhance the morphology- and DNA-based identification techniques for immatures (e.g., eggs, larvae, and pupae) and adults to encompass a broader spectrum of insect species. This expansion is indispensable before these methods can be effectively employed in real cases to evaluate their feasibility and limitations. The acquisition of comprehensive data on the developmental growth rates of various forensically significant insect species and the patterns of insect succession on decomposing cadavers across different environments and regions is of utmost importance. The availability of the data holds the potential to reinforce the reliability, efficiency, and expediency of forensic investigations. It is imperative to acknowledge that the utilization of insect evidence in forensic investigations constitutes a multifaceted domain requiring expertise in both entomology and forensic science. Consequently, the establishment of appropriate operational guidelines becomes essential to mitigate the risk of misinterpretations.

10.7 Mongolia

Khatanbaatar Igori

Mongolia, located in East North Asia with a population of 3.4 million, is the 18th largest (Gonchigjav et al. 2022) and the least densely populated country in the world, and one-third of the population lives in rural areas with a traditional semi-nomadic lifestyle (NSOM 2023; Chimed-Ochir et al. 2022). Mongolia is a land-locked country located in central-east Asia between Russia and China, characterized by long, cold winters, during which night temperatures can drop to −40 °C (Farukh et al. 2009).

10.7.1 Vector-Borne Diseases

Ticks and tick-borne diseases (TB) cause substantial economic, veterinary, and major public health concerns in Mongolia. The distribution of ticks and tick-borne diseases in Mongolia is broadly related to local climatic conditions, which are considered to be mostly arid. Nomadic pastoralists, which make up ~26% of Mongolia's population, are at an increased risk of both tick bite exposure and economic loss associated with clinical disease in nomadic herders (Lkhagvatseren et al. 2019; Altantogtokh et al. 2022).

The various vegetation zones from boreal forests to steppes and deserts in Mongolia allow for the existence of many different tick species adapted to completely different biotopes. Eighteen species of ticks from the genera *Ixodes*, *Haemaphysalis*, *Dermacentor*, *Hyalomma*, and *Rhipicephalus* have been reported in Mongolia (Dash 1969). *Ixodes persulcatus* and *D. nuttalli* have been studied intensively, as they seem to be the two most important species for human and animal health (Černý et al. 2019).

In addition to tick-borne infectious diseases, vector-borne diseases such as tick-borne encephalitis virus, rickettsial diseases, and other diseases, leishmaniasis, Surra (trypanosomiasis) in animals, plague, Lyme borreliosis have been reported in Mongolia (Table 10.3) (Fig. 10.11).

Table 10.3 The list of vector-borne diseases in Mongolia

Diseases name	Causative agents (serotype/genotype/lineage detected)	Areas of distribution (Provinces/aimag)	Year of detection	Imported cases to other countries	(Putative) vectors involved	References
Tick-borne encephalitis virus	TBEV strains Zausaev (AF527415) and Vasilchenko (AF069066). Siberian subtype of TBEV.	Selenge and Bulgan	Human case since 1980	Russia	*Ixodes persulcatus*	Frey et al. (2012) Muto et al. (2015)
Rickettsial diseases and other diseases	*Rickettsia sibirica* *Candidatus Rickettsia tarasevichiae* *R. raoultii*	Umnugobi, Selenge Arkhangai, Khentii	Human case since 1984		*Dermacentor nuttalli* *D. daghestanicus*, *Hyalomma asiaticum* *I. persulcatus*	Dash et al. (1988) Boldbaatar et al. (2017) Speck et al. (2012) Fischer et al. (2020)
Leishmaniasis	*Leishmania turanica* *L. gerbilli*	Dornogobi, Umnugobi Bayankhongor Uvurkhangai	Unknown		*Phlebotomus alexandri*, *Ph. andrejevi*, *Ph. mongolensis*, *Ph. wui*, *Sergentomyia murgabiensis murgabiensis*, *S. gobica*	Artemyev and Neronov (1982, 1984) Neronov et al. (1987) Gantuya (2010)
Surra (Trypanosomiasis) in animals	*T. equiperdum*	Nationwide	Unknown		Unknown	Suganuma et al. (2016)
Plague	*Yersinia pestis*	Nationwide	Human case since 1704–1788		*O. silantiewi*	Fijn and Terbish (2021)
Lyme Borreliosis	*B. burgdorferi* sensu lato	Zavkhan Selenge	Unknown		*I. persulcatus*	Von Fricken et al. (2019)

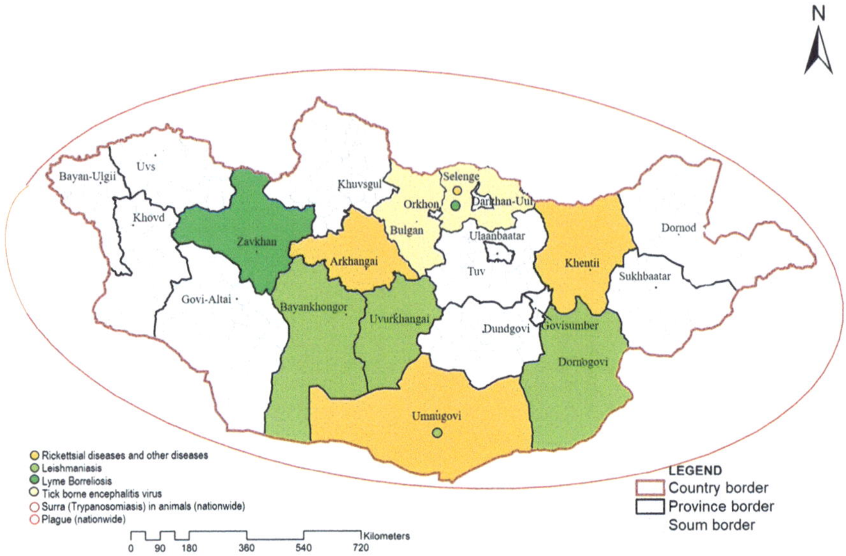

Fig. 10.11 The distribution of vector-borne diseases in Mongolia

10.7.2 Tick-Borne Encephalitis Virus

Tick-borne encephalitis virus (TBEV) is a zoonotic virus belonging to the genus *Flavivirus* in the family Flaviviridae. The virus, which is endemic in Europe and northern parts of Asia, causes severe encephalitis. Tick-borne encephalitis (TBE) has been reported in Mongolia since the 1980s, but details about the biological characteristics of the endemic virus are lacking. Mongolian TBEV strains were isolated from *I. persulcatus* ticks, and phylogenetic analysis of the E-gene classified this isolate as a Siberian subtype of TBEV (Frey et al. 2012; Muto et al. 2015).

10.7.3 Rickettsial Diseases and Other Diseases

In Mongolia, *Dermacentor nuttalli* collected from the steppe and forested steppe and examined for *Rickettsia sibirica,* 0.9–5.4% were infected. *Dermacentor nuttalli, D. daghestanicus*, and *H. asiaticum* from the Gobi Desert had infection rates of 0.4%, 0.5%, and 0.8%, respectively (Dash et al. 1988).

Rickettsia raoultii was the predominant Rickettsia (82% prevalence) found in *D. nuttalli* and was also detected in *I. persulcatus* (0.8%). In addition, *R. sibirica*

was identified in approximately 4% of *D. nuttalli* (Speck et al. 2012). *Candidatus Rickettsia tarasevichiae* (*Candidatus R. tarasevichiae*), an emerging tick-borne human pathogen, is an obligate intracellular gram-negative bacterium belonging to the ancestral group of the genus *Rickettsia* (Walker et al. 2008). More recently, *Candidatus R. tarasevichiae* has been identified from *I. persulcatus* in northern Mongolia (Boldbaatar et al. 2017).

All adult ticks were identified as *D. nuttalli*. DNA of *R. raoultii* (252/310; 81.3%), an uncharacterized *Anaplasma* species preliminarily named *Anaplasma* sp. Mongolia (26/310; 8.4%), *Candidatus* Midichloria sp. (18/310; 5.8%), *Theileria equi* (16/310; 5.2%), *Babesia caballi* (5/310; 1.6%), *T. orientalis* (1/310; 0.3%), *Borrelia afzelii* (1/310; 0.3%) and *Candidatus Neoehrlichia mikurensis* (1/310; 0.3%), was detected in adult ticks collected from the field (Fischer et al. 2020).

10.7.4 Leishmaniasis

Haematophagous insects of the family Psychodidae are mostly found in the aptly named subfamily Phlebotominae. This subfamily contains several hundred species, and many of them transmit protozoan pathogens in the genus *Leishmania* to humans and other mammals.

A total of six species of sandflies have been detected in Mongolia. *Phlebotomus alexandri, Ph. andrejevi, Ph. mongolensis, Ph. wui, Sergentomyia murgabiensis,* and *S. gobica* were investigated by Artemyev and Neronov (1982, 1984).

In exploring Phlebotominae, Mongolia is of special interest to us for the relative abundance of host–vector–pathogen interactions seen in southwestern Mongolia and because its location is central to the ranges of several Old World groups. This highlights the need for taxonomic clarity and information about diversification dynamics in this group.

In the natural foci of the Zoonotic cutaneous leishmaniasis (ZCL) on the territory of Mongolia, great gerbils are the main natural hosts of *Leishmania* spp. discussed above (Neronov et al. 1987; Gantuya 2010).

Subcutaneous inoculation of human volunteers with two strains of *L. turanica* (one from a Mongolian great gerbil and one of the sand flies from Uzbekistan) resulted in mild cutaneous manifestations. The Mongolian isolate infection manifested in a small nodule, which persisted for 2 weeks and resolved with no ulceration. Infection with a strain originating from a sand fly led to a prolonged incubation period and ulceration. The lesions persisted for approximately 2.5 months and healed spontaneously, leaving small scars (Strelkova et al. 1990).

10.7.5 Surra (Trypanosomiasis) in Animals

Forty-nine species of Tabanidae, 48 of Muscidae, 39 of Sarcophagidae, and 20 of Calliphoridae were recorded from Mongolia. The infection rates of *Hybomitra expollicata*, *H. morgan*, and *Atylotus quadrifarius* with Trypanosoma parasites in Khovd Province in Mongolia were 6.5%, 14.9%, and 2.2%, respectively (Myagmarsuren 1996). Some scientific reports have also shown that the seroprevalence of trypanosomosis in equines in Mongolia is 6–8% (Clausen et al. 2003) and 4.1–5.5% (Mizushima et al. 2020). However, these studies did not identify the causative species, as it is difficult to distinguish between *Trypanosoma evansi* and *T. equiperdum* using serological diagnostic techniques. *Tripanosoma equiperdum* was isolated from the urethral tract of a horse showing dourine symptoms in Mongolia, characterized as a true *T. equiperdum* strain, and named the IVM-t1 strain (Suganuma et al. 2016). More recently, the prevalence of animal trypanosomosis (21.27%) in Bayan-Ulgii and Khovd provinces, which are located in western Mongolia, was reported (Musinguzi et al. 2022).

10.7.6 Plague

Epidemiologists recognize that close contact with infected Siberian marmots (*Marmota sibirica*) is a significant route of transmission for humans. The plague is caused by the bacteria *Yersinia pestis* and exhibits two highly lethal forms, bubonic and pneumonic (Fijn and Terbish 2021). The traumatic penetration of pseudotuberculosis microbes into the blood of Mongolian marmots for the whole population during hibernation created conditions for the implementation of stable bacteremia and the development of the original epizootic *M. sibirica*, *Oropsylla silantiewi,* and *Y. pestis* plague system (Suntsov 2017). Between 2012 and 2019, plague bacteria were prevalent in different kinds of rodents in 137 districts across Mongolia (Kehrmann et al. 2020).

10.7.7 Lyme Borreliosis

A total of 150 suspected clinical cases of Lyme borreliosis (LB) originated from 13 different provinces in Mongolia, 78.7% (118/150) of whom recalled experiencing a tick bite before the onset of symptoms. Information on cases and symptoms by province can be found in Zavkhan and Selenge, located in northern Mongolia, which had the highest number of suspected LB cases, with 25% of all reported cases originating in Zavkhan and 20% in Selenge. The most frequently reported symptoms include fever, rash, headache, and enlarged lymph glands (Von Fricken et al. 2019). Prevalence of Lyme borreliosis antibody detected in horses in the Selenge and Orkhon River basins of Mongolia (Sugar 2010).

10.8 Nepal

Basu Dev Pandey and Kouichi Morita

Nepal is a Himalayan landlocked country with largely mountainous terrain located in South Asia that is bordered by China and India and contains eight of the world's ten highest peaks (Fig. 10.12). The population of Nepal is almost 30 million people. Administratively, the country is divided into 7 provinces, 77 districts, and 856 municipalities. Kathmandu is the capital city of Nepal and has a population of over 1.4 million. The altitudes vary from the lowest 100 m above sea level to the highest 8848 m.

10.8.1 Vector-Borne Diseases

Nepal is endemic for five major vector-borne diseases (VBDs), namely, malaria, dengue fever, lymphatic filariasis, Japanese encephalitis (JE), and visceral leishmaniasis (Dhimal et al. 2015). Chikungunya and scrub typhus have also emerged in different areas of the country. There is increasing evidence about the impacts of climate change on VBDs, especially in tropical highlands and temperate regions,

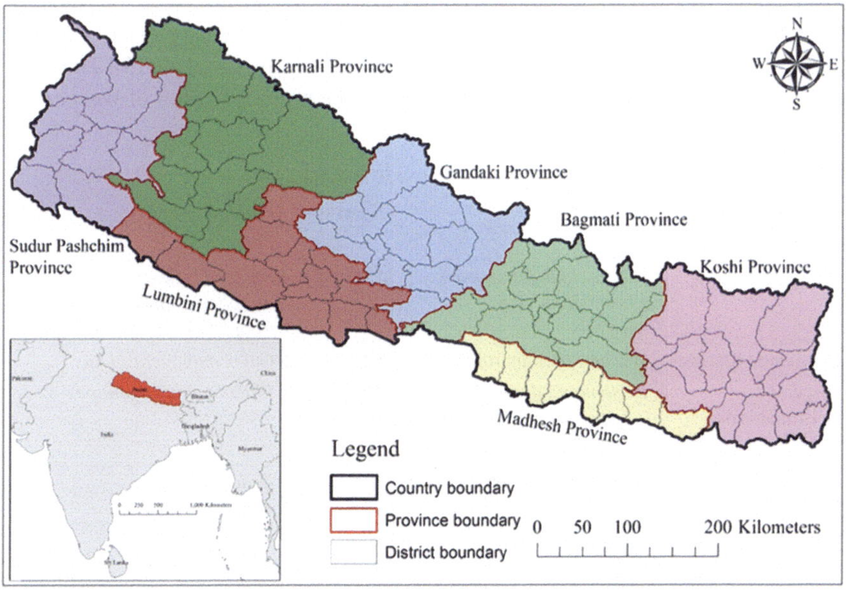

Fig. 10.12 Political map of Nepal and its location in Asia showing provinces and districts. The map is referred to Khanal et al. (2020)

although urbanization, water scarcity, frequent movement, improper waste disposal, and poor sanitation may fuel the increase in VBDs in Nepal. Pronounced warming in the mountains, increasing temperatures by more than 0.056 °C over the past 40 years, and an expansion of autochthonous cases of VBDs to nonendemic areas, including mountain regions, has led to a major epidemic of dengue in all 7 provinces and 77 districts in 2022 (Pandey et al. 2022).

10.8.2 Vectors

Forty-four species of medically important *Anopheles* mosquitoes have been recorded in Nepal, and of these, seven have been implicated in malaria transmission (Dhimal et al. 2015). The major species recorded are *Anopheles minimus* Theobald, *An. fluviatilis* James, *An. maculatus* Theobald, *An. dravidicus* Christophers, *An. pseudowillmori* (Theobald), *An. willmori* Theobald, and *An. annularis* van der Wulp (Table 10.4). On the other hand, the principal vector responsible for dengue and chikungunya transmission is *Aedes aegypti* (L.), which was identified up to a 2000-m altitude, and the secondary vector for dengue is *Aedes albopictus* (Skuse). The other common VBD is visceral leishmaniasis, first identified in the lowland Terai region of Nepal and now reported in more than 65 districts. It is transmitted by the vector, the sand fly, *Phlebotomus argentipes* Annandale & Brunetti, and the parasite *Leishmania donovani* (Laveran et Mesnil) (Uranw et al. 2013). Japanese encephalitis virus (JEV) was first reported in the Rupendehi district of Nepal in 1978. It became endemic in the lowlands and is transmitted by the mosquito species *Culex tritaeniorhynchus* Giles as a principal vector (Dhimal et al. 2015). The mosquito species *Cx. quinquefasciatus* Say, which is the principal vector of *Wuchereria*

Table 10.4 Characteristics of vector-borne diseases in Nepal[a]

Disease	Pathogen	Reservoirs	Principle vectors
Parasitic			
Malaria	*Plasmodium falciparum* *Plasmodium vivax*	Humans	*Anopheles fluviatilis* *Anopheles annularis* *Anopheles maculatus* group
Lymphatic filariasis	*Wuchereria bancrofti*	Humans	*Culex quinquefasciatus*
Visceral Leishmaniasis	*Leishmania donovani*	Mammals	*Phlebotomus argentipes*
Viral			
Dengue fever	Dengue virus (Flaviviridae)	Humans	*Aedes aegypti* *Aedes albopictus*
Chikungunya	Chikungunya virus (Alphavirus)	Humans	*Aedes aegypti* *Aedes albopictus*
Japanese encephalitis	Japanese encephalitis virus (Flaviviridae)	Birds, pigs	*Culex tritaeniorhynchus*

[a]This table is modified from Dhimal et al. (2015)

bancrofti (Cobbold) microfilaria, was first recorded in Nepal in 1956 and found to occur within the lymphatic filariasis endemic zones of this country.

10.8.3 Future Challenges for VBDs Control in Nepal

Recently, the outbreak of scrub typhus was reported after the devastating earthquake of 2015 in Nepal and was caused by *Orientia tsutsugamushi,* which is transmitted by the vector mites *Leptotrombidin deliense* Walch and *L. akamushi* (Brumpt) (Dhimal et al. 2021). It is difficult to say precisely what the real situation of vectors is due to a shortage of entomologists and the absence of regular vector surveillance in Nepal. However, VBDs have been expanding their geographical ranges, especially in mountain regions of the country. Urgent mapping is necessary for the identification, distribution, and biology of vectors since there has been no nationwide survey of vectors in Nepal since 1990 (Darsie and Pradhan 1990).

Acknowledgments We would like to acknowledge Dr. Laxman Khanal, Associate Professor, Central Department of Zoology, Tribhuban University, for preparing the map of Nepal.

10.9 People's Republic of China

Jiamei Sun

China is a vast territory country located in East Eurasia and on the west coast of the Pacific Ocean, stretching from the equatorial zone to the cold temperate zone, covering five time zones, posing the monsoon climate, the temperate continental climate, and the alpine climate. The widely varying geographical landscapes, complex climate, and ecological environment in China have nurtured a diversity of pathogens, host animals, and arthropods, leading to the widespread distribution and frequent occurrence of arthropod-borne diseases (ABDs) in the past, not only damaging public health but also threatening social and economic development. Through unremitting efforts, China has rapidly decreased ABDs in incidence and mortality (Yang 2019) and has eliminated malaria and lymphatic filariasis (Sun et al. 2013; Cao et al. 2021). However, influenced by multiple natural and social factors, such as globalization, climate change, urbanization, insecticide resistance, and ecological environment changes, ABDs in China have shown new epidemic trends, such as the recurrence and emergence of neglected, unknown ABDs. Maintaining a sturdy surveillance system and developing new risk assessment and prewarning methods for recurrent and emerging ABDs is crucial to effective public health prevention and control of ABDs. Creative approaches may help control ABDs, such as using application programs to promote public participation (Huang et al. 2023).

10.9.1 Vectors and ABDs

In China, six ABDs are listed as notifiable infectious diseases, including malaria, Japanese encephalitis, dengue fever in Class B, and leishmaniasis, filariasis, and typhus in Class C. Here some ABDs of public health importance in mainland China are described. Among these ABDs, typhus and dengue fever are high in incidence, and Japanese encephalitis and SFTS are high in mortality (Liu 2022). Nearly all arboviruses are zoonotic diseases, some of which are important emerging diseases.

10.9.1.1 Malaria

The primary vector of malaria transmission in mainland China is *Anopheles*. Innovative tools and strategies play crucial roles in the fight against malaria, including discovering the effective anti-malaria drug artemisinin from Chinese herbal medicines (Chen et al. 2021b). In 2017, China achieved the goal of reducing the number of annual cases from 30 million to zero. In 2021, the World Health Organization announced that China had eliminated malaria (Cao et al. 2021). Afterward, localized outbreaks caused by imported factors occurred in some places. Postelimination surveillance to consolidate the achievements in eliminating malaria has become the main task (Feng et al. 2022).

10.9.1.2 Lymphatic Filariasis

Lymphatic filariasis (LF) is caused by *Wuchereria bancrofti* and *Brugia malayi*, with the primary vector of *Culex quinquefasciatus*, *Cx. pipiens pallens*, and *An. sinensis*, *An. anthropophagus*, respectively. China used to be one of the world's most heavily endemic countries for LF. Through treatment of LF patients, general administration of fortified diethylcarbamazine salt, and other means, China has achieved the goal of eliminating LF nationwide and was recognized by the WHO in 2007 (Sun et al. 2013; Sun et al. 2020). Currently, there are occasional reports of imported cases.

10.9.1.3 Japanese Encephalitis

Japanese encephalitis (JE) is a natural zoonotic disease caused by the Japanese encephalitis virus, mainly transmitted through *Cx. tritaeniorhynchus*, with pigs and birds as reservoirs. JE was first discovered in Japan and spread to China in the 1930s. In the 1960s and 1970s, epidemic outbreaks of JE hit China with high morbidity and mortality (Chen et al. 2021c). The epidemic season is primarily from May to October, peaking from July to September. Except for Xinjiang, Tibet, and Qinghai, there is a risk of transmission in all other regions. Predominant types of JE virus are genotypes 1 and 3 (Wang et al. 2007). With the application of vaccines, socioeconomic development, and improvement of healthcare conditions, significant achievements have been made in JE prevention and control. The annual incidence has continued to decline. The age composition of the disease changes, with pediatric cases decreasing while adult cases increasing. In recent years, the transmission of JE has rebounded in northern China (Wang et al. 2021).

10.9.1.4 Dengue Fever

Dengue fever is caused by the Dengue virus, with *Aedes aegypti* and *Ae. albopictus* being vectors in mainland China. *Ae. aegypti* is found in the southern region, while *Ae. albopictus* is widely distributed. Indigenous transmission caused by imported cases characterizes the dengue fever epidemic in China. Dengue fever cases are reported almost nationwide except in Tibet, mainly in Guangdong and Yunnan Provinces (Yue et al. 2022; Liu et al. 2019; Wu et al. 2022) (Fig. 10.13). The dengue fever epidemic in Southeast Asia, South Asia, South America, and other countries and regions continues to be high. With globalization and the advancement of the "Belt and Road" initiative, imported cases of dengue fever may increase. Since 2020, in response to the global COVID-19 pandemic, China's restrictions on international flights and entry quarantine controls have significantly reduced the number of imported cases of dengue fever (Tan et al. 2023; Guo et al. 2022). With the resumption of regular national communication, the risk of dengue fever entering and causing local transmission must rebound.

10.9.2 Other Mosquito-Borne Diseases

Other mosquito-borne viruses, including Chikungunya fever (Chen et al. 2021a), West Nile virus disease (Zhang et al. 2021), Rift Valley fever (Liu et al. 2021), and Zika virus disease (Zhou et al. 2020), which are caused by Chikungunya virus, West Nile virus, Rift Valley fever virus, and Zika virus, respectively, are emerging zoonoses. Imported cases of these diseases have been reported in China. The first imported case of Chikungunya fever was found in 2008 and has increased since then. There have been outbreaks of local transmission caused by imported Chikungunya fever cases in Guangdong, Zhejiang, and Yunnan, but there is no evidence of a stable natural epidemic.

10.9.2.1 SFTS

Severe fever with thrombocytopenia syndrome (SFTS) is an emerging infectious disease first discovered in China and reported in surrounding countries. In 2011, Chinese researchers identified the pathogenic virus as a Bunia virus (Yu et al. 2011), renamed *Dabie bandavirus* by the International Committee on Taxonomy of Viruses in 2019. Currently, except in Hainan, Tibet, Qinghai, and Ningxia, SFTS has been widely reported in mainland China. In 2011–2021, a total of 18,902 confirmed cases and 966 deaths from SFTS were reported in 27 provinces across the country, concentrated in seven areas, including Henan, Shandong, Hubei, Anhui, Jiangsu, Zhejiang, and Liaoning, with a trend of spreading from central to the outside (Chen et al. 2022b; Huang et al. 2021a) (Fig. 10.14). It often occurs from April to October,

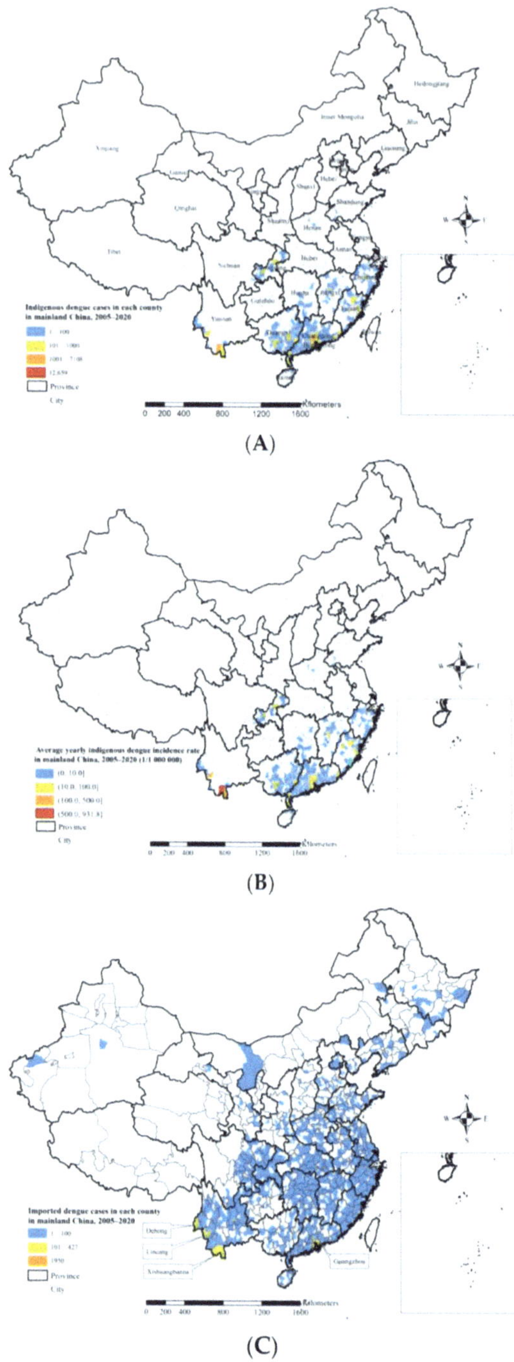

Fig. 10.13 Spatial distribution of dengue fever in mainland China during 2005–2020. (**a**) Indigenous cases. (**b**) Indigenous incidence rate. (**c**) Imported cases (Yue et al. 2022, quoted under an open access Creative Commons CC BY 4.0 license)

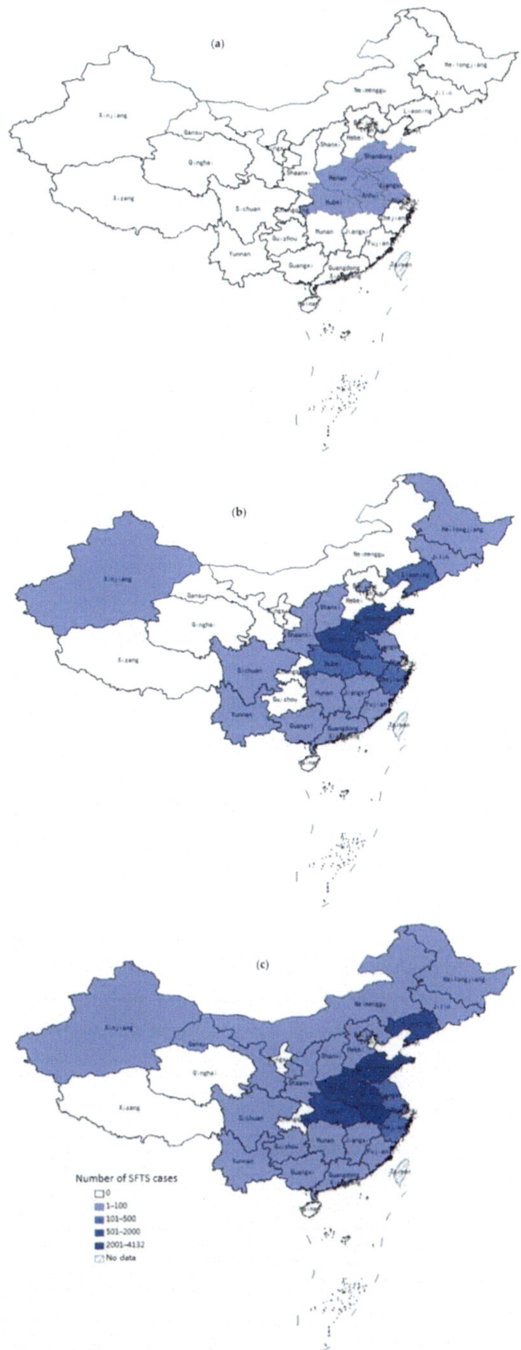

Fig. 10.14 Regional distribution of SFTS cases in mainland China: (**a**) SFTS distribution in 2010, (**b**) the accumulative SFTS distribution from 2010 to 2014, and (**c**) the accumulative SFTS distribution from 2010 to 2019 (Huang et al. 2021a, quoted under an open access Creative Commons CC BY 4.0 license)

peaking from May to June. The elderly population has a high risk of morbidity and mortality. The clear transmission of SFTS is through tick bites. Contact with patients' blood, body fluids, secretions, or excreta may also cause infection. The *Haemaphysalis longicornis* tick is a proven effective vector for SFTS and is widely distributed. SFTS has also been detected in other ticks. Livestock such as cattle and sheep are natural hosts with high infection rates (Li et al. 2021). The migration of ticks and host animals promotes the formation of new epidemic areas. Modern logistics and transportation have accelerated the passive migration of hosts, and factors such as global warming have increased the speed of diffusion. There is no effective vaccine or specific therapeutic drug. Cutting off the transmission route is a relatively effective prevention and control measure. Public health education is essential to promote tick prevention and control knowledge, take personal protective measures, and seek medical attention promptly after tick bites and illnesses. Not listed as a national notifiable infectious disease, SFTS is currently regulated according to Class B, decided by the National Health Commission. Establishing mechanisms for severe illness identification and referral and improving diagnostic and medical treatment capabilities are crucial to reducing mortality rates.

10.9.3 Other Tick-Borne Diseases

Tick-borne diseases are a class of diseases transmitted by bloodsucking vector ticks. In addition to SFTS, there are Crimean-Congo hemorrhagic fever (Li et al. 2020), forest encephalitis (Shi et al. 2022), relapsing fever (Zhang et al. 2022), Q fever (Huang et al. 2021b), spotted fever (Teng et al. 2023), Lyme disease (Stark et al. 2022), human granulocytic anaplasmosis (Dumic et al. 2022), and babesiosis (Dumic et al. 2022; Chen et al. 2020). These tick-borne diseases are natural foci and zoonoses, and there are sporadic reports in China. Forest encephalitis and Lyme disease are listed as occupational infectious diseases. Ticks in China are rich in species and widely distributed. The migration of ticks and their hosts drives the spread of tick-borne pathogens. Population and distribution monitoring of pathogens and vectors and infection risk assessment are essential.

10.9.4 Leishmaniasis

In China, leishmaniasis is mainly visceral leishmaniasis (VL), also known as kala-azar, caused by *Leishmania donovani* and *L. infantum*. There are occasional reports of cutaneous leishmaniasis, mostly imported. The main vector species is *Phlebotomus chinensis*, which is widely distributed. Others are *Ph. longiductus, Ph. Alexandri*, and *Ph. wui* (controversial, a synonym of *Ph. smirnovi*) (Zhang and Leng 1997; Tian et al. 2023). The presence of infected dogs and other potential host animals makes it difficult to control leishmaniasis (Liao et al. 2021; Zhen et al. 2023). VL spread primarily north of the Yangtze River throughout history. Currently, VL is under effective control. It is generally scattered with low-level endemics in the northwest and southwest, often in clusters (Yang et al. 2023; Zhou et al. 2023; Li et al. 2022). The incidence in Shanxi, Gansu, and Henan rebounded. Cases in Gansu, Xinjiang, and Sichuan, the three historically high-incidence areas, have declined owing to government policy in poverty alleviation. Strengthening environmental governance in the central and western regions has reduced sandfly breeding grounds and density. Due to the existence of natural epidemic sources, it is challenging to eliminate leishmaniasis.

10.9.5 Typhus

Typhus is a zoonotic disease that includes epidemic typhus and endemic typhus. The pathogens were *Rickettsia prowazekii* and *R. mooseri*. They are transmitted by body lice and rat fleas, respectively. The incidence was high in southern China and has decreased significantly since the 1990s. After falling to the lowest level in 2017, it rebounded in 2018–2021 (Han et al. 2023). Local epidemics have appeared in the north.

10.9.6 Chagas Disease

American trypanosomiasis, also known as Chagas disease, is a tropical parasitic disease caused by *Trypanosoma cruzi*, transmitted by triatomines. Chagas disease has become one of the most transmitted diseases, spreading worldwide in recent decades. China is not an endemic area of the disease. Although there have not yet been any recorded cases of Chagas disease, one of the potential vectors, *Triatoma rubrofasciata*, is widely recorded in southern China (Guo et al. 2021). Surveillance and early warning of imported *T. cruzi* carriers, especially at frontier ports, and risk assessment of outbreaks are essential measures to prevent the invasion of American trypanosomiasis into China.

10.10 Republic of Indonesia

Astri Nur Faizah

Indonesia is an archipelago country located in Southeast Asia and Oceania with 16,671 named islands in total (UNGEGN 2019) and is the fourth most populated nation behind China, India, and the United States (UNSD 2022). With vast land and seas of approximately 12 km^2, Indonesia has some of the world's highest concentrations of neglected tropical diseases, some of which are vector-borne (Tan et al. 2014; Wibawa and Satoto 2016). Vector-borne diseases (VBDs), such as dengue fever, chikungunya, and malaria, are major public health concerns in Indonesia, posing significant risks to the population. Efforts to control vector-borne diseases have included the use of insecticides, mosquito nets, and vaccines and the establishment of special applications for *Stegomyia* index reporting, but challenges including access to healthcare and limited funding for control programs persist. Here, we will describe prevalent vector-borne diseases (VBDs) that are prioritized by the Ministry of Health of the Republic of Indonesia (MOH), along with the vectors involved in transmitting them and any information regarding imported infection cases in international travelers returning from Indonesia. Some other VBDs with a history of presence in the country will also be discussed, along with the challenges to control efforts and the future directions for research and control, and finally conclude with the impact of vector-borne diseases on public health in Indonesia and neighboring countries.

10.10.1 Vector-Borne Diseases (VBDs): A High Priority for Disease Elimination in Indonesia

Indonesia is the world's largest archipelago with diverse geographic, climatic, and demographic conditions that may impact the dynamics of disease transmission (Sasmono et al. 2020). The MOH regularly reports four mosquito-borne diseases, dengue fever, chikungunya, filariasis, and malaria, which are endemic and currently prioritized for disease elimination (MOH 2022).

10.10.1.1 Dengue Fever (DF)

Dengue fever (DF), dengue hemorrhagic fever (DHF), and dengue shock syndrome (DSS) depending on the severity of clinical manifestation, are caused by the dengue virus (DENV) belonging to the family Flaviviridae. The first emergence of DENV transmission in human populations was said to be a few hundred years ago from nonhuman primates in a sylvatic cycle in Africa and Asia (Holmes and Twiddy 2003). Malaysia is suggested to be the ancient shelter for the DENV sylvatic cycle for DENV serotypes that were cross-species transferred separately to humans (Messina et al. 2014). In Asia, DENV serotype 2 (DENV-2) was first reported to emerge in 1944 in Papua New Guinea and Indonesia, while DENV-3 and DENV-4 were first reported in 1953 in the Philippines and Thailand (Harapan et al. 2020). In Indonesia, DENV was thought to first emerge in 1968 in Jakarta and Surabaya (Sumarmo 1987), where outbreaks were documented. Currently, four DENV serotypes and few respected genotypes are present in Indonesia (DENV-1 G-I, DENV-1 G-IV, DENV-2 G-Cosmopolitan, DENV-3 G-I, DENV-4 G-II) (Sasmono et al. 2015), and based on the number of annual cases reported to the WHO, Indonesia was ranked second with the highest number of cases per year worldwide (Tan et al. 2014). Anamnesis of the phase of the illness, clinical symptoms, physical examination including vital signs and signs of bleeding, and supporting examinations are used to confirm the diagnosis of severe dengue (DHF) cases. In 2021 alone, there were 73,518 cases of DHF with 705 deaths recorded. Those figures are slightly below the total number of cases and deaths in 2020, i.e., 108,303 and 747, respectively (MOH 2022). The highest incidence rates (IR) of dengue fever in Indonesia were recorded in 2016 and 2019, with 79 and 52 incidences per 100,000 population, respectively.

Dengue virus is present in 34 provinces in Indonesia, and the country has been repeatedly identified as the source of DENV infections in travelers returning to their home countries. Studies in Australia and Taiwan reported that approximately a quarter of the total dengue cases imported into Queensland (Warrilow et al. 2012) and Taiwan (Yang et al. 2018) originated from Indonesia. A study even referred to Indonesia as the top country that frequently imported all four serotypes to Taiwan during 2008–2010 (Huang et al. 2012). Furthermore, most DF cases that were brought into Western Australia specifically pointed to Bali as the source of infection (Ernst et al. 2015). Presumably being the central hotspot that can mix and at the same time import multiple DENV serotypes/genotypes to other countries (Ernst et al. 2015), Indonesia, or Bali Island in particular, may need rigorous surveillance to determine timely information on circulating DENV serotypes and genotypes before human cases in nonendemic countries.

The method that was mainly employed in preventing dengue infections is by controlling the disease vectors, e.g., biological (such as Wolbachia and paratransgenesis) and chemical control (such as insecticides and larvicides). Despite a weak effect (Buhler et al. 2019), chemical control remains the primary vector control strategy, particularly during disease outbreaks (Harapan et al. 2020). The primary vectors for DENV transmission in residential areas are *Aedes aegypti* (L.) and *Ae. albopictus* (Skuse), which feeds during the daytime. Therefore, multiple layers of vector control strategies, such as insecticide-treated windows, door screens, and curtains, may help prevent mosquito bites, whereas insecticide/larvicide-treated water container covers, and insecticide-treated bed nets may help reduce vector abundance. Although biological control of DENV vectors is generally underway, a recent successful report of wMel-infected *Ae. aegypti* populations in Yogyakarta in attenuating DENV transmission (Utarini et al. 2021) implied the potential of implementing the same program in other provinces in Indonesia.

In Indonesia, epidemiological investigation (Penyelidikan Epidemiologi, PE) is usually performed within 24 h of receiving a confirmed or suspected dengue case report in an area. A specialized PE team will investigate any human cases and *Stegomyia* indices within a 100-m radius of where the patient resides. The aim of the PE strategy is to determine the transmission potential and further spread of dengue as well as the countermeasures that need to be taken in the area (MOH 2022). Recently, available tetravalent, live-attenuated, dengue vaccines, Dengvaxia® (CYD-TDV) and QDENGA® (TAK-003) have been approved by the Indonesia National Agency for Drug and Food Control, BPOM, for use in 9–45-year-olds

(Takeda 2022; Zeng et al. 2018). A study found that implementing vaccination and prevaccination screening programs in Indonesia would be cost-effective yet unaffordable to implement as national routine immunization (Suwantika et al. 2021).

10.10.1.2 Chikungunya Fever

Chikungunya (CHIK) is caused by the chikungunya virus (CHIKV) belonging to the family Alphaviridae. The first CHIKV-like disease was reported to infect a physician working in Jakarta in 1779 (Halstead 2015). Serological studies conducted between 1969 and 1972 confirmed anti-CHIKV antibodies in most Indonesian islands except for Java Island (Kanamitsu et al. 1979; Tesh et al. 1975). The MOH documented the first Chikungunya cases on Kalimantan Island in 1973, followed by reported cases in 1982 on Sumatra Island and several outbreaks in 1983 and 1984 (Harapan et al. 2019). The outbreaks reemerged two decades later in South Sumatra, Aceh, and West Java between 2001 and 2003 (Laras et al. 2005). Subsequent Chikungunya outbreaks were indicated in the west and central Indonesian islands in 2009 and 2010 (Arif et al. 2020). In 2021, 241 cases were found in West Java, Central Java, and Bali Province, with no deaths reported (MOH 2022). Due to the diagnosis method being based solely on unspecific clinical presentation, the incidence rate of Chikungunya is likely an underestimate, and it remains an underdiagnosed acute febrile illness in Indonesia (Arif et al. 2020). One suggestion for public health measures for Chikungunya will be the establishment of easy access to CHIKV diagnostic tests to avoid unnecessary treatments or antibiotics and provide an early response to vector control implementation in the area.

Two genotypes of CHIKV have been detected in Indonesia, with most of the identified viruses belonging to the Asian genotype (Harapan et al. 2019). The other genotype, East/Central/South African (ECSA), was identified domestically and outside Indonesia between 2008 and 2011. Singapore, Taiwan, and the Netherlands reported multiple imported CHIKV infections in travelers returning from Indonesia (Ng et al. 2009; Scholte et al. 2013; Yang et al. 2016). Only the CHIKV West African genotype has yet to be identified in Indonesia. The Asian genotype may have been circulating in Indonesia for at least three decades. The epidemiological investigation that follows the discovery of a Chikungunya case is the same as that of Dengue measures.

10.10.1.3 Lymphatic Filariasis

Lymphatic filariasis (LF), also known as elephantiasis, is considered a neglected tropical disease (NTD) caused by parasitic worms. The total number of LF cases in 2021 in Indonesia was 9354, with details as follows: Papua (3629), East Nusa Tenggara (1307), West Papua (620), Gorontalo, Yogyakarta, Bali, and North Kalimantan (<5 combined cases in the last four provinces) (MOH 2022). There are five provinces reported to frequently show high LF cases: East Nusa Tenggara, Aceh, West Papua, Papua, and West Java.

Indonesia is the only country in the world with various types of LF parasites: *Brugia malayi*, *B. timori*, *Wuchereria bancrofti* (Lee and Ryu 2019), and *B. pahangi* (Palmieri et al. 1985). The distribution of *B. malayi* in Indonesia is much wider compared to the other two and can be found on Sumatra, Kalimantan, Sulawesi, and Maluku Island. *Brugia malayi* can be transmitted by *Anopheles barbirostris* van der Wulp and Mansonian species, including *Mansonia uniformis* (Theobald), *Ma. annulifera* (Theobald), *Ma. dives* (Schiner), *Ma. bonneae* Edwards, *Ma. annulata* Leicester, and *Ma. indiana* Edwards (Rahayu et al. 2020). A recent study conducted in South Borneo Province found that *B. malayi* may also be transmitted by *M. annulifera, Ae. albopictus*, and *Culex tritaeniorhynchus* Giles (Supriyono and Tan 2020).

Brugia timori is found in restricted areas of eastern Indonesia and is transmitted by *An. barbirostris*. Urban *W. bancrofti* is found on Java Island, whereas rural *W. bancrofti* is found outside Java, particularly in Papua and East Nusa Tenggara. In urban areas, *W. bancrofti* is transmitted by *Cx. quinquefasciatus* Say, while in rural areas, it is transmitted by *Culex* sp. and *Anopheles* sp. (*An. gambiae* Giles, *An. funestus* Giles, *An. scapularis* (Rondani)) (Lee and Ryu 2019). *Brugia pahangi* is generally found in cats, but studies have shown that it can also infect humans, with clinical symptoms similar to the usual LF (Palmieri et al. 1985; Tan et al. 2011). *Brugia pahangi* was found in South Kalimantan (Palmieri et al. 1985), and a recent study in Malaysia demonstrated the possible role of *Armigeres subalbatus* (Coquillett) as the vector (Muslim et al. 2013).

The direct economic loss due to productivity loss in Indonesia is significant (Aisyah et al. 2022). The global program to eradicate filariasis (GPELF) initiated by the WHO has been running since 2000, and Indonesia has been implementing it fully. The GPELF approach combined mass drug administration (diethylcarbamazine citrate and albendazole in a single dose) for high-risk populations and simultaneously reduced LF morbidity in infected patients (WHO 2010b). There has been an imported case of filariasis in Singapore (Chavatte and Jureen 2017), which mentioned Indonesia as a potential source of infection.

10.10.1.4 Malaria

Because of its geography, the presence of certain types of mosquitoes, a large population with high levels of migration, and socioeconomic and political factors, Indonesia is largely at risk for malaria (Lee and Ryu 2019; Sitohang et al. 2018). Indonesia is one of nine malaria-endemic countries in Southeast Asia, accounting for 21% of reported cases and 16% of malaria deaths in the region (WHO 2019). By 2018, 285 districts in Indonesia had accomplished their goal of eliminating malaria, thus clearly showing significant progress towards the regional goal of malaria elimination by 2030 (Sitohang et al. 2018). The two provinces with the highest number of malaria cases in Indonesia are Papua (42.7%) and West Papua (38.4%). Determination of endemicity stratification is determined based on the annual parasite incidence (API) value: low endemic area (API < 1), medium (API = 1–5), and high (API > 5). In the 2021 report, there were 347 malaria-free districts/cities

(67.5%), 124 low-malaria districts/cities (24.2%), 17 medium-malaria districts/cities (3.3%), and 26 high-endemic districts/cities (5%) (MOH 2022).

Five species of Plasmodia are present in Indonesia: *Plasmodium falciparum*, *P. vivax*, *P. malariae*, *P. ovale*, and *P. knowlesi*. The approximate human cases of *P. falciparum* and *P. vivax* are nearly on par, with *P. falciparum* dominating certain highly endemic areas. In 2019, 57% of nonzoonotic malaria cases were caused by *P. falciparum*, 35% by *P. vivax*, 7% by coinfections, 1% by *P. malariae*, and very few cases were caused by *P. ovale* (bin Said et al. 2022). *Plasmodium knowlesi* is the cause of significant zoonotic malaria that infects macaques and was recently recognized as the fifth human malaria parasite (Naik 2020; Singh and Daneshvar 2013). There has been very limited information regarding *P. knowlesi* due to its diagnostic difficulties and resemblance to other species. One review article thoroughly compiled a detection report during the 2017–2021 timeframe, informing the distribution of *P. knowlesi* in Aceh, North Sumatra, Jambi, Central Kalimantan, and South Kalimantan Provinces (bin Said et al. 2022).

Known anopheline malaria vectors that are present in Indonesia are *An. aconitus* Dönitz, *An. balabacensis* Baisas, *An. bancroftii* Giles, *An. barbirostris*, *An. farauti* Laveran, *An. flavirostris* (Ludlow), *An. karwari* (James), *An. kochi* Dönitz, *An. koliensis* Owen, *An. leucosphyrus* (Dönitz), *An. maculatus* Theobald, *An. nigerrimus* Giles, *An. parangensis* (Ludlow), *An. punctulatus* Dönitz, *An. sinensis* Wiedemann, *An. subpictus* Grassi, *An. sundaicus* (Rodenwaldt), *An. tessellatus* Theobald, *An. vagus* Dönitz, *An. letifer* Sandosham, *An. ludlowi* (Theobald), *An. minimus* Theobald, *An. umbrosus* (Theobald), *An. peditaeniatus* (Leicester), and *An. annullaris* (Lee and Ryu 2019). For *P. knowlesi*, of several known *P. knowlesi* vectors, five are distributed in Indonesia: *An. latens* Sallum & Peyton (formerly *An. leucosphyrus* species A), *An. balabacensis*, *An. leucosphyrus*, *An. introlatus* Baisas, and *An. cracens* (formerly *An. dirus* species B) (bin Said et al. 2022).

To date, the standardized and effective treatment for malaria in Indonesia is Artemicinin-based Combination Therapy (ACT) administration given in the first 24 h of a fever patient (MOH 2022). Indonesia is one of the suspected countries for any imported malaria cases into Japan (Taniguchi et al. 2008).

10.10.2 Concerned Vector-Borne Diseases in Indonesia

Mosquito-borne diseases remain a public health concern in Indonesia, impacting millions of people living in the country. Aside from the four diseases mentioned above, other vector-borne diseases, including tick-borne and biting midge-borne diseases, which were detected and found to be infecting humans and/or animals in Indonesia are also discussed. There were at least 11 other diseases confirmed to be present despite nascent or sporadic independent detections/surveillances (Table 10.5).

Table 10.5 List of vector-borne diseases in Indonesia

Disease name	Causative agents (serotype/genotype/lineage detected)	Areas of distribution	Year of detection	Imported cases to other countries	(Putative) vectors involved[c]	References
Dengue fever[a]	*Dengue virus* Four serotypes, five different genotypes: DENV-1 G-I, DENV-1 G-IV, DENV-2 G-Cosmopolitan, DENV-3 G-I, DENV-4 G-II	Nationwide (all 34 provinces)	Annual human cases reported (since 1968)	Australia (Ernst et al. 2015; Warrilow et al. 2012), Taiwan (Huang et al. 2012; Yang et al. 2018)	*Ae. aegypti, Ae. albopictus*	Ministry of Health of the Republic of Indonesia (2022)
Chikungunya fever[a]	*Chikungunya virus* Two genotypes: CHIKV Asian and East/Central/South African (ECSA)	Java and Bali Island (during the last two decades)	Annual human cases reported (since 1779)	Japan (Kobashi et al. 2010)	*Ae. aegypti, Ae. albopictus*	Arif et al. (2020), Ministry of Health of the Republic of Indonesia, (2022)
Lymphatic filariasis[a]	*Wuchereria bancrofti, Brugia malayi, Brugia timori, Brugia pahangi*	Nationwide	Annual human cases reported	Singapore (Chavatte and Jureen 2017)	*An. barbirostris, An. gambiae, An. funestus, An. scapularis, Ar. subalbatus, Ae. albopictus*, Cx. *tritaeniorhynchus, M. uniformis, M. annulifera, M. dives, M. bonneae, M. annulata, M. indiana,*	Ministry of Health of the Republic of Indonesia (2022)

(continued)

Table 10.5 (continued)

Disease name	Causative agents (serotype/genotype/lineage detected)	Areas of distribution	Year of detection	Imported cases to other countries	(Putative) vectors involved[c]	References
Malaria[a]	*Plasmodium falciparum, P. vivax, P. knowlesi, P. ovale*	Nationwide, with various API level (low, medium, high-endemic areas). Most provinces in Eastern Indonesia are highly endemic	Annual human cases reported	Japan (Taniguchi et al. 2008)	*An. aconitus, An. balabacensis, An. bancroftii, An. barbirostris, An. farauti, An. flavirostris, An. karwari, An. kochi, An. koliensis, An. leucosphyrus, An. maculatus, An. nigerrimus, An. parangensis, An. punctulatus, An. sinensis, An. subpictus, An. sundaicus, An. tessellatus, An. vagus, An. lebifer, An. ludlowi, An. minimus, An. umbrosus, An. peditaeniatus, An. annullaris, An. latens, An. introlatus, An. cracens*	Ministry of Health of the Republic of Indonesia (2022)
Japanese encephalitis[b]	*Japanese encephalitis virus (JEV)* Four genotypes; G-I, G-II, G-III, G-IV	Nearly nationwide (29 provinces), with Bali, West Kalimantan, East Nusa Tenggara, West Java, and East Java are the areas with high incidences	Sporadic JEV detections in pig/bat/mosquito/macaque/human were reported (since 1960)	Australia (Hills et al. 2010; Macdonald et al. 1989), Sweden(Buhl and Lindquist 2009; Hills et al. 2010; Wittesjö et al. 1995), Germany (Caramello et al. 2007), Denmark (Buhl and Lindquist 2009; Hills et al. 2010), France (Hills et al. 2010), and Netherlands (Buhl and Lindquist 2009; Hills et al. 2010).	*Cx. tritaeniorhynchus, Cx. vishnui, Cx. gelidus, Ar. subalbatus, Cx. fuscocephala, Cx. bitaeniorhynchus, Cx. quinquefasciatus, An. vagus, An. kochi, An. annularis*	Diptyanusa et al. (2022), Faizah et al. (2021), Garjito et al. (2018, 2019), Schuh et al. (2013, 2014), van Peenen et al. (1975)

Schistosomiasis	*Schistosoma japonicum*	Restricted to three villages in Central Sulawesi Province	Annual human cases reported (since 1937)	n/a	Water snail (*Oncomelania hupensis lindoensis*)	Azhari (2022), Garjito et al. (2008), Nurwidayati et al. (2019), Tan et al. (2014)
Leishmaniasis	*Leishmania* spp.	Not endemic in Indonesia	One cutaneous Leishmaniasis-like disease presentation in Bali in 2016, No autochthonous cases reported	Russia (Nagorny et al. 2019)	*Phlebotomus* and *Lutzomyia*	Ekawasti and Martindah (2020), Ekawati and Saputra (2016)
Zika disease	*Zika virus (ZIKV)* Asian/American lineage	North Sumatra, West Sumatra, Jambi, Lampung, Banten, DKI Jakarta, West Java, Central Java, East Java, Bali, East Kalimantan, Southeast Sulawesi, West Nusa Tenggara Province	1977, 1978, 1983, 2013, 2014, 2016 (Human sera)	Australia (Leder et al. 2013; Leung et al. 2015)	*Aedes* spp.	Olson et al. 1981, (1983), Perkasa et al. (2016), Sasmono et al. (2018, 2021)

(continued)

Table 10.5 (continued)

Disease name	Causative agents (serotype/genotype/lineage detected)	Areas of distribution	Year of detection	Imported cases to other countries	(Putative) vectors involved[c]	References
West Nile fever	*West Nile virus (WNV)* Lineage 1 Kunjin strain Lineage 2	Bali, West Nusa Tenggara, West Java Province	1979 (cattle), 1983, 2004–2005 (human)	n/a	unknown	Miura et al. (1982), Myint et al. (2014), Olson et al. (1983)
Bluetongue disease[b]	*Bluetongue virus (BTV)* serotype 1, 3, 6, 7, 9, 10, 12, 16, 20, 21, 23	Nationwide	Sporadic BTV detections in sentinel cattle/biting midge/mosquito (Since 1981)	Australia (Firth et al. 2017)	*Culicoides fulvus, C. peregrinus, C. shortii, C. orientalis, Aedes* sp., *Anopheles* sp.	Daniels et al. (1995), Miura et al. (1982), Sendow (2002), Sendow et al. (1991, 1994), Sendow et al. (1993a, b)
Bovine ephemeral fever[b]	*Bovine ephemeral fever virus (BEFV)*	Nationwide	1919–1993 (bovine sera)	n/a	Mosquito	Daniels et al. (1995)
Banna disease	*Banna virus*	Central Java and West Java	1980, 1981, 2020 (mosquitoes)	n/a	*Cx. vishnui, An. vagus, An. subpictus, Cx, tritaeniorhynchus Ae. albopictus, Ae. aegypti*	Attoui et al. (1998, Brown et al. (1993), Supriyono et al. (2020)
Akabane fever[b]	*Akabane virus (AKAV)* genogroup Ib	West Java, Bali Province	1979, 2014 (Bovine sera)	n/a	*Culicoides*	Purnomo Edi et al. (2017)

Rickettsial diseases	*Rickettsia typhi* (TGR), *Orientia tsutsugamushi* (STGR)	Jakarta, West Java, Central Java, East Java, Bali, West Papua, Northeastern Papua	Sporadic serological screening in human/wild rodents (since 1976)	Japan (Kato et al. 2014; Takeshita et al. 2010), Europe (Kiely et al. 2011; Parola 1998)	Flea: *Xenopsylla cheopis* Chigger/Trombiculidae: *Leptotrombidium (Leptotrombidium) delicense, L. (L.) fletcheri* (Hadi et al. 1980)	Capeding et al. (2013), Gasem et al. (2009), Lokida et al. (2020), Punjabi et al. (2012), Putri et al. (1997)
Surra (Trypanosomiasis)[b]	*Trypanosoma evansi*	Nationwide (Sumba Island was free until 2009)	Annual cases reported (since 1937)	n/a	Mechanical vectors; *Tabanus* spp., *Stomoxys* spp.	Dewi et al. (2020), Reid and Copeman (2000), Sawitri et al. (2019), Syah et al. (2020)

[a] Prioritized vector-borne diseases regularly monitored by the Ministry of Health of the Republic of Indonesia
[b] Vector-borne diseases of veterinary significance
[c] Genus abbreviation, *Ae Aedes, Cx Culex, An Anopheles, C Culicoides, M Mansonia*

10.10.2.1 Japanese Encephalitis

Japanese encephalitis (JE) is caused by Japanese encephalitis virus (JEV) belonging to the genus *Flavivirus*, which is transmitted to humans through the bites of infected mosquitoes. JEV antibodies in humans were first detected in Lombok in 1960, while the virus was successfully isolated for the first time from mosquitoes in Jakarta and West Java Province in 1972 (Garjito et al. 2018). Since then, JE cases have been discovered from pig, bat, mosquito, and even macaque samples collected in 29 provinces (Diptyanusa et al. 2022; Faizah et al. 2021; Garjito et al. 2018; Olson et al. 1985; Putra et al. 2022; Schuh et al. 2013; van Peenen et al. 1975). Those sporadic discoveries were based on multiple independent surveys rather than national routine surveillance (Garjito et al. 2018). At least ten mosquito species belonging to three genera were involved in JEV transmission in Indonesia: *Cx. tritaeniorhynchus*, *Cx. gelidus* Theobald, *Cx. vishnui* Theobald, *Cx. fuscocephala* Theobald, *Cx. bitaeniorhynchus* Giles, *Cx. quinquefasciatus* Say, *An. vagus*, *An. kochi*, *An. annularis* van der Wulp, and *Ar. subalbatus*. On the other hand, four JEV genotypes have been detected in Indonesia (Faizah et al. 2021; Garjito et al. 2018, 2019; Schuh et al. 2013), leaving G-V as the only undetected genotype.

The endemicity of JEV in Indonesia affects not only naïve local populations but also international travelers visiting Indonesia. Travel-associated human cases, either fatal or nonfatal encephalitis, have been reported in Australia (Hills et al. 2010; Macdonald et al. 1989), Sweden (Buhl and Lindquist 2009; Hills et al. 2010; Wittesjö et al. 1995), Germany (Caramello et al. 2007), Denmark (Buhl and Lindquist 2009; Hills et al. 2010), France (Hills et al. 2010), and the Netherlands (Buhl and Lindquist 2009; Hills et al. 2010). Bali Province, in particular, was the travel destination for most of the imported cases. Recent serological surveys reported that up to 92% of the children in Bali had antibodies against JEV, with an annual incidence estimated to be 7.1 per 100,000 children (Kardena et al. 2021). Hence, Bali was chosen as the first area in Indonesia to implement a JEV vaccine program (Chengdu SA-14-14-2, live-attenuated) for children. Despite the lack of a complete evaluation of the vaccination program, JEV infection in Bali children appears to be decreasing (Kardena et al. 2021). Implementation of the JEV vaccination program in children has also been expanded to North Sulawesi (Yosephine 2017) and further will be included in the basic immunization for children aged 9 months throughout Indonesia upon program completion in hyperendemic areas.

10.10.2.2 Schistosomiasis

Schistosomiasis, also known as snail fever or bilharzia, is caused by parasitic worms (*Schistosoma japonicum*) transmitted by aquatic snails (*Oncomelania hupensis lindoensis*). The parasite is found in only three areas of Indonesia's Central Sulawesi Province: Lindu, Napu, and Bada Valleys (Budiono et al. 2019; Tan et al. 2014).

Schistosomiasis is a neglected tropical disease in Indonesia; nevertheless, the disease has been successfully controlled for the past seven decades (Garjito et al. 2008). Rigorous elimination programs implemented by its government may now be nearing completion, as disease prevalence has been reported to be less than 1% since 2016 (Azhari 2022). The figure contrasts with the disease prevalence reported in the same area in the 1970s, i.e., more than 37% (Oemijati et al. 1974). Aside from humans, schistosomiasis has also been found to infect other mammalian species: wild rodents, wild pigs, wild deer, wild shrews, wild civet cats, cattle, water buffalo, horses, and dogs in endemic areas (Carney et al. 1978; Lee and Ryu 2019). Except for rodents, the role of these animals in disease transmission remains unknown.

One study confirmed the decline in disease prevalence observed since 2016; however, screening results in snails and rats showed a trend of increasing infection rates (Nurwidayati et al. 2019). Thus, the complete eradication of schistosomiasis may also be dependent on the reduction in intermediate vectors and mammal reservoirs. Other schistosomiasis control activities that have been carried out by the Central Sulawesi Provincial Health Office are mass drug treatment administration (MDA, using Praziquantel) if the prevalence is above 1% and selective treatment if the prevalence in humans is less than 1% (Nurwidayati et al. 2019).

10.10.2.3 Leishmaniasis

Leishmaniasis is caused by parasites of the Leishmania species, which are transmitted to humans and animals through the bites of infected sandflies. Although this disease is not endemic in Indonesia, a case of an HIV-infected patient with a cutaneous leishmaniasis-like presentation was found in Bali in 2016 (Ekawati and Saputra 2016). Moreover, the main vectors involved in transmission, sandflies in the genera Phlebotomus and Lutzomyia, are abundant in Indonesia. Leishmaniasis has received WHO attention because the disease is expected to continue to expand and afflict the global population; thus, there is an urgency for prevention in countries with suitable vector habitats.

A survey of autochthonous leishmaniasis cases in Russia implied that some infections were acquired from Indonesia (Nagorny et al. 2019), despite no autochthonous human cases ever reported in Indonesia. In another case report in Thailand, autochthonous visceral leishmaniasis was found in a patient with no travel history except for Northern Indonesia during his lifetime (Siriyasatien et al. 2010). The report concluded that the parasite may be acquired in Thailand due to the absence of cases in Indonesia as well as the occurrence of prior leishmaniasis cases in Thailand.

Leishmaniasis has been endemic in the Indian subcontinent and East Africa, with several outbreaks occurring in the Middle East because of war-related population migration. With Indonesian troops being sent multiple times for a peace mission on the Lebanese-Israeli border where cutaneous leishmaniasis is endemic, there is a risk of cases being imported into Indonesia. (Alawieh et al. 2014; Ekawasti and Martindah 2020; Widana and Hilmawan 2019).

10.10.2.4 Zika Virus Infection

Zika virus infection is caused by the Zika virus (ZIKV) of the *Flavivirus* genus, which is transmitted to humans through the bite of *Aedes* spp. mosquito or *Ae. aegypti* in particular. There have been ZIKV detections in Asia, with the first 2 cases reported in Malaysia and Central Java Province in Indonesia in 1966 and 1977, respectively (Olson et al. 1981). Serological surveys were also conducted in Lombok in 1983, resulting in the detection of ZIKV antibodies in human sera (Olson et al. 1983). In more extensive surveillance in 2014, ZIKV antibodies could be detected in children living in various provinces in Indonesia: North Sumatra, West Sumatra, Jambi, Lampung, Banten, DKI Jakarta, West Java, Central Java, East Java, East Kalimantan, and Southeast Sulawesi, thus indicating widespread Zika virus transmission and achieved immunity in Indonesian populations (Sasmono et al. 2018, 2021). The first autochthonous ZIKV infection was confirmed in Sumatra between December 2014 and April 2015 (Perkasa et al. 2016), highlighting the identification of Asian/American lineage ZIKV in Indonesia. Travel-associated ZIKV cases were discovered in Australian travelers returning from Jakarta and Bali in 2012 and 2015, respectively (Leder et al. 2013; Leung et al. 2015). Indonesia, along with India, the Philippines, Pakistan, and Bangladesh, may have a high risk of ZIKV importation, with a possible significant health impact on the population (Duong et al. 2017). Nonetheless, due to difficulties in obtaining confirmed laboratory diagnosis, the true number of ZIKV infections is likely to be underestimated, necessitating continued surveillance and robust knowledge dissemination to local public health workers.

10.10.2.5 West Nile Fever

West Nile fever is caused by West Nile virus (WNV) and is considered a serious public health threat because of the large outbreaks of epidemic encephalitis that it may cause. The virus can be categorized phylogenetically into at least five lineages, in which Lineage 1 and Lineage 2 are dominant. Lineage 1 has three sub-lineages, a, b, and c, where the latter (1c) has been reclassified as Lineage 5. Lineage 1a has been found in Africa, the Middle East, Europe, and America, whereas Lineage 1b (known as Kunjin strain) is present in Australia. Strains of Lineage 2 have been identified in sub-Saharan Africa, Madagascar, Europe (Greece, Hungary, and Italy), and Russia (McMullen et al. 2013). Mosquito–bird transmission cycle maintains the natural cycle of WNV, with the northern house mosquito, *Cx. pipiens* L., and the freshwater mosquito, *Cx. annulirostris* Skuse, were reported to act as the primary vectors in the United States and Australia, respectively. *Culex annulirostris* is also distributed in the Indonesian islands. The first detection of WNV (Kunjin strain) in Indonesia was from bovine sera collected on the island of Bali in 1979 (Miura et al. 1982) and then followed by detections in human sera in Lombok, West Nusa Tenggara Province, in 1983 (Olson et al. 1983). Another WNV detection was from a positive sample collected in 2004–2005 in West Java Province. The patient was a 15-year-old boy who presented with systemic febrile illness but no neurological symptoms (Myint et al. 2014). However, the latter study identified the presence of WNV Lineage 2 in Indonesia, which is related to other isolates causing outbreaks in Europe, rather than Lineage 1 (Kunjin strain), which is circulating in Australia and was previously detected in bovine sera in 1979. The presence of WNV in Indonesia is currently unknown, due to limited laboratory capacity in local health care services to detect arboviruses other than DENV. Increased human, vector, and animal surveillance studies, as well as the establishment of dedicated research facilities in Indonesia, may help prevent emerging pathogens.

10.10.2.6 Bluetongue Disease

Bluetongue disease (BTD), which infects ruminants worldwide and is not known to infect humans, is caused by bluetongue virus (BTV) of the genus *Orbivirus* and is transmitted primarily by biting midges (*Culicoides* spp.) (Firth et al. 2017). The first BTV outbreak in Indonesia was reported in 1981 in imported Suffolk sheep (Saminathan et al. 2020). Since clinical BTD has not been reported in indigenous ruminants, added to the fact that the infected animals recovered and no deaths were reported, BTV strains in Indonesia do not appear to be highly virulent (Daniels et al. 1995; Sendow 2005). Nonetheless, the detection of BTV serotypes 1, 12, and 20 from cattle in Java and Bali Island (Miura et al. 1982) and the isolation of BTV serotypes 1, 7, 9, 21, and 23 from sentinel cattle in various locations across Indonesian provinces have confirmed the presence of the virus nationwide (Sendow et al. 1991; Sendow et al. 1993a).

Bluetongue virus serotypes 1, 6, 10, and 21 were isolated from *Culicoides* spp. (*Cu. fulvus* Sen & Das Gupta, *Cu. peregrinus* Kieffer, *Cu. shortti* Smith and Swaminath, and *Cu. orientalis* Kieffer), *An. vagus* and mosquito pools consisted of *Aedes* and *Anopheles* (Brown et al. 1992; Sendow 2002, 2005; Sendow et al. 1993b). Unsuccessful infection and transmission of BTV by *Ae. albopictus* in laboratory experimental conditions has ruled out the species as a potential BTV vector (Ledda et al. 2022). Four confirmed biting midge vectors in Australia, *Cu. actoni* Smith, *Cu. brevitarsis* Kieffer, *Cu. fulvus*, and *Cu. wadai* Kitaoka, are widely distributed in Indonesia. Together with *Cu. orientalis* and *Cu. nudipalpis* Delfinado, these were vector candidates for transmitting BTV in Indonesia. However, the dominant BTV vector in Africa and the Middle East, *Cu. imicola* Kieffer, has yet to be found in Indonesia (Daniels et al. 1995).

At least 11 BTV serotypes have been detected in Indonesia (Daniels et al. 1995; Saminathan et al. 2020), that is, 73% of the 15 serotypes are present in Southeast Asia or 39% of the 28 available serotypes worldwide. Regardless of the segment examined, the Indonesian BTV isolates were more closely related to viruses circulating in Australia than to those circulating in China, India, or Taiwan (Firth et al. 2017). Infected insects carried by wind from infected areas to the north and northwest are the most likely mode of disease introduction. Molecular analyses showed that Southeast Asian strains of the disease have moved into northern Australia, with differences in observations between India and eastern Australia suggesting movement along a gradient through Southeast Asia (Daniels et al. 1995). Even though the BTV serotypes circulating in Indonesia were deemed nonpathogenic, Sendow (2005) suggested the importance of vaccination in all imported sheep, especially those from BTV-free areas, along with the extension of the quarantine period for better acclimation before being distributed to any areas in Indonesia.

10.10.2.7 Bovine Ephemeral Fever

Bovine ephemeral fever (BEF), or 3-day fever, is caused by the bovine ephemeral fever virus (BEFV) belonging to the family Rhabdoviridae and is not known to be a human disease agent. The virus has been isolated from various mosquito species (*An. bancroftii* Giles and *Cx. annulirostris*) and biting midges (*Cu. arakawae, Cu. bedfordi* Ingram and Macfie, *Cu. brevitarsis* Kieffer, *Cu. coarctatus, Cu. cornutus, Cu. kingi, Cu. imicola, Cu. nivosus*) (Lee 2019). In Indonesia, clinical BEF presentation in cattle was first observed in West Java Province at approximately 1919. Later, outbreaks were reported on Sumatra Island between 1928 and 1931, in East Java Province in 1978, and in East Java again in 1985, with high mortality rates and over 78% seroprevalence. In 1991, another outbreak was recorded in South Kalimantan Province (Daniels et al. 1995).

A high prevalence of BEFV infection was recorded in bovine sera collected in Java and Bali Island in 1979 (up to 79%) (Miura et al. 1982). Extensive serological surveys conducted during 1989–1993 suggested that the disease was already widespread and endemic in Indonesia, as 12 out of 29 provinces showed BEFV seroconversions in sentinel cattle (Daniels et al. 1995).

The disease can lead to significant economic losses in livestock production. Since there is no specific treatment for BEF, control measures will focus on preventing the spread of the virus through the use of insecticides and good hygiene practices.

10.10.2.8 Banna Disease

In 1987, Banna virus (BAV) was first isolated from encephalitis patients in Xishuangbanna Prefecture in southern China and later isolated from pigs, cattle, ticks, mosquitoes, and human patients with febrile illness (Liu et al. 2010). Banna virus (BAV) belongs to the genus *Seadornavirus* family Reoviridae and has been identified in China and Indonesia, with the latter only identified from mosquitoes. BAV was isolated from mosquitoes; *Cx. vishnui*, *An. vagus*, and *An. subpictus* collected in Central Java in 1980–1981 (Brown et al. 1993), and *Ae. albopictus*, *Ae. aegypti*, and *Cx. tritaeniorhynchus* collected in Bogor, West Java Province (Supriyono et al. 2020).

The full range of diseases caused by BAV and the mechanisms by which it is transmitted to humans remains unclear, and the scope and impact of BAV presence in Asia require further investigation (Liu et al. 2010). In Indonesia, the prevalence of BAV infection in mammals and its association with human diseases have not been established, thus warranting further investigation. BAV may circulate in areas with JEV and may be misdiagnosed as JE due to similar symptoms, potentially underestimating its public health impact and highlighting the need for further surveillance and improved diagnostic systems (Attoui et al. 1998; Liu et al. 2010).

10.10.2.9 Rickettsial Diseases

Rickettsial diseases or rickettsiosis are caused by intracellular bacteria (*Rickettsia* and *Orientia* pathogenic species) and are transmitted to humans through arthropod vectors such as fleas, ticks, and mites. These diseases are the second most common cause of nonmalarial febrile illness in Southeast Asia after dengue and are generally divided into three major groups: the typhus group (TGR), scrub typhus group (STGR), and spotted fever group (SFGR) (Bhengsri et al. 2016). The most prevalent findings in Southeast Asia are flea-borne endemic (murine) typhus (TGR) caused by Rickettsia typhi and scrub typhus (STGR) related to *O. tsutsugamushi* infection (Satjanadumrong et al. 2019).

Rickettsioses in Indonesian populations are not well known, and very few studies have conducted specific pathogen screening in patients with febrile illness. One extensive study conducted in 2010–2011 revealed that Rickettsia was responsible for 7.6% of febrile illness cases in children in Jakarta, West Java, and Bali Province (Capeding et al. 2013). Separate studies identified murine typhus in Jayapura (in Northeastern Papua, during 210–2011), Semarang (Central Java 2005–2006), and Malang (East Java); spotted fever in Jayapura; and scrub typhus cases in Jayapura and Malang (Gasem et al. 2009; Punjabi et al. 2012; Putri et al. 1997). A recent study conducted in 2013–2016 found a much higher prevalence of murine typhus as a significant cause of fever in Indonesia, i.e., as high as 30.8% (Lokida et al. 2020). Furthermore, the study found that acute rickettsial infection cases were misdiagnosed, which may be due to overlap with other infectious etiologies, lack of inclusion in differential diagnoses, and poor access to diagnostic tests.

Rickettsia and *Orientia* species persist in animal and arthropod reservoirs. In Indonesia, rodents were found to be important reservoirs for rickettsial diseases, as seroconversion was observed in wild rats collected in Jakarta and Central Java (15% for murine typhus TGR and 39.1% for SFGR) (Ibrahim et al. 1999), in Northeastern Papua (11% for *R. typhi* and 12.5% for *O. tsutsugamushi*) (Richards et al. 2002), and on Biak and Owi Island, Papua Province, in August 1976 (for *O. tsutsugamushi*) (Hadi et al. 1980). The latter study found *O. tsutsugamushi*, the causal agent of scrub typhus, in the spleens of wild rats infested by two chigger (Tsutsugamushi) species: *Leptotrombidium* (*Leptotrombidium*) *deliense* Walch and *L.* (*L.*) *fletcheri* Womersley and Heaslip. While the presence of *R. typhi* (TGR) and *O. tsutsugamushi* (STGR) has been confirmed, the Rickettsia species of SFGR in Indonesia remain inconclusive (Richards 2022; Satjanadumrong et al. 2019) and require further investigation.

More than 450 Rickettsiosis cases in international travelers were reported in 2004 alone, with evidence of *R. typhi* infection from tropical and subtropical countries and *O. tsutsugamushi* infection from Asia-Pacific countries (Ericsson et al. 2004). Murine typhus cases have been identified in Japanese (Kato et al. 2014; Takeshita et al. 2010) and European travelers (Kiely et al. 2011; Parola 1998) returning from Indonesia. Given the high prevalence observed, Rickettsial diseases should also be prioritized for diagnosis, treatment, and prevention in Indonesia. Improved detection and management can improve patient outcomes and reduce healthcare costs. While PCR and IFA are the most effective diagnostic methods, they require expensive equipment and trained technicians, making ELISA a good alternative in resource-limited settings (Lokida et al. 2020). It is critical to provide appropriate empiric treatment and implement prevention guidelines for at-risk local populations to reduce the impact of rickettsioses.

10.10.2.10 Akabane Disease

Akabane disease is caused by Akabane virus (AKAV), which belongs to the genus *Orthobunyavirus* and was first isolated from *Ae. vexans* (Meigen) and *Cx. tritaeniorhynchus* in Japan in 1959 (Oya et al. 1961). The primary vectors for AKAV, however, are the *Culicoides* biting midges, which have a wide distribution worldwide (Yanase et al. 2005). AKAV infection in pregnant cattle, sheep, and goats may cause abortion, stillbirth, and congenital abnormalities. AKAV has caused many large-scale epidemics in East Asia and Australia since 1972 (Zeller and Bouloy 2000) and has also been detected in Vietnam, Australia, South Africa, the Middle East, Argentina, and Indonesia (Edmondson et al. 2012; Miura et al. 1982; Purnomo Edi et al. 2017).

In Indonesia, AKAV was first detected in bovine serum collected in Java and Bali in 1979 (Miura et al. 1982) and was recently isolated from bovine sera collected in West Java in 2014 (Purnomo Edi et al. 2017), revealing the close relationship of the Indonesian isolate to other AKAV genogroup Ib isolates found in Turkey in 2015. There is no information regarding the AKAV vector in Indonesia, despite the presence of *Cu. brevitarsis*, which is the confirmed vector for AKAV in Australia (Bellis et al. 2014; Sick et al. 2019). In addition, the nationwide spread of BTV, another *Culicoides*-borne virus that infects cattle and sheep, suggests the possible circulation of AKAV in Indonesia (Purnomo Edi et al. 2017).

10.10.2.11 Surra (Trypanosomiasis)

Surra is caused by the blood protozoan parasite *Trypanosoma evansi* and spread by mechanical transmission of hematophagous flies (*Tabanus* spp., *Stomoxys* spp., *Haematopota* spp., *Lyperosia* spp., and *Chrysops* spp.) (Desquesnes et al. 2013). *Trypanosoma evansi* was thought to first spread to Indonesia via livestock imports from India in the early nineteenth century and currently gained endemic status throughout all provinces in Indonesia, excluding Sumba Island, until 2009. In

2010–2012, the most severe Surra outbreak in Indonesia occurred on Sumba Island, killing over 2000 livestock and causing an IDR of 25.7 billion in economic losses (Dewi et al. 2020). The high cost of the outbreak was caused by the lack of a control program in place and farmers' lack of knowledge about Surra diseases at the start of the outbreak. Surra control is primarily based on the identification of infested animals, so farmers or livestock keepers should observe clinical signs and report them to the local veterinarians to receive proper treatment. Surra's high prevalence in Indonesian provinces necessitates increased local government attention, such as optimizing livestock traffic control, developing a Surra disease reporting system to increase public participation, and conducting regular community socialization (Syah et al. 2020).

Aside from infecting water buffaloes, cattle, dogs, and horses, the causal pathogen *T. evansi* infects endangered wild animals such as Asian elephant (*Elephas maximus*), leopard (*Panthera pardus*), tiger (*P. tigris*), jaguar (*P. onca*), and Sumatran rhinoceros (*Dicerorhinus sumatrensis*) (Aregawi et al. 2019). Furthermore, the parasite was reported to have infected an Indian cattle farmer in 2004 (Joshi et al. 2005), followed by other human infection reports in Egypt and Vietnam (Lai et al. 2010; Lun et al. 2009). Serological assays on Sumba Island, Indonesia, during the Surra outbreak in 2012 found human infections in farmers, albeit in the absence of clinical manifestations (Sawitri et al. 2019) Notwithstanding, *T. evansi* infection in humans warrants attention due to its emerging disease potential.

10.10.3 Other Vector-Borne Diseases

Indonesia's diverse ecological, cultural, and economic factors make it a unique area for the study of vector-borne diseases, which are influenced by both natural and human-related factors (Rodhain 2000). Owing to the role of the country as home to a diverse array of plants and animals, pathogens that endanger the lives of wild and endangered animals have recently started to be studied. Some examples are avian haemosporidian (*Plasmodium*, *Haemoproteus*, and *Leucocytozoon*) parasites in wild birds (Yuda 2019) and *P. pitheci* malaria in Bornean orang-utans, which serves as an endemic threat to the health of orang-utans in rescue and rehabilitation settings or even in the wild (Sanchez et al. 2022).

Many other vector-borne diseases are understudied and may need wider attention. A study on arthropod-virus screening of bovine sera collected in 1979 on Java and Bali Island revealed not only JEV, WNV, BTV, and BEFV but also other Flaviviruses (Apoi, St. Louis encephalitis, Murray Valley encephalitis, Langat, Stratford, Kokobera virus), Alphavirus (Getah virus), Bunyaviruses (Aino, Tinaroo, Peaton, Douglas, Wongal, Ilesha, Batai, Kairi, Wyeomyia, Trubanaman, Kowanyama, Belmont virus), and Orbiviruses (Ibaraki, Mitchell River, Warrego virus) (Miura et al. 1982). The results from the study implied the presence of a much larger number of yet-to-be-discovered arboviruses.

10.10.4 Challenges to Control Efforts and Future Directions for Research and Control in Indonesia

Indonesia is home to a diverse array of plants, animals, and microorganisms, many of which play important roles in the ecosystem and contribute to the overall health and well-being of the country. However, some of these organisms can also pose a threat to human health when they act as disease agents. Vector-borne diseases, which are transmitted by insects or other arthropods, are a major public health concern in Indonesia. Examples of control measures that have thus far been implemented in Indonesia to control vector-borne diseases include the traditional use of insecticides and mosquito nets, and the recently implemented methods to control and prevent VBD transmission in Indonesia include the use of vaccines, e.g., Dengue and Japanese encephalitis vaccines (Kardena et al. 2021; Yosephine 2017; Zeng et al. 2018), and the release of Wolbachia-infected mosquitoes in Yogyakarta (Utarini et al. 2021). Despite many drawbacks, such as resistance development towards insecticides, traditional methods are the backbone of controlling vectors in rural and remote areas.

The large and geographically dispersed population of Indonesia is undoubtedly a challenge to control efforts, given the nationwide distribution of most VBDs (Fig. 10.15). Control measures can be difficult to implement and enforce in areas

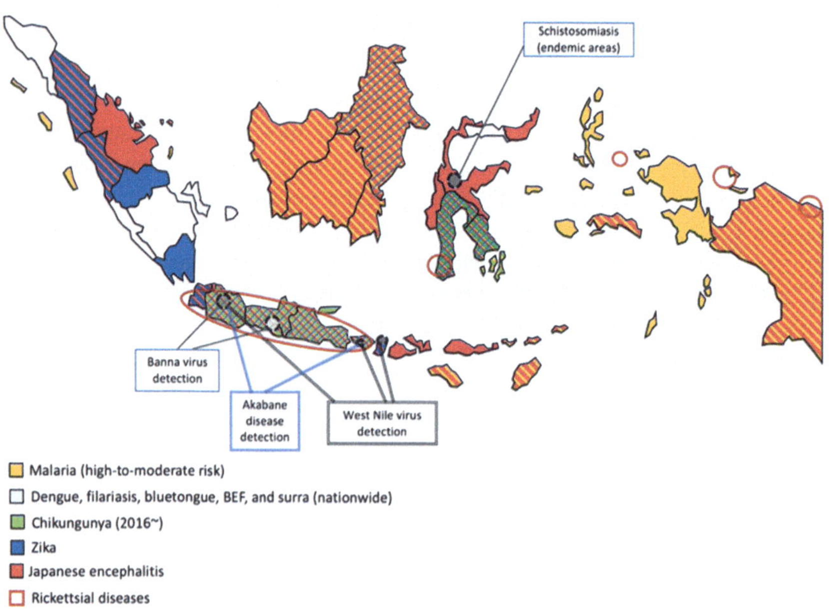

Fig. 10.15 Distribution map of vector-borne diseases in the Indonesian archipelago. Five diseases were considered nationwide, while the rest have sporadic distribution, albeit requiring further and regular investigations

with over 270 million people and 16,000 islands, particularly in rural and remote areas, due to the lack of resources and infrastructure and inadequate data collection systems. To overcome these challenges, various approaches can be employed. One effective method is to improve disease surveillance and data collection systems to identify patterns, trends, and risk factors within the local population. This information can then be used to deploy targeted prevention and control measures. Another key approach is to form partnerships and collaborations between government agencies, nongovernmental organizations (NGOs), and local communities. These collaborations can facilitate the sharing of resources, data, and expertise, which can be limited in remote areas. Involving the community in implementing control measures can also lead to more sustainable and effective solutions in the long term. Additionally, it's feasible to improve the capacity and preparedness of local and regional public health agencies to respond better to disease outbreaks. This can involve providing training and accessible equipment and infrastructure and improving communication strategies between all levels of health and medical facilities to help achieve the intended impacts.

It is also important to note that the community's vulnerability to incoming health threats can be defined by variations in underlying social and economic determinants of health, such as poverty and extreme environmental degradation. Addressing these underlying factors can hopefully reduce the disease burden and improve the overall health of the local population.

10.11 Republic of Türkiye

Yusuf Ozbel

Medical entomology is very important for Türkiye because Türkiye is a country where many vector arthropods and vector-borne diseases are found due to its geographical location, climate, and biodiversity. Various control programs are carried out in Türkiye to combat vectors that threaten public health. These programs are organized by the Ministry of Health and related institutions.

Türkiye is a transcontinental country located mainly on the Anatolian Peninsula in West Asia, with a small portion on the Balkan Peninsula in Southeast Europe. It borders the Black Sea to the north; Georgia to the northeast; Armenia, Azerbaijan, and Iran to the east; Iraq to the southeast; Syria and the Mediterranean Sea to the south; the Aegean Sea to the west; and Greece and Bulgaria to the northwest. Cyprus is off the south coast.

Türkiye is subdivided into 81 provinces for administrative purposes. Each province is divided into districts, for a total of 922 districts. Türkiye is also subdivided into 7 regions and 21 subregions for geographic, demographic, and economic measurements, surveys, and classifications; this does not refer to an administrative division.

Türkiye is a transcontinental country bridging Southeastern Europe and Western Asia. Asian Türkiye, which includes 97% of the country's territory, is separated from European Türkiye by the Bosporus, the Sea of Marmara, and the Dardanelles. European Türkiye comprises only 3% of the country's territory. The country is encircled by seas on three sides: the Aegean Sea to the west, the Black Sea to the north, and the Mediterranean Sea to the south.

The coastal areas of Türkiye bordering the Aegean and Mediterranean Seas have a temperate Mediterranean climate, with hot, dry summers and mild to cool, wet winters. The coastal areas bordering the Black Sea have a temperate oceanic climate with warm, wet summers and cool to cold, wet winters. The Turkish Black Sea coast receives the most precipitation and is the only region of Türkiye that receives high precipitation throughout the year. The coastal areas bordering the Sea of Marmara, which connects the Aegean Sea and the Black Sea, have a transitional climate between a temperate Mediterranean climate and a temperate oceanic climate with warm to hot, moderately dry summers and cool to cold, wet winters.

10.11.1 Education on Medical Entomology in Türkiye

In higher education institutions, since there are no training programs focused only on medical entomology in Türkiye, but it is generally included as a department in more comprehensive undergraduate and graduate programs, the number of specialist medical entomologists in the country is low. Researchers working in universities and some state institutions generally turn to the field of entomology with theoretical and practical individual training at home or abroad after attracting their attention. In the universities where these researchers are located, education and research activities on parasitology and entomology are carried out. Scientific studies in this field are published in national and international journals.

Medical entomology education in Türkiye is generally given in the form of theoretical and some practical courses in the parasitology departments of medical and veterinary faculties. In addition, some biology departments also have entomology courses. However, since medical entomology is a specific specialty, it is usually discussed in more detail in advanced (MSc or Ph.D.) education, but there is no graduate program titled "Medical Entomology."

In addition, the Ministry of Health organizes various online and in person training programs and seminars to protect public health. Within the scope of these programs, medical entomology subjects are also included.

Professional organizations such as the Parasitological Society of Türkiye, Microbiological Society of Türkiye, and Clinical Microbiology Specialty Society organize special courses, seminars, and conferences on medical entomology or dedicate a section within national congresses.

10.11.2 *Vectors and Vector-Borne Diseases*

The main vectors and vector-borne diseases seen in Türkiye are:

10.11.2.1 Mosquito-Borne Diseases (*Anopheles*, *Aedes*, *Culex*)

Some mosquito-borne diseases, such as malaria, have been an important public health problem throughout history. However, in recent years, the spread of this disease has been largely controlled because of malaria control programs. As a result of intensive control programs in Turkey, malaria is a disease coming at the stage of elimination and no local cases have been reported since 2010. Only 250–300 imported malaria cases, mainly falciparum malaria, have been reported each year.

Two outbreaks, one in 2010 and one in 2018, caused by West Nile virus (WNV) occurred with 35 and 26 cases, respectively (Touray et al. 2023). Serological surveillance data suggested human and animal exposure to WNV in some provinces, and an outbreak in 2010 involving individuals as well as sporadic cases since 2009, indicate symptomatic infections (Fig. 10.16) (Ergunay et al. 2014). Turkey has had 2 cases of dengue and 1 case of Chikungunya, both imported, while no Zika case has been detected so far. At present, a total of 64 mosquito species have been detected in Türkiye.

Another mosquito-borne infection for dogs, dirofilariasis cases, is also common among the dog population.

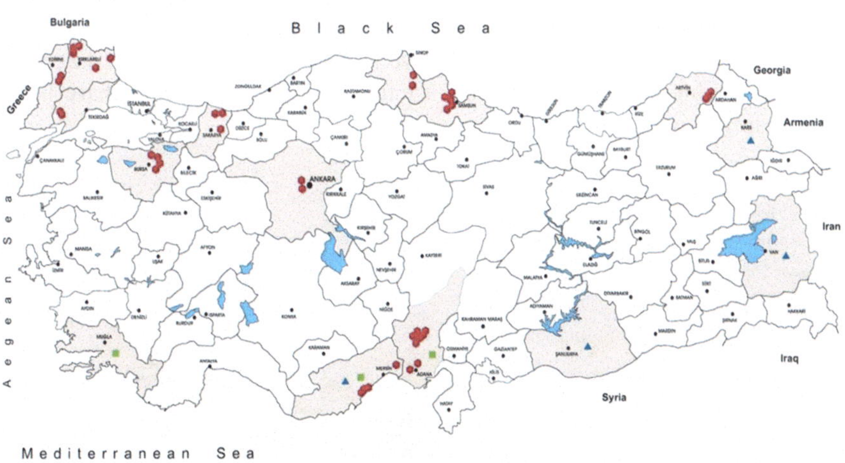

Fig. 10.16 Illustrated map of provinces showing WNV activity. Red circle: entomological screening, blue triangle: serological screening; green square: viral RNA screening

10.11.2.2 Tick-Borne Diseases (Ixodidae)

Ticks, which are vectors of some diseases such as Crimean-Congo hemorrhagic fever (CCHF), are an important public health problem, especially in areas where livestock is concentrated. In addition, the cases of anaplasmosis (*Anaplasma phagocytophilum*), babesiosis (*Babesia microti, B. divergens, B. bovis*), ehrlichiosis (human and canine monocytotropic ehrlichiosis; *Ehrlichia chaffeensis, E. ewingii, E. canis*), Lyme disease (*Borrellia burgdorferi*), Q fever (*Coxiella burnetii*), and tularemia (*Francisella tularensis*) were also seen. A total of 55 tick species have been detected in Türkiye so far (Touray et al. 2023).

10.11.2.3 Sandfly-Borne Diseases (Phlebotomine)

Leishmaniases (kala-azar/visceral leishmaniasis, cutaneous leishmaniasis) caused by four Old World species are endemic diseases reporting around 1000 cases each year. In humans, leishmaniasis is seen in two clinical forms as visceral leishmaniasis (VL, kala-azar) and cutaneous leishmaniasis (CL). *Leishmania tropica, L. infantum, L. major,* and as sporadic *L. donovani* are responsible for CL while *Leishmania infantum* is responsible for VL. Cutaneous leishmaniasis has been reported from 45 provinces in Turkey so far (Fig. 10.17). The total number of cases in 2014 was 1.678 (9.884 in the last 5 years), and a significant portion (almost 50%) of cases are seen in Şanlıurfa Province located in the southeastern region of the country (Ozbel 2013). The morbidity rate is calculated as 3.04 for the year 2017, with 2460 recorded local cases. Only 25–30 visceral leishmaniasis cases have been officially reported per year. Canine leishmaniasis is also very common in Türkiye, with an average 16% seroprevalence rate. There are comprehensive vector studies performed in the country covering almost 45 provinces, and a total of 28 sand fly species have been detected in Türkiye so far.

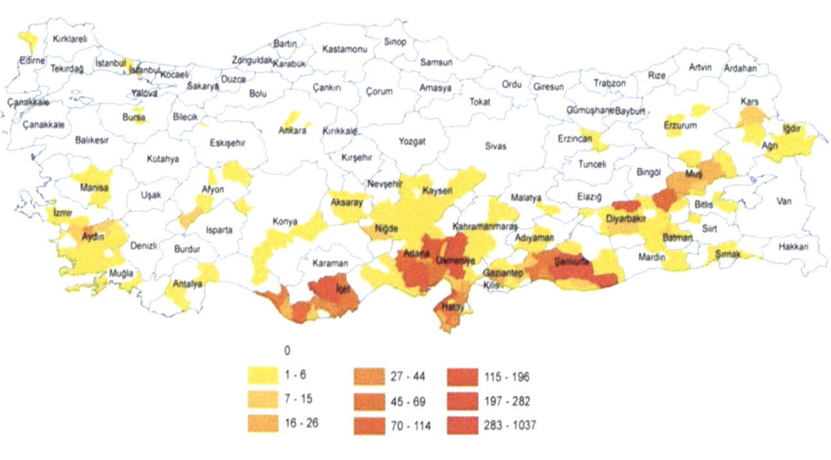

Fig. 10.17 The endemicity of cutaneous leishmaniasis in various provinces of Türkiye

In addition, sandfly fever caused by different phleboviruses including several local and novel ones [Sandfly fever Turkiye virus (SFTV), Adana virus (ADAV), Toros virus (TRV), Zerdali virus (ZERV)] is also present and caused outbreaks in the summer of 2007 in three different areas. But due to notification is not mandatory there are no official reports for understand its epidemiology (Touray et al. 2023).

10.11.2.4 Black Fly-Borne Diseases (Simulidae)

The first human case caused by *Onchocerca caecutienus* was reported in 1976. In 2012, a case of *O. lupi* was reported from Edirne province, located in the European part of Türkiye (Duzlu et al. 2020).

10.11.3 Control Programs at the National Level in Türkiye

In Türkiye, there are vector control programs at the national level carried out for the control and prevention of vector-borne diseases. The Ministry of Health of the Republic of Türkiye is taking the lead in this regard and coordinating the relevant programs.

10.11.3.1 Malaria Control Program

Türkiye has conducted malaria control programs since the mid-twentieth century. Because of the successful application of the programs, malaria cases were greatly reduced and by the late 1960s, malaria was brought under control in Türkiye.

Türkiye was included in the malaria elimination program by the World Health Organization (WHO), and no indigenous cases have been seen since 2010, but import or induced cases have been reported each year (WHO 2023d). Within the scope of this program, activities such as malaria case follow-up, mosquito population control, and public awareness are carried out. Notification of the cases is mandatory in Türkiye.

10.11.3.2 Crimean-Congo Hemorrhagic Fever (CCHF) Control Program

Control programs are carried out at the national level for CCHF disease, which has been seen in Türkiye since the early 2000s. Within the scope of this program, tick population control, health workers, and public guidelines are prepared in risky areas to raise public awareness, and early diagnosis and treatment activities are carried out. The number of cases decreased from 1315 in 2008 to 343 in 2017. Notification of the cases is mandatory in Türkiye.

10.11.3.3 Leishmaniasis Control Program

In Türkiye, cutaneous leishmaniasis is endemic, and visceral leishmaniasis is seen sporadically, and there are control programs for these diseases. These programs include free diagnosis, treatment, and control of sand flies, which are vectors of the disease, and public awareness activities. In Türkiye, leishmaniasis is known as an endemic disease, especially in Southeastern Anatolia and Mediterranean regions. The cutaneous form of leishmaniasis is more common. Notification of the cases is mandatory in Türkiye.

10.11.3.4 Dengue, Zika, and Chikungunya Control Program

The first resident populations of *Aedes albopictus*, the vector of these diseases, *Aedes* mosquitoes, which invaded Türkiye from the northeast and Thrace, were detected in the Eastern Black Sea Region in 2015. After covering the entire Black Sea coast, the increase in rainfall in the spring and summer months of 2022 and 2023 due to climate change facilitated the spread of this species to the south (Akıner et al. 2018). *Aedes aegypti*, the main vector of the Zika virus, was first detected in Türkiye in 2015, but its spread area was limited to the Eastern Black Sea. In addition to the Ministry of Health, the Veterinary and Health units of the provincial and district municipalities are engaged in larval adult struggle with mosquitoes. The spreading of the most invasive species, *Aedes albopictus*, is now continuing throughout Türkiye from north to south, especially following the coastal areas.

These programs are coordinated and carried out by the relevant units of the Ministry of Health. In addition, the active participation of local governments, non-governmental organizations, and the public is required for these programs to be successful.

10.12 Socialist Republic of Viet Nam

Tran Vu Phong and Vu Sinh Nam

Vietnam is a country located at the easternmost tip of the Indochina Peninsula in Southeast Asia, bordering Laos, Cambodia, China, the East Sea, and the Gulf of Thailand. The population in 2020 was 98 million people belonging to 54 ethnic groups, predominantly the Kinh group (86%). The long country is affected by two climate zones: subtropical (northern part) with four seasons and tropical (southern part) with dry and wet seasons. Vietnam is located in the Indomalaya ecoregion. According to the 2005 National State of the Environment Report, Vietnam is among the 25 countries with a high degree of biodiversity, ranked 16th worldwide in terms of biodiversity, and is home to approximately 16% of the world's largest biodiversity species. There are also many types of ecosystems with a high diversity of natural biometrics, especially invertebrate animals. Most of them are good for agriculture, but some are pests for the field and human life. Recently, vector-borne diseases have become a threat to public health and spread with the increasing human population, civilization, rapid transportation, and climate change. In the twenty-first century, when malaria and plague thresholds were reduced by successful prevention and control programs within 30 years, dengue fever became the most important public health problem in Vietnam, followed by Japanese encephalitis (JE) and Zika virus infection. Other unknown pathogens of encephalitis syndromes have also been found in some mosquito species in resident and natural foci (Duong et al. 2021). Recently, the abundance and vectorial capability of the sandfly with leishmania (Nam et al. 2019) and flavivirus have been revealed throughout the country. In this part, we focus on the mosquito vector belonging to three groups, *Anopheles*, *Culex*, and *Aedes,* that exist in different ecosystems in Vietnam. Anopheline mosquitoes (some of these are vectors of malaria parasites) mostly inhabit mountainous areas and some mangrove forests. Some *Culex* mosquitoes have been recognized as the most important vectors of JE virus (JEV) in endemic areas and are dominant in rural areas although their distribution is affected by agricultural activities. The *Aedes* mosquitoes (some of these are the vectors of dengue virus: DENV, chikungunya virus: CHIKV, and Zika virus: ZIKV) prefer urban dwellings due to the abundance of human hosts (Duong et al. 2021).

10.12.1 Aedes-Borne Diseases

10.12.1.1 Dengue Fever

In Vietnam, the dengue epidemic was first reported in 1958 and has now become an endemic disease, with a high number of cases within 40 years (Fig. 10.18). Most dengue cases have been reported in the southern part (tropical area) due to the high abundance of the mosquito vector *Aedes aegypti* (L.) Now when malaria is gradually being eliminated, dengue has become a leading public health problem in Vietnam. Although there was a National Dengue Prevention and Control Project from 1999 to 2020 with many efforts by the public health system and the community, the success was only in reducing the number of cases and reducing the mortality rate from dengue temporarily. The reasons for the increase in the number of cases and the expansion of dengue-affected areas are debated, but some reliable hypotheses are as follows: (1) human population growth also overurbanization increase the favorable conditions for mosquito production, and many sources of food for mosquitoes develop in urban areas lacking infrastructure; (2) increased human activity by modern means of transport globally and across regions has spread the disease-transmitting *Aedes* mosquito, which spreads the disease rapidly everywhere; and (3) climate change with warming manifestation makes the survival conditions of mosquitoes far beyond the old limits, and the density of vector populations also increases dramatically with extreme climate events (WHO 2011).

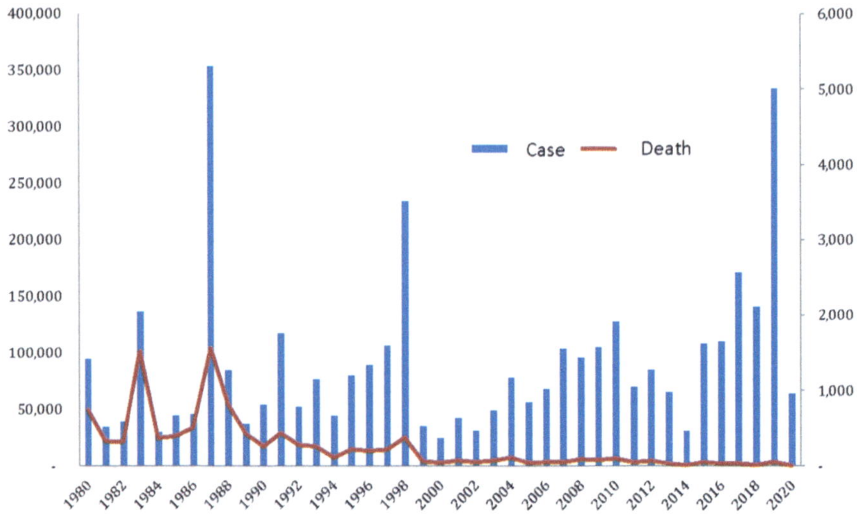

Fig. 10.18 Changes in dengue cases and deaths in Vietnam from 1980 to 2020. (Data from the Vietnam Ministry of Health (MOH) 2021)

Although both mosquito species, *Ae. aegypti* and *Ae. albopictus* (Skuse), are considered dengue vectors and exist throughout the country, *Ae. aegypti* is the main vector of dengue and is found in most epidemic areas. The geographic distribution of each species is opposite from the north to the south. *Aedes albopictus* was dominant in the north (subtropical climate), *Ae. aegypti* was dominant in the south, and their population abundance was much higher than that in the north (Kawada et al. 2009). This is very consistent with the difference in climate in the north, where the winter season is not suitable for *Ae. aegypti* to grow. In contrast, the southern equatorial climate is hot all year round, so this mosquito species is not restricted by temperature but only affected by seasonal rainfall in the year. Therefore, the population density of *Ae. aegypti* is always much higher than that in the north, and this mosquito species is present in both urban and rural areas. That is why the number of dengue cases is mainly recorded in the South (Fig. 10.19).

10.12.1.2 Zika Virus Infection

Accumulating the first 11 months of 2016, Vietnam recorded 136 positive cases of ZIKV in nine provinces and cities. One case of microcephaly was recorded, most likely due to ZIKV. None of these patients had a history of returning from the epidemic area. There were 20 cases reported to be pregnant among the female cases of ZIKV (accounting for 22%). Cases of ZIKV infection continued to be recorded in the following years but in much lower numbers. The total number of cases recorded by the end of 2019 was 287.

Aedes aegypti mosquito was confirmed as a vector of Zika virus in most of the epidemics in Vietnam (Duong et al. 2021; WHO 2016).

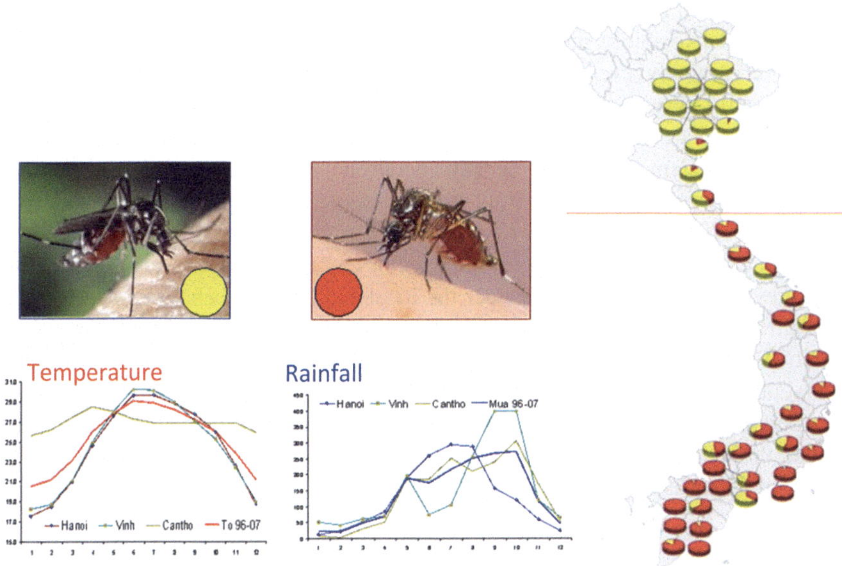

Fig. 10.19 The geographic distribution of *Aedes aegypti* and *Aedes albopictus* in Vietnam. (Data from Kawada et al. 2009)

10.12.1.3 Chikungunya Fever

In Vietnam, the CHIKV epidemic has been detected since 1975, but the sentinel surveillance system only recorded a few sporadic cases of the disease, which did not break out into an epidemic. A study to determine the presence of the virus in mosquitoes and suspected patients in 5 provinces bordering Laos and Cambodia did not detect any CHIKV by PCR in the period 2012–2014 (Phong et al. 2015). Recently, the surveillance system recorded six patients and one female *Ae. aegypti* mosquito with CHIKV in an outbreak in the border area bordering Cambodia in An Giang Province in 2020 (Quang et al. 2022). CHIKV is primarily transmitted by the *Aedes* mosquitoes, similar to the transmission of DENV and ZIKV.

10.12.2 Prevention and Control for the Aedes Mosquitoes

Since *Ae. aegypti* can transmit DENV, ZIKV, and CHIKV, we would like to take it as a common target for vector control of these diseases. There are no vaccines or specific treatments for all three viruses and their interactions with the host individually. Therefore, vector control is the most effective measure for the prevention and control of dengue fever, Zika virus infection, and chikungunya fever.

Measures to control disease vectors are expected to meet three criteria: effectiveness, economy, and safety for humans, other organisms, and the environment. However, mosquitoes and other disease-carrying insects are also common groups of organisms living in ecosystems with humans and other species. Because of their high reproductive ability, mosquito species that live near humans can recover populations very quickly, even when killed up to 90%. Therefore, it is very difficult to have a control measure with high killing efficiency and long-term effects that only affects the target mosquito population without negatively affecting other objects in the same ecosystem. Therefore, the battle between humans and disease-transmitting insects is always tough, and there are many unwanted losses to species and habitats (Clark et al. 2013; WHO 2016).

When choosing the most appropriate vector control measure or combination of measures, it is imperative to consider the local ecology and behavior of the target mosquito species, the resources available to implement it, the cultural context in which the interventions are implemented, the feasibility of applying appropriate interventions, and the extent to which they should be covered. On a national scale, the socialization of vector control has been applied by many developed countries around the world to mobilize the resources of the entire society, not just the health sector or the government (Phong et al. 2022).

10.12.2.1 Copepods

Vietnam has conducted many studies on the application of *Mesocyclops*, the carnivorous crustacean copepod species, as the biological agent for *Aedes* mosquito control programs, including thorough and careful studies in the laboratory and the field. From 1989, copepods were applied in the field on a small scale and then expanded to localities, regions, and areas throughout the country (Nam et al. 2005).

These projects have been implemented in 12 provinces with high rates of dengue patients. The purpose of choosing this implementation site is both to solve the local dengue problem in particular and to build appropriate models for dengue prevention in general. In addition, we assess the sustainability of the active control program after intensive support for 3 years by small loan investment to maintain the initial activities in the community (Nam et al. 2009).

10.12.2.2 *Wolbachia*

Wolbachia is a bacterium that lives in insect cells and is passed from one generation to the next through eggs. In nature, up to 60% of insect species that carry *Wolbachia* bacteria, such as butterflies, dragonflies, fruit flies, and some mosquitoes, live near humans. However, this bacterium does not exist in the mosquito *Ae. aegypti*, the main vector of dengue hemorrhagic fever (DHF) caused by DENV. *Wolbachia pipientis* Hertig was first found on the ovaries and testes of the mosquito, *Culex pipiens* L., in 1920. The first studies on *Wolbachia* demonstrated that it is not a mammalian pathogen; it exists naturally in insects and is harmless to humans and pets. Subsequent studies proved that this is a very common bacterium in natural insects, estimated to be up to 60% of insect species in nature (between 2 and 5 million different species of insects on earth) carrying *Wolbachia* bacteria. This is a very contagious bacterium that spreads easily in the insect population (Hoffmann et al. 2014).

In Vietnam, research on *Wolbachia* has been carried out since 2006 with the support of international experts. The National Institute of Hygiene and Epidemiology team successfully transplanted and bred local midges carrying *Wolbachia* in the laboratory and used them for mosquito release in the field on Tri Nguyen Island in 2013 and 2014. This was followed by a series of studies conducted to apply them in phases in Nha Trang City. After the release of mosquitoes, the population of mosquitoes carrying *Wolbachia* was maintained quite well, and the number of dengue cases in the trial commune decreased significantly compared to other communes/wards of Nha Trang City in 2018–2019. The measure was supported and responded to by the people and the government (Duong et al. 2021).

10.12.2.3 Eco Health

The strategy to contribute to improved dengue prevention uses transdisciplinary analysis to better understand the biology, ecology, and social factors associated with dengue, thereby developing and evaluating preventive measures. Ecosystem- and community-center interdisciplinary management approach to reduce dengue vector breeding sites. Cat Ba Island was selected as one of the six well-known tourist destinations in six Southeast Asian countries to participate in this study.

The pretrial survey showed that tourism development in Cat Ba (including land use change, occupational restructuring, increase in tourist facilities and services, number of tourists, temperature, and rainfall conditions) has increased the risk of dengue. There is a high correlation between the number of tourists ($R = 0.63$, $p < 0.05$), the total number of tourist establishments ($R = 0.21$, $p < 0.05$), and the number of dengue cases by month in Cat Ba (Fig. 10.20). There are two species of dengue vectors, *Ae. aegypti* and *Ae. albopictus*, with high density. After 2 years of implementation, the ecological health approach through multisectoral, community-based coordination was highly effective in the active prevention of dengue in Cat Ba. 97.8% to 100% reduction in *Ae. aegypti* compared with before the intervention. People's knowledge, attitudes, and behaviors increased compared to before the intervention ($p < 0.05$). Methods of prevention of dengue and ecological health are accepted by the people and the government. No dengue cases were recorded after 2 years of applying prevention methods in the intervention area (Duong et al. 2021).

10.12.2.4 Community Mobilization

The active dengue prevention model combines two-pronged approaches—top-down and bottom-up—including mobilizing the real participation of the community, local authorities, schools, and health staff (technical advice). In this model, the health sector needs to mobilize the participation of the community as well as the support

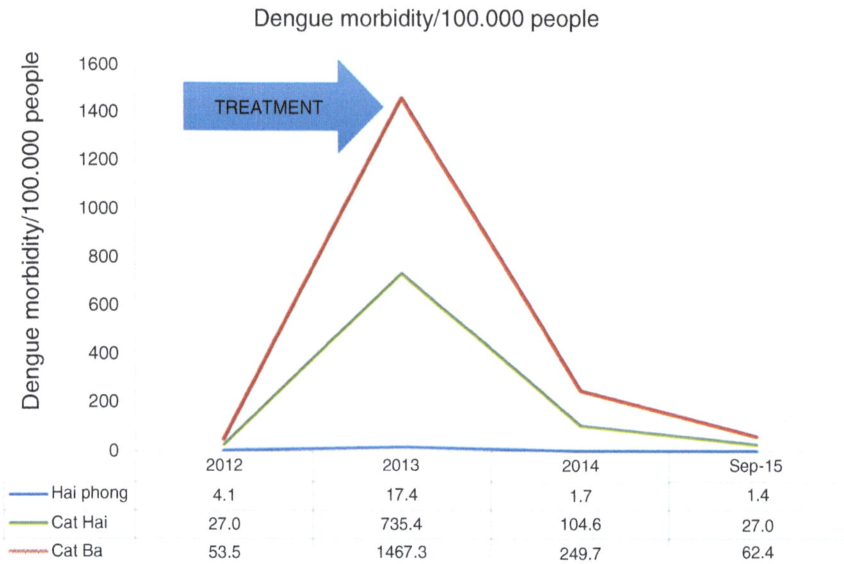

Fig. 10.20 Dengue morbidity/100.000 of Cat Ba town, Cat Hai district, and Hai Phong Provinces in 2013–2015

and direct participation of the local government, relevant sectors, and the school. The establishment of a steering committee at the grassroots level is very important to ensure real implementation, management, and monitoring of activities and, at the same time, promptly adjust and change plans to suit specific circumstances. Of the locality, the three main components are government, health care, and schools (Nam et al. 2005).

10.12.2.5 Socialization

The concept of "socialization" has been used in all socioeconomic activities, including public health issues and disease prevention. Today, as Vietnam's economy has shifted to a market mechanism and the country has emerged from poverty, organizations and individuals are forced to shift from subsidies to economic accounting, including public health programs. Mobilizing community participation is even more important in health programs that require the use of internal resources. In addition, concerns about livelihood in daily life also affect each person, family, and community. Although the mosquitoes that transmit dengue are generated in people's homes, few people in areas at risk areas of this disease (crowded residential areas, poor infrastructure) have the ability to control insects at home. Even if there are a few houses that are able to do so, the risk from surrounding houses is still high. Therefore, mobilizing financial resources to use services from professional pest control companies is a trend in residential areas with high economic potential and population,

Fig. 10.21 Establishment of the Vietnam Pest Management Association (VPMA) in 2019

such as urban areas. These areas are at low risk of dengue because of good infrastructure, few water containers that are breeding grounds for mosquitoes, and especially because they use pest control services regularly in public areas and in each household. Moreover, some families in these areas used proactive methods to protect themselves avoid mosquitoes frequently (Phong et al. 2022).

Developed countries around the world are all operating under this mechanism, and the pest control industry is a potential field of activity. Currently, there are similar professional associations in the world such as the United States (National Pest Management Association), Singapore (Singapore Pest Management Association), the United Kingdom (British Pest Control Association), Australia (Australian Pest Control Association), Indonesia (Indonesia Pest Control Association), Thailand (Thailand Pest Control Association), and Japan (Japan Pest Control Association). In 2019, Vietnam established the Vietnam Pest Management Association (VPMA) with the operational criteria of capacity building; experience sharing for colleagues; unitedly striving to contribute to the achievement of the goal of reducing risk factors affecting public health; and reasonable, effective, and safe pest control for humans and the environment (Fig. 10.21). The long-term outcome contributes to physical and mental development, increases longevity, and improves quality of life (Phong et al. 2022).

10.12.3 Anopheles-Borne Diseases

10.12.3.1 Malaria

In Vietnam, all four types of human malaria parasites (mostly *Plasmodium falciparum* and *P. vivax*) have been circulated in endemic areas for a long time. To reduce malaria morbidity and mortality, Vietnam's National Malaria Control Program has implemented many appropriate strategies. As a result, from more than 1 million cases of malaria with 4646 deaths in 1991, by 2018, there were only 6870 cases and only 1 death. Many provinces and cities no longer transmit malaria. However, technical challenges such as *Anopheles* spp. resistant to insecticides, the malaria parasite *Plasmodium* spp. effective insecticide resistance in use, malaria in the group of people who go to the forest, sleep in the fields, and frequent exchanges across the Vietnam-Laos-Cambodia border are very complicated.

The major malaria vectors in Vietnam are *An. minimus* Theobald, *An. dirus* Peyton & Harrison, and *An. epiroticus* Linton & Harbach. For example, the population density of *An. minimus* increases at the beginning and end of the rainy season. Thus, the incidence of malaria cases due to *An. minimus* bites shows two peaks, at the beginning and end of the rainy season. In areas with *An. dirus* malaria occurs during the rainy season and in areas with *An. epiroticus* occurs throughout the year. At present, the transmission season may change due to climate change and seasonal human activities by activities that increase or decrease exposure to malaria vector mosquitoes (MOH 2021).

10.12.3.2 Malaria Prevention and Control

Vietnam has successfully implemented the Malaria Prevention Program for the past 30 years with the help of the WHO and international organizations. To date, the endemic area has been reduced, but the number of malaria cases and deaths has increased by more than 90% compared to the 1980s. However, malaria surveillance and control activities are still continuously carried out by the preventive health system from the central level to the provinces and districts to the local health system. Proactive control programs in endemic areas using a combination of mosquito control, early detection, and thorough treatment for communities in high-risk areas.

10.12.4 Culex-Borne Diseases

10.12.4.1 Japanese Encephalitis (JE)

Although the JE vaccine has been available and applied in vaccination programs for more than 20 years, the risk of Japanese encephalitis infection is high in some remote areas (Lai Châu, Sơn La, Điện Biên Provinces) due to low vaccination coverage of children and virus circulation in natural and domestic animals (Fig. 10.22) (Quoc et al. 2018).

In Vietnam, three mosquito species have been considered vectors of JEV: *Culex tritaeniorhynchus* Giles, *Cx. vishnui* Theobald and *Cx. gelidus* Theobald. JEV could be preserved in birds, swine, and mosquitoes for a long time as a natural reservoir. The *Culex* mosquitoes mostly bite domestic animals such as buffalo, swine, chickens, and dogs but sometimes also humans. Thus, JEV has been transmitted from the reservoir to humans by the vector mosquitoes. Climate and rice harvesting season have a strong influence on *Culex* populations.

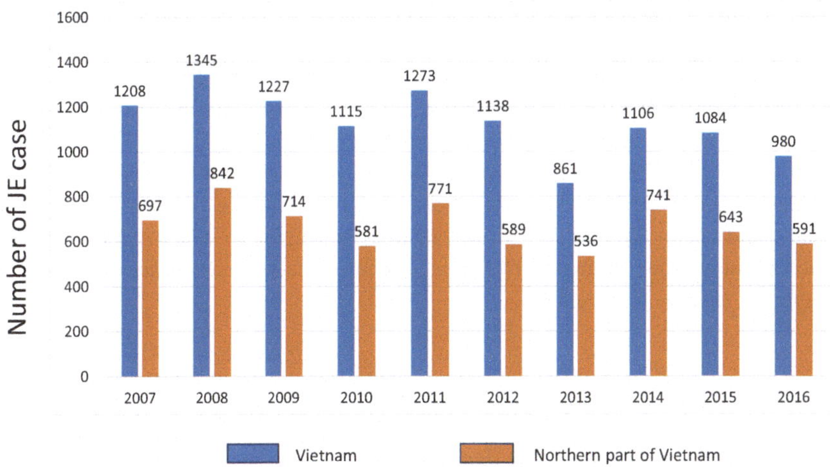

Fig. 10.22 Comparison of the number of encephalitis cases in Vietnam national and northern region in 2007–2016. (Created with data from WHO 2016)

10.12.4.2 JE Prevention and Control

JE vaccines have been available and applied in national immunization programs in Vietnam for more than 20 years in Vietnam. However, *Culex* mosquitoes are still common in all localities, especially in rural areas with wet rice cultivation and domestic livestock and poultry farming. These mosquito species have wild behavior, high dispersal and adaptability, and the ability to develop large populations by sucking the blood of many species of animals and poultry. Therefore, it is difficult to apply measures to control *Culex* mosquitoes, so the main preventive measures are to reduce the risk of mosquito bites. In modern urban areas and industrial zones, the periodic proactive control of mosquitoes and disease-carrying insects is mostly done through services provided by private companies. This minimizes the risk of virus transmission and the spread of vector-borne diseases. This is also the trend of socializing disease vector control, which has been implemented in many countries around the world and in Southeast Asia.

10.12.5 Other Vector-Borne Diseases

10.12.5.1 Plague

The plague has been a horror throughout human history and is known as the Black Death, an infectious disease that is internationally quarantined at airports and global border gates. The disease is caused by the bacterium *Yersinia pestis*, which is transmitted by flea bites and is spread worldwide by the rat and rodent population that is the host of the flea.

In Vietnam, the plague was first recorded in 1898 in Nha Trang by ships and boats from Hong Kong. Period of outbreak, widespread circulation 1961–1990. The period of narrowing and circulation of some persistent drives from 1991 to 2002. Controlled period 2003–present. Since March 2003, no human cases have been recorded. Since April 2004, it has not been possible to isolate *Yersinia pestis* from the host and vector. To date, 34 species of fleas belonging to 18 genera have been identified in various animals. The most common vector is the flea, especially *Xenopsylla cheopis* (Rothschild).

10.12.5.2 Plague Prevention and Control

China borders Vietnam by more than 1300 km, and there are still outbreaks of plague in wild animals in nature. Therefore, this animal-derived disease can follow rats and fleas into Vietnam through extensive trade and travel. In addition, the circulation of rats and fleas in Vietnam is quite common, so when pathogens enter the community, outbreaks can occur.

Vietnam is still maintaining an active plague surveillance system from the central level to the system of the International Health Quarantine Centers of provinces bordering China, Laos, and Cambodia. The International Health Quarantine Centers regularly inspect and closely supervise people, vehicles, and goods imported into our country, especially for means of transport originating from epidemic-endemic areas. Lymph nodes; improve environmental sanitation at the border gate (such as removing waste, garbage, and rodent food supplies); and kill rats and fleas on ships, airplanes, airports, and harbors according to regulations.

10.13 Taiwan

Hwa-Jen TengHan-Hsuan Chung, Pai-Shan Chiang and Shiu-Ling Chen

Taiwan is an island country in East Asia that is surrounded by Japan in the northeast and the Philippines in the south. The climates of this country are subtropical and tropical, which are suitable for vectors to breed. In Taiwan, vector-borne diseases under surveillance by the Taiwan CDC are Japanese encephalitis (JE), dengue fever, chikungunya fever, Zika virus infection, malaria, West Nile fever, yellow fever, severe fever with thrombocytopenia syndrome (SFTS), scrub typhus, endemic typhus, epidemic typhus, Lyme diseases, and hantavirus infection, which are transmitted by mosquitoes, ticks, mites, fleas, and rats. Among these diseases, Japanese encephalitis, scrub typhus, murine typhus, and hantavirus infection are known as indigenous diseases, while others are related to traveling diseases with infections or the introduction of pathogens from other countries.

Due to global warming and the rapid movement of the human population, the introduction of vector-borne pathogens through travelers and migratory birds is very likely in Taiwan. Therefore, reducing local vector population densities and surveillance of the pathogens in vectors have become very important. Here, we described the occurrence of the indigenous diseases and their vectors in Taiwan by vector types (i.e. mosquito-borne diseases, chigger mite-borne diseases, fleas-borne diseases, and rodent-borne diseases). In addition, the newly found tick-borne disease, SFTS, was also included.

10.13.1 Mosquito-Borne Diseases

In Taiwan, 132 mosquito species have been identified in 17 genera (Lien 2004). Among them, *Aedes aegypti* (L.) and *Ae. albopictus* (Skuse) are the main vectors of dengue virus (DENV), chikungunya virus, Zika virus, and yellow fever virus (Chen et al. 2019). *Culex tritaeniorhynchus* Giles, *Cx. annulus* Theobald, and *Cx. fuscocephala* Theobald transmit the Japanese encephalitis virus (JEV) every year. The vector of malaria is *Anopheles minimus* Theobald, and the secondary vector is *An. sinensis* Wiedemann. DENVs are introduced to Taiwan by travelers in early summer (Fig. 10.23a) and then transmitted to the local vector population. When the environment is favorable, a local outbreak occurs. Generally, this outbreak has ended in winter, except for a few years. In these years (1987–1988 (DENV-1), 2001–2002 (DENV-2), 2009–2010 (DENV-3), and 2014–2015 (DENV-1)), the viruses overwintered and caused a large outbreak in the following year (Fig. 10.23b). *Aedes aegypti* is distributed in southern Taiwan year-round, with the population peaking in summer (Wu et al. 2013b). In addition, this species is also found in Taitung City in eastern Taiwan and Penghu Island. *Ae. albopictus* is found on the island at sea level below 1760 m (Lin et al. 2014). JE is an indigenous disease in which viruses are active in May and June (Lin et al. 2010). Therefore, the peak number of human cases was found between June and July (Fig. 10.24), as was the mosquito population. In 1968, a JE vaccine with three doses was launched for children. Therefore, only sporadic cases were found each year between 16 and 37. For malaria, in 2001 and 2003, 1 and 2 introduced cases were found in Taitung Counties, respectively. Since 2004, only imported malaria cases were found between 2 and 26. *Anopheles minimus* used to be found island-wise (DOH 1991) and is now distributed only in four counties in southern and eastern Taiwan.

Three long-term surveys of mosquitoes were conducted in Taiwan to provide information for mosquito control strategies. One surveillance started in 2004 and was conducted by quarantine personnel, Taiwan CDC, using ovitraps, sentinel traps, or light traps in airports and harbors to identify the invading mosquito species every 3 months. For example, *Cx. gelidus* Theobald, a vector of JEV in Asia, was detected occasionally. The invading species *Cx. pipiens* form *molestus* Forskal was first documented in 1996 (Lien et al. 1996) and was found at all sites in this surveillance. The second surveillance was launched in 2004 in response to the introduction of

(a)

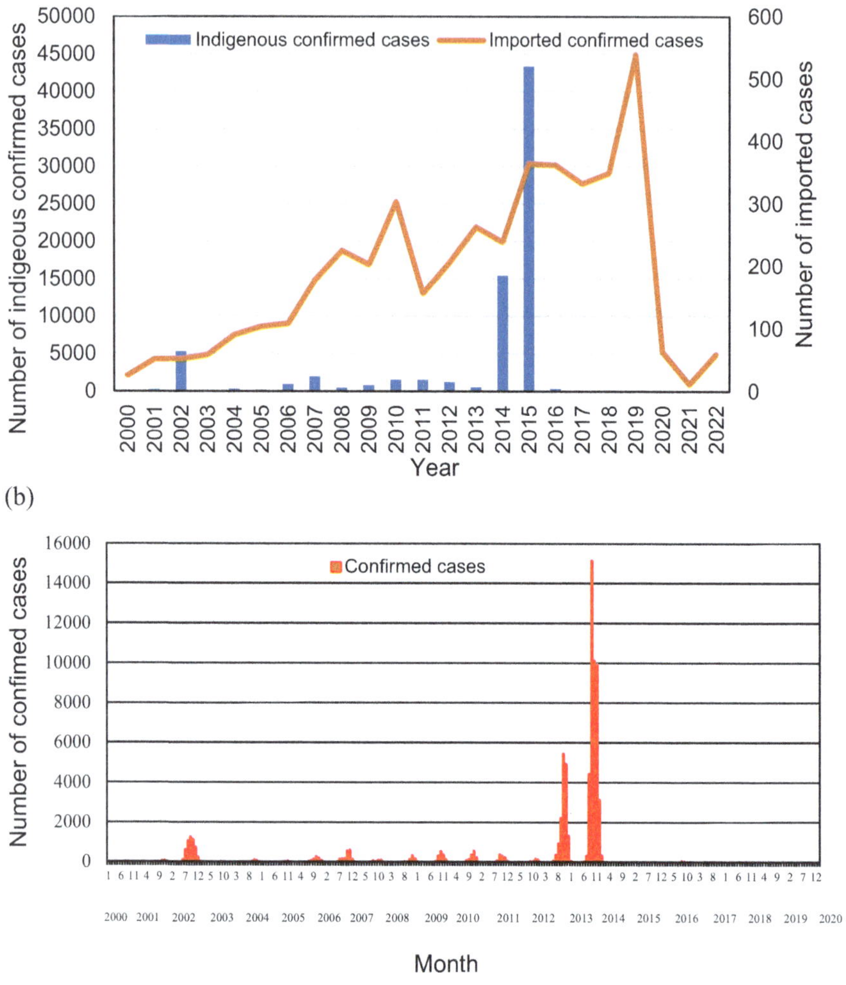

Fig. 10.23 The annual number of indigenous and imported confirmed cases (**a**) and the monthly number of dengue confirmed cases (**b**) between 2000 and 2022. (Data downloaded from the Taiwan CDC website)

malaria cases. This surveillance is conducted by the local health bureau using light traps to monitor the distribution of *An. minimus* as well as the nocturnal mosquito species (e.g., JE vectors). The third surveillance was launched in 2003 in response to the large 2002 outbreak. This surveillance, also conducted by the local health bureau, monitors the *Aedes* mosquitoes with ovitraps and household surveys of containers for breeding *Aedes* immatures and adults with sweeping nets.

The application of insecticides for vector control is necessary, especially during an *Aedes*-borne viral disease outbreak. In Taiwan, insecticide insensitivity has been

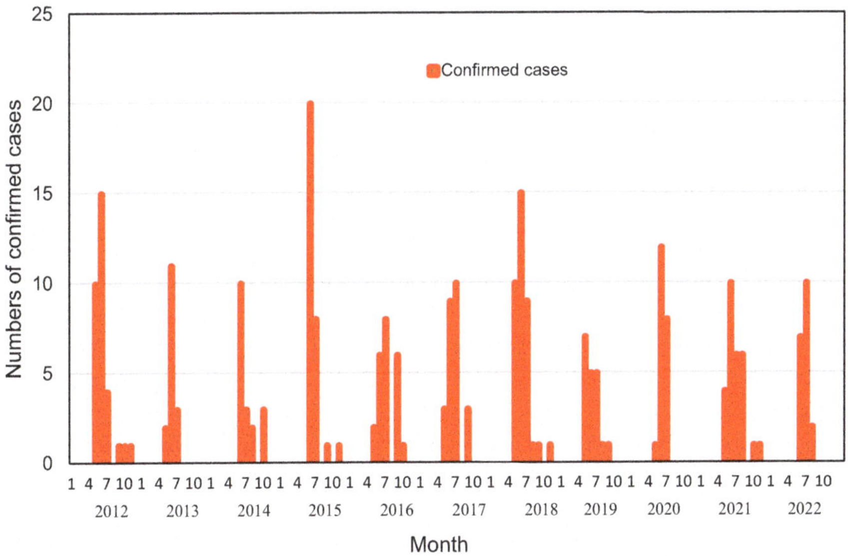

Fig. 10.24 The monthly number of JE-confirmed cases between 2012 and 2022. (Data downloaded from the Taiwan CDC website)

observed in the primary dengue vector *Ae. aegypti* (Lin et al. 2012; Wu et al. 2013a). In 2011, bioassays to evaluate the effectiveness of commercial insecticides with different active ingredients and formulations were launched by the Taiwan CDC and the National Health Research Institute. On the other hand, the mechanism of resistance in *Ae. aegypti* was also discovered. Both receptor insensitivity and overproduction of detoxification enzymes are involved (Wu et al. 2013a). In the haplotype with S989P and V1016G, mutations of the voltage-gated sodium channel (VGSC) were associated with resistance (Chung et al. 2019). Therefore, routine monitoring of VGSC mutations in *Ae. aegypti* was implemented by the Taiwan CDC in 2016 to detect insecticide resistance early. Information on insecticide efficacy and insecticide resistance was provided to the local government for adjusting the vector control strategy. More recently, the resistance-related four-locus VGSC genotypes in *Ae. aegypti* have also been proposed for routine resistance monitoring (Chung et al. 2022). In contrast, insecticide resistance was not common in *Ae. albopictus* in Taiwan (Lin et al. 2012).

A new mosquito biological control strategy, *Wolbachia*-mediated population suppression, has also been evaluated in Taiwan. *Wolbachia* is a maternally inherited bacterial endosymbiont exit in many arthropod species. The cytoplasmic incompatibility (CI) and pathogen blocking (PB) properties of *Wolbachia* are potential tools for the control of vector-borne pathogens. A *Wolbachia*-transinfected *Ae. aegypti*, *w*AlbB-Tw, has been established to effectively reduce the density of *Ae. aegypti*. Now, lab-scale characterization and semi-field evaluation of *w*AlbB-Tw revealed significant potential for controlling *Ae. aegypti* population in Taiwan (Liu et al. 2022).

10.13.2 Chigger Mite (Tsutsugamushi)-Borne Diseases

Scrub typhus is the most common endemic mite-borne disease in Taiwan. This disease is an acute febrile illness caused by the obligate intracellular bacterium *Orientia tsutsugamushi* (also known as *Rickettsia tsutsugamushi*) contracted from the bite of an infected larval stage trombiculid mite (chigger). *Orientia tsutsugamushi* is preserved in reservoirs (e.g., rodents, mammals, and birds) and transmitted to chiggers when they bite (Elliott et al. 2019). The infectious chiggers were transovarially transferred to their eggs, and then, bacteria remained in chiggers, larvae, and adults. Almost all confirmed cases are indigenous, and this disease occurs year-round with one or two peaks. The increased case number usually starts in April–May and reaches its highest value in June–July, and the second peak may be found in September–October (Fig. 10.25) (Chen et al. 2022). In Taiwan, most of the cases occurred on offshore islands and in eastern Taiwan, including Penghu (archipelago), Kinmen (island), Taitung, and Hualien Counties (Fig. 10.26a). This may be related to the distribution of its vectors and reservoirs in Taiwan. *Rattus* species, especially *R. losea* (Swinhoe), is an important reservoir that sustains the transmission of *O. tsutsugamushi* with competent vectors, and *Leptotrombidium deliense* Walch is a pivotal vector for summer scrub typhus in Taiwan. In addition, *L. scutellare* Nagayo et al. is a principal vector for the transmission of winter-type scrub typhus found in Taiwanese offshore islands (Kinmen and Lienchiang counties) (Kuo et al. 2015). In

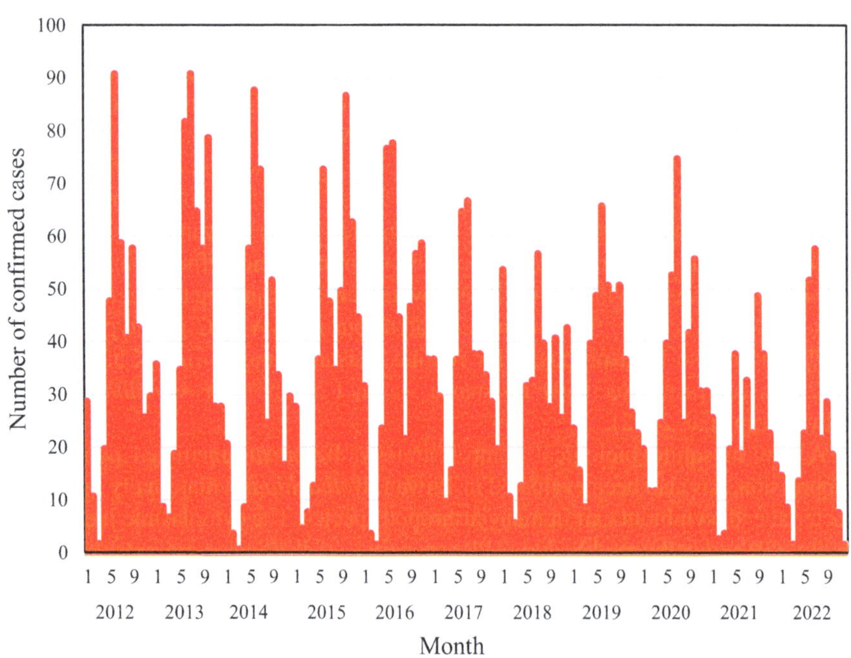

Fig. 10.25 The monthly number of scrub typhus confirmed cases between 2012 and 2022. (Data downloaded from the website of the Taiwan CDC)

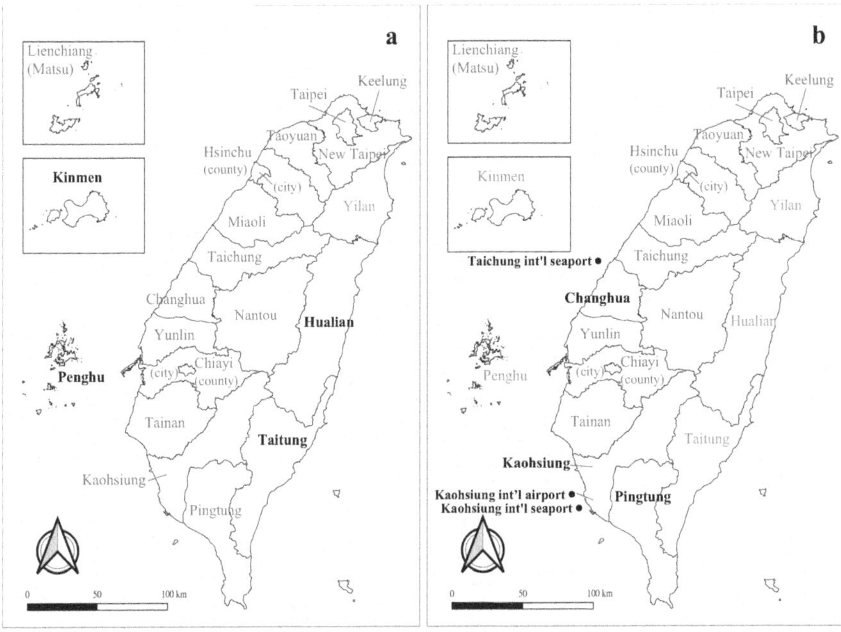

Fig. 10.26 Administrative districts with the highest number of scrub typhus cases (**a**) and murine typhus cases and ports with high seropositive rates for anti-*R. typhi* responses in small mammals (**b**) in Taiwan. The name of each administrative district is labeled in gray. The district with the most cases and the port with a high seropositivity rate of anti-*R. typhi* responses in small mammals are labeled in black and bold font

addition, according to the long-term surveillance of *O. tsutsugamushi* in Taiwan, the genotype of the pathogen and the species of vector showed great diversity in genetic traits (Chen et al. 2022). Worldwide, most cases of scrub typhus occur in rural areas of Southeast Asia, Indonesia, China, Japan, India, and northern Australia. Taiwan is located in the center of the Tsutsugamushi triangle, harboring abundant rodents and migratory birds from the East Asia/Australasia Flyway that might promote host diversity and the spectrum and expansion of dominant vectors (Kim et al. 2017).

10.13.3 Flea-Borne Diseases

Murine typhus (also known as endemic typhus fever) is a flea-borne zoonotic disease that is caused by the bacterium *Rickettsia typhi*. The vector species of this disease is the rat flea, *Xenopsylla cheopis,* or the cat flea, *Ctenocephalides felis*. The reservoirs of *R. typhi* include rats (especially *R. rattus* and *R. norvegicus*) and some other small mammals. In nature, the maintenance cycle was among rat-flea-rat. In Taiwan, indigenous murine typhus human cases have been continuously detected since 2007. The annual number of indigenous infections ranged from 13 to 44 cases. Most cases were detected in May, June, and July (late spring/early summer), and the other months still had a few confirmed cases. According to the results of small mammal surveillance conducted at 25 international airports and ports from 2007 to 2019 in Taiwan, *Rattus norvegicus, Suncus murinus,* and *R. losea* were the major abundant rodents in ports. Overall anti-*R. typhi* seropositive rate in small mammals was 8.2%. Graphically, the ports detected with the highest seropositive rates were Kaohsiung International Seaport (28.3%), Taichung International Seaport (20.2%), and Kaohsiung International Airport (6.2%), while all other international ports revealed overall *R. typhi* seropositivity rates between 0% and <5%. The ports with the highest seropositivity are inside/near the areas with the greatest number of indigenous murine typhus (Fig. 10.26b). This observation in Taiwan is consistent with the international situation in which seaports have been regarded as the primary focus of murine typhus transmission. In addition, human incidence rates were positively correlated with the 1–2 months prior seropositive rates in small mammals (Chiang et al. 2022). Surveillance in small mammals may be useful for early reminders for physicians of sporadic murine typhus cases based on the delayed correlation observed.

10.13.4 Tick-Borne Diseases

Severe fever with thrombocytopenia syndrome (SFTS) is a tick-borne viral disease that was reported for the first time in 2019 on a hill in Taipei City, Taiwan. SFTS was designated as a notifiable infectious disease in 2020. In the following period, an infected patient was not detected in Taiwan until another case, who lived in Lanyu Township (Orchid Island, in the Pacific Ocean to the east of Taiwan), was reported in 2022. Overall, the disease occurred domestically but sporadically in Taiwan. To evaluate the risk to the environment, the Taiwan CDC launched a series of environmental investigations in the nearby surroundings of these confirmed cases. Nevertheless, by the end of 2023, the SFTS virus (SFTSV) was not detected in all ticks collected annually. The genetic analysis shows that the SFTSV detected from the first case in Taiwan was not closely related to Chinese or Korean strains but to Japanese strains collected in Kyushu, Japan, in 2016 (Peng et al. 2020; Sato et al. 2021; Yoshikawa et al. 2014). The ticks collected in the nearby surroundings of the first case were all *Haemaphysalis hystricis* Supino, which has also been observed with SFTSV in Kyushu, Japan (Sato et al. 2021). These observations in Taiwan on both molecular virology and vectors suggest an exchange of ticks or tick-borne pathogens between different regions of East Asia.

10.13.5 Rodent-Borne Diseases

Most cases of Hantavirus syndrome were found in spring in the Taipei metropolitan area and the rural areas of Kaohsiung and Pingtung. However, the number of hantavirus cases per year was limited to 0–4 between 2001 and 2019 from the web database of Taiwan CDC (Hsieh et al. 2020). Occasional hantavirus infections have also been detected in the local rat population.

References

Abdullah SR, Omar B, Salleh AFM, Marwi MA, Othman H, Wahid SA (2016) Fly colonization and carcass decomposition in a high-rise building in Malaysia. Int Med J 23(1):94–99

Abdullah SR, Omar B, Salleh AFM, Othman H, Wahid SA (2017) Comparative fly species composition on indoor and outdoor forensic cases in Malaysia. J Forensic Leg Med 45:41–46

Abdullah SR, Shafie SS, Wahid SA (2022) Forensically important fly larvae on floating corpses in Malaysia: three case reports. Egypt J Forensic Sci 12:18

AbuBakar S, Sam IC, Wong PF, MatRahim N, Hooi PS, Roslan N (2007) Reemergence of endemic Chikungunya, Malaysia. Emerg Infect Dis 13(1):147–149

Ahmad Firdaus MS, Marwi MA, Syamsa RA, Zuha RM, Ikhwan Z, Omar B (2010) Morphological descriptions of second and third instar larvae of *Hypopygiopsis violacea* Macquart (Diptera: Calliphoridae), a forensically important fly in Malaysia. Trop Biomed 27(1):134–137

Ahmad NW, Lee HL, Chen CD, Heo CC, Abdullah AG, Wan-Norjuliana WM, Chew WK, Jeffery J, Rosli H, Sofian-Azirun M (2011) Comparative insect fauna succession on indoor and outdoor monkey carrions in a semi-forested area in Malaysia. Asian Pac J Trop Biomed 1(2):S232–S238

Ahmad R, Suzilah I, Wan Najdah WMA, Topek O, Mustafakamal I, Lee HL (2018) Factors determining dengue outbreak in Malaysia. PLoS One 13(2):e0193326

Ahmad NA, Endersby-Harshman NM, Mohd Mazni NR, Mohd Zabari NZA, Amran SNS, Ridhuan Ghazali MK, Abdul Karim MA, Cheong YL, Sinkins SP, Ahmad NW, Hoffmann AA (2020) Characterization of sodium channel mutations in the dengue vector mosquitoes *Aedes aegypti* and *Aedes albopictus* within the context of ongoing *Wolbachia* releases in Kuala Lumpur, Malaysia. Insects 11(8):529

Aisyah DN, Kozlakidis Z, Diva H, Trimizi SN, Sianipar LR, Wijayanti E, Avicena AM, Adisasmito W (2022) The spatial-temporal distribution of chronic lymphatic filariasis in Indonesia: an 18-year registry-based analysis. Microbiol Res 13(4):681–690

Akıner MM, Demirci B, Bedir H, Öztürk M, Demirtaş R, Doğan AF, Gökdemir A, Topluoğlu S, Altuğ Ü, Kurtcebe ZÖ, Irmak H (2018) Surveillance and control of invasive *Aedes* species in the Eastern Black Sea area of Türkiye. Turk Bull Hyg Exp Biol 75(3):225–238

Alawieh A, Musharrafieh U, Jaber A, Berry A, Ghosn N, Bizri AR (2014) Revisiting leishmaniasis in the time of war: the Syrian conflict and the Lebanese outbreak. Int J Infect Dis 29:115–119

Algalil FMA, Zambare SP (2017) Molecular identification of forensically important flesh flies (Diptera: Sarcophagidae) using COI gene. J Entomol Zool Stud 5(2):263–267

Altantogtokh D, Lilak AA, Takhampunya R, Sakolvaree J, Chanarat N, Matulis G, Poole-Smith BK, Boldbaatar B, Davidson S, Hertz J, Bolorchimeg B, Tsogbadrakh N, Fiorenzano JM, Lindroth EJ, von Fricken ME (2022) Metagenomic profiles of *Dermacentor* tick pathogens from across Mongolia, using next generation sequencing. Front Microbiol 13:946631

Anderson JF, Andreadis TG, Main AJ, Kline DL (2004) Prevalence of West Nile virus in tree canopy inhabiting *Culex pipiens* and associated mosquitoes. Am J Trop Med Hyg 71(1):112–119

Arai S, Kuwata R, Higa Y, Maekawa Y, Tsuda Y, Roychoudhury S, Bertuso AG, Tran VP, Nguyen TY, Etoh T, Otuka A, Matsumura M, Nabeshima T, Tanaka-Taya K, Okabe N, Kobayashi M, Sawabe K (2022) Two hidden taxa in the Japanese encephalitis vector mosquito, *Culex tritaeniorhynchus*, and the potential for long-distance migration from overseas to Japan. PLoS Negl Trop Dis 16(6):e0010543

Aregawi WG, Agga GE, Abdi RD, Büscher P (2019) Systematic review and meta-analysis on the global distribution, host range, and prevalence of *Trypanosoma evansi*. Parasit Vectors 12(1):67

Arif M, Tauran P, Kosasih H, Pelupessy NM, Sennang N, Mubin RH, Sudarmono P, Tjitra E, Murniati D, Alam A, Gasem MH, Aman AT, Lokida D, Hadi U, Parwati KTM, Lau CY, Neal A, Karyana M (2020) Chikungunya in Indonesia: epidemiology and diagnostic challenges. PLoS Negl Trop Dis 14(6):e0008355

Artemyev MM, Neronov VM (1982) Sandfly fauna (Diptera, Phsychodidae, Phlebotominae) of the Mongolian People's Republic and peculiarities of its formation. In: Zoologicheskiye issledovaniya v MNR (Zoological studies in the Mongolian People's Republic). Nauka, Moscow, pp 108–119. (in Russian)

Artemyev MM, Neronov VM (1984) Distribution and ecology of sandflies of the Old World (genus *Phlebotomus*). Institute of Evolutionary Morphology and Animal Ecology, USSR Academy of Sciences, Moscow, 207 pp. (in Russian)

Attoui H, Charrel RN, Billoir F, Cantaloube JF, de Micco P, de Lamballerie X (1998) Comparative sequence analysis of American, European and Asian isolates of viruses in the genus Coltivirus. J Gen Virol 79(10):2481–2489

Azhari AN (2022) The last miles in Indonesia's schistosomiasis elimination programme. WHO News. https://www.who.int/indonesia/news/detail/26-06-2022-the-last-miles-in-indonesia-s-schistosomiasis-elimination-programme. Accessed 04 Jan 2023

Bellis G, Dyce A, Gopurenko D, Yanase T, Garros C, Labuschagne K, Mitchell A (2014) Revision of the *Culicoides* (Avaritia) Imicola complex Khamala & Kettle (Diptera: Ceratopogonidae) from the Australasian region. Zootaxa 3768(4):401–427

Bhattacharya S, Rinzin N, Chusak P, Dash AP, Chowdhury R, Tobgay T, Narain JP (2010) Occurrence & significance of kala-azar in Bhutan. Indian J Med Res 132:337–338

Bhengsri S, Baggett HC, Edouard S, Dowell SF, Dasch GA, Fisk TL, Raoult D, Parola P (2016) Sennetsu neorickettsiosis, spotted fever group, and typhus group rickettsioses in three provinces in Thailand. Am J Trop Med Hyg 95(1):43–49

bin Said I, Kouakou YI, Omorou R, Bienvenu AL, Ahmed K, Culleton R, Picot S (2022) Systematic review of *Plasmodium knowlesi* in Indonesia: a risk of emergence in the context of capital relocation to Borneo? Parasit Vectors 15(1):258

Boldbaatar B, Jiang RR, von Fricken ME, Lkhagvatseren S, Nyamdawa P, Baigalmaa B, Wang YW, Anderson BD, Jiang JF, Gray GC (2017) Distribution and molecular characteristics of rickettsiae found in ticks across Central Mongolia. Parasit Vectors 10(1):61

Brown SE, Gorman BM, Tesh RB, Knudson DL (1992) Isolation of bluetongue and epizootic hemorrhagic disease viruses from mosquitoes collected in Indonesia. Vet Microbiol 32(3–4):241–252

Brown SE, Gorman BM, Tesh RB, Knudson DL (1993) Coltiviruses isolated from mosquitoes collected in Indonesia. Virology 196(1):363–367

Budiono NG, Satrija F, Ridwan Y, Handharyani E, Murtini S (2019) The contribution of domestic animals to the transmission of schistosomiasis japonica in the Lindu Subdistrict of the Central Sulawesi Province, Indonesia. Vet World 12(10):1591–1598

Buhl MR, Lindquist L (2009) Japanese encephalitis in travelers: review of cases and seasonal risk. J Travel Med 16(3):217–219

Buhler C, Winkler V, Runge-Ranzinger S, Boyce R, Horstick O (2019) Environmental methods for dengue vector control – a systematic review and meta-analysis. PLoS Negl Trop Dis 13(7):e0007420

Byrd JH, Castner JL (2010) Insects of forensic importance. In: Byrd JH, Castner JL (eds) Forensic entomology – the utility of arthropods in legal investigations, 2nd edn. CRC Press, Boston, pp 39–126

Cao J, Newby G, Cotter C, Hsiang MS, Larson E, Tatarsky A, Gosling RD, Xia Z, Gao Q (2021) Achieving malaria elimination in China. Lancet Public Health 6(12):e871–e872

Capeding MR, Chua MN, Hadinegoro SR, Hussain IIHM, Nallusamy R, Pitisuttithum P, Rusmil K, Thisyakorn U, Thomas SJ, Huu Tran N, Wirawan DN, Yoon IK, Bouckenooghe A, Hutagalung Y, Laot T, Wartel TA (2013) Dengue and other common causes of acute febrile illness in Asia: an active durveillance dtudy in children. PLoS Negl Trop Dis 7(7):e2331

Caramello P, Canta F, Balbiano R, Lipani F, Ariaudo S, de Agostini M, Calleri G, Boglione L, di Caro A (2007) A case of imported JE acquired during short travel in Vietnam. Are current recommendations about vaccination broader? J Travel Med 14(5):346–348

Carney WP, van Peenen PF, Sudomo M (1978) A mammalian reservoir of *Schistosoma japonicum* in the Napu Valley, Central Sulawesi, Indonesia. J Parasitol 64(6):1138–1139

Centers for Disease Control and Prevention (CDC) (2022) Dirofilariasis. https://www.cdc.gov/dpdx/dirofilariasis/index.html. Accessed 22 Dec 2022

Černý J, Buyannemekh B, Needham T, Gankhuyag G, Oyuntsetseg D (2019) Hard ticks and tick-borne pathogens in Mongolia – a review. Ticks Tick Borne Dis 10(6):101268

Chanmol W, Jariyapan N, Somboon P, Bates MD, Bates PA (2019) Development of *Leishmania orientalis* in the sand fly *Lutzomyia longipalpis* (Diptera: Psychodidae) and the biting midge *Culicoides soronensis* (Diptera: Ceratopogonidae). Acta Trop 199:105157

Chansiri K, Tejangkura T, Kwaosak P, Sarataphan N, Phantana S, Sukhumsirichart W (2002) PCR based method for identification of zoonotic *Brugia malayi* microfilariae in domestic cates. Mol Cell Probes 16(5):129–135

Chavatte JM, Jureen R (2017) Imported asymptomatic bancroftian filariasis discovered from a *Plasmodium vivax* infected patient: a case report from Singapore. Case Rep Infect Dis 2017:1972587, 8 pp

Chen CD, Nazni WA, Lee HL, Jeffery J, Wan-Norjuliana WM, Abdullah AG, Sofian-Azirun M (2008) Larval growth parameters and growth rates of forensically important flies, *Hypopygiopsis violacea* Macquart, 1835 and *Chrysomya rufifacies* Macquart, 1842. Proc Asean Congr Trop Med Parasitol 3:97–100

Chen H, Hu K, Zou J, Xiao J (2013) A cluster of cases of human-to-human transmission caused by severe fever with thrombocytopenia syndrome bunyavirus. Int J Infect Dis 17(3):e206–e208

Chen IC, Chen YC, Teng HJ, Shu PY, Li SY (2019) Ecological characteristics and viral transmission capability of *Aedes aegypti* and *Ae. albopictus*. Taiwan Epidemiol Bull 35(13):73–74

Chen M, Liu Q, Xue J, Chen S, Huang D, Yu Y, Cai Y, Lu Y, Song P, Zhang R, Ai L, Chen J (2020) Spreading of human babesiosis in China: current epidemiological status and future challenges. China CDC Wkly 2(33):634–637

Chen B, Chen Q, Li Y, Mu D, Wang Z, Zhu M, Chen N, Yin W (2021a) Epidemiological characteristics of imported Chikungunya fever cases in China, 2010–2019. Dis Surveill 36(6):539–543. (in Chinese)

Chen JH, Fen J, Zhou XN (2021b) From 30 million to zero malaria cases in China: lessons learned for China-Africa collaboration in malaria elimination. Infect Dis Poverty 10(1):51

Chen XJ, Wang HY, Li XL, Gao XY, Li MH, Fu SH, He Y, Li F, Xu ST, Wu D, Li YX, Yin ZD, Yang G, Liang GD (2021c) Japanese encephalitis in China in the period of 1950–2018: from discovery to control. Biomed Environ Sci 34(3):175–183

Chen HF, Peng SH, Tsai KH, Yang CF, Chang MC, Hsueh YL, Su CL, Wang RY, Shu PY, Yang SL (2022a) Molecular epidemiology of scrub typhus in Taiwan during 2006–2016. PLoS Negl Trop Dis 16(4):e0010369

Chen QL, Zhum MT, Chen N, Yang D, Yin WW, Mu D, Li Y, Zhang YP, Zainawudong Y (2022b) Epidemiological characteristics of severe fever with thrombocytopenia syndrome in China, 2011–2021. Chin J Epidemiol 43(6):852–859. (in Chinese)

Chiang PS, Su SW, Yang SL, Shu PY, Lee WP, Li SY, Teng HJ (2022) Delayed correlation between the incidence rate of indigenous murine typhus in humans and the seropositive rate of *Rickettsia typhi* infection in small mammals in Taiwan from 2007–2019. PLoS Negl Trop Dis 16(4):e0010394

Chimed-Ochir O, Delgermaa V, Takahashi K, Purev O, Sarankhuu A, Fujino Y, Bayarmagnai N, Dugee O, Erkhembayar R, Lkhagvaa B, Ochir C, Sosorburam T, Naghavi M (2022) Mongolia health situation: based on the Global Burden of disease study 2019. BMC Public Health 22(1):5

Choochote W, Somboon P, Kamboonruang C, Suwanpanit P (1986) A survey for natural vectors of *Dirofilaria immitis* in Chiang Mai Province, Northern Thailand. Southeast Asian J Trop Med Public Heath 17(1):146–147

Choochote W, Sukhavat K, Keha P, Somboon P, Khamboonruang C, Suwanpanit P (1987) The prevalence of *Dirofilaria immitis* in stray dog and its potential vector in Amphur Muang Chiang Mai, Northern Thailand. Southeast Asian J Trop Med Public Health 18(1):131–134

Choovattanapakorn N, Yanola J, Lumjuan N, Saingamsook J, Somboon P (2017) Characterization of metabolic detoxifying enzymes in an insecticide resistant strain of *Aedes aegypti* harboring homozygous S989P and V1016G *Kdr* mutations. Med Entomol Zool 68(1):1–8

Chung HH, Cheng IC, Chen YC, Lin C, Tomita T, Teng HJ (2019) Voltage-gated sodium channel intron polymorphism and four mutations comprise six haplotypes in an *Aedes aegypti* population in Taiwan. PLoS Negl Trop Dis 13(3):e0007291

Chung HH, Tsai CH, Teng HJ, Tsai KH (2022) The role of voltage-gated sodium channel genotypes in pyrethroid resistance in *Aedes aegypti* in Taiwan. PLoS Negl Trop Dis 16(9):e0010780

Chusri S, Thammapalo S, Silpapojakul K, Siriyasatien P (2014) Animal reservoirs and potential vectors of *Leishmania siamensis* in southern Thailand. Southeast Asian J Trop Med Public Health 45(1):13–19

Clark GG, Golden FV, Allan SA, Cooper Band MF, McNelly JR (2013) Behavioral responses of two dengue virus vectors, *Aedes aegypti* and *Aedes albopictus* (Diptera: Culicidae), to DUET and its components. J Med Entomol 50(5):1059–1070

Clausen PH, Chuluun S, Sodnomdarjaa R, Greiner M, Noeckler K, Staak C, Zessin KH, Schein E (2003) A field study to estimate the prevalence of *Trypanosoma equiperdum* in Mongolian horses. Vet Parasitol 115(1):9–18

Counotte MJ, Kim CR, Wang J, Bernstein K, Deal CD, Broutet NJN, Low N (2018) Sexual transmission of Zika virus and other flaviviruses: a living systematic review. PLoS Med 15(7):e1002611

Daniels PW, Sendow I, Soleha E, Sukarsih, Hunt NT, Bahri S (1995) Australian-Indonesian collaboration in veterinary arbovirology – a review. Vet Microbiol 46(1–3):151–174

Darsie RF Jr, Pradhan SP (1990) The mosquitoes of Nepal: their identification, distribution and biology. Mosq Syst 22(2):69–130

Dash M (1969) To the fauna of ixodid ticks of Mongolia. Folia Parasitol 16(1):18

Dash M, Biambaa B, Neronov VM (1988) The ixodid tick fauna of the Mongolian People's Republic. I. The species distribution. Med Parazitol (Mosc) 3:37–42 (in Russian)

Department of Health (DOH) (1991) Malaria eradication in Taiwan. Department of Health, The Executive Yuan, Republic of China, Taipei. 300 pp

Desquesnes M, Dargantes A, Lai DH, Lun ZR, Holzmuller P, Jittapalapong S (2013) *Trypanosoma evansi* and Surra: a review and perspectives on transmission, epidemiology and control, impact, and zoonotic aspects. Biomed Res Int 2013:1–20

Dewi RS, Damajanti R, Wardhana AH, Mulatsih S, Poetri ON, Steeneveld W, Hogeveen H (2020) The economic losses of Surra outbreak in Sumba Timur, Nusa Tenggara Timur-Indonesia. Trop Anim Sci J 43(1):77–85

Dhimal M, Ahrens B, Kuch U (2015) Climate change and spatiotemporal distributions of vector-borne diseases in Nepal – a systematic synthesis of literature. PLoS One 10(6):e0129869

Dhimal M, Dumre SP, Sharma G, Khanal P, Ranabhat K, Shah LP, Lal BK, Jha R, Upadhyaya BP, Acharya B, Shrestha SK, Davidson SA, Charoensinphon P, Karki KB (2021) An outbreak investigation of scrub typhus in Nepal: confirmation of local transmission. BMC Infect Dis 21:193

Dickson BFR, Graves PM, McBride WJ (2017) Lymphatic filariasis in Mainland Southeast Asia: a systematic review and meta-analysis of prevalence and disease burden. Trop Med Infect Dis 2(3):32

Dieng H, Saifur RGM, Hassan AA, Salmah MRC, Boots M, Satho T, Jaal Z, AbuBakar S (2010) Indoor-breeding of *Aedes albopictus* in Northern Peninsular Malaysia and its potential epidemiological implications. PLoS One 5(7):e11790

Diptyanusa A, Herini E, Indarjulianto S, Satoto TBT (2022) Estimation of Japanese encephalitis virus infection prevalence in mosquitoes and bats through nationwide sentinel surveillance in Indonesia. PLoS One 17(10):e0275647

Dorji T, Yoon IK, Holmes EC, Wangchuk S, Tobgay T, Nisalak A, Chinnawirotpisan P, Sangkachantaranon K, Gibbons RV, Jarman RG (2009) Diversity and origin of dengue virus serotypes 1, 2, and 3. Bhutan Emerg Infect Dis 15(10):1630–1632

Dorji K, Phuentshok Y, Zangpo T, Dorjee S, Dorjee C, Jolly P, Morris R, Marquetoux N, McKenzie J (2019) Clinical and epidemiological patterns of scrub typhus, an emerging disease in Bhutan. Trop Med Infect Dis 4(2):56

Dumic I, Jevtic D, Veselinovic M, Nordstrom CW, Jovanovic M, Mogulla V, Veselinovic EM, Hudson A, Simeunovic G, Petcu E, Ramanan P (2022) Human Granulocytic anaplasmosis – a systematic review of published cases. Microorganisms 10(7):1433

Duong V, Dussart P, Buchy P (2017) Zika virus in Asia. Int J Infect Dis 54:121–128

Duong TN, Phong TV, Nam VS, Duoc VT, Tu TC, Yen NT, Son TH, Thuy VN, Anh NTM, Anh PT, Cuong TC, Thinh ND, Giang NTT, Anh NT, Lieu VT (2021) Surveillance and control of medical insects and animals for a number of common human diseases. Medical Publishing House (Vietnam), Hanoi, 182 pp

Düzlü Ö, İnci A, Yıldırım A, Doğanay M, Özbel Y, Aksoy S (2020) Vector-borne zoonotic diseases in Türkiye: rising threats on public health. Türkiye Parazitol Derg 44(3):168–175

Edmondson MA, Roberts JF, Baird AN, Bychawski S, Pugh DG (2012) Theriogenology of sheep and goats. In: Sheep and goat medicine, 2nd edn. Elsevier, B.V, Amsterdam, pp 150–230

Ekawasti F, Martindah E (2020) Awareness of the existence of leishmaniasis as protozoan zoonosis in Indonesia. WARTAZOA Indones Bull Anim Vet Sci 30(2):79–90

Ekawati N, Saputra H (2016) Cutaneous Leishmaniasis with differential diagnosis of histoplasmosis in an HIV patient. In: Laporan Sebuah Kasus. Bagian/Smf Patologi Anatomi, F.K. UNUD/RSUP Sanglah Denpasar, Bali. (in Indonesian)

Elliott I, Pearson I, Dahal P, Thomas NV, Roberts T, Newton PN (2019) Scrub typhus ecology: a systematic review of *Orientia* in vectors and hosts. Parasit Vectors 12(1):513

Ergunay K, Gunay F, Erisoz Kasap O, Oter K, Gargari S, Karaoglu T, Tezcan S, Cabalar M, Yildirim Y, Emekdas G, Alten B, Ozkul A (2014) Serological, molecular and entomological surveillance demonstrates widespread circulation of West Nile virus in Turkey. PLoS Negl Trop Dis 8(7):e3028

Ericsson CD, Jensenius M, Fournier PE, Raoult D (2004) Rickettsioses and the international traveler. Clin Infect Dis 39(10):1493–1499

Ernst T, McCarthy S, Chidlow G, Luang-Suarkia D, Holmes EC, Smith DW, Imrie A (2015) Emergence of a new lineage of dengue virus type 2 identified in travelers entering Western Australia from Indonesia, 2010–2012. PLoS Negl Trop Dis 9(1):e0003442

Faizah AN, Kobayashi D, Amoa-Bosompen M, Higa Y, Tsuda Y, Itokawa K, Miura K, Hirayama K, Sawabe K, Isawa H (2020) Evaluating the competence of the primary vector, *Culex tritaeniorhynchus*, and the invasive mosquito species, *Aedes japonicus japonicus*, in transmitting three Japanese encephalitis virus genotypes. PLoS Negl Trop Dis 14(12):e000898

Faizah AN, Kobayashi D, Maekawa Y, Amoa-Bosompem M, Fauziyah S, Mulyatno KC, Subekti S, Rohmah EA, Lusida MI, Mori Y, Miura K, Hirayama K, Isawa H, Sawabe K (2021) Identification and isolation of Japanese encephalitis virus genotype IV from *Culex vishnui* collected in Bali, Indonesia in 2019. Am J Trop Med Hyg 105(3):813–817

Farukh MA, Hayasaka H, Mishigdor O (2009) Recent tendency of Mongolian wildland fire incidence: analysis using MODIS hotspot and weather data. J Nat Disaster Sci 31(1):23–33

Feng X, Feng J, Zhang L, Tu H, Xia Z (2022) Vector control in China, from malaria endemic to elimination and challenges ahead. Infect Dis Poverty 11(1):54

Fernandes RS, O'Connor O, Bersot MIL, Girault D, Dokunengo MR, Pocquet N, Dupont-Rouzeyrol M, Lourenço-de-Oliveira R (2020) Vector competence of *Aedes aegypti, Aedes albopictus* and *Culex quinquefasciatus* from Brazil and New Caledonia for three Zika virus lineages. Pathogens 9(7):575

Fijn N, Terbish B (2021) The multiple faces of the Marmot: associations with the plague, hunting, and cosmology in Mongolia. Hum Ecol 49:539–549

Firth C, Blasdell KR, Amos-Ritchie R, Sendow I, Agnihotri K, Boyle DB, Daniels P, Kirkland PD, Walker PJ (2017) Genomic analysis of bluetongue virus episystems in Australia and Indonesia. Vet Res 48(1):82

Fischer T, Myalkhaa M, Krücken J, Battsetseg G, Batsukh Z, Baumann MPO, Clausen PH, Nijhof AM (2020) Molecular detection of tick-borne pathogens in bovine blood and ticks from Khentii, Mongolia. Transbound Emerg Dis 67(Suppl 2):111–118

Frey S, Mossbrugger I, Altantuul D, Battsetseg J, Davaadorj R, Tserennorov D, Buyanjargal T, Otgonbaatar D, Zoller L, Speck S, Wolfel R, Dolber G, Essbauer S (2012) Isolation, preliminary characterization, and full-genome analyses of tick-borne encephalitis virus from Mongolia. Virus Genes 45(3):413–425

Froeschl G, Beissner M, Huber K, Bretzel G, Hoelscher M, Rothe C (2018) *Plasmodium knowlesi* infection in a returning German traveler from Thailand: a case report on an emerging malaria pathogen in a popular low-risk travel destination. BMC Infect Dis 18(1):148

Fukunaga T, Phommasack B, Bounlu K, Saito M, Tadano M, Makino Y, Kanemura K, Arakaki S, Shinjo M, Insisiengmay S (1994) Epidemiological situation of dengue infection in Lao PDR. Trop Med 35(4):219–227

Fungfuang W, Udom C, Tongthainan D, Kadir KA, Singh B (2020) Malaria parasites in macaques in Thailand: stump-tailed macaques (*Macaca arctoides*) are new natural hosts for *Plasmodium knowlesi*, *Plasmodium inui*, *Plasmodium coatneyi* and *Plasmodium fieldi*. Malar J 19(1):350

Gantuya S (2010) Studies on zoonotic Leishmaniasis caused by *Leishmania major* sensu lato. PhD thesis, The University of Tokyo, Tokyo, Japan

Garjito TA, Sudomo M, Abdullah, Dahlan M, Nurwidayati A (2008) Schistosomiasis in Indonesia: past and present. Parasitol Int 57(3):277–280

Garjito TA, Widiarti, Anggraeni YM, Alfiah S, Tunggul Satoto TB, Farchanny A, Samaan G, Afelt A, Manguin S, Frutos R, Aditama TY (2018) Japanese encephalitis in Indonesia: an update on epidemiology and transmission ecology. Acta Trop 187:240–247

Garjito TA, Prihatin MT, Susanti L, Prastowo D, Sa'adah SR, Taviv Y, Satoto TBT, Waluyo J, Manguin S, Frutos R (2019) First evidence of the presence of genotype-1 of Japanese encephalitis virus in *Culex gelidus* in Indonesia. Parasit Vectors 12(1):19

Garros C, Van Bortel W, Trung HD, Coosemans M, Manguin S (2006) Review of the *minimus* complex of *Anopheles*, main malaria vector in Southeast Asia: from taxonomic issues to vector control strategies. Trop Med Int Health 11(1):102–114

Gasem MH, Wagenaar JFP, Goris MGA, Adi MS, Isbandrio BB, Hartskeerl RA, Rolain JM, Raoult D, van Gorp ECM (2009) Murine typhus and leptospirosis as causes of acute undifferentiated fever, Indonesia. Emerg Infect Dis 15(6):975–977

Godsey MS Jr, Burkhalter K, Young G, Delorey M, Smith K, Townsend J, Levy C, Mutebi JP (2012) Entomologic investigations during an outbreak of West Nile virus disease in Maricopa County, Arizona, 2010. Am J Trop Med Hyg 87(6):1125–1131

Gonchigjav B, Bayaraa A, Enkhee B, Purevjal M, Pandaan B (2022) Organ donation and transplantation in Mongolia. Transplantation 106(1):1–3

Gosselin M, Wille SM, Fernandez Mdel M, Di Fazio V, Samyn N, De Boeck G, Bourel B (2011) Entomotoxicology, experimental set-up and interpretation for forensic toxicologists. Forensic Sci Int 208(1–3):1–9

Gould DJ, Bailey CL, Vongpradist S (1982) Implication of forest mosquitoes in the transmission of *Wuchereria bancrofti* in Thailand. Mosq News 42:560–564

Gould DJ, Scanlon JE, Ward RA (1966) *Anopheles* vectors of malaria in Southeast Asia. Army Sci Conf Proc 1:361–373

Greenberg B, Kunich C (2002) Entomology and the law – flies as forensic indicators. Cambridge University Press, Cambridge, 315 pp

Guan LR, Zhou ZB, Jin CF, Fu Q, Chai JJ (2016) Phlebotomine sand flies (Diptera: Psychodidae) transmitting visceral leishmaniasis and their geographical distribution in China: a review. Infect Dis Poverty 5:15

Guo YD, Cai JF, Chang YF, Li X, Liu QL, Xue XH, Wang X, Zhong M, Wen JF, Wang JF (2011) Identification of forensically important sarcophagid flies (Diptera: Sarcophagidae) in China, based on COI and 16S rDNA gene sequences. J Forensic Sci 56(6):1534–1540

Guo YD, Cai JF, Meng FM, Chang YF, Gu Y, Lan LM, Liang L, Wen JF (2012a) Identification of forensically important flesh flies based on a shorter fragment of the cytochrome oxidase subunit I gene in China. Med Vet Entomol 26(3):307–313

Guo YD, Cai JF, Xiong F, Wang HJ, Wen JF, Li JB, Chen YQ (2012b) The utility of mitochondrial DNA fragments for genetic identification of forensically important sarcophagid flies (Diptera: Sarcophagidae) in China. Trop Biomed 29(1):51–60

Guo Y, Zha L, Yan W, Li P, Cai J, Wu L (2014) Identification of forensically important sarcophagid flies (Diptera: Sarcophagidae) in China based on COI and period gene. Int Leg Med 128(1):221–228

Guo Y, Shi Y, Zhang Y, Li Y, Wang M, Liu Q (2021) Modelling the ecological suitability of the Chagas disease vector *Triatoma rubrofasciata*- China. China CDC Wkly 3(35):736–740

Guo X, Ma C, Wang L, Zhao N, Liu S, Xu W (2022) The impact of COVID-19 continuous containment and mitigation strategy on the epidemic of vector-borne diseases in China. Parasit Vectors 15(1):78

Hadi TR, Nalim S, Sukaeri S, Dennis DT (1980) Scrub typhus survey of Biak and Owi islands: ectoparasites of small mammals and rickettsial isolations. Southeast Asian J Trop Med Public Health 11(2):220–226

Hall RD (1990) Medicocriminal entomology. In: Catts EP, Haskell NH (eds) Entomology and death: a procedural guide. Joyce's Print Shops, Clemson, pp 1–8

Hall RD (2001) Introduction: perceptions and status of forensic entomology. In: Byrd JH, Castner JL (eds) Forensic entomology: the utility of arthropods in legal investigations. CRC Press, Boston, pp 1–15

Halstead SB (2015) Reappearance of chikungunya, formerly called dengue, in the Americas. Emerg Infect Dis 21(4):557–561

Hamid NA, Omar B, Marwi MA, Ahmad FMS, Mansar AH, Feng SS, Moktar NA (2003) A review of forensic specimens sent to forensic entomology laboratory Universiti Kebangsaan Malaysia for the year 2001. Trop Biomed 20(1):27–31

Hammon WM, Rudnick A, Sather GE (1960) Viruses associated with epidemic hemorrhagic fevers of The Philippines and Thailand. Science 131(3407):1102–1103

Han L, Zhang YF, Teng ZQ, Kan B, Qin T (2023) Epidemiological characteristics of typhus in China, 1950–2021. Chin J Epidemiol 44(3):430–437. (in Chinese with English abstract)

Harapan H, Michie A, Mudatsir M, Nusa R, Yohan B, Wagner AL, Sasmono RT, Imrie A (2019) Chikungunya virus infection in Indonesia: a systematic review and evolutionary analysis. BMC Infect Dis 19(1):243

Harapan H, Michie A, Sasmono RT, Imrie A (2020) Dengue: a minireview. Viruses 12(8):829

Harinasuta C, Sucharit S, Deesin T, Surathin K, Vutikes S (1970) Bancroftian filariasis in Thailand, a new endemic area. Southeast Asian J Trop Med Public Health 1(2):233–245

Harinasuta C, Sucharit S, Choochote W (1981) The susceptibility of leaf monkeys to bancroftian filariasis in Thailand. Southeast Asian J Trop Med Public Health 12(4):581–589

Hebert PDN, Gregory TR (2005) The promise of DNA barcoding for taxonomy. Syst Biol 54(5):852–859

Hemingway J, Ranson H (2000) Insecticide resistance in insect vectors of human disease. Annu Rev Entomol 45:371–391

Heo CC, Marwi MA, Jeffery J, Omar B (2008a) Insect succession on a decomposing piglet carcass placed in a man-made freshwater pond in Malaysia. Trop Biomed 25(1):23–29

Heo CC, Marwi MA, Ahmad FMS, Jeffery J, Kurahashi H, Omar B (2008b) Study of insect succession and rate of decomposition on a partially burned pig carcass in an oil palm plantation in Malaysia. Trop Biomed 25(3):202–208

Heo CC, Kurahashi H, Marwi MA, Jeffery J, Omar B (2011) Opportunistic insects associated with pig carrions in Malaysia. Sains Malays 40(6):601–604

Heo CC, Latif B, Silahuddin SA, Kurahashi H (2015a) Morphology of third instar *Boettcherisca highlandica* Kurahashi & Tan, 2009 (Diptera: Sarcophagidae). Halteres 6:121–131

Heo CC, Latif B, Silahuddin SA, Nazni WA, Omar B (2015b) Morphological descriptions on the larvae of *Hypopygiopsis fumipennis* (Walker, 1856) (Diptera: Calliphoridae). Trop Biomed 32(1):151–159

Heo CC, Rahimi R, Mengual X, Isa MSM, Zainal S, Khofar PN, Nazni WA (2019) *Eristalinus arvorum* (Fabricius, 1787) (Diptera: Syrphidae) in human skull: a new fly species of forensic importance. J Forensic Sci 65(1):276–282

Hills SL, Griggs AC, Fischer M (2010) Japanese encephalitis in travelers from non-endemic countries, 1973–2008. Am J Trop Med Hyg 82(5):930–936

Hiscox A, Kaye A, Vongphayloth K, Banks I, Piffer M, Khammanithong P, Sananikhom P, Kaul S, Hill N, Lindsay SW, Brey PT (2013) Risk factors for the presence of *Aedes aegypti* and *Aedes albopictus* in domestic water-holding containers in areas impacted by the Nam Theun 2 hydroelectric project. Laos Am J Trop Med Hyg 88(6):1070–1078

Hoffmann AA, Iturbe-Ormaetxe I, Callahan AG, Phillips BL, Billington K, Axford JK, Montgomery B, Turley AP, O'Neill SL (2014) Stability of the wMel *Wolbachia* infection following invasion into *Aedes aegypti* populations. PLoS Negl Trop Dis 8(9):e3115

Holmes E, Twiddy S (2003) The origin, emergence and evolutionary genetics of dengue virus. Infect Genet Evol 3(1):19–28

Hsieh CJ, Li CW, Cheng CA, Wu DC, Wu WC, Lin FH, Chou YC, Yu CP (2020) Epidemiologic characteristics of domestic patients with hemorrhagic fever with renal syndrome in Taiwan: a 19-year retrospective study. Int J Environ Res Public Health 17(15):5291

Huang JH, Su CL, Yang CF, Liao TL, Hsu TC, Chang SF, Lin CC, Shu PY (2012) Molecular characterization and phylogenetic analysis of dengue viruses imported into Taiwan during 2008–2010. Am J Trop Med Hyg 87(2):349–358

Huang YJ, Ayers VB, Lyons AC, Unlu I, Alto BW, Cohnstaedt LW, Higgs S, Vanlandingham D (2016) *Culex* species mosquitoes and Zika virus. Vector Borne Zoonotic Dis 6(16):673–676

Huang X, Li J, Li A, Wang S, Li D (2021a) Epidemiological characteristics of severe fever with thrombocytopenia syndrome from 2010 to 2019 in Mainland China. Int J Environ Res Public Health 18(6):3092

Huang M, Ma J, Jiao J, Li C, Chen L, Zhu Z, Ruan F, Xing L, Zheng X, Fu M, Ma B, Gan C, Mao Y, Zhang C, Sun P, Liu X, Lin Z, Chen L, Lu Z, Zhou D, Wen B, Chen W, Xiong X, Xia J (2021b) The epidemic of Q fever in 2018 to 2019 in Zhuhai city of China determined by metagenomic next-generation sequencing. PLoS Negl Trop Dis 15(7):e0009520

Huang Y, Shen H, Wang F, Qian W, Dong S, Qi X (2023) Design and implementation of WeChat Mini Program for invasive insect recognition. J Anhui Agric Sci 51(1):222–230

Hustedt J, Prasetyo DB, Fiorenzano JM, von Fricken ME, Hertz JC (2022) Phlebotomine sand flies (Diptera: Psychodidae) and sand fly-borne pathogens in the Greater Mekong Subregion: a systematic review. Parasit Vectors 15(1):355

Ibrahim IN, Okabayashi T, Ristiyanto, Lestari EW, Yanase T, Muramatsu Y, Ueno H, Morita C (1999) Serosurvey of wild rodents for Rickettsioses (spotted fever, murine typhus and Q fever) in Java Island, Indonesia. Eur J Epidemiol 15(1):89–93

Imanishi N, Higa Y, Teng HJ, Sunahara T, Minakawa N (2018) Identification of three distinct groups of *Anopheles lindesayi* in Japan by morphological and genetic analyses. Jpn J Infecrt Dis 71(6):427–435

Infectious Diseases Weekly Report (IDWR) (2022) IDWR surveillance data table 2022 week 52. National Institute of Infectious Diseases. https://www.niid.go.jp/niid/ja/idwr-dl/2022.html. Accessed 27 Dec 2023

Intarapuk A, Bhumiratana A (2021) Investigation of *Armigeres subalbatus*, a vector of zoonotic *Brugia pahangi* filariasis in plantation areas in Suratthani, Southern Thailand. One Health 13:100261

Iwasa M, Makino S, Asakura H, Kobori H, Morimoto Y (1999) Detection of *Escherichia coli* O157: H7 from *Musca domestica* (Diptera: Muscidae) at a cattle farm in Japan. J Med Entomol 36(1):108–112

Iyengar MO (1953) Filariasis in Thailand. Bull World Health Organ 9(6):731–799

Jafari S, Oshaghi MA, Akbarzadeh K, Abai MR, Koosha M, Mohtarami F (2019) Identification of forensically important flesh flies using the cytochrome c oxidase subunits I and II genes. J Med Entomol 56(5):1253–1259

Jariyapan N, Daroontum T, Jaiwong K, Chanmol W, Intakhan N, Sor-Suwan S et al (2018) *Leishmania (Mundinia) orientalis* n. sp. (Trypanosomatidae), a parasite from Thailand responsible for localised cutaneous leishmaniasis. Parasit Vectors 11(1):351

Jeslyn WPS, Huat TC, Vernon L, Irene LMZ, Sung LK, Jarrod LP et al (2011) Molecular epidemiological investigation of *Plasmodium knowlesi* in humans and macaques in Singapore. Vector Borne Zoonotic Dis 11(2):131–135

Joshi PP, Shegokar VR, Powar RM, Herder S, Katti R, Salkar HR, Dani VS, Bhargava A, Jannin J, Truc P (2005) Human trypanosomiasis caused by *Trypanosoma evansi* in India: the first case report. Am J Trop Med Hyg 73(3):491–495

Kaikuntod M, Arijkumpa O, Kladkempetch D, Fukumoto S, Thongkorn K, Boonyapakorn C et al (2021) Geographic spatial distribution patterns of *Dirofilaria immitis* and *Brugia pahangi* infection in commnity dogs in Chiang Mai, Thailand. Animals (Basel) 11(1):33

Kamyingkird K, Junsiri W, Chimnoi W, Kengradomkij C, Saengow S, Sangchuto K et al (2017) Prevalence and risk factors associated with *Dirofilaria immitis* infection in dogs and cats in Songkhla and Satun provinces, Thailand. Agric Nat Resour 51:299–302

Kanamitsu M, Taniguchi K, Urasawa S, Ogata T, Wada Y, Wada Y, Saroso JS (1979) Geographic distribution of arbovirus antibodies in indigenous human populations in the Iando-Australian archipelago. Am J Trop Med Hyg 28(2):351–363

Kanjanopas K, Siripattanapipong S, Ninsaeng U, Hitakarun A, Jitkaew S, Kaewtaphaya P et al (2013) *Sergentomyia (Neophlebotomus) gemmea*, a potential vector of *Leishmania siamensis* in southern Thailand. BMC Infect Dis 13:333

Kano R, Field G, Shinonaga S (1967) Fauna Japonica – Sarcophagidae (Insecta: Diptera). Biogeographical Society of Japan, Tokyo, 168 pp

Kardena IM, Adi AAAM, Astawa NM, O'Dea M, Laurence M, Sahibzada S, Bruce M (2021) Japanese encephalitis in Bali, Indonesia: ecological and sociocultural perspectives. Int J Vet Sci Med 9(1):31–43

Kato H, Yanagisawa N, Sekiya N, Suganuma A, Imamura A, Ajisawa A (2014) Murine typhus in a Japanese traveler returning from Indonesia: a case report. Kansenshogaku Zasshi 88(2):166–170. (in Japanese)

Kavitha R, Nazni WA, Tan TC, Lee HL, Azirun MS (2013) Review of forensically important entomological specimens collected from human cadavers in Malaysia (2005–2010). J Forensic Leg Med 20(5):480–482

Kawada H, Higa Y, Nguyen TY, Tran SH, Nguyen HT, Takagi M (2009) Nationwide investigation of the pyrethroid susceptibility of mosquito larvae collected from used tires in Vietnam. PLoS Med 9(3):391–397

Kehrmann J, Popp W, Delgermaa B, Otgonbayar D, Gantumur T, Buer J, Tsogbadrakh N (2020) Two fatal cases of plague after consumption or raw marmot organs. Emerg Microbes Infect 9(1):1878–1880

Khalid MZ, Shariff AM, Ismail WNW (2015) Developmental stages of *Hypopygiopsis violacea* (Family: Calliphoridae), a forensically important blowfly on rat carcass. Borneo J Resour Sci Technol 5(1):26–33

Khamboonruang J, Thitasut P, Pan-In S, Morakote N, Choochote W, Somboon P et al (1987) Filariasis in Tak Province, northwest Thailand: the presence of subperiodic variant *Wuchereria bancrofti*. Southeast Asian J Trop Med Public Health 18(2):218–222

Khanal L, Paudel BK, Acharya BK (2020) Community vulnerability to epidemics in Nepal: a high-resolution spatial assessment amidst COVID-19 pandemic. Npl J Zool 4(1):23–35

Khemrattakool P, Yanola J, Lumjuan N, Somboon P (2019) Pyriproxyfen-treated polypropylene sheets and resting boxes for controlling mosquitoes in livestock operations. Insects 10(2):55

Khongwichit S, Chansaenroj J, Thongmee T, Benjamanukul S, Wanlapakorn N, Chirathaworn C, Poovorawan Y (2021) Large-scale outbreak of Chikungunya virus infection in Thailand, 2018–2019. PLoS One 16(3):e0247314

Khoso FN, Wong SK, Chia SL, Lau WH (2015) Molecular identification of synanthropic flies in Malaysia. Aust J Basic & Appl Sci 9(5):469–476

Kiely B, Weekes MP, Stockdale AJ, Lever AML (2011) Severe typhus group rickettsiosis complicated by pulmonary edema in a returning traveler from Indonesia. Am J Trop Med Hyg 85(6):1121–1123

Kilpatrick AM, Kramer LD, Jones MJ, Marra PP, Daszak P (2006) West Nile virus epidemics in North America are driven by shifts in mosquito feeding behavior. PLoS Biol 4(4):e82

Kim G, Ha NY, Min CK, Kim HI, Yen NT, Lee KH, Oh I, Kang JS, Choi MS, Kim IS, Cho NH (2017) Diversification of *Orientia tsutsugamushi* genotypes by intragenic recombination and their potential expansion in endemic areas. PLoS Negl Trop Dis 11(3):e0005408

Kintz P, Godelar B, Tracqui A, Mangin P, Lugnier AA, Chaumont AJ (1990a) Fly larvae: a new toxicological method of investigation in forensic medicine. J Forensic Sci 35(1):204–207

Kintz P, Tracqui A, Mangin P (1990b) Toxicology and fly larvae on a putrefied cadaver. J Forensic Sci Soc 30(4):243–246

Kittayapong P, Kaeothaisong NO, Ninphanomchai S, Limohpasmanee W (2018) Combined sterile insect technique and incompatible insect technique: sex separation and quality of sterile *Aedes aegypti* male mosquitoes released in a pilot population suppression trial in Thailand. Parasit Vectors 11(Suppl 2):657

Kittayapong P, Ninphanomchai S, Limohpasmanee W, Chansang C, Chansang U, Mongkalangoon P (2019) Combined sterile insect technique and incompatible insect technique: the first proof-of-concept to suppress *Aedes aegypti* vector populations in semi-rural settings in Thailand. PLoS Negl Trop Dis 13(10):e0007771

Knight KL (1978) A new *Aedes* (*Finlaya*) mosquito from Thailand. Mosq Syst 10:106–116

Kobashi K, Kobayashi T, Nakamura-Uchiyama F, Ohnishi K (2010) A Japanese patient with chikungunya fever returning from Flores Island, Indonesia. Kansenshogaku Zasshi 84(4):457–459

Kobayashi M, Sasaki T, Saito N, Tamura K, Suzuki K, Watanabe H, Agui N (1999) Houseflies: not simple mechanical vectors of enterohemorrhagic *Escherichia coli* O157:H7. Am J Trop Med Hyg 61(4):625–629

Kobayashi M, Nihei N, Kurihara T (2002) Analysis of northern distribution of *Aedes albopictus* (Diptera: Culicidae) in Japan by geographical information system. J Med Entomol 39(1):4–11

Kobayashi D, Faizah AN, Amoa-Bosompem M, Watanabe M, Maekawa Y, Hayashi T, Higa Y, Sawabe K, Isawa H (2020) Analysis of *Trypanosoma* sequences from *Haemaphysalis flava* (Acari: Ixodidae) and *Tabanus rufidens* (Diptera: Tabanidae) collected in Ishikawa, Japan. Med Entomol Zool 71(4):279–288

Kobayashi D, Kuwata R, Kimujra T, Shimoda H, Fujita R, Faizah AN, Kai I, Matsumura R, Kuroda Y, Watanabe S, Kuniyoshi S, Yamauchi T, Watanabe M, Higa Y, Hayashi T, Shinomiya H, Maeda K, Kasai S, Sawabe K, Isawa H (2021a) Detection of Jingmenviruses in Japan with evidence of vertical transmission in ticks. Viruses 13(12):2547

Kobayashi D, Watanabe M, Faizah AN, Amoa-Bosompem M, Higa Y, Tsuda Y, Sawabe K, Isawa H (2021b) Discovery of a novel flavivirus (Flaviviridae)from the horse fly, *Tabanus rufidens* (Diptera: Tabanidae): the possible coevolutionary relationships between the classical insect specific flaviviruses and host dipteran insects. J Med Entomol 58(2):880–890

Kodama F, Yamaguchi H, Park E, Tatemoto K, Sashika M, Nakao R, Terauchi Y, Mizuma K, Orba Y, Kariwa H, Hagiwara K, Okazaki K, Goto A, Komagome R, Miyoshi M, Ito T, Yamano K, Yoshii K, Funaki C, Ishizuka M, Shigeno A, Itakura Y, Bell-Sakyi L, Edagawa S, Nagasaka A, Sakoda Y, Sawa H, Maeda K, Saijo M, Matsuno K (2021) A novel nairovirus associated with acute febrile illness in Hokkaido, Japan. Nat Commun 12(1):5539

Kumara TK, Abu HA, Che Salmah MR, Bhupinder S (2009a) The infestation of *Dermestes ater* (De Geer) on a human corpse in Malaysia. Trop Biomed 26(1):73–79

Kumara TK, Abu Hassan A, Che Salmah MR, Bhupinder S (2009b) Larval growth of the muscid fly, *Synthesiomyia nudiseta* (Wulp), a fly of forensic importance, in the indoor fluctuating temperatures of Malaysia. Trop Biomed 26(2):200–205

Kumara TK, Abu Hassan A, Che Salmah MR, Bhupinder S (2010) A report on the pupae of *Desmometopa* sp. (Diptera: Milichiidae) recovered from a human corpse in Malaysia. Trop Biomed 27(1):131–133

Kumara TK, Disney RH, Hassan AA, Flores M, Tan SH, Mohamed Z, Chesalmah MR, Bhupinder S (2012) Occurrence of oriental flies associated with indoor and outdoor human remains in the tropical climate of North Malaysia. J Vector Ecol 37(1):62–68

Kumara TK, Hassan AA, Salmah MR, Bhupinder S (2013) Larval growth of *Liosarcophaga dux* Thompson (Diptera: Sarcophagidae) under uncontrolled indoor temperatures in Malaysia. Southeast Asian J Trop Med Public Health 44(2):182–187

Kuo CC, Lee PL, Chen CH, Wang HC (2015) Surveillance of potential hosts and vectors of scrub typhus in Taiwan. Parasit Vectors 8:611

Kurahashi H, Leh MU (2007) The flies from Sarawak, East Malaysia (Diptera: Muscidae, Calliphoridae, Sarcophagidae and Tachinidae). Med Entomol Zool 58(4):261–273

Kurahashi H, Leh MU (2009) The blow flies from Sarawak, East Malaysia (Diptera: Calliphoridae), with practical keys and a checklist. Sarawak Mus J 66(87):277–303

Kurahashi H, Tan SH (2009) The sarcophagid flies from Peninsular Malaysia (Diptera: Sarcophagidae). Med Entomol Zool 60(4):283–296

Kurahashi H, Benjaphong N, Omar B (1997) Blow flies (Insecta: Diptera: Calliphoridae) of Malaysia and Singapore. Raffles Bul Zool 5(Suppl):1–88

Kurahashi H, Tan SH, Leh MU (2021) Keys to the flesh flies of Sarawak, East Malaysia, with description of two genera and eight new species (Diptera: Sarcophagidae). Med Entomol Zool 72(3):119–175

Lai DH, Wang QP, Li Z, Luckins AG, Reid SA, Lun ZR (2010) Investigations into human serum sensitivity expressed by stocks of *Trypanosoma brucei evansi*. Int J Parasitol 40(6):705–710

Laras K, Sukri NC, Larasati RP, Bangs MJ, Kosim R, Djauzi, Wandra T, Master J, Kosasih H, Hartati S, Beckett C, Sedyaningsih ER, Beecham HJ, Corwin AL (2005) Tracking the re-emergence of epidemic chikungunya virus in Indonesia. Trans R Soc Trop Med Hyg 99(2): 128–141

Lataillade LG, Vazeille M, Obadia T, Madec Y, Mousson L, Kamgang B, Chen CH, Failloux AB, Yen PS (2020) Risk of yellow fever virus transmission in the Asia-Pacific region. Nat Commun 11(1):5801

Ledda S, Foxi C, Puggioni G, Bechere R, Rocchigiani AM, Scivoli R, Coradduzza E, Cau S, Vento L, Satta G (2022) Experimental infection of *Aedes* (*Stegomyia*) *albopictus* and *Culex pipiens* mosquitoes with bluetongue virus. Med Vet Entomol 37(1):105–110

Leder K, Druce JD, Kwong JC (2013) Zika virus infection acquired during brief travel to Indonesia. Am J Trop Med Hyg 89(3):516–517

Lee HL (1989) Recovery of forensically important entomological specimens from human cadavers in Malaysia – an update. Malays J Pathol 11:33–36

Lee HL (1994) Larvae of *Eristalis* spp (Family: Syrphidae) found in human cadaver in Malaysia. J Biosci 5(1–2):67–68

Lee HL (1996a) Forensically important fly maggots recovered from human cadavers in Malaysia. Trop Biomed 13:93–94

Lee HL (1996b) Recovery of forensically important insect larvae from human cadavers in Malaysia (1993–1996). Malays J Pathol 18(2):125–127

Lee F (2019) Bovine ephemeral fever in Asia: recent status and research gaps. Viruses 11(5):412

Lee HL, Cheong WH (1982) Larva of *Hermetia* species from a dead human female. Southeast Asian J Trop Med Public Health 13(2):289–290

Lee HL, Marzuki T (1993) Preliminary observation of arthropods on carrion and its application to forensic entomology in Malaysia. Trop Biomed 10:5–8

Lee J, Ryu JS (2019) Current status of parasite infections in Indonesia: a literature review. Korean J Parasitol 57(4):329–339

Lee HL, Abdullah AG, Cheong WH (1984) The use of fly larvae from human corpses in determining the time of death: a review and some technical considerations. J Med Hlth Lab Tech Malaysia 9:15–17

Lee HL, Krishnasamy M, Abdullah AG, Jeffery J (2004) Review of forensically important entomological specimens in the period of 1972–2002. Trop Biomed 21(2):69–75

Leelayoova S, Siripattanapipong S, Manomat J, Piyaraj P, Tan-Ariya P (2017) Leishmaniasis in Thailand: a review of causative agents and situations. Am J Trop Med Hyg 96(3):534–542
Leng YJ (1997) Eighty-year research of Phlebotomine sandflies (Diptera: psychodidae) in China (1915–1995). I. Taxonomy and zoogeographical distribution. Parasite 4(2):107–126
Leung GHY, Baird RW, Druce J, Anstey NM (2015) Zika virus infection in Australia following a monkey bite in Indonesia. Southeast Asian J Trop Med Public Health 46(3):460–464
Li Y, Yan C, Liu D, He B, Tu C (2020) Seroepidemiological investigation of Crimean-Congo hemorrhagic fever virus in sheep and camels of Inner Mongolia of China. Vector Borne Zoonotic Dis 20(6):461–467
Li JC, Wang YN, Zhao J, Li H, Liu W (2021) A review on the epidemiology of severe fever with thrombocytopenia syndrome. Chin J Epidemiol 42(12):2226–2233. (in Chinese with English abstract)
Li Y, Luo Z, Hao Y, Zhang Y, Yang L, Li Z, Zhou Z, Li S (2022) Epidemiological features and spatial-temporal clustering of visceral leishmaniasis in mainland China from 2019 to 2021. Front Microbiol 13:959901
Liao LF, Simayi A, Xu YM, Wusobate, Yan SS (2021) Investigation on natural host animals of desert type of kala-azar in Tarim Basin. Bull Dis Cont Prev 36(1):1–6+33
Lien JC (2004) Pictorial keys to the mosquitoes of Taiwan. Yi Hsien Publication Co. Ltd, New Taipei, 178 pp
Lien JC, Wu TN, Lin CH, Lin CC, Weng MH (1996) Occurrence of *Culex pipiens* issp. *molestus* Forskal, 1775 in northern Taiwan. Chin J Parasitol Parasit Dis 9:19–26
Lin C, Jian SW, Teng HJ (2010) Surveys on Japanese encephalitis vectors in Taiwan area during 2004–2008. Taiwan Epidemiol Bull 26:202–218
Lin YH, Wu HH, Hsu EL, Chang NT, Luo YP (2012) Insecticide resistance in *Aedes aegypti* (L.) and *Aedes albopictus* (Skuse) larvae in southern Taiwan. Formos Entomol 32:107–121
Lin C, Wang CY, Teng HJ (2014) The study of dengue virus distribution in Taiwan from 2009 to 2011. Taiwan Epidemiol Bull 30(15):150
Liu QY (2022) Reported cases of vector-borne diseases in China, 2005–2020: epidemic trend, challenges in prevention and control, and related coping strategies. Chin J Vector Biol Control 33(1):1–7
Liu H, Li MH, Zhai YG, Meng WS, Sun XH, Cao YX, Fu SH, Wang HY, Xu LH, Tang Q, Liang GD (2010) Banna virus, China, 1987–2007. Emerg Infect Dis 16(3):514–517
Liu K, Hou X, Wang Y, Sun J, Xiao J, Li R, Lu L, Xu L, Sang S, Hu J, Wu H, Song X, Zhao N, Yan D, Li J, Liu X, Liu Q (2019) The driver of dengue fever incidence in two high-risk areas of China: a comparative study. Sci Rep 9(1):19510
Liu B, Ma J, Jiao Z, Gao X, Xiao J, Wang H (2021) Risk assessment for the Rift Valley fever occurrence in China: special concern in south-west border areas. Transbound Emerg Dis 68(2):445–457. (in Chinese)
Liu WL, Yu HY, Chen YX, Chen BY, Leaw SN, Lin CH, Su M-P, Tsai LS, Chen Y, Shiao SH, Xi Z, Jang ACC, Chen CH (2022) Lab-scale characterization and semi-field trials of *Wolbachia* Strain wAlbB in a Taiwan *Wolbachia* introgressed *Ae. aegypti* strain. PLoS Negl Trop Dis 16(1):e0010084
Lkhagvatseren S, Hogan KM, Boldbaatar B, von Fricken ME, Anderson BD, Pulscher LA, Caddell L, Nyamdawa P, Gray GC (2019) Discrepancies between self-reported tick bites and evidence of tick-borne disease exposure among nomadic Mongolian herders. Zoonoses Public Health 66(5):480–486
Lokida D, Hadi U, Lau CY, Kosasih H, Liang CJ, Rusli M, Sudarmono P, Lukman N, Laras K, Asdie RH, Murniati D, Utama IMS, Mubin RH, Karyana M, Gasem MH, Alisjahbana B (2020) Underdiagnoses of Rickettsia in patients hospitalized with acute fever in Indonesia: observational study results. BMC Infect Dis 20(1):364
Lumjuan N, Wicheer J, Leelapat P, Choochote W, Somboon P (2014) Identification and characterisation of *Aedes aegypti* aldehyde dehydrogenases involved in pyrethroid metabolism. PLoS One 9(7):e102746

Lun ZR, Reid SA, Lai DH, Li FJ (2009) Atypical human trypanosomiasis: a neglected disease or just an unlucky accident? Trends Parasitol 25(3):107–108

Macdonald WBG, Tink AR, Ouvrier RA, Menser MA, Silva LM, Nairn H, Hawkes RA (1989) Japanese encephalitis after a two-week holiday in Bali. Med J Aust 150(6):334–339

Maekawa Y, Higa Y, Sawabe K, Kasai S (2020) Expansion of the distribution of the *Aedes albopictus* in Japan. IASR 41:92–93 (in Japanese)

Magnus MM, Espósito DLA, Costa VAD, Melo PS, Costa-Lima C, Fonseca BALD, Addas-Carvalho M (2018) Risk of Zika virus transmission by blood donations in Brazil. Hematol Transfus Cell Ther 40(3):250–254

Mahara F, Koga K, Sawada S, Taniguchi T, Shigemi F, Tsuboi Y, Ooya A, Koyama H, Uchiyama T, Uchida T (1985) The first report of the rickettsial infections of spotted fever group in Japan: three clinical cases. J Asoc Infect Dis 59(8):1165–1172

Mahat NA, Zafarina Z, Jayaprakash PT (2009) Influence of rain and malathion on the oviposition and development of blowflies (Diptera: Calliphoridae) infesting rabbit carcasses in Kelantan, Malaysia. Forensic Sci Int 192:19–28

Mahat NA, Jayaprakash PT, Zafarina Z (2012) Malathion extraction from larvae of *Chrysomya megacephala* (Fabricius) (Diptera: Calliphoridae) for determining death due to malathion. Trop Biomed 29(1):9–17

Mahat NA, Yin CL, Jayaprakash PT (2014) Influence of paraquat on *Chrysomya megacephala* (Fabricius) (Diptera: Calliphoridae) infesting minced-beef substrates in Kelantan, Malays. J Forensic Sci 59(2):529–532

Malaysian Society of Infectious Diseases and Chemotherapy (MSIDC) (2020) Upcoming vaccines – chikungunya. In: Lim VKE (ed) Guidelines for adult immunization, 3rd edn. Percetakan Skyline Shd. Bhd, Kuala Lumpur, pp 186–237

Maramat R, Rahim NAA (2015) Forensically important flies associated with decomposing rabbit carcasses in mangrove forests in Kuching, Sarawak, Malaysia. Malays J Forensic Sci 6(1):79–83

Maramat R, Ibrahim N, Othman M, Rahim NAA (2021) Larvae of forensically important flies associated with rabbit carcasses placed in a closed room in Kuching, Sarawak. J Clin Health Sci 6(1):137–147

Matsumoto-Takahashi ELA, Iwagami M, Oyoshi K, Sasaki Y, Hongvanthong B, Kano S (2023) Deforestation inhibits malaria transmission in Lao PDR: a spatial epidemiology using Earth observation satellites. Trop Med Health 51(1):60

Mayxay M, Douangdala P, Vilayhong C, Phommasone K, Chansamouth V, Vongsouvath M, Rattanavong S, Chang K, Sengvilaipaseuth O, Chanthongthip A, Thongpaseuth S, Newton PN, Dubot-Pérès A (2020) Outcome of Japanese encephalitis virus (JEV) infection in pediatric and adult patients at Mahosot Hospital, Vientiane, Lao PDR. Am J Trop Med Hyg 104(2): 567–575

McMullen AR, Albayrak H, May FJ, Davis CT, Beasley DWC, Barrett ADT (2013) Molecular evolution of lineage 2 West Nile virus. J Gen Virol 94(Pt 2):318–325

Mead PS, Hills SL, Brooks JT (2018) Zika virus as a sexually transmitted pathogen. Curr Opin Infect Dis 31(1):39–44

Meiklejohn KA, Wallman JF, Dowton M (2011) DNA-based identification of forensically important Australian Sarcophagidae (Diptera). Int J Legal Med 125(1):27–32

Messina JP, Brady OJ, Scott TW, Zou C, Pigott DM, Duda KA, Bhatt S, Katzelnick L, Howes RE, Battle KE, Simmons CP, Hay SI (2014) Global spread of dengue virus types: mapping the 70 year history. Trend Microbiol 22(3):138–146

Minard G, Tran Van V, Tran FH, Melaun C, Klimpel S, Koch LK, Ly Huynh KK, Huynh TTT, Huu TN, Potier P, Mavingui P, Valiente Moro C (2017) Identification of sympatric cryptic species of *Aedes albopictus* subgroup in Vietnam: new perspectives in phylosymbiosis of insect vector. Parasit Vectors 10(1):276

Ministry of Health (MOH), Malaysia (2010) Clinical practice guidelines on management of dengue infection in adults, 2nd edn. https://www.moh.gov.my/moh/attachments/5502.pdf. Accessed 03 Mar 2023

Ministry of Health (MOH), Vietnam (2021) Guidelines for surveillance and prevention of malaria. https://thuvienphapluat.vn/van-ban/The-thao-Y-te/Quyet-dinh-4922-QD-BYT-2021-Huong-dan-giam-sat-benh-sot-ret-492499.aspx. Accessed 05 Jan 2024

Ministry of Health and Welfare (MHW), Japan (1999) Annual report on health and welfare. Social Security and National Life. https://www.mhlw.go.jp/www1/english/wp_5/vol1/p2c6s2.html. Accessed 27 Dec 2023

Ministry of Health of the Republic of Indonesia (MOH) (2022) Indonesia health profile 2021 (eds: Sibuea F, Hardhana B, Widiantini W). https://www.kemkes.go.id/id/profil-kesehatan-indonesia-2021. Accessed 03 Oct 2023

Ministry of Public Health (MOPH), Thailand (2022) Title. http://malaria.ddc.moph.go.th/malariaR10/index_v2.php. Accessed 20 Dec 2022

Miura Y, Inaba Y, Tsuda T, Tokuhisa S, Sato K, Akashi H, Matumoto M (1982) A survey of antibodies to arthropod-borne viruses in Indonesian cattle. Jpn J Vet Sci 44(6):857–863

Mizushima D, Amgalanbaatar T, Davaasuren B, Kayano M, Naransatsral S, Myagmarsuren P, Otgonsuren D, Enkhtaivan B, Davkharbayar B, Mungun-Ochir B, Baatarjargal P, Nyamdolgor U, Soyolmaa G, Altanchimeg A, Zoljargal M, Nguyen TT, Battsetseg B, Battur B, Inoue N, Yokoyama N, Suganuma K (2020) Nationwide seroprevalence surveillance of non-tsetse-transmitted horse trypanosomes in Mongolia. Parasite Epidemiol Control 10:e00158

Motoki MT, Fonseca DM, Miot EF, Demari-Silva B, Thammavong P, Chonephetsarath S, Phommavanh N, Hertz JC, Kittayapong P, Brey PT, Marcombe S (2019a) Population genetics of *Aedes albopictus* (Diptera: Culicidae) in its native range in Lao People's Democratic Republic. Parasit Vectors 12(1):477

Motoki MT, Vongphayloth K, Rueda LM, Miot EF, Hiscox A, Hertz JC, Brey PT (2019b) New records and updated checklist of mosquitoes (Diptera: Culicidae) from Lao People's Democratic Republic, with special emphasis on adult and larval surveillance in Khammuane Province. J Vector Ecol 44(1):76–88

Musinguzi SP, Suganuma K, Davaasuren B, Battsetseg B, Battur B, Inoue N (2022) A PCR-based survey of animal trypanosomes among domestic animals herded together in the Bayan-Ulgii and Khovd provinces of Mongolia. J Protozool Res 32(1–2):1–10

Muslim A, Fong MY, Mahmud R, Lau YL, Sivanandam S (2013) *Armigeres subalbatus* incriminated as a vector of zoonotic *Brugia pahangi* filariasis in suburban Kuala Lumpur, Peninsular Malaysia. Parasit Vectors 6(1):219

Musso D, Stramer SL, AABB Transfusion-Transmitted Diseases Committee, Busch MP, International Society of Blood Transfusion Working Party on Transfusion-Transmitted Infectious Diseases (2016) Zika virus: a new challenge for blood transfusion. Lancet 387(10032):1993–1994

Muto M, Boldbaatar B, Tsevel B, Dashzevge E, Yoshii K, Kariwa H (2015) Isolation and characterization of tick-borne encephalitis virus from *Ixodes persulcatus* in Mongolia in 2012. Ticks Tick Borne Dis 6(5):623–629

Myagmarsuren D (1996) Zoopilic and pasture flies of Mongolia, their agricultural importance. ScD thesis, Institute of Zoology, USSR Academy of Sciences, Sankt Petersburg, Russia. (in Russian)

Myint KSA, Kosasih H, Artika IM, Perkasa A, Puspita M, Ma'roef CN, Antonjaya U, Ledermann JP, Powers AM, Alisjahbana B (2014) West Nile virus documented in Indonesia from acute febrile illness specimens. Am J Trop Med Hyg 90(2):260–262

Nachaiwieng W, Yanola J, Chamnanya S, Lumjuan N, Somboon P (2021) Efficacy of five commercial household insecticide aerosol sprays against pyrethroid resistant *Aedes aegypti* and *Culex quinquefasciatus* mosquitoes in Thailand. Pestic Biochem Phys 178:104911

Nagorny SA, Ermakova LA, Golovchenko NV, Pshenichnaya NY, Zhuravlev AS, di Muccio T, Gradoni L (2019) Recent autochthonous cases of leishmaniasis in residents of the Republic of Dagestan, Russian Federation. Int J Infect Dis 86:171–174

Naik DG (2020) *Plasmodium knowlesi*-mediated zoonotic malaria: a challenge for elimination. Trop Parasitol 10(1):3–6

Nam VS, Yen NT, Phong TV, Ninh TU, Mai LQ, Lo LV, Nghia LT, Bektas A, Briscombe A, Aaskov JG, Ryan PA, Kay BH (2005) Elimination of dengue by community programs using *Mesocyclops* (copepoda) against *Aedes aegypti* in central Vietnam. Am J Trop Med Hyg 72(1):67–73

Nam VS, Phong TV, Yên NT, Kay BH (2009) Evaluation of the effectiveness and sustainability of the community model using the biological agent Mesocyclops in the prevention of dengue fever in Vietnam. Vietnam J Prev Med 101(2):90–101

Nam VS, Son TH, Phong TV, Tu TC, Duong TN, Anh DD, Yen NT, Lieu VT, Phuong NK, Hoang NV, Cécile C, Chau NV, Rahola N, Bañuls AL (2019) Diversity and ecology of sand flies (Diptera: Psychodidae), potential vectors of *Leishmania* in the Quang Ninh Province, Vietnam. J Med Entomol 57(1):259–265

Nam SJ, Oh YI, Kim HJ, Cheon DS, Noh SJ, Hong YJ (2020) Unusual case of severe fever with thrombocytopenia syndrome showing clinical manifestations in a companion dog. Vet Med Sci 6(3):353–358

Namgay R, Drukpa T, Wangdi T, Pemo D, Harbach RE, Somboon P (2018) A checklist of the *Anopheles* mosquito species (Diptera: Culicidae) in Bhutan. Acta Trop 188:206–212

Nandi BC (2002) Fauna of India and the adjacent countries – Diptera (Volume X) Sarcophagidae. Zoological Survey of India, Ministry of Environment, Forest and Climate Change (MoEFCC), Government of India, Kolkata, 608 pp

Naszeerah WN (2015) Characterization of anopheline species composition along the Bhutan-India border region. PhD thesis, Yale University, Public Health Theses, New Haven, Connecticut. http://elischolar.library.yale.edu/ysphtdl/1205. Accessed 25 Dec 2023

National Institute of Infectious Diseases (NIID) (2023a) Oz virus infection in human. https://www.niid.go.jp/niid/ja/diseases/a/ozv/2629-cepr/12134-ozv-eng.html. Accessed 27 Dec 2023

National Institute of Infectious Diseases (NIID) (2023b) Severe fever with thrombocytopenia syndrome (SFTS). https://www.niid.go.jp/niid/ja/diseases/sa/sfts.html. Accessed 27 Dec 2023

National Statistics Office of Mongolia (NSOM) (2023). https://www.1212.mn/mn/statistic/statcate/573051/table-view/DT_NSO_0300_071V3. Accessed 15 Dec 2023

Nazni WA, Jeffery J, Sa'diyah I, Noorjuliana WM, Chen CD, Rohayu SA, Hafizam AH, Lee HL (2008) First report of maggots of family: *Piophilidae* recovered from human cadavers in Malaysia. Trop Biomed 25(2):173–175

Neronov VM, Strelkova MV, Shurkhal AA, Luschekina AA, Artemyev MM (1987) Natural focality of zoonotic cutaneous leishmaniasis in the Mongolian People's Republic; results and objectives of integrated research. Folia Parasitol 34(1):1–9

Ng LC, Tan LK, Tan CH, Tan SSY, Hapuarachchi HC, Pok KY, Lai YL, Lam-Phua SG, Bucht G, Lin RTP, Leo YS, Tan BH, Han HK, Ooi PLS, James L, Khoo SP (2009) Entomologic and virologic investigation of chikungunya, Singapore. Emerg Infect Dis 15(8):1243–1249

Nurwidayati A, Frederika PP, Sudomo M (2019) Schistosomiasis fluctuations in endemic areas of Central Sulawesi Province in 2011–2018. Health Res Bull 47(3):199–206

O'Neill S (2018) The use of *Wolbachia* by the world mosquito program to interrupt transmission of *Aedes aegypti* transmitted viruses. Adv Exp Med Biol 1062:355–360

O'Neill SL, Ryan PA, Turley AP, Wilson G, Retzki K, Iturbe-Ormaetxe I et al (2019) Scaled deployment of *Wolbachia* to protect the community from dengue and other *Aedes* transmitted arboviruses. Gates Open Res 2:36

Obsomer V, Defourny P, Coosemans M (2007) The *Anopheles dirus* complex: spatial distribution and environmental drivers. Malar J 6:26

Oemijati S, Carney WP, Clarke MD, Joesoef A, Cross JH, Hadidjaja P (1974) Schistosomiasis and other human parasitoses of Lake Lindu in Central Sulawesi (Celebes), Indonesia. Am J Trop Med Hyg 23(3):385–392

Olson JG, Ksiazek TG, Suhandiman, Triwibowo (1981) Zika virus, a cause of fever in Central Java, Indonesia. Trans R Soc Trop Med Hyg 75(3):389–393

Olson JG, Ksiazek TG, Gubler DJ, Lubis SI, Simanjuntak G, Lee VH, Nalim S, Juslis K, See R (1983) A survey for arboviral antibodies in sera of humans and animals in Lombok, Republic of Indonesia. Ann Trop Med Parasitol 77(2):131–137

Olson JG, Ksiazek TG, Lee VH, Tan R, Shope RE (1985) Isolation of Japanese encephalitis virus from *Anopheles annularis* and *Anopheles vagus* in Lombok, Indonesia. Trans R Soc Trop Med Hyg 79(6):845–847

Omar B (2002) Key to third instar larvae of flies of forensic importance in Malaysia. In: Greenberg B, Kunich JC (eds) Entomology and law: flies as forensic indicators. Cambridge University Press, Cambridge, pp 143–147

Omar B, Marwi MA, Sulaiman S, Oothuman P (1994a) Dipteran succession in monkey carrion at a rubber tree plantation in Malaysia. Trop Biomed 11:77–82

Omar B, Marwi MA, Mansar AH, Rahman MS, Oothuman P (1994b) Maggots of *Synthesiomyia nudiseta* (Wulp) (Diptera: Muscidae) as decomposers of corpses found indoors in Malaysia. Trop Biomed 11:145–148

Oya A, Okuno T, Ogata T, Kobayashi I, Matsuyama T (1961) Akabane, a new arbovirus isolated in Japan. Jpn J Med Sci Biol 14(3):101–108

Ozbel Y (2013) The infections transmitted by sand flies in Turkey. Ankara Univ Vet Fak Derg 60:225–228

Palmieri JR, Ratiwayanto S, Masbar S, Tirtokusumo S, Rusch J, Marwoto HA (1985) Evidence of possible natural infections of man with *Brugia pahangi* in South Kalimantan (Borneo), Indonesia. Trop Geogr Med 37(3):239–244

Pandey BD, Karabatsos N, Cropp B, Tagaki M, Tsuda Y, Ichinose A, Igarashi A (1999) Identification of a flavivirus isolated from mosquitoes in Chiang Mai Thailand. Southeast Asian J Trop Med Public Health 30(1):161–165

Pandey BD, Pandey K, Dumre SP, Morita K, Costello A (2022) Struggling with a new dengue epidemic in Nepal. Lancet Infect Dis 23(1):16–17

Parola P (1998) Murine typhus in travelers returning from Indonesia. Emerg Infect Dis 4(4):677–680

Pastorino B, Sengvilaipaseuth O, Chanthongthip A, Vongsouvath M, Souksakhone C, Mayxay M, Thirion L, Newton PN, de Lamballerie X, Dubot-Peres A (2019) Low Zika virus seroprevalence in Vientiane, Laos, 2003–2015. Am J Trop Med Hyg 100(3):639–642

Pearce JC, Learoyd TP, Langendorf BJ, Logan JG (2018) Japanese encephalitis: the vectors, ecology and potential for expansion. J Travel Med 25(Suppl 1):S16–S26

Peng SH, Yang SL, Tang SE, Wang TC, Hsu TC, Su CL, Chen MY, Shimojima M, Yoshikawa T, Shu PY (2020) Human case of severe fever with thrombocytopenia syndrome virus infection, Taiwan, 2019. Emerg Infect Dis 26(7):1612–1614

Perkasa A, Yudhaputri F, Haryanto S, Hayati RF, Ma'roef CN, Antonjaya U, Yohan B, Myint KSA, Ledermann JP, Rosenberg R, Powers AM, Sasmono RT (2016) Isolation of Zika virus from febrile patient, Indonesia. Emerg Infect Dis 22(5):924–925

Phong TV, Duoc VT, Duc HM, Le NH, Yen NT, Dong TD, Cuong TC, Anh DTV, Tu TC, Soai NV, Son TH, Anh NTM, Minh HD, Hang NM, Duong TN (2015) Chikungunya virus and it's vector at 5 cross-border provinces between Vietnam, Laos and Cambodia, 2012–2014. Vietnam J Prev Med 25(12+13):172–173

Phong TV, Dung NT, Tam NL, Tu TC, Cuong TC, Nam VS (2022) Community mobilization and socialization for vector control. Vietnam J Prev Med 32(2):218–224

Phongmany S, Rolain JM, Phetsouvanh R, Blacksell SD, Soukkhaseum V, Rasachack B, Phiasakha K, Soukkhaseum S, Frichithavong K, Chu V, Keolouangkhot V, Martinez-Aussel B, Chang K, Darasavath C, Rattanavong O, Sisouphone S, Mayxay M, Vidamaly S, Parola P, Thammavong C, Heuangvongsy M, Syhavong B, Raoult D, White NJ, Newton PN (2006) Rickettsial infections and fever, Vientiane, Laos. Emerg Infect Dis 12(2):256–262

Phuentshok Y, Dorji K, Zangpo T, Davidson SA, Takhampunya R, Tenzinla T, Dorjee C, Morris RS, Jolly PD, Dorjee S, McKenzie JS (2018) Survey and phylogenetic analysis of rodents and important rodent-borne zoonotic pathogens in Gedu, Bhutan. Korean J Parasitol 56(5):521–525

Plernsub S, Stenhouse SA, Tippawangkosol P, Lumjuan N, Yanola J, Choochote W, Somboon P (2013) Relative developmental and reproductive fitness associated with F1534C homozygous knockdown resistant gene in *Aedes aegypti* from Thailand. Trop Biomed 30(4):621–630

Plernsub S, Saingamsook J, Yanola J, Lumjuan N, Tippawangkosol P, Sukontason K, Walton C, Somboon P (2016a) Additive effect of knockdown resistance mutations, S989P, V1016G and F1534C, in a heterozygous genotype conferring pyrethroid resistance in *Aedes aegypti* in Thailand. Parasit Vectors 9(1):417

Plernsub S, Saingamsook J, Yanola J, Lumjuan N, Tippawangkosol P, Walton C, Somboon P (2016b) Temporal frequency of knockdown resistance mutations, F1534C and V1016G, in *Aedes aegypti* in Chiang Mai city, Thailand and the impact of the mutations on the efficiency of thermal fogging spray with pyrethroids. Acta Trop 162:125–132

Pothikasikorn J, Bangs MJ, Boonplueang R, Chareonviriyaphap T (2008) Susceptibility of various mosquitoes in Thailand to nocturnal subperiodic *Wuchereria bancrofti*. J Vector Ecol 33(2):313–320

Prapanthadara L, Reunkum W, Leelapat P, Suwan W, Yanola J, Somboon P (2005) Glutathione S-transferase isoenzymes and the DDTase activity in two DDT-resisttant strains of *Aedes aegypti*. Dengue Bull 29:183–191

Punjabi NH, Taylor WRJ, Murphy GS, Purwaningsih S, Picarima H, Sisson J, Olson JG, Baso S, Wangsasaputra F, Lesmana M, Oyofo BA, Simanjuntak CH, Subekti D, Corwin AL, Richie TL (2012) Etiology of acute, non-malaria, febrile illnesses in Jayapura, Northeastern Papua, Indonesia. Am J Trop Med Hyg 86(1):46–51

Purnomo Edi S, Ibrahim A, Sukoco R, Bunali L, Taguchi M, Kato T, Yanase T, Shirafuji H (2017) Molecular characterization of an Akabane virus isolate from West Java, Indonesia. J Vet Med Sci 79(4):774–779

Putaporntip C, Kuamsab N, Pattanawong U, Yanmanee S, Seethamchai S, Jongwutiwes S (2021) *Plasmodium cynomolgi* co-infections among symptomatic malaria patients, Thailand. Emerg Infect Dis 27(2):590–593

Putaporntip C, Kuamsab N, Seethamchai S, Pattanawong U, Rojrung R, Yanmanee S, Cheng CW, Jongwutiwes S (2022) Cryptic *Plasmodium inui* and *Plasmodium fieldi* infections among symptomatic malaria patients in Thailand. Clin Infect Dis 75(5):805–812

Putra IGAA, Adi AAAM, Astawa INM, Kardena IM, Wandia IN, Soma IG, Brotcorne F, Fuentes A (2022) First survey on seroprevalence of Japanese encephalitis in long-tailed macaques (*Macaca fascicularis*) in Bali, Indonesia. Vet World 15:1341–1346

Putri MP, Olson JG, Corwin AL, Rahardjo E, Baskoro AD, Hernowati TE, Church CJ, Dasch GA, Soendoro M, Kelly DJ, Yanuwiadi B, Sardjono TW, Soeatmadji DW, Strickman D, Saragih JM, Hakim L, Richards AL, Widodo MA, Roebiyoso (1997) Seroepidemiologic evidence for murine and scrub typhus in Malang, Indonesia. Am J Trop Med Hyg 57(1):91–95

Quang LC, Thao NTT, Hai DT, Quoc DK, Huy LH, Ly LTT, Tung LT, Khanh LHK, Manh DH, Tam PT, My NTK, Hoang PH, Dung NH, Dung MV, Thuong NV (2022) Characteristics of cases infected with chikungunya virus in An Phu district, An Giang province in 2020. Vietnam J Prev Med 32(2):86–92

Quoc VV, Duoc VT, Phong TV, Duc HM, Nam VS, Yen NT, Son TH, Tu TC, Cuong TC, Dung NT, Hung DN, Suu NV, Duong TN (2018) Vectors of Japanese encephalitis in some target North-west mountainous region. Vietnam J Prev Med 28(7):96 (in Vietnamese with English summary)

Rahayu N, Suryatinah Y, Ridha MR, Kusumaningtyas H, Ipa M, Dhewantara PW (2020) Detecting *Brugia malayi* in lymphatic filariasis mosquito vector in North Hulu Sungai District, South Kalimantan, Indonesia. Adv Health Sci Res 22:179–184

Rajagopal K, Dhang CC, Lee HL, Ahmad NW, Ibrahim S, Aris EM (2008) Estimated post-mortem intervals (PMI) of pathologist and entomologist in Malaysia: a comparison. Proc Asean Congr Trop Med Parasitol 3:21–27

Rashid RA, Osman K, Ismail MI, Zuha RM, Hassan RA (2008) Determination of malathion levels and the effect of malathion on the growth of *Chrysomya megacephala* (Fabricius) in malathion-exposed rat carcass. Trop Biomed 25(3):184–190

Rashid RA, Khairul O, Zuha RM, Heo CC (2010) An observation on the decomposition process of gasoline-ingested monkey carcasses in a secondary forest in Malaysia. Trop Biomed 27(3):373–383

Rashid RA, Zulkifli NF, Rashid RA, Rosli SF, Sulaiman SHS, Ahmad NW (2012) Effects of Ketum extract on blowfly *Chrysomya megacephala* development and detection of mitragynine in larvae sample. In: 2012 IEEE symposium on business, engineering and industrial application, Bandung, Indonesia. https://doi.org/10.1109/ISBEIA.2012.6422898. Accessed 03 Mar 2023

Rashid AR, Siti AS, Siti FR, Reena AR, Sharifah HSS, Nurul FZ, Nazni WA (2013) Forensic implications of blowfly *Chrysomya rufifacies* (Calliphoridae: Diptera) development rates affected by Ketum extract. World Acad Sci Eng Technol Int J Med Health Sci 7(7): 332–336

Ratcliffe ST, Webb DW, Weinzievr RA, Robertson HM (2003) PCR-RFLP identification of Diptera (Calliphoridae, Muscidae and Sarcophagidae) – a generally applicable method. J Forensic Sci 48(4):783–785

Rattanarithikul R, Harbach RE, Harrison BA, Panthusiri P, Jones JW, Coleman RE (2005) Illustrated keys to the mosquitoes of Thailand, II Genera *Culex* and *Lutzia*. Southeast Asian J Trop Med Public Health 36(Suppl 2):1–97

Rattanarithikul R, Harrison BA, Harbach RE, Panthusiri P, Coleman R (2006) Illustrated keys to the mosquitoes of Thailand, IV. *Anopheles*. Southeast Asian J Trop Med Public Health 37(Suppl 2):1–128

Reid JA (1953) Notes on house-flies and blowflies in Malaya. Bull Inst Med Res Kuala Lumpur 7:1–26

Reid SA, Copeman DB (2000) Surveys in Papua New Guinea to detect the presence of Trypanosoma evansi infection. Aust Vet J 78(12):843–845

Richards AL (2022) Available data do not suggest *Rickettsia rickettsii* has been found in Indonesia. EcoHealth 19(1):3–4

Richards AL, Rahardjo E, Rusjdi AF, Kelly DJ, Dasch GA, Church CJ, Bangs MJ (2002) Evidence of *Rickettsia typhi* and the potential for murine typhus in Jayapura, Irian Jaya, Indonesia. Am J Trop Med Hyg 66(4):431–434

Riyong D, Choochote W, Jitpakdi A, Suvannadabba S, Leemingsawat S, Chaitong U (2000) Autogenous *Aedes togoi* sub-colony (Chanthaburi, Thailand strain), an efficient laboratory vector in study of filariasis. Southeast Asian J Trop Med Public Health 31(2):246–251

Rodhain F (2000) The state of vector-borne diseases in Indonesia. Bull Soc Pathol Exot 93(5):348–352

Rohdendorf BB (1937) Fauna U.S.S.R., insects, diptera, vol 19 No 1, Family Sarcophagidae, part 1. Academy of Sciences of the U.S.S.R, Moscow. (in Russian and German)

Rojanapanus S, Toothong T, Boondej P, Thammapalo S, Khuanyoung N, Santabutr W et al (2019) How Thailand eliminated lymphatic filariasis as a public health problem. Infect Dis Poverty 8(1):38

Rojanapithayakorn W (1998) Dengue haemorrhagic fever in Thailand. Dengue Bull 22:60–72

Roziah A, Tan SH, Lee HL, Mohamed Z (2015) Mitochondrial and nuclear DNA for identification of forensically important flesh flies (Sarcophagidae: *Boettcherisca* spp). Entomol Ornithol Herpetol 4:163

Ruchusatsawat K, Wongjaroen P, Posanacharoen A, Rodriguez-Barraquer I, Sangkitporn S, Derek AT, Cummings DAT, Salje H (2019) Long-term circulation of Zika virus in Thailand: an observational study. Lancet Infect Dis 19(4):439–446

Sadanandane C, Gunasekaran K, Panneer D, Subbarao SK, Rahi M, Vijayakumar B et al (2022) Studies on the fitness characteristics of *w*Mel- and *w*AlbB-introgressed *Aedes aegypti* (Pud) lines in comparison with *w*Mel- and *w*AlbB-transinfected *Aedes aegypti* (Aus) and wild-type *Aedes aegypti* (Pud) lines. Front Microbiol 13:947857

Saeung A, Hempolchom C, Baimai V, Thongsahuan S, Taai K, Jariyapan N et al (2013) Susceptibility of eight species members in the *Anopheles hyrcanus* group to nocturnally subperiodic *Brugia malayi*. Parasit Vectors 6:5

Saigusa K, Takamiya M, Aoki Y (2005) Species identification of the forensically important flies in Iwate prefecture, Japan based on mitochondrial cytochrome oxidase gene subunit I (COI) sequences. Legal Med 7(3):175–178

Saingam P, Pidtana K, Suida P, Poramathikul K, Lertsethtakarn P, Kuntawunginn W et al (2022) Case series of three malaria patients from Thailand infected with the simian parasite, *Plasmodium cynomolgi*. Malar J 21(1):142

Saingamsook J, Saeung A, Yanola J, Lumjuan N, Walton C, Somboon P (2017) A multiplex PCR for detection of knockdown resistance mutations, V1016G and F1534C, in pyrethroid-resistant *Aedes aegypti*. Parasit Vectors 10(1):465

Saingamsook J, Yanola J, Lumjuan N, Walton C, Somboon P (2019) Investigation of relative development and reproductivity fitness cost in three insecticide-resistant strains of *Aedes aegypti* from Thailand. Insects 10(9):265

Salleh AFM, Marwi MA, Jeffery J, Hamid NAA, Zuha RM, Omar B (2007) Review of forensic entomology cases from Kuala Lumpur Hospital and Hospital Universiti Kebangsaan Malaysia, 2002. J Trop Med Parasitol 30:51–54

Salleh AFM, Marwi MA, Jeffery J, Hamid NAA, Zuha RM, Omar B (2008) A review of forensic entomology cases at Ipoh Hospital and Hospital Universiti Kebangsaan Malaysia for the year 2003. Indones J Leg Forensic Sci 1(1):1–4

Salleh AFM, Marwi MA, Abdullah SR, Azil AH, Omar B (2014) Effects of temperatures on development of a forensically important fly *Chrysomya villeneuvi* Patton (Diptera: Calliphoridae). Sains Malays 43(10):1537–1542 (in Malay with English abstract)

Samerjai C, Sukontason KL, Sontigun N, Sukontason K, Klong-klaew T, Chareonviriyaphap T, Kurahashi H, Klimpel S, Kochmann J, Saeung A, Somboon P, Wannasan A (2020) Mitochondrial DNA-based identification of forensically important flesh flies (Diptera: Sarcophagidae) in Thailand. Insects 11(1):2

Saminathan M, Singh KP, Khorajiya JH, Dinesh M, Vineetha S, Maity M, Rahman AF, Misri J, Malik YS, Gupta VK, Singh RK, Dhama K (2020) An updated review on bluetongue virus: epidemiology, pathobiology, and advances in diagnosis and control with special reference to India. Vet Q 40(1):258–321

Sanchez KL, Greenwood AD, Nielsen A, Nugraha RTP, Prameswari W, Nurillah A, Agustina F, Campbell-Smith G, Dharmayanthi AB, Pratama R, Exploitasia I, Baird JK (2022) *Plasmodium pitheci* malaria in Bornean orang-utans at a rehabilitation Centre in West Kalimantan, Indonesia. Malar J 21(1):280

Sasmono RT, Wahid I, Trimarsanto H, Yohan B, Wahyuni S, Hertanto M, Yusuf I, Mubin H, Ganda IJ, Latief R, Bifani PJ, Shi PY, Schreiber MJ (2015) Genomic analysis and growth characteristic of dengue viruses from Makassar, Indonesia. Infect Genet Evol 32:165–177

Sasmono RT, Dhenni R, Yohan B, Pronyk P, Hadinegoro SR, Soepardi EJ, Ma'roef CN, Satari HI, Menzies H, Hawley WA, Powers AM, Rosenberg R, Myint KSA, Soebandrio A (2018) Zika virus seropositivity in 1–4-year-old children, Indonesia, 2014. Emerg Infect Dis 24(9):1740–1743

Sasmono RT, Santoso MS, Pamai YWB, Yohan B, Afida AM, Denis D, Hutagalung IA, Johar E, Hayati RF, Yudhaputri FA, Haryanto S, Stubbs SCB, Blacklaws BA, Myint KSA, Frost SDW (2020) Distinct dengue disease epidemiology, clinical, and diagnosis features in Western, Central, and Eastern regions of Indonesia, 2017–2019. Front Med (Lausanne) 7:582235

Sasmono RT, Johar E, Yohan B, Ma'roef CN, Pronyk P, Hadinegoro SR, Soepardi EJ, Bouckenooghe A, Hawley WA, Rosenberg R, Powers AM, Soebandrio A, Myint KSA (2021) Spatiotemporal heterogeneity of Zika virus transmission in Indonesia: serosurveillance data from a pediatric population. Am J Trop Med Hyg 104(6):2220–2223

Satjanadumrong J, Robinson MT, Hughes T, Blacksell SD (2019) Distribution and ecological drivers of spotted fever group rickettsia in Asia. EcoHealth 16(4):611–626

Satjawongvanit H, Phumee A, Tiawsirisup S, Sungpradit S, Brownell N, Siriyasatien P et al (2019) Molecular analysis of canine filaria and its *Wolbachia* endosymbionts in domestic dogs collected from two animal university hospitals in Bangkok Metropolitan Region, Thailand. Pathogens 8(3):114

Sato T, Matsumoto F, Abe T, Nihei N, Kobayashi M (2012) Reacent distribution of *Aedes albopictus* in Iwate Prefecture, Northern Japan and analysis of climatic conditions for northern limit of geographic distribution by geographic information systems. Med Entomol Zool 63(3):195–204

Sato Y, Mekata H, Sudaryatma PE, Kirino Y, Yamamoto S, Ando S, Sugimoto T, Okabayashi T (2021) Isolation of severe fever with thrombocytopenia syndrome virus from various tick species in Area with human severe fever with thrombocytopenia syndrome cases. Vector Borne Zoonotic Dis 21(5):378–384

Sawabe K, Takagi M, Tsuda Y, Tuno N (2003) Molecular variation and phylogeny of the *Anopheles minimus* complex (Diptera: Culicidae) inhabiting Southeast Asian countries, based on ribosomal DNA internal transcribed spacers, ITS1 and 2, and the 28S D3 sequences. Southeast Asian J Trop Med Public Health 34(4):771–780

Sawabe K, Hoshino K, Isawa H, Sasaki T, Hayashi T, Tsuda Y, Kurahashi H, Tanabayashi K, Hotta A, Saito T, Yamada A, Kobayashi M (2006) Detection and isolation of highly pathogenic H5N1 avian influenza A viruses from blow flies collected in the vicinity of an infected poultry farm in Kyoto, Japan, 2004. Am J Trop Med Hyg 75(2):327–332

Sawabe K, Tanabayashi K, Hotta A, Isawa H, Sasaki T, Yamada A, Kurahashi H, Shudo C, Kobayashi M (2009) Survival of avian H5N1 influenza A viruses in *Calliphora nigribarbis* (Diptera: Calliphoridae). J Med Entomol 46(4):852–855

Sawabe K, Isawa H, Hoshino K, Sasaki T, Roychoudhury S, Higa Y, Kasai S, Tsuda Y, Nishiumi I, Hisai N, Hamao S, Kobayashi M (2010) Host-feeding habits of *Culex pipiens* and *Aedes albopictus* (Diptera: Culicidae) collected at the urban and suburban residential areas of Japan. J Med Entomol 47(3):442–450

Sawabe K, Imanishi-Kobayashi N, Maekawa Y, Higa Y, Kim KS, Hoshino K, Tsuda Y, Hayashi T, Nihei N, Takai K, Kurihara T, Kobayashi M (2021) Updated distribution of anopheline mosquitoes in Hokkaido, Japan, and the first evidence of *Anopheles belenrae* in Japan. Parisit Vectors 14:494

Sawitri DH, Wardhana AH, Sadikin M, Wibowo H (2019) Detection of Surra (trypanosomiasis) positivity in humans in an outbreak area of Indonesia. Med J Indones 28(2):196–202

Scholte FEM, Tas A, Martina BEE, Cordioli P, Narayanan K, Makino S, Snijder EJ, van Hemert MJ (2013) Characterization of synthetic chikungunya viruses based on the consensus sequence of recent E1-226 V isolates. PLoS One 8(8):e71047

Schuffenecker I, Iteman I, Michault A, Murri S, Frangeul L, Vaney MC, Lavenir R, Pardigon N, Reynes JM, Pettinelli F, Biscornet L, Diancourt L, Michel S, Duquerroy S, Guigon G, Frenkiel MP, Bréhin AC, Cubito N, Desprès P, Kunst F, Rey FA, Zeller H, Brisse S (2006) Genome microevolution of chikungunya viruses causing the Indian Ocean outbreak. PLoS Med 3(7):e263

Schuh AJ, Guzman H, Tesh RB, Barrett ADT (2013) Genetic diversity of Japanese encephalitis virus isolates obtained from the Indonesian Archipelago between 1974 and 1987. Vector Borne Zoonotic Dis 13(7):479–488

Schuh AJ, Ward MJ, Leigh Brown AJ, Barrett ADT (2014) Dynamics of the emergence and establishment of a newly dominant genotype of Japanese encephalitis virus throughout asia. J Virol 88(8):4522–4532

Sendow I (2002) Isolasi virus bluetongue serotipe 1, 6 dan 21 dari Serangga Nyamuk di Jawa Barat (Isolation of bluetongue virus serotype 1, 6, and 21 from mosquitoes in West Java). Jurnal Ilmu Ternak Dan Veteriner (Indones J Anim Vet Sci) 7(4):272–278. (in Indonesian)

Sendow I (2005) Studi Patogenitas Isolat Lokal Virus *Bluetongue* pada Domba Lokal dan Impor [Pathogenicity study of local bluetongue virus isolates in local and imported sheep]. Jurnal Ilmu Ternak Dan Veteriner (Indones J Anim Vet Sci) 10(1):51–62 (in Indonesian)

Sendow I, Daniels P, Soleha E, Hunt N, Ronohardjo P (1991) Isolation of bluetongue viral serotypes 7 and 9 from healthy sentinel cattle in West Java, Indonesia. Aust Vet J 68(12):405–406

Sendow I, Soleha E, Daniels P, Achdiyati J, Sebayang D, Karma K, Erasmus B (1993a) Isolation of bluetongue virus serotypes 1, 21 and 23 from healthy sentinel cattle in Irian Jaya, Indonesia. Aust Vet J 70(6):229–230

Sendow I, Sukarsih, Soleha E, Daniels PW (1994) Isolation of bluetongue virus serotype 21 from mosquitoes in West Java, Indonesia. In Penyakit Hewan, 26(47, 21–25). Balai Penelitian Veteriner, Bogor

Sendow I, Sukarsih, Soleha E, Erasmus BJ, Daniels PW (1993b) Isolation of bluetongue virus serotype 21 from *Culicoides* spp. in Indonesia. Vet Microbiol 36(3–4):349–353

Senior-White RA, Aubertin D, Smart J (1940) The fauna of British India, including the remainder of the oriental region. In: Diptera vol 6, Family Calliphoridae. Taylor and Francis, London, 288 pp

Sharma M, Singh D, Sharma AK (2015) Molecular identification of forensically important Indian species of flesh flies (Diptera: Sarcophagidae) by using COI gene of mitochondrial DNA. J Forensic Res 6:316

Shi DM, Song L, Sun ML, Li DD, Wang J, Zou HL, Pan XB (2022) Analysis of 1153 cases of forest encephalitis reported by domestic documents. Chin J Ind Hyg Occup Dis 40(11):841–845 (in Chinese)

Sick F, Beer M, Kampen H, Wernike K (2019) *Culicoides* biting midges – underestimated vectors for arboviruses of public health and veterinary importance. Viruses 11(4):376

Silahuddin SA, Latif B, Kurahashi H, Heo CC (2015) The importance of habitat in the ecology of decomposition on rabbit carcasses in Malaysia: implications in forensic entomology. J Med Entomol 52(1):9–23

Singh B, Daneshvar C (2013) Human infections and detection of *Plasmodium knowlesi*. Clin Microbiol Rev 26(2):165–184

Singh I, Kurahashi H, Kano R (1979) A preliminary key to the common calliphorid flies of Peninsular Malaysia (Insecta: Diptera). Bull Tokyo Med Dent Univ 26(1):5–24

Siriyasatien P, Suwanpimolkul G, Wilde H, Suankratay C (2010) Autochthonous visceral leishmaniasis in a human immunodeficiency virus (HIV)-infected patient: the first in Thailand and review of the literature. Am J Trop Med Hyg 82(1):4–8

Sitohang V, Sariwati E, Fajariyani SB, Hwang D, Kurnia B, Hapsari RK, Laihad FJ, Sumiwi ME, Pronyk P, Hawley WA (2018) Malaria elimination in Indonesia: halfway there. Lancet Glob Health 6(6):e604–e606

Smith KGV (1986) A manual of forensic entomology. Trustees of the British Museum (Natural History), London, 250 pp

Somboon P, Aramrattana A, Lines J, Webber R (1998) Entomological and epidemiological investigations of malaria transmission in relation to population movements in forest areas of north-west Thailand. Southeast Asian J Trop Med Public Health 29(1):3–9

Somboon P, Choochote W, Khamboonruang C, Keha P, Suwanphanit P, Sukontasan K et al (1989) Studies on the Japanese encephalitis vectors in Amphoe Muang, Chiang Mai, Northern Thailand. Southeast Asian J Trop Med Public Health 20(1):9–17

Somboon P, Lines J, Aramrattana A, Chitprarop U, Prajakwong S, Khamboonruang C (1995) Entomological evaluation of community-wide use of lambdacyhalothrin-impregnated bed nets against malaria in a border area of north-west Thailand. Trans R Soc Trop Med Hyg 89(3):248–254

Somboon P, Prapanthadara L, Suwonkerd W (2003) Insecticide susceptibility tests of *Anopheles minimus s.l., Aedes aegypti, Aedes albopictus,* and *Culex quinquefasciatus* in northern Thailand. Southeast Asian J Trop Med Public Health 34(1):87–93

Somboon P, Phanitchakun T, Namgay R, Harbach RE (2022a) The *Anopheles baileyi* species complex (Diptera: Culicidae: Anophelinae) in Bhutan. Acta Trop 226:106241

Somboon P, Phanitchakun T, Namgay R, Harbach RE (2022b) The *Anopheles lindesayi* species complex (Diptera: Culicidae) in Bhutan. J Med Entomol 59(4):1236–1251

Somboon P, Rory A, Tsuda Y, Takagi M, Harbach RE (2010) Systematics of *Anopheles* (*Cellia*) *yaeyamaensis* sp. n. alias species E of the *An. minimus* complex in southeastern Asia (Diptera: Culicidae). Zootaxa 2651:43–51

Somboon P, Saeung A, Saingamsook J, Namgay R, Harbach RE (2024) A new species of the *Anopheles barbirostris* complex (Diptera: Culicidae) from Bhutan. J Med Entomol 61(2):377–388

Somboon P, Walton C, Sharpe RG, Higa Y, Tuno N, Tsuda Y, Takagi M (2001) Evidence of new sibling species *Anopheles minimus* from the Ryukyu Archipelago, Japan. J Am Mosq Control Assoc 17(2):98–113

Somlor S, Vongpayloth K, Diancourt L, Buchy P, Duong V, Phonekeo D, Ketmayoon P, Vongphrachanh P, Brey PT, Caro V, Buisson Y, Grandadam M (2017) Chikungunya virus emergence in the Lao PDR, 2012–2013. PLoS One 12(12):e0189879

Somwang P, Yanola J, Suwan W, Walton C, Lumjuan N, Prapanthadara L, Somboon P (2011) Enzymes-based resistant mechanism in pyrethroid resistant and susceptible *Aedes aegypti* strains from northern Thailand. Parasitol Res 109(3):531–537

Song Z, Wang X, Liang G (2008) Species identification of some common necrophagous flies in Guangdong province, southern China based on the rDNA internal transcribed spacer 2 (ITS2). Forensic Sci Int 175(1):17–22

Speck S, Derschum H, Damdindorj T, Dashdavaa O, Jiang J, Kaysser P, Jigjav B, Nyamdorj E, Baatar U, Munkhbat E, Choijilsuren O, Gerelchuluun O, Romer A, Richards AL, Kiefer D, Scholz H, Wolfel R, Zoller L, Dobler G, Essbauer S (2012) *Rickettsia raoultii,* the predominant *Rickettsia* found in Mongolian *Dermacentor nuttalli*. Ticks Tick Borne Dis 3(4):227–321

Stark JH, Li X, Zhang JC, Burn L, Valluri SR, Liang J, Pan K, Fletcher MA, Simon R, Jodar L, Gessner BD (2022) Systematic review and meta-analysis of Lyme disease data and seropositivity for *Borrelia burgdorferi*, China, 2005–2020. Emerg Infect Dis 28(12):2389–2397

Stenhouse SA, Plernsub S, Yanola J, Lumjuan N, Dantrakool A, Choochote W, Somboon P (2013) Detection of the V1016G mutation in the voltage-gated sodium channel gene of *Aedes aegypti* (Diptera: Culicidae) by allele-specific PCR assay, and its distribution and effect on deltamethrin resistance in Thailand. Parasit Vectors 6(1):253

Strelkova MV, Shurkhal AV, Kellina OI, Eliseev LN, Evans DA, Peters W, Chapman CJ, Leblancq SM, Van Eys GJJM (1990) A new species of Leishmania isolated from the great gerbil *Rhombomys opimus*. Parasitology 101(3):327–335

Sudjaritruk T, Kaewpoowat Q, Prasarakee C, Sarachai S, Taure AF, Sricharoen N et al (2022) Seroepidemiological study of Japanese encephalitis virus in Chiang Mai: immunity and susceptibility 28 years after introduction of a vaccination programme. PLoS Negl Trop Dis 16(8):e0010674

Suganuma K, Narantsatsral S, Battur B, Yamasaki S, Otgonsuren D, Musinguzi SP, Davaasuren B, Battsetseg B, Inoue N (2016) Isolation, cultivation and molecular characterization of a new *Trypanosoma equiperdum* strain in Mongolia. Parasit Vectors 9(1):481

Sugar S (2010) New and re-emerging infectious diseases in livestock of Mongolia – review (2000–2010 year). In: Proceedings 2010, Central Veterinary laboratory, Ulaanbaatar, pp 56–61 (in Mongolian)

Sukehiro N, Kida N, Umezaki M, Murakami T, Arai N, Jinnai T, Inagaki S, Tsuchida H, Murayama H, Tsuda Y (2013) First report on invasion of yellow fever mosquito, *Aedes aegypti*, at Narita International Airport, Japan in August 2012. Jpn J Infect Dis 66(3):189–194

Sumarmo (1987) Dengue haemorrhagic fever in Indonesia. Southeast Asian J Trop Med Public Health 18(3):269–274

Sumarnrote A, Overgaard HJ, Marasri N, Fustec B, Thanispong K, Chareonviriyaphap T, Corbel V (2017) Status of insecticide resistance in *Anopheles* mosquitoes in Ubon Ratchathani province, Northeastern Thailand. Malar J 16(1):299

Sun DJ, Deng XL, Duan JH (2013) The history of the elimination of lymphatic filariasis in China. Infect Dis Poverty 2(1):30

Sun DJ, Fang Y, Huang Y, Zhang Y (2020) Chapter 4: Contributions to the lymphatic filariasis elimination programme and postelimination surveillance in China by NIPD-CTDR. Adv Parasitol 110:145–183

Sunantaraporn S, Sor-Suwan S, Boonserm R, Thepparat A, Phumee A, Bellis G et al (2021) *Culicoides latreille* (Diptera: Ceratopogonidae) as potential vectors for *Leishmania martiniquensis* and *Trypanosoma* sp. in northern Thailand. PLoS Negl Trop Dis 15(12):e0010014
Suntsov VV (2017) Recent speciation of plague microbe *Yersinia pestis* in the heterothermal (heteroimmune) environment of marmot–flea (*Marmota sibirica–Oropsylla silantiewi*): biogeocenotic preconditions and preadaptations. Biol Bull Rev 7(4):299–311
Supriyono S, Tan S (2020) DNA of *Brugia malayi* detected in several mosquito species collected from Balangan District, South Borneo Province, Indonesia. Vet World 13(5):996–1000
Supriyono, Kuwata R, Torii S, Shimoda H, Ishijima K, Yonemitsu K, Minami S, Kuroda Y, Tatemoto K, Tran NTB, Takano A, Omatsu T, Mizutani T, Itokawa K, Isawa H, Sawabe K, Takasaki T, Yuliani DM, Abiyoga D, Hadi UK, Setiyono A, Hondo E, Agungpriyono S, Maeda K (2020) Mosquito-borne viruses, insect-specific flaviviruses (family Flaviviridae, genus Flavivirus), Banna virus (family Reoviridae, genus Seadornavirus), Bogor virus (unassigned member of family Permutotetraviridae), and alphamesoniviruses 2 and 3 (family Mesoniviridae, genus Aphamesonivirus) isolated from Indonesian mosquitoes. J Vet Med Sci 82(7):1030–1041
Suttana S, Yanola J, Lumjuan N, Somboon P (2019) The efficacy of pyriproxyfen-treated resting boxes on the reproductivity of *Aedes aegypti* (Diptera: Culicidae) in the laboratory. Trop Biomed 36(2):514–530
Suwantika AA, Supadmi W, Ali M, Abdulah R (2021) Cost-effectiveness and budget impact analyses of dengue vaccination in Indonesia. PLoS Negl Trop Dis 15(8):e0009664
Syah A, Basri C, Wicaksono A (2020) Incidence of Surra disease in beef cattle in Barru District, South Sulawesi Province in 2015–2017. J Med Vet 3(2):145–153 (in Indonesian)
Syamsa RA, Ahmad FMS, Marwi MA, Zuha RM, Omar B (2010) An analysis of forensic entomological specimens by Universiti Kebangsaan Malaysia. Med J Malaysia 65(3):185–188
Syamsa RA, Ahmad FMS, Zuha RM, Khairul AZ, Marwi MA, Shahrom AW, Omar B (2012) An occurrence of *Synthesiomyia nudiseta* (Wulp) (Diptera: Muscidae) from a human corpse in a high-rise building in Malaysia: a case report. Trop Biomed 29(1):107–112
Syamsa RA, Omar B, Zuha RM, Faridah MN, Swarhib MS, Hidayatulfathi O, Shahrom AW (2015) Forensic entomology of high-rise buildings in Malaysia: three case reports. Trop Biomed 32(2):291–299
Takahashi K, Takahashi Y, Kobayashi S, Ishizuka M, Hirano M, Muto M, Nishiyama S, Kariwa H, Yoshii K (2020) Characterization of tick-borne encephalitis virus isolated from a tick in central Hokkaido in 2017. J Gen Virol 101(5):497–509
Thengchaisri N, Inpankaew T, Arthitwong S, Steiner JM, Sattasathuchana P (2022) Molecular prevalence of *Dirofilaria immitis* and *Wolbachia infections* in pet and semi-domesticated cats in Bangkok, Thailand. Vet World 15(2):239–243
Takahashi T, Maeda K, Suzuki T, Ishido A, Shigeoka T, Tominaga T, Kamei T, Honda M, Ninomiya D, Sakai T, Senba T, Kaneyuki S, Sakaguchi S, Satoh A, Hosokawa T, Kawabe Y, Kurihara S, Izumikawa K, Kohno S, Azuma T, Suemori K, Yasukawa M, Mizutani T, Omatsu T, Katayama Y, Miyahara M, Ijuin M, Doi K, Okuda M, Umeki K, Saito T, Fukushima K, Nakajima K, Yoshikawa T, Tani H, Fukushi S, Fukuma A, Ogata M, Shimojima M, Nakajima N, Nagata N, Katano H, Fukumoto H, Sato Y, Hasegawa H, Yamagishi T, Oishi K, Kurane I, Morikawa S, Saijo M (2014) The first identification and retrospective study of severe fever with thrombocytopenia syndrome in Japan. J Infect Dis 209(6):816–827
Takashima I, Morita K, Chiba M, Hayasaka D, Sato T, Takezawa C, Igarashi I, Kariwa H, Yoshimatsu K, Arikawa J, Hashimoto N (1997) A case of tick-borne encephalitis in Japan and isolation of the virus. J Clin Microbiol 35(8):1943–1947
Takeda (2022) QDENGA summary of product characteristics. https://www.takeda.com/4a410b/siteassets/system/what-we-do/areas-of-focus/vaccines/pdf/acc_qdenga_smpc.pdf. Accessed 04 Jan 2023
Takeda T, Ito T, Chiba M, Takahashi K, Niioka T, Takashima I (1998) Isolation of tick-borne encephalitis virus from *Ixodes ovatus* (Acari: Ixodidae) in Japan. J Med Entomol 35(3):227–231

Takeda T, Ito T, Osada M, Takahashi K, Takashima I (1999) Isolation of tick-borne encephalitis virus from wild rodents and a seroepizootiologic survey in Hokkaido, Japan. Am J Trop Med Hyg 60(2):287–291

Takeshita N, Imoto K, Ando S, Yanagisawa K, Ohji G, Kato Y, Sakata A, Hosokawa N, Kishimoto T (2010) Murine typhus in two travelers returning from Bali, Indonesia: an underdiagnosed disease. J Travel Med 17(5):356–358

Tan SH, Mohd Aris E, Surin J, Omar B, Kurahashi H, Mohamed Z (2009) Sequence variation in the cytochrome oxidase subunit I and II genes of two commonly found blow fly species, *Chrysomya megacephala* (Fabricius) and *Chrysomya rufifacies* (Macquart) (Diptera: Calliphoridae) in Malaysia. Trop Biomed 26(2):173–181

Tan SH, Mohd Aris E, Kurahashi H, Mohamed Z (2010a) A new record of *Iranihindia martellata* (Senior-White, 1924) (Diptera: Sarcophagidae) from Peninsular Malaysia and female identification using both morphology and DNA-based approaches. Trop Biomed 27(2):287–293

Tan SH, Rizman-Idid M, Mohd-Aris E, Kurahashi H, Mohamed Z (2010b) DNA-based characterisation and classification of forensically important flesh flies (Diptera: Sarcophagidae) in Malaysia. Forensic Sci Int 199:43–49

Tan LH, Fong MY, Mahmud R, Muslim A, Lau YL, Kamarulzaman A (2011) Zoonotic *Brugia pahangi* filariasis in a suburbia of Kuala Lumpur City, Malaysia. Parasitol Int 60(1):111–113

Tan M, Kusriastuti R, Savioli L, Hotez PJ (2014) Indonesia: an emerging market economy beset by neglected tropical diseases (NTDs). PLoS Negl Trop Dis 8(2):e2449

Tan XH, Deng AP, Zhang YT, Luo M, Deng H, Yang YW, Duan JH, Peng ZQ, Zhang M (2023) Analysis of the impact of health management measures for entry personnel on imported Dengue fever in Guangdong Province, 2020–2022. Chin J Epidemiol 44(6):954–959. (in Chinese)

Tanaka K, Mizusawa K, Saugstad ES (1979) A revision of the adult and larval mosquitoes of Japan (including the Ryukyu Archipelago and the Ogasawara Islands) and Korea (Diptera: Culicidae). Contrib Am Entomol Inst 16:1–987

Taniguchi K, Yoshida M, Sunagawa T, Tada Y, Okabe N (2008) Imported infectious diseases and surveillance in Japan. Travel Med Infect Dis 6(6):349–354

Taylor AJ, Vongphayloth K, Vongsouvath M, Grandadam M, Brey PT, Newton PN, Sutherland IW, Dittrich S (2016) Large-scale survey for tickborne bacteria, Khammouan Province, Laos. Emerg Infect Dis 22(9):1635–1639

Teng Z, Gong P, Wang W, Zhao N, Jin X, Sun X, Zhou H, Lu J, Lin X, Wen B, Kan B, Xu J, Qin T (2023) Clinical forms of Japanese spotted fever from case-series study, Zigui County, Hubei Province, China, 2021. Emerg Infect Dis 29(1):202–206

Tenzin W, Sennanayake S, Dorjee S, Tobgay T, Zhou G, Piyasiri SB, Lakshika WM, Weerasinghe I, Karunaweera N (2024) Distribution and seasonal variation of *Phlebotomus* and *Sergentomyia* sand fly populations in Bhutan. Am J Trop Med Hyg 110(6):1117

Tesh RB, Gajdusek DC, Garruto RM, Cross JH, Rosen L (1975) The distribution and prevalence of group a arbovirus neutralizing antibodies among human populations in Southeast Asia and the Pacific Islands. Am J Trop Med Hyg 24(4):664–675

The United Nations Group of Experts on Geographical Names (UNGEGN) (2019) Country report of the Republic of Indonesia. https://unstats.un.org/unsd/ungegn/sessions/2019-new-york-ungegn-1st-session/documents/GEGN.2_2019_CRP.132_Country_Report_Indonesia_2019.pdf. Accessed 04 Jan 2023

The United Nations Statistics Division (UNSD) (2022) Population and vital statistics report: vol LXXIV. https://unstats.un.org/unsd/demographic-social/products/vitstats/Sets/Series_A_2022.pdf. Accessed 04 Jan 2023

Thevan K, Disney RHH, Ahmad AH (2010) First records of two species of oriental scuttle flies (Diptera: Phoridae) from forensic cases. Forensic Sci Int 195:e5–e7

Thisyakorn U, Jongwutiwes S, Vanichsetakul P, Lertsapcharoen P (1999) Visceral leishmaniasis: the first indigenous case report in Thailand. Trans R Soc Trop Med Hyg 93(1):23–24

Tian XD, Dai PF, Zhao JY, Dong HY, Cheng JX (2023) Distribution and seasonal fluctuation of sandflies in Shanxi province, China. Chin J Vector Biol Control 34(3):417–421 (in Chinese)

Tiawsirisup S, Nithiuthai S (2006) Vector competence of *Aedes aegypti* (L.) and *Culex quinquefasciatus* (Say) for *Dirofilaria immitis* (Leidy). Southeast Asian J Trop Med Public Health 37(Suppl 3):110–114

Touray M, Bakirci S, Ulug D, Gulsen SH, Cimen H, Yavasoglu SI, Simsek FM, Ertabaklar H, Ozbel Y, Hazir S (2023) Arthropod vectors of disease agents: their role in public and veterinary health in Turkiye and their control measures. Acta Trop 243:106893

Tsetsarkin KA, Vanlandingham DL, McGee CE, Higgs S (2007) A single mutation in chikungunya virus affects vector specificity and epidemic potential. PLoS Pathog 3(12):2201

Tsheten T, Mclure A, Clements ACA, Gray DJ, Wangdi T, Wangchuk S, Wangdi K (2021) Epidemiological analysis of the 2019 dengue epidemic in Bhutan. Int J Environ Res Public Health 18(1):354

Tshokey T, Graves S, Tshering D, Phuntsho K, Tshering K, Stenos J (2017) Scrub typhus outbreak in a remote primary school, Bhutan, 2014. Emerg Infect Dis 23(8):1412–1414

Tsuru M, Suzuki M, Suzuki T, Murakami T, Matsui K, Maeda Y, Yoshikawa T, Kurosu T, Shimojima M, Shimada T, Hasegawa H, Maeda K, Morikawa S, Saijo M (2021) Pathological characteristics of a patient with severe fever with thrombocytopenia syndrome (SFTS) infected with SFTS virus through a sick cat's bite. Viruses 13(2):204

Uranw S, Hasker E, Roy L, Meheus F, Das ML, Bhattarai NR, Rijal S, Boelaert M (2013) An outbreak investigation of visceral leishmaniasis among residents of Dharan town, eastern Nepal, evidence for urban transmission of *Leishmania donovani*. BMC Infect Dis 13:21

Utarini A, Indriani C, Ahmad RA, Tantowijoyo W, Arguni E, Ansari MR, Supriyati E, Wardana DS, Meitika Y, Ernesia I, Nurhayati I, Prabowo E, Andari B, Green BR, Hodgson L, Cutcher Z, Rancès E, Ryan PA, O'Neill SL, Dufault SM, Tanamas SK, Jewell NP, Anders KL, Simmons CP (2021) Efficacy of wolbachia-infected mosquito deployments for the control of dengue. N Engl J Med 384(23):2177–2186

van de Straat B, Sebayang B, Grigg MJ, Staunton K, Garjito TA, Vythilingam I et al (2022) Zoonotic malaria transmission and land use change in Southeast Asia: what is known about the vectors. Malar J 21(1):109

van Peenen PFD, Joseph PL, Atmosoedjono S, Irsiana R, Saroso JS (1975) Japanese encephalitis virus from pigs and mosquitoes in Jakarta, Indonesia. Trans R Soc Trop Med Hyg 69(5–6):477–479

Von Fricken ME, Rolomjav L, Illar M, Altantogtokh D, Hogan KM, Uyanga B, Ganbold D, Tsogbadrakh N (2019) Geographic range of Lyme borreliosis in Mongolia. Vector Borne Zoonotic Dis 19(9):658–661

Vongphayloth K, Apanaskevich DA, Lakeomany K, Phommavanh N, Sinh NV, Fiorenzano JM, Hertz JC, Sutherland IW, Brey PT, Robbins RG (2022) The genus *Haemaphysalis* (Acari: Ixodidae) in Laos: an update of species records and a review of taxonomic challenges. J Med Entomol 59(6):1986–1992

Walker DH, Paddock CD, Dumler JS (2008) Emerging and re-emerging tick-transmitted -rickettsial and ehrlichial infections. Med Clin North Am 92(6):1345–1361

Wang HY, Takasaki T, Fu SH, Sun XH, Zhang HL, Wang ZX, Hao ZY, Zhang JK, Tang Q, Kotaki A, Tajima S, Liang XF, Yang WZ, Kurane I, Liang GD (2007) Molecular epidemiological analysis of Japanese encephalitis virus in China. J Gen Virol 88(Pt 3):885–894

Wang ZH, Li W, Guo YH, Yue YJ, Wang J, Ma XR, Xu MF, Wu HX, Liu QY, Meng FX (2021) Epidemiological characteristics of Japanese encephalitis in eight provinces and municipalities of northern China, 2005–2019. Chin J Vector Biol Control 32(2):197–203 (in Chinese)

Wangchuk S, Lhazeen K, Dorji T (2012) Clinical characteristics of scrub typhus in Gedu and Mongar (Bhutan). http://www.rcdc.gov.bt/web/wp-content/uploads/2012/07/Scrub-Typhus-in-Gedu-and-Mongar.pdf. Accessed 01 Dec 2023

Wangchuk S, Gyeltshen S, Dorji K, Wangdi T, Dukpa T, Namgay R, Dorjee S, Tobgay T, Chaijaroenkul W, Na-Bangchang K (2019) Malaria elimination in Bhutan: asymptomatic malaria cases in the Bhutanese population living in malaria-risk areas and in migrant workers from India. Rev Inst Med Trop Sao Paulo 61:e52

Suwonkerd W, Ritthison W, Ngo CT, Tainchum K, Bangs MJ, Chareonviriyaphap T (2013) Vector biology and malaria transmission in Southeast Asia, Chapter 10. In: Sylvie M (ed) Anopheles mosquitoes. IntechOpen, Rijeka
Warrilow D, Northill JA, Pyke AT (2012) Sources of dengue viruses imported into Queensland, Australia, 2002–2010. Emerg Infect Dis 18(11):1850–1857
Wells JD, Pape T, Sperling FAH (2001) DNA-based identification and molecular systematics of forensically important Sarcophagidae (Diptera). J Forensic Sci 46(5):1098–1102
Wharton RH (1962) The biology of *Mansonia* mosquitoes in relation to the transmission of filariasis in Malaya. Bulletin No 11. Institute for Medical Research Federation of Malaya, Kuala Lumpur, pp 1–114
Wibawa T, Satoto TBT (2016) Magnitude of neglected tropical diseases in Indonesia at postmillennium development goals era. J Trop Med 2016:1–9
Widana IDKK, Hilmawan A (2019) Urgensi Pencegahan dan Pengendalian Risiko Infeksi Leishmaniasis atas Kontingen Garuda di Lebanon. J Kesehat Lingkung Indones 18(1):34–41 (in Indonesian)
Wittesjö B, Eitrem R, Niklasson B, Vene S, Mangiafico JA (1995) Japanese encephalitis after a 10-day holiday in Bali. Lancet 345(8953):856
Wongkamchai S, Monkong N, Mahannol P, Taweethavonsawat P, Loymak S, Foongladda S (2013) Rapid detection and identification of *Brugia malayi*, *B. pahangi*, and *Dirofilaria immitis* by high-resolution melting assay. Vector Borne Zoonotic Dis 13(1):31–36
Wongsathuaythong S, Punyagupta S, Sirivorasarn P, Chularek P (1963) Filariasis in Thailand, a preliminary report. Roy Thai Army Med J 16:101
World Bank (2020) Transforming Lao PDR from a land-locked to a land-linked economy. https://www.worldbank.org/en/country/lao/publication/transforming-lao-pdr-from-a-land-locked-to-a-land-linked-economy. Accessed 27 Dec 2023
World Bank (2023) GNI per capita, atlas method (current US$) – Lao PDR. https://data.worldbank.org/indicator/NY.GNP.PCAP.CD?locations=LA. Accessed 27 Dec 2023
World Health Organization (WHO) (2010a) Bhutan malaria control programme review: a report. https://apps.who.int/iris/bitstream/handle/10665/204831/B4669.pdf;sequence=1. Accessed 25 Dec 2023
World Health Organization (WHO) (2010b) Progress report 2000–2009 and strategic plan 2010–2020 of the global programme to eliminate lymphatic filariasis: halfway towards eliminating lymphatic filariasis. https://iris.who.int/handle/10665/44473. Accessed 04 Jan 2023
World Health Organization (WHO) (2016) Emerge Zika virus situation reports. https://www.who.int/health-topics/zika-virus-disease#tab=tab_1. Accessed 05 Jan 2024
World Health Organization (WHO) (2019) Joint malaria programme review reveals Indonesia is on track for malaria elimination. https://www.who.int/indonesia/news/detail/25-11-2019-title-joint-malaria-programme-review-reveals-indonesia-is-on-track-for-malaria-elimination. Accessed 04 Jan 2023
World Health Organization (WHO) (2020) World malaria report 2020. https://www.who.int/publications/i/item/9789240015791. Accessed 20 Dec 2022
World Health Organization (WHO) (2022) WHO guidelines for malaria. WHO, Geneva. (WHO/UCN/GMP/2022.01 Rev.2). https://iris.who.int/bitstream/handle/10665/354781/WHO-UCN-GMP-2022.01-Rev.2-eng.pdf?isAllowed=y&sequence=1. Accessed 20 Dec 2022
World Health Organization (WHO) (2023a) Dengue and severe dengue. https://www.who.int/news-room/fact-sheets/detail/dengue-and-severe-dengue. Accessed 27 Dec 2023
World Health Organization (WHO) (2023b) Eliminating malaria in Lao PDR: 460000 to 2300 cases in 25 years. https://www.who.int/laos/news/spotlights/460000-to-2300-eliminating-malaria-in-lao-pdr#:~:text=Driven%20by%20new%20'accelerator'%20strategies,education%20brought%20significant%20economic%20consequences. Accessed 27 Dec 2023
World Health Organization (WHO) (2023c) WHO announces the elimination of lymphatic filariasis as a public health problem in the Lao People's Democratic Republic. https://www.who.int/news/item/16-10-2023-who-announces-the-elimination-of-lymphatic-filariasis-as-a-public-health-problem-in-the-lao-people-s-democratic-republic#:~:text=The%20World%20

Health%20Organization%20(WHO,as%20a%20public%20health%20problem. Accessed 27 Dec 2023

World Health Organization (WHO) (2023d) World malaria report 2023. https://www.who.int/teams/global-malaria-programme/reports/world-malaria-report-2023. Accessed 17 Jan 2024

World Health Organization (WHO): South-East Asia Regional Office (2011) Comprehensive guidelines for prevention and control of dengue and dengue haemorrhagic fever. WHO Regional Office for South East Asia, Singapore, 210 pp

Wu HH, Lin YH, Pai HH, Hsu EL, Chang NT, Luo YP (2013a) Insecticide resistance status in *Aedes aegypti* (L.) adults from Southern Taiwan. Formos Entomol 33(4):253–270

Wu HH, Wang CY, Teng HJ, Lin C, Lu LC, Jian SW, Chang NT, Wen TH, Wu JW, Liu DPLJ, Norris D, Wu HS (2013b) A dengue vector surveillance by human population-stratified ovitrap survey for *Aedes* (Diptera: Culicidae) adult and egg collections in high dengue-risk areas of Taiwan. J Med Entomol 50(2):261–269

Wu T, Wu Z, Li YP (2022) Dengue fever and dengue virus in the People's Republic of China. Rev Med Virol 32(1):e2245

Yanase T, Kato T, Kubo T, Yoshida K, Ohashi S, Yamakawa M, Miura Y, Tsuda T (2005) Isolation of bovine arboviruses from *Culicoides* biting midges (Diptera: Ceratopogonidae) in Southern Japan: 1985–2002. J Med Entomol 42(1):63–67

Yang WZ (2019) Dramatic achievements in infectious disease prevention and treatment in China during the past 70 years. Chin J Epidemiol 40(12):1493–1498 (in Chinese)

Yang CF, Su CL, Hsu TC, Chang SF, Lin CC, Huang JC, Shu PY (2016) Imported chikungunya virus strains, Taiwan, 2006–2014. Emerg Infect Dis 22(11):1981–1984

Yang CF, Chang SF, Hsu TC, Su CL, Wang TC, Lin SH, Yang SL, Lin CC, Shu PY (2018) Molecular characterization and phylogenetic analysis of dengue viruses imported into Taiwan during 2011–2016. PLoS Negl Trop Dis 12(9):e0006773

Yang H, Sun J, Zheng C, Shi Y, Yin W, Zhou S (2023) Epidemiological characteristics of visceral leishmaniasis in China, 2017–2021. Dis Surveill 38(6):676 (in Chinese)

Yangzom T, Cruz I, Bern C, Argaw D, den Boer M, Vélez ID, Bhattacharya SK, Molina R, Alvar J (2012) Endemic transmission of visceral leishmaniasis in Bhutan. Am J Trop Med Hyg 87(6):1028–1037

Yanola J, Somboon P, Walton C, Nachaiwieng W, Prapanthadara L (2010) A novel F1552/C1552 point mutation in the *Aedes aegypti* voltage-gated sodium channel gene associated with permethrin resistance. Pestic Biochem Phys 96(3):127–131

Yanola J, Somboon P, Walton C, Nachaiwieng W, Somwang P, Prapanthadara L (2011) High throughput assays for detection of the F1534C mutation in the voltage-gated sodium channel gene in permethrin-resistant *Aedes aegypti* and the distribution of this mutation throughout Thailand. Trop Med Inter Health 16(4):501–509

Yosephine P (2017) Indonesia national immunization program. In: 3rd International symposium for Asia Pasific experts. https://www.fondation-merieux.org/wp-content/uploads/2017/10/vaccinology-2017-prima-yosephine.pdf. Accessed 04 Jan 2023

Yoshikawa T, Fukushi S, Tani H, Fukuma A, Taniguchi S, Toda S, Shimazu Y, Yano K, Morimitsu T, Ando K, Yoshikawa A, Kan M, Kato N, Motoya T, Kuzuguchi T, Nishino Y, Osako H, Yumisashi T, Kida K, Suzuki F, Takimoto H, Kitamoto H, Maeda K, Takahashi T, Yamagishi T, Oishi K, Morikawa S, Saijo M, Shimojima M (2014) Sensitive and specific PCR systems for detection of both Chinese and Japanese severe fever with thrombocytopenia syndrome virus strains and prediction of patient survival based on viral load. J Clin Microbiol 52(9):3325–3333

Yu XJ, Liang MF, Zhang SY, Liu Y, Li JD, Sun YL, Zhang L, Zhang QF, Popov VL, Li C, Qu J, Li Q, Zhang YP, Hai R, Wu W, Wang Q, Zhan FX, Wang XJ, Kan B, Wang SW, Wan KL, Jing HQ, Lu JX, Yin WW, Zhou H, Guan XH, Liu JF, Bi ZQ, Liu GH, Ren J, Wang H, Zhao Z, Song JD, He JR, Wan T, Zhang JS, Fu XP, Sun LN, Dong XP, Feng ZJ, Yang WZ, Hong T, Zhang Y, Walker DH, Wang Y, Li DX (2011) Fever with thrombocytopenia associated with a novel bunyavirus in China. N Engl J Med 364(16):1523–1532

Yuda P (2019) Detection of avian malaria in wild birds at Trisik Beach of Yogyakarta, Java (Indonesia). Ann Parasitol 65(2):171–175

Yue Y, Liu Q, Liu X, Zhao N, Yin W (2022) Dengue fever in mainland China, 2005–2020: a descriptive analysis of dengue cases and *Aedes* data. Int J Environ Res Public Health 19(7):3910

Zangmo S, Klungthong C, Chinnawirotpisan P, Tantimavanich S, Kosoltanapiwat N, Thaisomboonsuk B, Phuntsho K, Wangchuk S, Yoon IK, Fernandez S (2015) Epidemiological and molecular characterization of dengue virus circulating in Bhutan, 2013–2014. PLoS Negl Trop Dis 9(8):e0004010

Zeller H, Bouloy M (2000) Infections by viruses of the families Bunyaviridae and Filoviridae. Rev Sci Tech 19(1):79–91

Zeng W, Halasa-Rappel YA, Baurin N, Coudeville L, Shepard DS (2018) Cost-effectiveness of dengue vaccination in ten endemic countries. Vaccine 36(3):413–420

Zhang LM, Leng YJ (1997) Eighty-year research of phlebotomine sandflies (Diptera: Psychodidae) in China (1915–1995). II. Phlebotomine vectors of leishmaniasis in China. Parasite 4(4):299–306

Zhang C, Fu X, Xie K, Yan W, Guo Y (2015) mtDNA analysis for genetic identification of forensically important sarcophagid flies (Diptera: Sarcophagidae) in China. J Med Entomol 52(6):1225–1233

Zhang Y, Lei W, Wang Y, Sui H, Liu B, Li F, He Y, Li Z, Fu S, Wang L, Xu L, Mahe M, Gao Z, Mamutijiang T, Lv Z, Xiang N, Zhou L, Ni D, Liang G, Li Q, Wang H, Feng Z (2021) Surveillance of West Nile virus infection in Kashgar Region, Xinjiang, China, 2013–2016. Sci Rep 11(1):14010

Zhang XA, Tian F, Li Y, Zhang XL, Jiang BG, Liu BC, Zhang JT, Tian S, Ding H, Li S, Li H, Fang LQ, Liu W (2022) Molecular detection and identification of relapsing fever Borrelia in ticks and wild small mammals in China. Emerg Microbes Infect 11(1):2632–2635

Zhen Y, Tie P, Bai Y, Yan C, Wang T, Wang J, Tian X, Dai P (2023) Investigation on visceral leishmaniasis in domestic dogs and sandfly density in epidemic area in Shanxi Province from 2021 to 2022. Chin J Parasitol Parasit Dis 41(4):470–475. (in Chinese)

Zhou CM, Liu JW, Qi R, Fang LZ, Qin XR, Han HJ, Mo RC, Yu H, Jiao YJ, Lin JY, Yu XJ (2020) Emergence of Zika virus infection in China. PLoS Negl Trop Dis 14(5):e0008300

Zhou Z, Pan G, Li Y, Liu Q, Yang L, Li Z, Ma Z, Zhang Y, Li S (2023) Prevalence of visceral leishmaniasis in China in 2021. Chin J Parasitol Parasit Dis 41(2):149–155. (in Chinese)

Zuha RM, Baba MH (2016) Sarcophaga dux Thomson (Diptera: Sarcophagidae) recovered from wrapped body of an infant: its evidential value. Serangga 21(2):163–170

Zuha RM, Razak TA, Nazni WA, Omar B (2012) Interaction effects of temperature and food on the development of forensically important fly, *Megaselia scalaris* (Loew) (Diptera: Phoridae). Parasitol Res 111(5):2179–2187

Zuha RM, Huong-Wen S, Disney RH, Omar B (2017) Scuttle flies (Diptera: Phoridae) inhabiting rabbit carcasses confined to plastic waste bins in Malaysia include new records and an undescribed species. Trop Life Sci Res 28(1):131–143

Zulfa R, Lo WC, Cheng PC, Martini M, Chuang TW (2022) Updating the insecticide resistance status of *Aedes aegypti* and *Aedes albopictus* in Asia: a systematic review and meta-analysis. Trop Med Infect Dis 7(10):306

Index

K. Sawabe et al. (eds.), *Medical Entomology in Asia*, Entomology Monographs, https://doi.org/10.1007/978-981-97-4406-0

D

Q

R

SPRINGER NATURE

GPSR Compliance

The European Union's (EU) General Product Safety Regulation (GPSR) is a set of rules that requires consumer products to be safe and our obligations to ensure this.

If you have any concerns about our products, you can contact us on ProductSafety@springernature.com

In case Publisher is established outside the EU, the EU authorized representative is:

Springer Nature Customer Service Center GmbH
Europaplatz 3
69115 Heidelberg, Germany

The manufacturer's authorised representative in the EU is Springer Nature Customer Service Centre GmbH, Europaplatz 3, 69115 Heidelberg, Germany. If you have any concerns regarding our products, please contact ProductSafety@springernature.com

Printed and bound by CPI Group (UK) Ltd, Croydon, CR0 4YY
01/10/2026
02225852-0019